T0401954

Plasticity and Creep of Metals

Andrew Rusinko and Konstantin Rusinko

Plasticity and Creep of Metals

Authors

Andrew Rusinko
Óbuda University
Dept. of Mechanical and
System Engineering
Népszínház St. 8
Budapest
Hungary

Konstantin Rusinko
Budapest University of Technology and
Economics
Budapest
Hungary

ISBN 978-3-642-21212-3 e-ISBN 978-3-642-21213-0

DOI 10.1007/978-3-642-21213-0

Library of Congress Control Number: 2011931537

Typeset & Cover Design: Scientific Publishing Services Pvt. Ltd., Chennai, India.

Printed on acid-free paper

9 8 7 6 5 4 3 2 1

springer.com

Contents

Chapter 1
Classical Theories of Plasticity

The macromechanical theories (also called the mathematical theories) of plasticity describe plastic deformations phenomenologically, on the macroscopic level, and establish relations among macroscopic mechanical quantities (such as stresses and strains). These theories are based on the notions of a yield surface and a hardening rule, governing the evolution of the yield surface (loading surface) and the stress-strain relations of the material. The analysis of classical theories of plasticity and their concurrence with experimental data are the focus of this chapter.

The classical theories considered here are based upon the following assumptions: (i) only small plastic strains are considered; (ii) the material is initially isotropic until an inelastic behavior occurs; (iii) work-hardening materials (except for Section 1.3 and 1.14 with perfectly plastic materials) are considered; (iv) the plastic strain involves only a change in shape but no change in volume; (v) materials behave in tension in a same manner to their behavior in compression; and (vi) temporary effects, influences of temperature and other actions (irradiation, magnetic field etc) are neglected [17,18,53,125,126].

1.1 Definition of Subject

Plasticity is a property of a material to undergo a non-reversible change of shape in response to an applied force. The theory of plasticity, being the section of continuum mechanics, is concerned with the analysis of stresses and elasto-plastic strains in a body.

A start point of classical theories of plasticity is the experimental macro-behavior of specimens during loading. Typical stress-strain curve of polycrystalline body has the form schematically shown in Fig. 1.1a. Such curve can be obtained from a tensile test, or at inner pressure of short cylinders as well as at torsion of thin-walled shafts.

Figure 1.1a shows the typical stress-strain curve for a simple tension specimen. One can point out characteristic points in this diagram. The proportional limit of a material is defined as a largest value of stress for which the stress is still proportional to the strain. The elastic limit of a material is defined as a largest value of stress that can be applied without causing any residual strain upon removal for the stress. If the material is unloaded from point M, beyond the elastic limit, and then loaded again, the unloading-loading portion $MO'M'$, hysteresis loop, appears as in Fig. 1.1a.

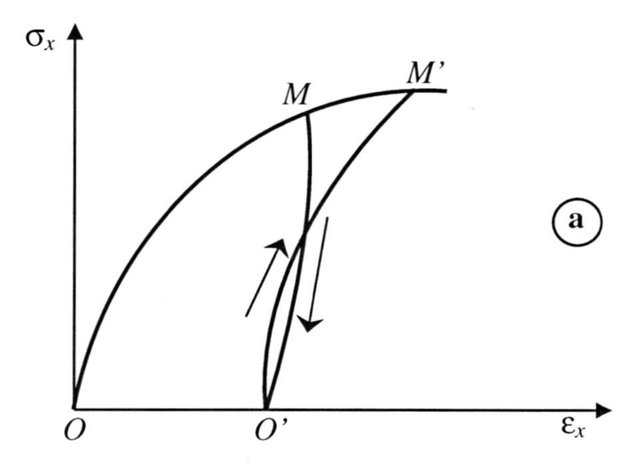

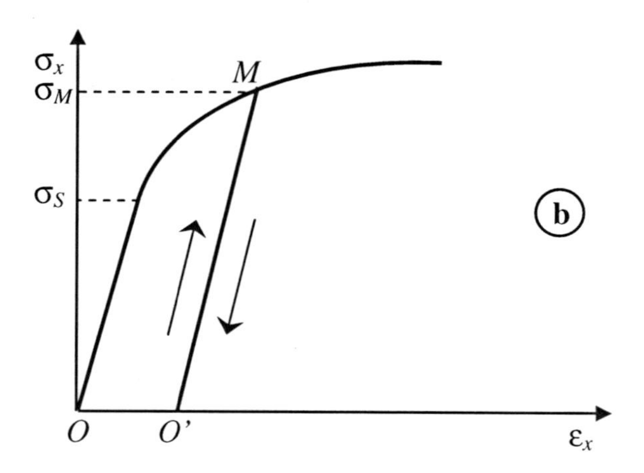

Fig. 1.1 Real (a) and idealized (b) stress-strain diagram.

In terms of classical theories, real diagrams are replaced by idealized ones (Fig. 1.1b) where there is only one characteristic point, yield strength (yield limit) of the material, σ_S, which is the stress required to produce a certain residual plastic strain, usually 0.2 %. For $\sigma < \sigma_S$, the loading portion is assumed to obey Hooke's law. Furthermore, the unloading MO' and the subsequent loading

$O'M$ are parallel to the initial elastic portion. The plastic (residual) strain (OO' in Fig. 1.1b) corresponding to a stress at M is defined as the difference between total and elastic strains.

The main task of the theory of plasticity consists in the determination of residual strains under any combined state of stress.

1.2 Hencky–Nadai Deformation Theory

The deformation theory was proposed by Hencky [38] for perfectly elastic-plastic materials and then generalized by Nadai to model the behavior of a work-hardening material. This theory assumes that the state of stress determines the state of strain uniquely as long as the plastic deformation continues.

Nadai [96-98] formulated the following laws of plasticity:

1. Stress-strain relation, deviators proportionality. As it is well known, stress/strain deviator tensor has the following components:

$$\overline{\sigma}_{ij} = \sigma_{ij} - \sigma\delta_{ij}, \quad \overline{\varepsilon}_{ij} = \varepsilon_{ij} - \varepsilon\delta_{ij}, \quad i, j = x, y, z, \tag{1.2.1}$$

where σ_{ij} and ε_{ij} are stress/strain tensor components, δ_{ij} is the Kronecker delta, and σ is the hydrostatic stress:

$$\sigma = \frac{1}{3}\left(\sigma_x + \sigma_y + \sigma_z\right), \tag{1.2.2}$$

ε is the volume-strain (dilatation):

$$\varepsilon = \frac{1}{3}\left(\varepsilon_x + \varepsilon_y + \varepsilon_z\right). \tag{1.2.3}$$

In Eqs. (1.2.2) and (1.2.3) and further we use σ_x, σ_y, and σ_z to denote stress components σ_{xx}, σ_{yy}, and σ_{zz}, respectively.

The main assumption is that the strain deviator components are proportionally related to the stress deviator components:

$$\overline{\varepsilon}_{ij} = \frac{\overline{\sigma}_{ij}}{2G_S}, \tag{1.2.4}$$

where G_S is the shear modulus of plasticity which will be concretized further. Equation (1.2.4) can be rewritten in the decompose form:

$$\varepsilon_x - \varepsilon = \frac{\sigma_x - \sigma}{2G_S}, \ \varepsilon_y - \varepsilon = \frac{\sigma_y - \sigma}{2G_S}, \ \varepsilon_z - \varepsilon = \frac{\sigma_z - \sigma}{2G_S}$$
$$\gamma_{xy} = \frac{\tau_{xy}}{G_S}, \ \gamma_{yz} = \frac{\tau_{yz}}{G_S}, \ \gamma_{xz} = \frac{\tau_{xz}}{G_S} \tag{1.2.5}$$

Equation (1.2.5) differs from Hooke's law only in that the constant module G is replaced by variable shear modulus of plasticity G_S, regulating the nonlinearity of the stress-strain relations.

Let us note that components ε_{ij} are related to displacement components, u_i, in the following way:

$$\varepsilon_{ij} = \frac{1}{2}\left(\frac{\partial u_i}{\partial x_j} + \frac{\partial u_j}{\partial x_i}\right). \tag{1.2.6}$$

Therefore, shear strain components (γ_{ij}) can be expressed through ε_{ij} as

$$\gamma_{ij} = 2\varepsilon_{ij}. \quad (i \neq j) \tag{1.2.7}$$

2. Hardening rule

The shear strain intensity (γ_0),

$$\gamma_0 = \frac{\sqrt{2}}{3}\left[\left(\varepsilon_x - \varepsilon_y\right)^2 + \left(\varepsilon_y - \varepsilon_z\right)^2 + \left(\varepsilon_x - \varepsilon_z\right)^2 + 6\left(\varepsilon_{xy}^2 + \varepsilon_{yz}^2 + \varepsilon_{zx}^2\right)\right]^{1/2}, \tag{1.2.8}$$

is assumed to be a universal function of the shear stress intensity (τ_0)[1],

$$\tau_0 = \frac{1}{\sqrt{2}}\left[\left(\sigma_x - \sigma_y\right)^2 + \left(\sigma_y - \sigma_z\right)^2 + \left(\sigma_x - \sigma_z\right)^2 + 6\left(\tau_{xy}^2 + \tau_{yz}^2 + \tau_{zx}^2\right)\right]^{1/2}. \tag{1.2.9}$$

Therefore, the hardening rule can be expressed as

$$\gamma_0 = \gamma_0(\tau_0), \tag{1.2.10}$$

where $\gamma_0(\tau_0)$ is an universal function of a material. This function can be obtained on the base of experimental diagrams in the following way. Consider the uniaxial tension when Eqs. (1.2.8) and (1.2.9) give

[1] We will use further throughout the notion "shear stress/strain intensity", which is the second invariant of stress/strain deviator tensor. In scientific literature, the notion "effective stress/strain" is also usable.

$$\gamma_0 = \frac{2}{3}\left(\varepsilon_x - \varepsilon_y\right), \qquad \tau_0 = \sigma_x . \tag{1.2.11}$$

Let us perform an experiment in uniaxial tension and measure the values of axial, ε_x, and transverse, $\varepsilon_y = \varepsilon_z$, strain components induced by $\sigma_x > 0$, $\sigma_y, \ldots, \tau_{yz} = 0$. The set of experimentally obtained values of ε_x, ε_y, and σ_x constitutes, through Eq. (1.2.11), the function $\gamma_0 = \gamma_0(\tau_0)$.

Since in terms of the theory of elasticity we have $\varepsilon_y = -\nu\varepsilon_x$ (ν is Poisson ratio [72]), Eq. (1.2.11) gives

$$\gamma_0 = \frac{2}{3}(1+\nu)\varepsilon_x . \tag{1.2.12}$$

Therefore, Hooke's law $\sigma_x = E\varepsilon_x$ (*E* is Young's modulus) in terms of τ_0 and γ_0 has the form:

$$\tau_0 = 3G\gamma_0 , \tag{1.2.13}$$

where *G* is the shear modulus of elasticity:

$$G = \frac{E}{2(1+\nu)} . \tag{1.2.14}$$

At significant plastic deformations, we have $\varepsilon_y = -0{,}5\varepsilon_x$, therefore, $\gamma_0 = \varepsilon_x$ and diagrams $\sigma_x - \varepsilon_x$ and $\tau_0 - \gamma_0$ are identical.

For the case of pure shear, when only one component γ_{xy} is nonzero, Eqs. (1.2.7)–(1.2.9) give

$$\tau_0 = \sqrt{3}\tau_{xy}, \qquad \gamma_0 = \frac{\gamma_{xy}}{\sqrt{3}} . \tag{1.2.15}$$

Therefore, the diagram $\tau_0 - \gamma_0$ is identical to that of $\tau_{xy} - \gamma_{xy}$ constructed for changed scale factors. The Hooke's law, $\tau_{xy} = G\gamma_{xy}$, also results in Eq. (1.2.13).

It is obvious that $\tau_0 - \gamma_0$ diagram can be constructed on the base of various experiments. By the universality of this diagram we mean that all experiments (tension, torsion, combined state of stress...) will result in one and the same $\tau_0 - \gamma_0$ curve shown in Fig. 1.2.

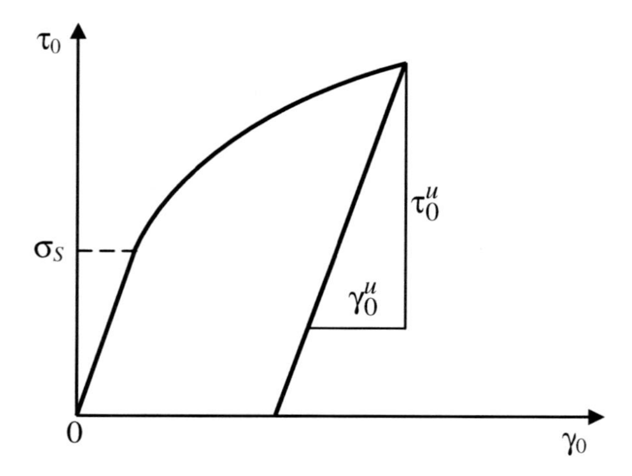

Fig. 1.2 Shear stress intensity-shear strain intensity curve.

From the existence of universal relation between τ_0 and γ_0, it follows that the deformation theory is based on Guber-Hencky-von Mises[2] yield criterion. It states that yielding begins when the shear stress intensity (effective stress) attains a critical value (constant of material).

On the base of Eq. (1.2.9), Guber-Hencky-von Mises yield criterion has the form

$$\left(\sigma_x - \sigma_y\right)^2 + \left(\sigma_y - \sigma_z\right)^2 + \left(\sigma_x - \sigma_z\right)^2 + 6\left(\tau_{xy}^2 + \tau_{yz}^2 + \tau_{zx}^2\right) = 2\sigma_S^2, \quad (1.2.16)$$

where σ_S is the yield limit (yield stress) in uniaxial tension. For the case of pure shear, Eq. (1.2.16) gives

$$\sigma_S = \sqrt{3}\tau_S, \quad (1.2.17)$$

where τ_S is the yield limit in pure shear.

It should be noted that Tresca was first to formulate the yield condition, which states that a metal yields plastically when maximum shear stress, τ_m, attains the critical value τ_S :

[2] Guber Maximilian prominent polish scientist in mechanics who was the rector of Lviv Polytechnica where the authors of this monograph worked.

$$\tau_m = \tau_S = \frac{\sigma_S}{2}. \tag{1.2.18}$$

However, numerous experiments showed that for many technically important materials the Guber criterion (1.2.16) better concurs with experiments than the Tresca criterion (1.2.18).

Let us study the relation of modulus G_S to τ_0 and γ_0. On the base of Eq. (1.2.5), the differences between normal strain components have the form:

$$\begin{aligned} \varepsilon_x - \varepsilon_y &= \frac{1}{2G_S}(\sigma_x - \sigma_y) \\ \varepsilon_y - \varepsilon_z &= \frac{1}{2G_S}(\sigma_y - \sigma_z) \\ \varepsilon_z - \varepsilon_x &= \frac{1}{2G_S}(\sigma_z - \sigma_x) \end{aligned} \tag{1.2.19}$$

and shear strain components are:

$$\gamma_{xy} = \frac{\tau_{xy}}{G_S}, \quad \gamma_{zy} = \frac{\tau_{zy}}{G_S}, \quad \gamma_{xz} = \frac{\tau_{xz}}{G_S}. \tag{1.2.20}$$

Substituting the left-hand side in Eqs. (1.2.19) and (1.2.20) into Eq. (1.2.8), we obtain

$$G_S = \frac{\tau_0}{3\gamma_0}. \tag{1.2.21}$$

For arriving at the result (1.2.21) we have taken into consideration the relation (1.2.9). Employing Eq. (1.2.21), we can reconstruct the $\tau_0 - \gamma_0$ diagram into the $G_S - \tau_0$ diagram (Fig. 1.3). This figure shows that $G_S = G$ in the elastic region ($\tau_0 < \sigma_S$) and the module G_S decreases for $\tau_0 > \sigma_S$. The decrease of G_S for $\tau_0 > \sigma_S$ is accounted for the fact that the fixed increment of τ_0 produces a larger increment in γ_0 in a plastic region than along the elastic (linear) potion of $\tau_0 - \gamma_0$ diagram (see Fig. 1.2).

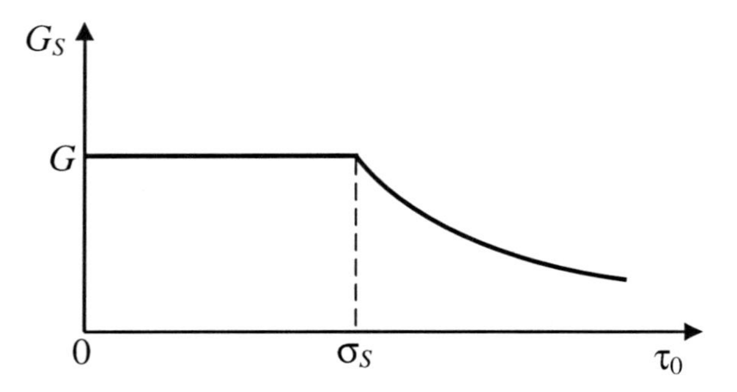

Fig. 1.3 Shear modulus G_S.

Substituting the value of G_S from Eq. (1.2.21) into Eq. (1.2.5), the deviators proportionality can be expressed as:

$$\begin{gathered}\varepsilon_x-\varepsilon=\frac{3\gamma_0}{2\tau_0}(\sigma_x-\sigma),\varepsilon_y-\varepsilon=\frac{3\gamma_0}{2\tau_0}(\sigma_y-\sigma),\varepsilon_z-\varepsilon=\frac{3\gamma_0}{2\tau_0}(\sigma_z-\sigma)\\ \gamma_{xy}=\frac{3\gamma_0}{2\tau_0}\tau_{xy},\gamma_{zy}=\frac{3\gamma_0}{2\tau_0}\tau_{zy},\gamma_{xz}=\frac{3\gamma_0}{2\tau_0}\tau_{xz}\end{gathered} \tag{1.2.22}$$

3. The law of the elasticity of volume deformation. According to numerous experiments [14], a hydrostatic compression does not result in a plastic flow. Therefore, the volumetric strain, 3ε, is assumed to be linearly related to hydrostatic stress:

$$3\varepsilon=\frac{\sigma}{K_0},\qquad K_0=const \tag{1.2.23}$$

where K_0 is the bulk modulus. According to the Hooke's law,

$$\varepsilon=\frac{\sigma(1-2\nu)}{E}. \tag{1.2.24}$$

Hence,

$$3K_0=\frac{E}{1-2\nu}. \tag{1.2.25}$$

Eqs. (1.2.5), (1.2.10), (1.2.21), and (1.2.23) constitute Hencky–Nadai deformation theory, and determine total (elastic and plastic) strain components.

4. A loading criterion can be expressed in the following way:

$$d\tau_0>0 \tag{1.2.26}$$

and if $d\tau_0<0$, an unloading takes place under elastic law.

Let us write Eq. (1.2.9) in another form. It must be noted that, if we substitute deviatoric components $\overline{\sigma}_{ij}$ for σ_{ij}, the value τ_0 will not change. In addition, it is possible to add the term $\overline{A}_1^2/2$ ($\overline{A}_1$ is the first invariant of stress deviator tensor) to the right-hand side in Eq. (1.2.9) because

$$\overline{A}_1 = \overline{\sigma}_x + \overline{\sigma}_y + \overline{\sigma}_z = 0\,. \tag{1.2.27}$$

This procedure gives a new expression for τ_0:

$$\tau_0^2 = \frac{3}{2}\left[\overline{\sigma}_x^2 + \overline{\sigma}_y^2 + \overline{\sigma}_z^2 + 2\left(\tau_{xy}^2 + \tau_{yz}^2 + \tau_{zx}^2\right)\right]. \tag{1.2.28}$$

If to use the summation convention, Eq. (1.2.28) can be rewritten in the form:

$$\tau_0^2 = \frac{3}{2}\overline{\sigma}_{ij}\overline{\sigma}_{ij}\,. \tag{1.2.29}$$

Now let us substantiate the correctness and general character of the loading criterion (1.2.26). Consider the case of uniaxial tension when the stress-strain relation during plastic loading takes the form

$$\sigma_x = E_S \varepsilon_x\,, \tag{1.2.30}$$

where E_S is a variable secant module and

$$d\sigma_x = E d\varepsilon_x \tag{1.2.31}$$

at the decrease of stress. At first sight it seems that Eq. (1.2.30) could be taken to be valid for $d\sigma_x > 0$, and Eq. (1.2.31) for $d\sigma_x < 0$. But this is incorrect for the negative values of σ_x in compression. It is easy to see that the product $\sigma_x d\sigma_x$ can be taken as the general loading criterion for uniaxial stretching-compressing: the positive sign of $\sigma_x d\sigma_x$ indicates the occurrence of plastic strain and its negative sign symbolizes an unloading. For the case of combined stress state it is natural to take the sign of expression $\overline{\sigma}_{ij} d\overline{\sigma}_{ij}$ as the loading criterion, which is the generalization of the product $\sigma_x d\sigma_x$ in tensile. On the other hand, Eq. (1.2.29) gives

$$\overline{\sigma}_{ij} d\overline{\sigma}_{ij} = \frac{2}{3}\tau_0 d\tau_0 \tag{1.2.32}$$

that leads to the criterion (1.2.26).

1.3 Hencky Relations

As it was pointed out earlier, Hencky [38] proposed the law of deviators proportionality, Eqs. (1.2.4) or (1.2.5), for perfectly-plastic materials. Let us analyze these equations for such materials.

In the case of perfectly-plastic materials, the stress-train diagram for uniaxial tension or compression is shown in Fig. 1.4a. Hencky supposed that, at the combined state of stress, the diagram is similar to 1.4a provided that the shear stress intensity of τ_0 is substituted for the component σ_x (Fig. 1.4b), and γ_0 for ε_x.

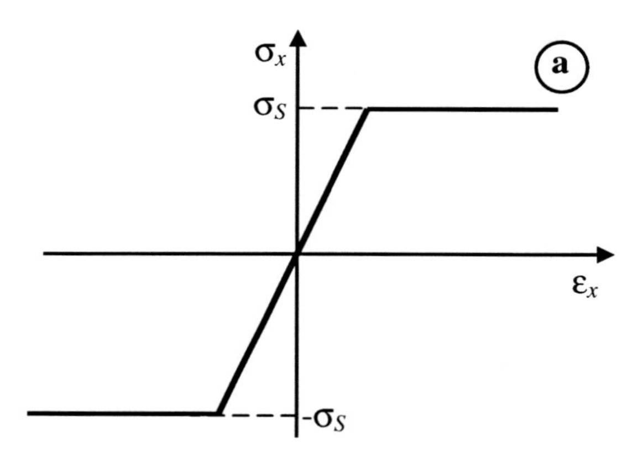

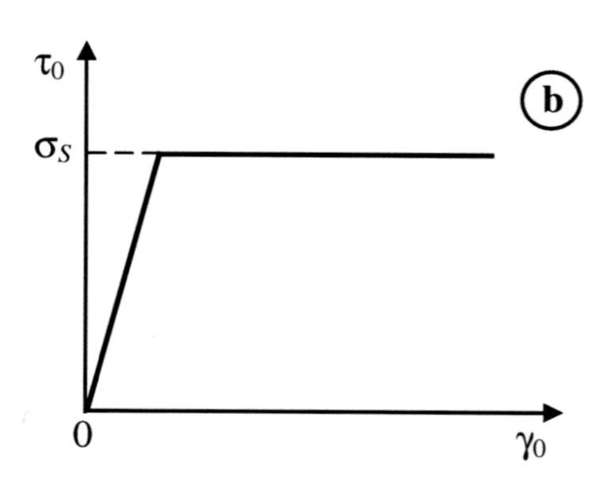

Fig. 1.4 Stress-strain curve in uniaxial stretching-compressing (a) and τ_0-γ_0 curve at arbitrary stress state (b) for ideal elastic-plastic material.

It is obvious that Eq. (1.2.21) is also valid for ideally elastic-plastic materials. Having substituted in this formula σ_S for τ_0, we obtain

$$G_S = \frac{\sigma_S}{3\gamma_0}. \tag{1.3.1}$$

Then, the deviators proportionality takes the form

$$\varepsilon_{ij} - \varepsilon\delta_{ij} = \frac{3\gamma_0}{2\sigma_S}\left(\sigma_{ij} - \sigma\delta_{ij}\right), \tag{1.3.2}$$

or, on the base of Eq. (1.2.23), we obtain Hencky equations:

$$\varepsilon_{ij} = \frac{\sigma}{3K_0}\delta_{ij} + \frac{3\gamma_0}{2\sigma_S}\left(\sigma_{ij} - \sigma\delta_{ij}\right). \tag{1.3.3}$$

The characteristic feature of Hencky equations is that fixed stress components do not determine strain components uniquely. Really, if to substitute strain components given by Eq. (1.3.3) into Eq. (1.2.8), we obtain that $\gamma_0 \equiv \gamma_0$. The uniqueness of strains is absolutely natural because plastic deformation is assumed to occur under a constant flow stress σ_S.

Equation (1.3.3) gives the stress components expressed in terms of ε_{ij}:

$$\sigma_{ij} = 3K_0\varepsilon\delta_{ij} + \left(\varepsilon_{ij} - \varepsilon\delta_{ij}\right)\frac{2\sigma_S}{3\gamma_0}, \tag{1.3.4}$$

and it is clear that stress components in this equation are the single-valued functions of strain components. If to neglect the volume change, then

$$\sigma_{ij} = \frac{2\sigma_S}{3\gamma_0}\varepsilon_{ij}. \tag{1.3.5}$$

By substituting the stress deviator components from Eq. (1.3.5) into Eq. (1.2.9), we obtain (together with Eq. (1.2.8)) that $\tau_0 = \sigma_S$, meaning that these stresses satisfy Guber's yield criterion. If loading is such that Guber's yield criterion is fulfilled, then $d\tau_0 = 0$, and Eqs. (1.3.2)–(1.3.5) are valid. If $d\tau_0 < 0$, unloading takes place under Hooke's law.

Let us show the application of Eq. (1.3.5) for the solution of boundary-value problems. Consider the circular cylinder of perfect-plastic material to which the following loading is applied: a) axial tension under stress $\sigma_z = \sigma_S$ which produces longitudinal strain ε_0, b) holding the value of ε_0, torsion with the relative angle of twist denoted by ϑ. Neglecting the volume change, let us find the distribution of acting stresses along the radius of the cylinder.

We choose a cylindrical coordinate system with the z-axis along the axis of the cylinder. The radial, circumferential, and axial strains can be determined in the following way:

$$\varepsilon_r = \frac{du}{dr}, \qquad \varepsilon_\varphi = \frac{u}{r}, \qquad \varepsilon_z = \varepsilon_0, \tag{1.3.6}$$

where u is the displacement in the radial direction. The condition of constant volume requires that

$$\frac{du}{dr} + \frac{u}{r} = -\varepsilon_0 . \tag{1.3.7}$$

The solution of the above equation gives:

$$u = \frac{C}{r} - \frac{1}{2}\varepsilon_0 r . \tag{1.3.8}$$

Since the displacement of the axis of the cylinder ($r = 0$) in the radial direction is equal to zero, Eq. (1.3.8) gives that $C = 0$ and Eqs. (1.3.6)–(1.3.8) lead to the relation

$$\varepsilon_r = \varepsilon_\varphi = -\frac{\varepsilon_0}{2} . \tag{1.3.9}$$

The shear strain is related to the relative angle of twist ϑ as

$$\gamma_{\varphi z} = \vartheta r . \tag{1.3.10}$$

The shear strain intensity, Eq. (1.2.8), can be rewritten as

$$\gamma_0 = \frac{1}{\sqrt{3}}\sqrt{3\varepsilon_0^2 + \vartheta^2 r^2} . \tag{1.3.11}$$

Eq. (1.3.5), on the base of Eq. (1.3.11) and taking into account that $\sigma = \sigma_z/3$, gives the formulas

$$\sigma_z = \frac{\sqrt{3}\sigma_S \varepsilon_0}{\sqrt{3\varepsilon_0^2 + \vartheta^2 r^2}}$$
$$\tau_{\varphi z} = \frac{1}{\sqrt{3}} \frac{\sigma_S \vartheta r}{\sqrt{3\varepsilon_0^2 + \vartheta^2 r^2}} \tag{1.3.12}$$

that determine the stresses distribution in the cylinder [98].

The component σ_z in Eq. (1.3.12) is maximal at $r = 0$, where $\sigma_z = \sigma_S$. In contrast to this, the component $\tau_{\varphi z}$ takes its maximal value at $r = R$ (*R* is the radius of the cylinder), and is equal to zero on the axis of cylinder. Stresses given by Eq. (1.3.12) satisfy Guber's yield criterion and the equilibrium equation. The compatibility conditions have no sense for perfectly-plastic materials.

The axial tensile force is calculated as

$$P = 2\pi \int_0^R r\sigma_z dr = \frac{2\pi\sqrt{3}R^2\sigma_S\varepsilon_0}{\sqrt{3\varepsilon_0^2 + \vartheta^2 R^2} + \sqrt{3}\varepsilon_0}. \tag{1.3.13}$$

It is obvious that the increase in angle ϑ in Eq. (1.3.13) results in the decrease of force *P* that is the condition of fulfillment of Guber's criterion.

1.4 Infinite Thin Plate with a Circular Hole: Comparison of Three Solutions

Let us consider the following boundary-value problem: an infinite thin plate with inner circular hole of radius *R* of an elasto-plastic work-hardening material is under the conditions of plane stress (p_1 is a radial stress as $r \to \infty$ and p_2 denotes a radial load at $r = R$). This section treats three solutions of this boundary-value problem obtained at various assumptions relative to the behavior of material during loading.

1. *The hole of plane, $r = R$, is stress free and constant stretching radial stress $\sigma_r = p_1$ as $r \to \infty$.* The solution of this problem [22,168] for the material of plate whose behavior is close to Hooke's law within the whole diapason of straining (including plastic deformation) has the form:

$$\begin{aligned} \sigma_r &= p_1\left(1 - \frac{R^2}{r^2}\right) - \frac{3}{2}kp_1^3\left(2\frac{R^2}{r^2} - \frac{R^4}{r^4} - \frac{R^6}{r^6}\right) \\ \sigma_\varphi &= p_1\left(1 + \frac{R^2}{r^2}\right) - \frac{3}{2}kp_1^3\left(-2\frac{R^2}{r^2} + 3\frac{R^4}{r^4} + 5\frac{R^6}{r^6}\right) \end{aligned} \tag{1.4.1}$$

where *k* is a small parameter characterizing the deviation of stress-strain relation from Hooke's law. The case $k = 0$ corresponds to the elastic solution. The fraction $\eta_0 = \sigma_\varphi / p_1$ at $r = R$ is called the stress concentration factor. From Eq. (1.4.1) we get

$$\eta_0 = 2 - 9kp_1^2 . \tag{1.4.2}$$

Therefore, in contrast to the elastic solution ($k = 0$), the stress concentration factor decreases with the increase in loading.

2. Let us present the other solution [48] of the considered problem obtained, in contrast to the previous case, *from the assumption of the incompressibility of metal*. For this case, on the base of Eq. (1.3.8), we have

$$u = \frac{C}{r}, \qquad \varepsilon_\varphi = -\varepsilon_r = \frac{C}{r^2}. \tag{1.4.3}$$

The shear strain intensity (1.2.8) now becomes

$$\gamma_0 = \frac{2}{\sqrt{3}}\frac{C}{r^2}. \tag{1.4.4}$$

The deviator proportionality gives

$$\sigma_r - \sigma_\varphi = 2G_S\left(\varepsilon_r - \varepsilon_\varphi\right) = -4G_S\frac{C}{r^2}, \tag{1.4.5}$$

and, on the base of Eq. (1.2.21), we find

$$\sigma_r - \sigma_\varphi = -\frac{4\tau_0(\gamma_0)}{3\gamma_0}\frac{C}{r^2}. \tag{1.4.6}$$

The equation of equilibrium is

$$r\frac{d\sigma_r}{dr} + \sigma_r - \sigma_\varphi = 0 . \tag{1.4.7}$$

Then, Eqs. (1.4.6) and (1.4.7) give the formulas determining the distribution of stress components in the plate:

$$\begin{aligned} \sigma_r &= \frac{4C}{3}\int\frac{\tau_0(\gamma_0)}{\gamma_0 r^3}dr + C_1, \\ \sigma_\varphi &= \frac{4C}{3}\int\frac{\tau_0(\gamma_0)}{\gamma_0 r^3}dr + C_1 + \frac{4C}{3}\frac{\tau_0(\gamma_0)}{\gamma_0 r^2}. \end{aligned} \tag{1.4.8}$$

The positive aspect of the obtained solution is that it is valid for any functional relation $\tau_0(\gamma_0)$. At the same time, this solution has important shortcoming that can be shown if we take the relation $\tau_0(\gamma_0)$ in the form:

$$\tau_0 = k_1 \gamma_0^k, \ (k_1, k = const, 0 < k \le 1). \qquad (1.4.9)$$

Then, Eqs. (1.4.8) and (1.4.9), with boundary conditions $\sigma_r = -p_2$ at $r = R$ and $\sigma_r = p_1$ as $r \to \infty$, give

$$\sigma_r = p_1 - (p_1 + p_2)\frac{R^{2k}}{r^{2k}},$$
$$\sigma_\varphi = p_1 + (p_1 + p_2)(2k - 1)\frac{R^{2k}}{r^{2k}} \qquad (1.4.10)$$

meaning that factor $\eta_0 = 2k$ is constant at $p_2 = 0$ and it does not depend on the value of p_1. This result originated from the assumption that constant volume is unacceptable. According to Eq. (1.4.2), obtained with the account of volume change, the stress concentration factor depends on external load and decreases with increase of p_1.

3. In contrast to above considered approaches, the solution of the problem (with boundary conditions $\sigma_r = -p_2$ at $r = R$ and the plate is stress free as $r \to \infty$) given below allows for an elastic volume change of material and is valid for any physical nonlinearity of material (i.e., for any deviation from Hooke's law on stress-strain diagram).

The deviator proportionality in a polar coordinate system takes the form

$$\varepsilon_\varphi = \frac{\sigma_\varphi - \sigma}{2G_S} + \frac{\sigma}{3K_0}, \quad \varepsilon_r = \frac{\sigma_r - \sigma}{2G_S} + \frac{\sigma}{3K_0},$$
$$\sigma = \frac{1}{3}(\sigma_\varphi + \sigma_r). \qquad (1.4.11)$$

Substituting these components in the equation of strain compatibility,

$$r\frac{d\varepsilon_\varphi}{dr} + \varepsilon_\varphi - \varepsilon_r = 0, \qquad (1.4.12)$$

we obtain

$$2\left(3\frac{d\sigma_r}{dr} + r\frac{d^2\sigma_r}{dr^2}\right)\left(\frac{1}{G_S} + \frac{1}{3K_0}\right) - \left(\sigma_r + 2r\frac{d\sigma_r}{dr}\right)\frac{1}{G_S^2}\frac{dG_S}{d\tau_0^2}\frac{d\tau_0^2}{dr} = 0 \cdot \qquad (1.4.13)$$

Let us find the solution of Eq. (1.4.13) for a particular form of $G_S(\tau_0)$ [143,148].

Let us set up the function $G_S(\tau_0)$ so that the following equation holds:

$$\frac{1}{G_S^2}\frac{dG_S}{d\tau_0^2}=-2k\left(\frac{1}{G_S}+\frac{1}{3K_0}\right), \tag{1.4.14}$$

where k is the constant of material. The solution of the above differential equation for $G_S(\tau_0)$, well-known Bernoulli equation, has the form

$$G_S=\frac{3K_0}{C\exp\left(2k\tau_0^2\right)-1}, \tag{1.4.15}$$

where C is the integration constant to be determined by the condition that $G_S=G$ at $\tau_0=0$:

$$G_S=\frac{3K_0G}{(3K_0+G)\exp\left(2k\tau_0^2\right)-G}. \tag{1.4.16}$$

By using of Eq. (1.4.16) it is possible to approximate the experimental diagrams $\tau_0-\gamma_0$ for a wide class of materials. From Eq. (1.4.16) it follows that the shear modulus G_S is smaller than G beginning from small values of τ_0, i.e., there is no elastic portion on the stress-strain diagram. Under the condition (1.4.14), Eq. (1.4.13) becomes simpler

$$3\frac{d\sigma_r}{dr}+r\frac{d^2\sigma_r}{dr^2}+k\left(\sigma_r+2r\frac{d\sigma_r}{dr}\right)\frac{d\tau_0^2}{dr}=0. \tag{1.4.17}$$

The shear stress intensity (1.2.9), in view of the equation of equilibrium (1.4.7), can be written as

$$\tau_0^2=\sigma_r^2+r\sigma_r\frac{d\sigma_r}{dr}+r^2\left(\frac{d\sigma_r}{dr}\right)^2. \tag{1.4.18}$$

Substituting τ_0^2 from the above equation into Eq. (1.4.17), we obtain the following differential equation for σ_r:

$$3r\frac{d\sigma_r}{dr}+r^2\frac{d^2\sigma_r}{dr^2}+k\left(2r\frac{d\sigma_r}{dr}+\sigma_r\right)\left[3r\sigma_r\frac{d\sigma_r}{dr}+3r^2\left(\frac{d\sigma_r}{dr}\right)^2+r^2\sigma_r\frac{d^2\sigma_r}{dr^2}+2r^3\frac{d\sigma_r}{dr}\frac{d^2\sigma_r}{dr^2}\right]=0. \tag{1.4.19}$$

The above equation is a generalized-homogeneous one and, therefore, it is possible to lower the order of derivative by means of the introduction of several new variables. The introduction of variable

$$r_1 = \ln r \tag{1.4.20}$$

gives

$$2\frac{d\sigma_r}{dr_1}+\frac{d^2\sigma_r}{dr_1^2}+\frac{4k}{9}\left(2\frac{d\sigma_r}{dr_1}+\sigma_r\right)\left[2\sigma_r\frac{d\sigma_r}{dr_1}+\left(\frac{d\sigma_r}{dr_1}\right)^2+\sigma_r\frac{d^2\sigma_r}{dr_1^2}+2\frac{d\sigma_r}{dr_1}\frac{d^2\sigma_r}{dr_1^2}\right]=0 \tag{1.4.21}$$

and this equation does not contain the variable r_1 in an explicit form. By employing variable

$$\mu=\frac{d\sigma_r}{dr_1}, \qquad \mu\frac{d\mu}{d\sigma_r}=\frac{d^2\sigma_r}{dr_1^2}, \tag{1.4.22}$$

we obtain

$$2+\frac{d\mu}{d\sigma_r}+\frac{4k}{9}(2\mu+\sigma_r)\left(2\sigma_r+\mu+\sigma_r\frac{d\mu}{d\sigma_r}+2\mu\frac{d\mu}{d\sigma_r}\right)=0. \tag{1.4.23}$$

Finally, by introducing variables

$$x=\sqrt{k}\sigma_r \text{ and } y=\sqrt{k}(\sigma_r+2\mu), \tag{1.4.24}$$

Eq. (1.4.23) becomes

$$\frac{dy}{dx}=-3\frac{1+xy}{1+y^2}. \tag{1.4.25}$$

Let us determine the boundary conditions in terms of variables x and y. The boundary conditions of problem, $\sigma_r \to 0$ and $\sigma_\varphi \to 0$ as $r \to \infty$, imply that the term $r\,d\sigma_r/dr$ in the equilibrium equation (1.4.7) tends to zero as $r \to \infty$. According to Eqs. (1.4.20) and (1.4.22), we have

$$r\frac{d\sigma_r}{dr}=\frac{d\sigma_r}{dr_1}=\mu. \tag{1.4.26}$$

Therefore, $\mu \to 0$ as $r \to \infty$ and Eq. (1.4.24) gives $x \to 0$ and $y \to 0$ as $r \to \infty$. Thus, the boundary conditions in terms of variables x and y are:

$$y=0 \text{ at } x=0. \tag{1.4.27}$$

If to take x in Eq. (1.4.25) as the function $x(y)$ to be determined, then this equation is the Abel equation of the second kind that has no solution in the closed form. Equation (1.4.25) is not integrable in elementary functions for $y(x)$ either.

Despite the fact that the solution $y(x)$ of differential equation (1.4.25) is not expressed in explicit form, we can indicate several properties of curve $y(x)$. The solution of Eq. (1.4.25) is a non-pair function because the replacements of x by $-x$ and y by $-y$ do not affect the derivative y' in Eq. (1.4.25). Further, let us consider an auxiliary hyperbola $y = -1/x$ shown by a dashed line in Fig. 1.5. Between the branches of this hyperbola, y' in Eq. (1.4.25) is negative, i.e., the integral of differential equation (1.4.25) is a descending function. Further, if to substitute $-1/x$ for y in the right-hand side in Eq. (1.4.25), then $y' = 0$. This means that the tangent line to the curve $y(x)$ is parallel to x-axis at the point of the intersection of hyperbola with curve $y(x)$. Derivative y' in Eq. (1.4.25) is positive for $|x| > a$ (see Fig. 1.5) and the solution $y(x)$ asymptotically approaches the hyperbola $y = -1/x$ at large values of x.

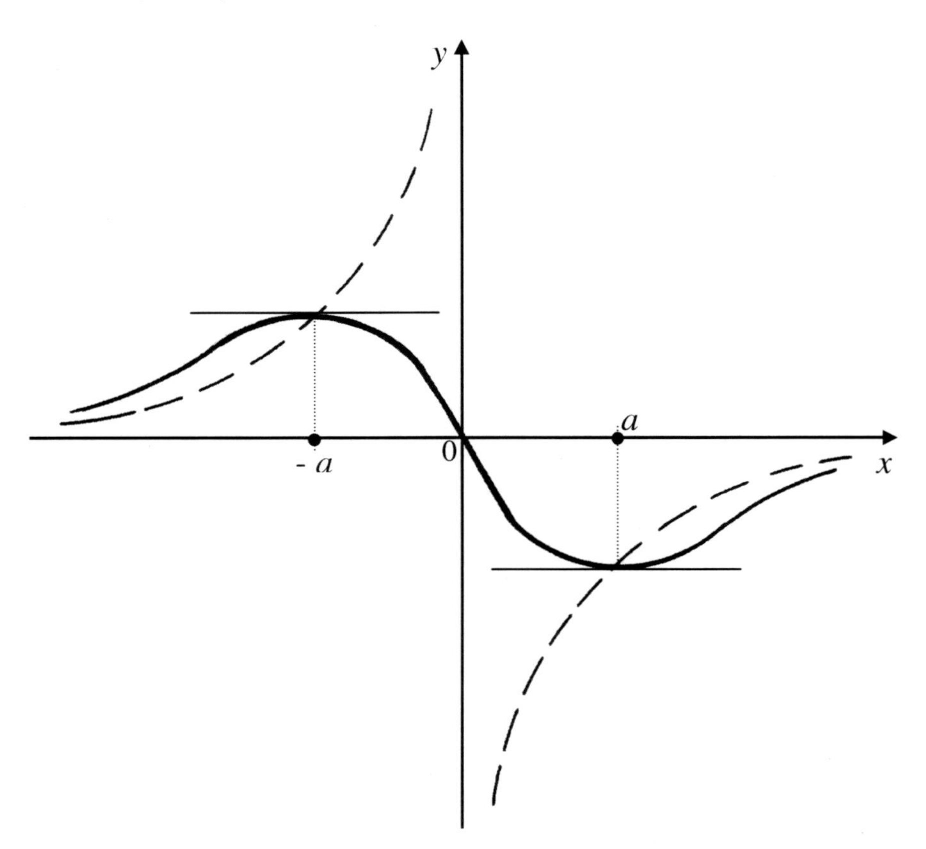

Fig. 1.5 Curve of solution of Eq. (1.4.25) (solid line).

The solution of Eq. (1.4.25) under condition (1.4.27), constructed by means of numerical methods on computer, is shown by solid line in Fig. 1.5.

By using Eqs. (1.4.20), (1.4.22), and (1.4.24), the solution of Eq. (1.4.25) expressed in terms of σ_r and r, $y\left(\sqrt{k}\sigma_r\right)$, can be found from the following equations:

$$2\sqrt{k}\int\frac{d\sigma_r}{y\left(\sqrt{k}\sigma_r\right)-\sqrt{k}\sigma_r}=\ln r+\ln C,$$
$$\sigma_\varphi=\frac{\sigma_r}{2}+\frac{y\left(\sqrt{k}\sigma_r\right)}{2\sqrt{k}}, \tag{1.4.28}$$

where σ_φ is obtained from the equilibrium equation (1.4.7).

At the vicinity of point $x=0$, the solution $y(x)$ can be evaluated by means of the McLauren polynomial series

$$y=\frac{y_0^{(j)}}{j!}x^j, \tag{1.4.29}$$

where $y_0^{(j)}$ is a j-order derivative with respect to x at $x=0$. The derivatives $y_0^{(j)}$ are calculated by differentiating Eq. (1.4.25). Dropping of the term x^5 and higher powers in x, we obtain

$$y=-3x+12x^3. \tag{1.4.30}$$

By using the transformations of variables in Eqs. (1.4.20), (1.4.22) and (1.4.24), Eq. (1.4.30) takes the form

$$\sigma_r=-p_2\frac{R^2}{r^2}-\frac{3}{2}kp_2^3\left(1-\frac{R^4}{r^4}\right)\frac{R^2}{r^2},$$
$$\sigma_\varphi=p_2\frac{R^2}{r^2}+\frac{3}{2}kp_2^3\left(1-5\frac{R^4}{r^4}\right)\frac{R^2}{r^2}. \tag{1.4.31}$$

These formulas were obtained in [168] by other method. Similar to Eqs. (1.4.1), Eqs. (1.4.31) are applicable at small parameter k. Eq. (1.4.31) gives the following value for η_0:

$$\eta_0 = 1 - 6kp_2^2 \tag{1.4.32}$$

and $\eta_0 = 1$ at elastic straining ($k = 0$).

While $y\left(\sqrt{k}\sigma_r\right)$ in Eq. (1.4.28) is unknown, the values of stress concentration factor can be evaluated not only at the vicinity of point $x = 0$. It is worthwhile to note that, as follows from Eq. (1.4.24), but also the magnitude of x is proportional to the boundary radial stress at $r = R$, $x = \sqrt{k}\, p_2$. Indeed, Eq. (1.4.28) at $r = R$ when $\sigma_r = -p_2$ gives

$$\sigma_\varphi = -\frac{p_2}{2} + \frac{y\left(-\sqrt{k}\, p_2\right)}{2\sqrt{k}} \tag{1.4.33}$$

and, as a consequence,

$$\eta_0 = \left.\frac{\sigma_\varphi}{p_2}\right|_{r=R} = -\frac{1}{2} + \frac{y\left(-\sqrt{k}\, p_2\right)}{2\sqrt{k}\, p_2}. \tag{1.4.34}$$

The above formula, in view of Eq. (1.4.30), gives that $\eta_0 \to 1$ as x or p_2 tends to zero. At the large values of p_2 the solution of the problem, $y(x)$, approaches the hyperbola implying that η_0 from Eq. (1.4.34) drops and tends to -1/2 as $p_2 \to \infty$. Therefore, the values of the η_0 lie in the range $-1/2 \le \eta_0 \le 1$ at $\infty \ge p_2 \ge 0$.

Summarizing the results of the point **3.**, Eq. (1.4.16) and Eqs. (1.4.28)–(1.4.34), we must emphasize that the obtained solution is applicable for compressible materials with any physical nonlinearity. This enables us to determine the range of stress concentration factor for all possible values of external load.

1.5 Ilyushin's Theorems

Ilyushin's first theorem on a proportional (simple) loading [48].

First of all, let us define the notion "proportional (simple) loading". The loading is simple or proportional if the current values of each stress-tensor component at each point in a body are related proportionally to their final values

$$\sigma_{ij} = \eta\sigma_{ij}^0, \tag{1.5.1}$$

where the parameter η monotonously grows from 0 to 1, and σ_{ij}^{0} denotes the final values of stress components ($\eta = 1$). In the case of proportional loading, the ratio

$$\frac{\sigma_{ij}}{\sigma_{kl}} = \frac{\sigma_{ij}^{0}}{\sigma_{kl}^{0}} = const \tag{1.5.2}$$

remains constant during the whole process of loading, i.e., Eqs (1.5.1) and (1.5.2) are equivalent. To show that Eq. (1.5.1) leads to Eq. (1.5.2) and vice versa, we write Eq. (1.5.2) in the form

$$\sigma_{ij} = \frac{\sigma_{kl}}{\sigma_{kl}^{0}} \sigma_{ij}^{0} \qquad \not{\Sigma} \tag{1.5.3}$$

The symbol $\not{\Sigma}$ means that there is no summation over dummy indexes. Denoting the fraction $\sigma_{kl} / \sigma_{kl}^{0}$ as η, we arrive at Eq. (1.5.1).

Let as examine whether Eq. (1.5.2) is valid for the following states of stresses. Calculate $\sigma_r / \sigma_\varphi$ ratio for the case of an infinite plate with a circular hole of radius R for the following boundary condition: $\sigma_r \to p_1$ as $r \to \infty$ and $\sigma_r = -p_2$ at $r = R$. The solution of this problem is given by Eqs. (1.4.31) for $p_2 > 0$ and $p_1 = 0$, and, in the case $p_2 = 0$ and $p_1 > 0$, the solution is given by Eqs. (1.4.1). Let us assume (approximately) that the solution for the case $p_1 > 0$ and $p_2 > 0$ can be expressed as the sum of the right-hand sides of Eqs. (1.4.1) and (1.4.31) which are obtained on the base of Hencky–Nadai stress-strain relations accounting for the volume change. On the base of stated above

$$\frac{\sigma_r}{\sigma_\varphi} = \frac{C_1 p_1 + C_2 k p_1^3 + C_3 p_2 + C_4 k p_2^3}{C_5 p_1 + C_6 k p_1^3 + C_7 p_2 + C_8 k p_2^3}, \tag{1.5.4}$$

where C_i are functions of r, whose forms are not important for current analysis.

Let us assume that p_2 changes proportionally to p_1 during loading, i.e.,

$$p_2 = \eta_1 p_1 \qquad (\eta_1 = const\,). \tag{1.5.5}$$

Then, Eq. (1.5.4) gives

$$\frac{\sigma_r}{\sigma_\varphi} = \frac{C_1 + C_2 k p_1^2 + C_3 \eta_1 + C_4 k \eta_1^3 p_1^2}{C_5 + C_6 k p_1^2 + C_7 \eta_1 + C_8 k \eta_1^3 p_1^2}. \tag{1.5.6}$$

From Eq. (1.5.6) we thus conclude that the ratio σ_r/σ_φ is not constant during loading (value of p_1 grows) although external forces p_1 and p_2 are related to each other proportionally.

The situation is different for the case of homogeneous state of stress: the proportional change of external forces automatically leads to Eq. (1.5.2). Consider, for example, thin-walled cylindrical cylinder subjected to tension by force *P* and torsion by moment M_k. Normal and shear stresses in the cylinder are

$$\sigma = \frac{P}{2\pi R h}, \quad \tau = \frac{M_k}{2\pi R^2 h}, \tag{1.5.7}$$

where *h* and *R* is the thickness and the average radius of cylinder, respectively. If moment M_k changes proportionally to force *P*,

$$M_k = P R \eta_1, \tag{1.5.8}$$

then, Eqs (1.5.7) gives

$$\frac{\tau}{\sigma} = \eta_1, \tag{1.5.9}$$

i.e., the loading is simple.

Numerous experiments show that Hencky–Nadai relations concur with experiments at simple loading at combined state of stress for many materials. But as it is seen from formula (1.5.6), proportional change of external forces does not mean automatically that the condition (1.5.1) is fulfilled.

Therefore, let us consider now the following problem:

a) Let external surface- and body-forces applied to a body (let they be denoted as ζ) vary proportionally with a common factor, η_1,

$$\zeta = \eta_1 \zeta^0 \quad (0 \le \eta_1 \le 1), \tag{1.5.10}$$

where ζ^0 is a maximal value of ζ.

b) Employing equilibrium equations, compatibility conditions and boundary condition (1.5.10) for $\eta_1 = 1$ as well as the Hencky–Nadai relations (1.2.4), some boundary-value (in a general case, inhomogeneous) problem has been solved, i.e., the stress and strain distribution in a body has been determined. Then, the following question arises: under what conditions the stresses induced by proportional external forces will also be proportional, i.e., governed by Eq. (1.5.1), at each point in a body?

The answer to the posed question gives ***Ilyushin's first theorem***:

In a view of the conditions a) and b), the power relation between shear-stress and shear-strain intensities given by Eq. (1.4.9):

$$\tau_0 = k_1 \gamma_0^k, \ (k_1, k = const, 0 < k \leq 1),$$

and an incompressibility of a body, $\varepsilon = 0$, are sufficient conditions of simple loading, i.e., fulfillment of Eq. (1.5.1), at each point in a body.

► Let us use a semi-inverse approach. Assume that Eq. (1.5.1) is valid at each point in the body, i.e., the parameter η does not vary with position in a body. Now, let us show that Eq. (1.4.9) and the condition $\varepsilon = 0$ lead to correct results. By correctness we mean the following. Let σ_{ij}^0, ε_{ij}^0 and $\sigma_{ij}, \varepsilon_{ij}$ denote stress and strain components corresponding to the external forces ς^0 and ς, respectively. According to the initial condition of the theorem, a boundary-value problem has been solved for external forces ς^0 that automatically implies that the components σ_{ij}^0 and ε_{ij}^0 satisfy equilibrium equations as well as boundary and compatibility conditions. The key question is that whether stress components σ_{ij} determined by Eq. (1.5.1) and strain components ε_{ij} induced by them also satisfy specified equations and conditions.

Due to the linearity of equilibrium equations and to that the coefficient η does not vary with position in body, components σ_{ij} determined by Eq. (1.5.1) also satisfy the equilibrium equations. On the base of Eq. (1.5.1), Eqs. (1.2.1), (1.2.9) and (1.2.24) give the following relationships:

$$\bar{\sigma}_{ij} = \eta \bar{\sigma}_{ij}^0, \quad \tau_0 = \eta \tau_0^0, \quad \varepsilon = \eta \varepsilon^0. \tag{1.5.11}$$

Therefore, Hencky–Nadai relations (1.2.4) give strain-deviator-tensor components,

$$\bar{\varepsilon}_{ij}^0 = \frac{\bar{\sigma}_{ij}^0}{2G_S\left(\tau_0^0\right)}, \tag{1.5.12}$$

for $\eta_1 = 1$ and, using Eq. (1.5.11), we get

$$\bar{\varepsilon}_{ij} = \frac{\bar{\sigma}_{ij}}{2G_S(\tau_0)} = \frac{\eta\bar{\sigma}_{ij}^0}{2G_S(\eta\tau_0^0)}. \tag{1.5.13}$$

Equations (1.5.11) and (1.5.13) give

$$\bar{\varepsilon}_{ij} = \eta_2\bar{\varepsilon}_{ij}^0, \ \eta_2 = \eta\frac{G_S(\tau_0^0)}{G_S(\eta\tau_0^0)} \tag{1.5.14}$$

and the strain-tensor components can be rewritten as

$$\varepsilon_{ij} = \eta\varepsilon^0\delta_{ij} + \eta_2\bar{\varepsilon}_{ij}^0 = (\eta - \eta_2)\varepsilon^0\delta_{ij} + \eta_2\varepsilon_{ij}^0. \tag{1.5.15}$$

Since $\varepsilon^0 = 0$,

$$\varepsilon_{ij} = \eta_2\varepsilon_{ij}^0. \tag{1.5.16}$$

The fact that components ε_{ij}^0 satisfy the compatibility conditions does not imply automatically that components ε_{ij} also satisfy these conditions because the value of η_2 varies with position in body (see Eq. (1.5.14)). But Eqs. (1.2.21), (1.4.9), and (1.5.14) give

$$G_S = \frac{k_1^{1/k}}{3}\tau_0{}^{1-1/k} \text{ and } \eta_2 = \eta^{1/k} = const, \tag{1.5.17}$$

which provides the fulfillment of the compatibility conditions for components ε_{ij} given by Eq. (1.5.16). To satisfy the boundary conditions, it is necessary to take that $\eta = \eta_1$.

Therefore, due to the theorem on the uniqueness of a solution of boundary-value problem, we have proved that it is the stresses given Eq. (1.5.1) that really act in a body because they satisfy all equations of continuous mechanics: equilibrium equations as well as boundary and compatibility conditions. Thus the assumption (1.5.1) is correct and the theorem is proved. ◀

As an illustration of the theorem, consider Eqs (1.4.10) obtained from the conditions identical to those of the theorem. Assuming that

$$p_1 = \eta p_1^0 \quad \text{and} \quad p_2 = \eta p_2^0, \tag{1.5.18}$$

Eqs (1.4.10) read:

$$\sigma_r = \eta \sigma_r^0, \quad \sigma_\varphi = \eta \sigma_\varphi^0, \tag{1.5.19}$$

where stresses σ_r^0 and σ_φ^0 correspond to the boundary values of p_1^0 and p_2^0 and stresses σ_r and σ_φ to p_1 and p_2.

Ilyushin's first theorem provides the legitimacy of Hencky–Nadai deformation theory. By the legitimacy we mean that this theorem provides the conditions for the realisation of proportional loading; for this type of loading, a huge amount of researches shows that Hencky–Nadai stress-strain relations concur with experimental data.

Let us note that the conditions of the first theorem are restrictive enough. In particular, a power relation between τ_0 and γ_0 implies that the stress-strain diagram has no linear portion. The incompressibility of material also leads to questionable results (see Sec. 1.4).

Ilyushin's second theorem on a proportional loading [48].

Consider symmetrical matrices, D_σ and D_ε, that represent the stress and strain tensor, respectively. Furthermore, we introduce the matrices, $D_{\bar{\sigma}/\tau_0}$ and $D_{\bar{\varepsilon}/\gamma_0}$

$$D_{\bar{\sigma}/\tau_0} = \begin{pmatrix} \bar{\sigma}_{11}/\tau_0 & \bar{\sigma}_{12}/\tau_0 & \bar{\sigma}_{13}/\tau_0 \\ \bar{\sigma}_{21}/\tau_0 & \bar{\sigma}_{22}/\tau_0 & \bar{\sigma}_{23}/\tau_0 \\ \bar{\sigma}_{13}/\tau_0 & \bar{\sigma}_{23}/\tau_0 & \bar{\sigma}_{33}/\tau_0 \end{pmatrix} \quad \text{and}$$
$$D_{\bar{\varepsilon}/\gamma_0} = \begin{pmatrix} \bar{\varepsilon}_{11}/\gamma_0 & \bar{\varepsilon}_{12}/\gamma_0 & \bar{\varepsilon}_{13}/\gamma_0 \\ \bar{\varepsilon}_{12}/\gamma_0 & \bar{\varepsilon}_{22}/\gamma_0 & \bar{\varepsilon}_{23}/\gamma_0 \\ \bar{\varepsilon}_{13}/\gamma_0 & \bar{\varepsilon}_{23}/\gamma_0 & \bar{\varepsilon}_{33}/\gamma_0 \end{pmatrix} \tag{1.5.20}$$

where $\bar{\sigma}_{ij}$ and $\bar{\varepsilon}_{ij}$ are stress and strain deviator tensor, and τ_0 and γ_0 are the second invariant of the stress and strain tensor, respectively (see Eqs. (1.2.8) and (1.2.9)).

Let us assume that the matrix D_σ is a function of some monotonously increasing parameter η_3 (for example, time):

$$D_\sigma = D_\sigma(\eta_3), \quad \sigma_{ij} = \sigma_{ij}(\eta_3). \tag{1.5.21}$$

The relation between D_σ and η_3 we call proportional (simple) if the matrix $D_{\overline{\sigma}/\tau_0}$ does not depend on η_3.

Further, let L denote some linear operator on D_σ, for example

$$L(D_\sigma) = A_1 D_\sigma + A_2 \frac{dD_\sigma}{d\eta_3} + A_3 \frac{d^2 D_\sigma}{d\eta_3^2} + \int_0^{\eta_3} A_4 D_\sigma d\eta_3, \tag{1.5.22}$$

where A_i are functions of invariants σ and τ_0. By "linearity" we mean that any operation in Eq. (1.5.22) (derivative, integral, etc.) appears linearly. Let us assume that the strain-stress relation is

$$D_\varepsilon = L(D_\sigma) \quad \text{or} \quad \varepsilon_{ij} = L(\sigma_{ij}). \tag{1.5.23}$$

Ilyushin's second theorem reads: **if D_σ-η_3 relation is simple and Eq. (1.5.23) holds, then a shear-strain intensity, γ_0, can be determined as**

$$\gamma_0 = \frac{2}{3} L(\tau_0). \tag{1.5.24}$$

In addition, matrices $D_{\overline{\varepsilon}/\gamma_0}$ and $D_{\overline{\sigma}/\tau_0}$ are identical (up to a constant):

$$D_{\overline{\varepsilon}/\gamma_0} = \frac{3}{2} D_{\overline{\sigma}/\tau_0}. \tag{1.5.25}$$

► Let us express D_σ and D_ε as the sum of hydrostatic and deviatoric parts

$$D_\sigma = \sigma E_0 + \tau_0 D_{\overline{\sigma}/\tau_0}, \tag{1.5.26a}$$

$$D_\varepsilon = \varepsilon E_0 + \gamma_0 D_{\overline{\varepsilon}/\gamma_0}, \tag{1.5.26b}$$

where E_0 is the unit matrix, and σ and ε are the first invariant of stress and strain tensor, respectively. Volumetric strain, ε, on the base of Eq. (1.5.23) takes the form

$$\varepsilon = \frac{1}{3}(\varepsilon_{11} + \varepsilon_{11} + \varepsilon_{11}) = \frac{1}{3}[L(\sigma_{11}) + L(\sigma_{22}) + L(\sigma_{33})] = \frac{1}{3}L(\sigma_{11} + \sigma_{22} + \sigma_{33}) = L(\sigma) \qquad (1.5.27)$$

that follows from the linearity of operator L. Further, let us act by the operator L on the left- and right-hand side of Eq.(1.5.26a). According to the condition of the theorem, the matrix $D_{\overline{\sigma}/\tau_0}$ does not depend on parameter η_3, therefore

$$D_\varepsilon = L(D_\sigma) = L(\sigma)E_0 + L(\tau_0)D_{\overline{\sigma}/\tau_0}\,. \qquad (1.5.28)$$

Comparing the right-hand sides of Eq. (1.5.26b) and (1.5.28) and taking into account Eq. (1.5.27), we obtain

$$\gamma_0 D_{\overline{\varepsilon}/\gamma_0} = L(\tau_0)D_{\overline{\sigma}/\tau_0}\,. \qquad (1.5.29)$$

Since the tensor $D_{\overline{\sigma}/\tau_0}$ does not depend on parameter η_3, i.e., its components $\overline{\sigma}_{ij}/\tau_0$ remain constant for any value of η_3, it can be stated that Eq. (1.5.1) holds if

$$\eta = \eta(\eta_3), \qquad (1.5.30)$$

where $\eta(\eta_3)$ is monotonously increasing function.

According to Eq. (1.5.22), we get

$$\frac{L(\sigma_{ij})}{L(\tau_0)} = \frac{A_1\sigma_{ij} + A_2\dfrac{d\sigma_{ij}}{d\eta_3} + \ldots + \int\limits_0^{\eta_3} A_4\sigma_{ij}d\eta_3}{A_1\tau_0 + A_2\dfrac{d\tau_0}{d\eta_3} + \ldots + \int\limits_0^{\eta_3} A_4\tau_0 d\eta_3}\,. \qquad (1.5.31)$$

The above formula, together with Eqs. (1.5.1) and (1.5.30), gives

$$\frac{L(\sigma_{ij})}{L(\tau_0)} = \frac{A_1\sigma_{ij}^0\eta + A_2\sigma_{ij}^0\dfrac{d\eta}{d\eta_3} + \ldots + \sigma_{ij}^0\int\limits_0^{\eta_3} A_4\eta d\eta_3}{A_1\tau_0^0\eta + A_2\tau_0^0\dfrac{d\eta}{d\eta_3} + \ldots + \tau_0^0\int\limits_0^{\eta_3} A_4\eta d\eta_3}, \qquad (1.5.32)$$

where τ_0^0 is the value of τ_0 corresponding to components σ_{ij}^0 ($\eta = 1$). Taking out σ_{ij}^0 and τ_0^0 of the brackets in the numerator and in denominator, and reducing the fraction, we obtain

$$\frac{L(\sigma_{ij})}{L(\tau_0)} = \frac{\sigma_{ij}^0}{\tau_0^0}. \tag{1.5.33}$$

Multiplying the numerator and denominator of the right-hand side in Eq. (1.5.33) by η and taking into account that $L(\sigma_{ij}) = \varepsilon_{ij}$, we obtain

$$\varepsilon_{ij} = \frac{L(\tau_0)}{\tau_0}\sigma_{ij}. \tag{1.5.34}$$

Substituting the above strain components into the expression for the second invariant (1.2.8), we arrive at Eq. (1.5.24). Eventually, Eqs. (1.5.24) and (1.5.29) give Eq. (1.5.25). The theorem is proved. ◄

On the base of the second theorem, general tensor-linear relations between stresses and strains can be studied under a simple load (Sec. 1.11).

Ilyushin's third theorem on unloading

Since small strains are only considered, it is possible to decompose them on elastic (ε_{ij}^e) and plastic (ε_{ij}^S) components:

$$\varepsilon_{ij} = \varepsilon_{ij}^e + \varepsilon_{ij}^S. \tag{1.5.35}$$

This formula is one of the basic relations of the theory of plasticity.

Lemma. If the deviator proportionality (1.2.4) is satisfied, then the shear strain intensity can also be decomposed on elastic (γ_0^e) and plastic (γ_0^S) components,

$$\gamma_0 = \gamma_0^e + \gamma_0^S, \tag{1.5.36}$$

where

$$\begin{aligned} \gamma_0^e &= \frac{\sqrt{2}}{3}\left[\left(\varepsilon_x^e - \varepsilon_y^e\right)^2 + \ldots 6\left(\varepsilon_{xy}^e\right)^2 + \ldots\right]^{1/2}, \\ \gamma_0^S &= \frac{\sqrt{2}}{3}\left[\left(\varepsilon_x^S - \varepsilon_y^S\right)^2 + \ldots 6\left(\varepsilon_{xy}^S\right)^2 + \ldots\right]^{1/2}. \end{aligned} \tag{1.5.37}$$

► To be convinced of the validity of Eq. (1.5.36), let us determine ε_{ij}^{e} and ε_{ij}^{S}. Substituting the elastic shear modulus G for modulus G_S in Eqs (1.2.4) and using Eqs. (1.2.1), (1.2.23), and (1.2.24), we find the elastic components ε_{ij}^{e}:

$$\varepsilon_{ij}^{e} = \frac{\sigma\delta_{ij}}{3K_0} + \frac{\sigma_{ij} - \sigma\delta_{ij}}{2G}. \tag{1.5.38}$$

Plastic components ε_{ij}^{S} can be obtained as the difference of the total (1.2.4) and elastic (1.5.38) components:

$$\varepsilon_{ij}^{S} = \frac{1}{2}\left(\sigma_{ij} - \sigma\delta_{ij}\right)\left(\frac{1}{G_S} - \frac{1}{G}\right). \tag{1.5.39}$$

By substituting the strain components from Eqs. (1.5.38) and (1.5.39) into the right-hand sides in Eqs.(1.5.37) and adding them together in Eq. (1.5.36), we will arrive at Eq. (1.2.8) (making use of Eq. (1.2.5)). ◄

Consider the case when a material is first deformed plastically and then unloaded. As the unloading is governed by the elastic law, $\Delta\gamma_0^S = 0$ during unloading, Eq. (1.5.36) takes the form:

$$\Delta\gamma_0 = \Delta\gamma_0^e, \tag{1.5.40}$$

i.e., the increment $\Delta\gamma_0$ does not depend on the value of plastic preloading. It must be noted that $\tau_0 - \gamma_0$ curve behaves in the same fashion as stress-strain curve during both loading and subsequent unloading: when the stress intensity τ_0 drops, the strain intensity γ_0 decreases along a straight line parallel to the initially linear portion of the curve, i.e.,

$$\tau_0^u = 3G\gamma_0^u, \tag{1.5.41}$$

where intensities τ_0^u and γ_0^u are shown in Fig. 1.2.

Let us assume that a solid body is first loaded by external surface and body forces, ς^0, which induce stresses σ_{ij}^0 and elasto-plastic strains ε_{ij}^0. Then, the external forces decreases to the value of $\varsigma < \varsigma^0$. As a result, stress and strain components drop by values of σ_{ij}^u and ε_{ij}^u to the residual values of σ_{ij} and ε_{ij}:

$$\sigma_{ij} = \sigma_{ij}^0 - \sigma_{ij}^u, \qquad (1.5.42)$$
$$\varepsilon_{ij} = \varepsilon_{ij}^0 - \varepsilon_{ij}^u,$$

Ilyushin's third theorem sets up the rule for the determination of the residual stresses and strains. Let the stress and strain components, σ_{ij}^0 and ε_{ij}^0, be the solution of elasto-plastic boundary-value problem for external forces ς^0 (i.e., σ_{ij}^0 and ε_{ij}^0 satisfy the equations of equilibrium and the conditions of the compatibility of deformations). We wish to determine the residual components σ_{ij} and ε_{ij}, which, naturally, satisfy specified equations for external forces ς. The formula (1.5.42), being linear, ensures that σ_{ij}^u and ε_{ij}^u satisfy the equations of equilibrium and the conditions of the compatibility of deformations as well. Since the unloading is governed by Eq. (1.5.41), we arrive at the following conclusion – **Ilyushin's third theorem** – *stresses* σ_{ij}^u *and strains* ε_{ij}^u *are obtained from the solution of* ***elastic*** *boundary-value problem for the external forces* $\varsigma^0 - \varsigma$. *For the case of complete unloading,* $\varsigma = 0$, *this* ***elastic*** *problem must be solved for the external forces* ς^0, *which acted at plastic preloading. Once the values of* σ_{ij}^u *and* ε_{ij}^u *are known, the residual stresses and strains are calculated by Eq. (1.5.42).*

As an example of the application of Ilyushin's third theorem, let us calculate residual stresses in a beam in pure bending after removing of bending moment. The material of beam is elastic-perfectly plastic with stress-strain diagram in tension (or compression) shown in Fig. 1.4a.

Under pure bending, a plane surface perpendicular to the longitudinal axis of the beam before bending will remain plane and perpendicular to the axis of the beam after bending. Therefore, the axial strain, ε_z, is

$$\varepsilon_z = cy. \qquad (1.5.43)$$

According to Eq. (1.5.43) and $\sigma - \varepsilon$ diagram showed in Fig. 1.4a, the normal stress distribution will be elastic over a central core region and plastic over an outer region, as shown in Fig. 1.6b, where y_0 defines the outer boundary of the elastic region and

$$\sigma_z = \begin{cases} \dfrac{y}{y_0}\sigma_S & \text{for } |y| \le y_0 \\ \pm\sigma_S & \text{for } y_0 \le |y| \le \dfrac{h}{2} \end{cases} \tag{1.5.44}$$

The bending moment M_x associated with the stress distribution given by Eq. (1.5.44) can be calculated as

$$\varsigma = M_x = \iint \sigma_z y dx dy\,; \tag{1.5.45}$$

the integral is taken over the cross-section of the beam. If the cross section is a rectangle of width b and of height h (Fig. 1.6a), Eq. (1.5.45) becomes simpler

$$M_x = 2b\int_0^{h/2} \sigma_z y dy = 2b\left(\int_0^{y_0} \sigma_z y dy + \int_{y_0}^{h/2} \sigma_z y dy\right) \tag{1.5.46}$$

and making use of Eq.(1.5.44) in Eq. (1.5.46), we will obtain

$$M_x = \frac{b\sigma_S}{12}\left(3h^2 - 4y_0^2\right). \tag{1.5.47}$$

For the case of elastic problem, the moment M_x given by Eq. (1.5.47) would produce the following normal stress distribution:

$$\sigma_z = \frac{12M_x y}{bh^3} = \frac{\left(3h^2 - 4y_0^2\right)\sigma_S}{h^3}y\,. \tag{1.5.48}$$

According to Ilyushin's third theorem, residual stresses acting in a solid after complete unloading can be found by subtracting of elastic solution from elastic-plastic one. Therefore, in order to determine residual stresses in the beam in bending, one must subtract the right-hand side of Eq. (1.5.48) from that of Eq. (1.5.44):

$$\sigma_z = \begin{cases} \left(\dfrac{1}{y_0} - \dfrac{3h^2 - 4y_0^2}{h^3}\right)y\sigma_S & \text{at } |y| \le y_0 \\ \pm\left(1 - \dfrac{3h^2 - 4y_0^2}{h^3}|y|\right)\sigma_S & \text{at } y_0 \le |y| \le \dfrac{h}{2} \end{cases} \tag{1.5.49}$$

For the case $y_0 = h/4$, the above formula takes the following form (the plot of σ_z is shown in Fig. 1.6c.):

$$\sigma_z = \begin{cases} \dfrac{5y}{4h}\sigma_S & \text{at } |y| \le \dfrac{h}{4} \\ \pm\left(1 - \dfrac{11|y|}{4h}\right)\sigma_S & \text{at } \dfrac{h}{4} \le |y| \le \dfrac{h}{2} \end{cases} \qquad (1.5.50)$$

If we substitute stresses (1.5.49) or (1.5.50) into Eq.(1.5.46), we obtain that bending moment $M_x = 0$, i.e., residual stresses statically equivalent to zero.

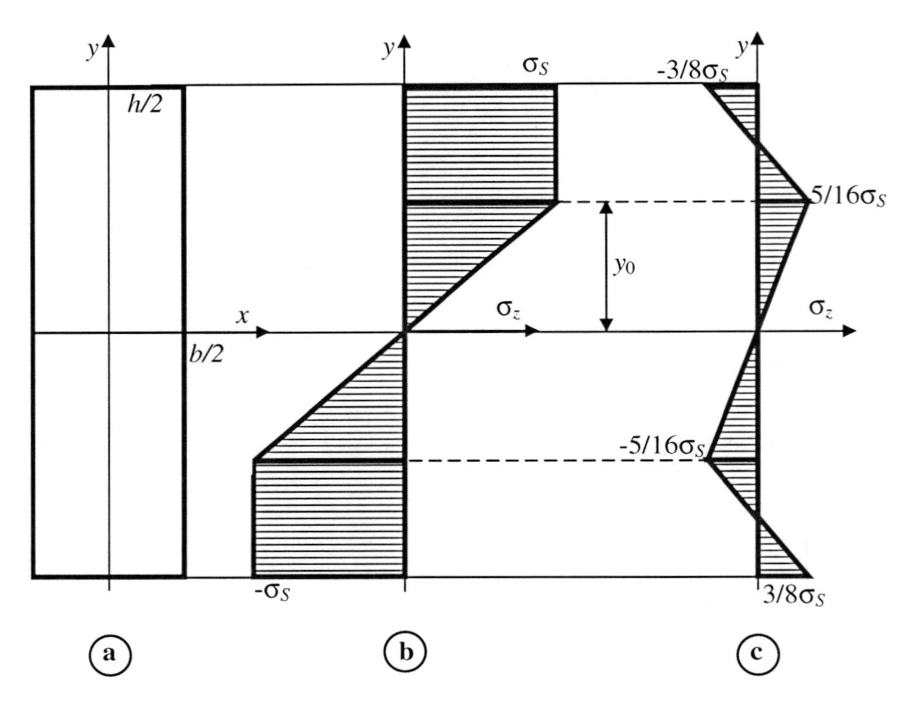

Fig. 1.6 Beam cross-section (a), distribution of normal stress across the section in elastic-plastic bending (b), distribution of residual stress (c).

1.6 Prager Deformation Theory

The Hencky–Nadai theory is a simplest theory among the most known theory of plasticity. In terms of this theory, a number of technically important problems are solved. At the same time, other deformation theories, giving more general stress-strain relations than the deviator proportionality [115], also exist.

Let us denote $D_{\overline{\sigma}}$ as the matrix with stress deviator components ($\overline{\sigma}_{ij} = \overline{\sigma}_{ji}$) and introduce the matrix $D_{\overline{\sigma}}^2 = D_{\overline{\sigma}} \cdot D_{\overline{\sigma}}$

$$D_{\overline{\sigma}}^2 = \begin{pmatrix} \overline{\sigma}_{11} & \overline{\sigma}_{12} & \overline{\sigma}_{13} \\ \overline{\sigma}_{21} & \overline{\sigma}_{22} & \overline{\sigma}_{23} \\ \overline{\sigma}_{31} & \overline{\sigma}_{32} & \overline{\sigma}_{33} \end{pmatrix} \begin{pmatrix} \overline{\sigma}_{11} & \overline{\sigma}_{12} & \overline{\sigma}_{13} \\ \overline{\sigma}_{21} & \overline{\sigma}_{22} & \overline{\sigma}_{23} \\ \overline{\sigma}_{31} & \overline{\sigma}_{32} & \overline{\sigma}_{33} \end{pmatrix} \tag{1.6.1}$$

whose entries d_{ij}:

$$d_{ij} = \overline{\sigma}_{im} \overline{\sigma}_{mj}. \tag{1.6.2}$$

It is easy to see that matrix $D_{\overline{\sigma}}^2$ is symmetrical. We may enter in analogical manner the following matrix

$$D_{\overline{\sigma}}^3 = D_{\overline{\sigma}} D_{\overline{\sigma}}^2 \tag{1.6.3}$$

with entries d_{ij}':

$$d_{ij}' = \overline{\sigma}_{im} \overline{\sigma}_{mk} \overline{\sigma}_{kj}. \quad \text{(1.6.4}$$

The sums in the right-hand side in this formula can be permuted, so that it is possible to write

$$D_{\overline{\sigma}}^3 = D_{\overline{\sigma}}^2 D_{\overline{\sigma}}. \tag{1.6.5}$$

Analogically, we may enter $D_{\overline{\sigma}}^4, D_{\overline{\sigma}}^5, \ldots$.

Let $D_{\overline{\varepsilon}}$ be the matrix to present a strain deviator tensor. We assume that the relation between the strain deviator tensor $D_{\overline{\varepsilon}}$ and $D_{\overline{\sigma}}^i$ can be written as infinite series

$$D_{\overline{\varepsilon}} = A_0' E_0 + A_1' D_{\overline{\sigma}} + A_2' D_{\overline{\sigma}}^2 + A_3' D_{\overline{\sigma}}^3 + \ldots \tag{1.6.6}$$

where $A_0', A_1', A_2' \ldots$ are functions of invariants of stress tensor, and E_0 is unit matrix.

To simplify Eq. (1.6.6), apply the Cayley–Hamilton theorem which states that every square matrix (tensor) satisfies its own characteristic equation. In particular, matrix $D_{\overline{\sigma}}$ must satisfy the following equation

$$D_{\bar{\sigma}}^{3} - \frac{\tau_0^2}{3} D_{\bar{\sigma}} + \frac{\overline{A}_3}{3} E_0 = 0\,, \tag{1.6.7}$$

where $\overline{A}_3$ is the third invariant of stress deviator tensor:

$$\overline{A}_3 = -\begin{vmatrix} \overline{\sigma}_x & \tau_{xy} & \tau_{xz} \\ \tau_{yx} & \overline{\sigma}_y & \tau_{yz} \\ \tau_{zx} & \tau_{zy} & \overline{\sigma}_z \end{vmatrix} \tag{1.6.8}$$

It is easy to see that the recurrent employing of Eq. (1.6.7) to the series (1.6.6) leads to the following relation

$$D_{\bar{\varepsilon}} = A_0 E_0 + A_1 D_{\bar{\sigma}} + A_2 D_{\bar{\sigma}}^{2}\,, \tag{1.6.9}$$

where A_0, A_1, and A_2 are functions of the invariant of stress.

Since the left-hand side in Eq. (1.6.9) is a deviator, therefore, its right-hand side must also be deviator, thus

$$3A_0 + A_2(d_{11} + d_{22} + d_{33}) = 0\,, \tag{1.6.10}$$

where d_{ij} are the components of the matrix $D_{\bar{\sigma}}^{2}$. On the basis of Eqs. (1.2.9) and (1.6.2), we obtain

$$A_0 = -\frac{2}{9} A_2 \tau_0^2\,, \tag{1.6.11}$$

i.e.,

$$D_{\bar{\varepsilon}} = A_1 D_{\bar{\sigma}} + A_2 \left(D_{\bar{\sigma}}^{2} - \frac{2}{9} \tau_0^2 E_0 \right). \tag{1.6.12}$$

The above formula proposed by Prager expresses a nonlinear relationship between strain deviator and stress deviator tensors.

V.V. Novozsilov expresses the parameters A_1 and A_2 through the invariants of stress and strain tensors [102]; here, we will not consider this case.

An inverse relation, to that in Eq. (1.6.9), must be written as

$$D_{\bar{\sigma}} = \tilde{A}_0' E_0 + \tilde{A}_1' D_{\bar{\varepsilon}} + \tilde{A}_2' D_{\bar{\varepsilon}}^{2} + \ldots. \tag{1.6.13}$$

Doing in a similar manner as in Eqs. (1.6.10)–(1.6.12), from Eq. (1.6.13) it follows that

$$D_{\overline{\sigma}} = \widetilde{A}_1 D_{\overline{\varepsilon}} + \widetilde{A}_2 \left(D_{\overline{\varepsilon}}^2 - \frac{2}{9}\gamma_0^2 E_0 \right), \tag{1.6.14}$$

where $\widetilde{A}_1$ and $\widetilde{A}_2$ are functions of invariants of strain and stress tensors. Equations (1.6.12) and (1.6.14) express relationship between the stress and strain deviatoric tensors. If $A_2 = \widetilde{A}_2 = 0$, then Eqs. (1.6.12) and (1.6.14) lead to the deviators proportionality.

1.7 Ilyushin's Space

The stress components appear in an unequal form in the expression for the shear stress intensity τ_0 (i.e., there are terms $2\sigma_x^2$, $2\sigma_x\sigma_y$, $6\tau_{xy}^2$, etc. in Eq. (1.2.9)). Furthermore, as it follows from Eq. (1.2.27), the stress deviator components $\overline{\sigma}_{ij}$ are dependent of each other. Instead of six dependent components $\overline{\sigma}_{ij}$, five independent components $S_1, S_2, \ldots, S_5$ are introduced [48] so that

(a) each of them appear identically in the formula for τ_0,

(b) transformations between $\overline{\sigma}_{ij}$ and S_k be one-to-one (single-valued) and linear function,

and (c):

$$\overline{\sigma}_{ij}\overline{\sigma}_{ij} = S_k S_k$$
$$(i, j = x, y, z; \quad k = 1, 2, \ldots, 5) \tag{1.7.1}$$

We set relations between normal stress deviator components and S_1 and S_2 in a such way

$$\begin{aligned} \overline{\sigma}_x &= a_x S_1 + b_x S_2 \\ \overline{\sigma}_y &= a_y S_1 + b_y S_2 \\ \overline{\sigma}_z &= a_z S_1 + b_z S_2 \end{aligned} \tag{1.7.2}$$

where $a_x, \ldots, b_z$ are coefficients to be determined further. We set up the shear stress components as

$$\sqrt{2}\tau_{xz} = S_3, \quad \sqrt{2}\tau_{xy} = S_4, \quad \sqrt{2}\tau_{yz} = S_5. \tag{1.7.3}$$

Substituting the values of $\overline{\sigma}_{ij}$ from Eq. (1.7.2) into Eq.(1.2.27), we obtain

$$\left(a_x + a_y + a_z\right)S_1 + \left(b_x + b_y + b_z\right)S_2 = 0. \tag{1.7.4}$$

Since S_1 and S_2 are independent variables, it follows that

$$\begin{aligned} a_x + a_y + a_z &= 0 \\ b_x + b_y + b_z &= 0 \end{aligned} \tag{1.7.5}$$

Then, in order to satisfy Eq. (1.7.1), the following relations must be fulfilled

$$\begin{aligned} a_x b_x + a_y b_y + a_z b_z &= 0, \\ a_x^2 + a_y^2 + a_z^2 &= 1, \\ b_x^2 + b_y^2 + b_z^2 &= 1. \end{aligned} \tag{1.7.6}$$

Let us consider two vectors, $\vec{\mathbf{a}}$ with coordinates a_x, a_y, a_z and $\vec{\mathbf{b}}$ with coordinates b_x, b_y, b_z. Then, it is possible to give the following geometrical interpretation to Eqs. (1.7.5) and (1.7.6). These equations imply that (a) vectors $\vec{\mathbf{a}}$ and $\vec{\mathbf{b}}$ are unit and mutually perpendicular; (b) vectors $\vec{\mathbf{a}}$ and $\vec{\mathbf{b}}$ lie in the plane equally inclined to x-, y-, and z-axes in three-dimensional Cartesian coordinate system.

Since there are only five equations (1.7.5) and (1.7.6) for six variables $a_x, \ldots, b_z$, the system is indeterminate. One of possible variants to set up the vectors $\vec{\mathbf{a}}$ and $\vec{\mathbf{b}}$ is shown in Fig. 1.7, where δ denotes an angle between the vector $\vec{\mathbf{a}}$ and the median MM_1 of triangle $M_1M_2M_3$, which is equally inclined to the x-, y-, and z-axes. Projecting the vectors $\vec{\mathbf{a}}$ and $\vec{\mathbf{b}}$ on these axes, we obtain

$$\begin{aligned} a_x &= \sqrt{\frac{2}{3}}\cos\delta, \; a_y = -\sqrt{\frac{2}{3}}\sin\left(\delta + \frac{\pi}{6}\right), \; a_z = \sqrt{\frac{2}{3}}\sin\left(\delta - \frac{\pi}{6}\right), \\ b_x &= \sqrt{\frac{2}{3}}\sin\delta, \; b_y = \sqrt{\frac{2}{3}}\cos\left(\delta + \frac{\pi}{6}\right), \; b_z = -\sqrt{\frac{2}{3}}\cos\left(\delta - \frac{\pi}{6}\right). \end{aligned} \tag{1.7.7}$$

where δ may take an arbitrary value. Hence,

$$\begin{aligned}
\bar{\sigma}_x \sqrt{\frac{3}{2}} &= S_1 \cos\delta + S_2 \sin\delta, \\
\bar{\sigma}_y \sqrt{\frac{3}{2}} &= -S_1 \sin\left(\delta + \frac{\pi}{6}\right) + S_2 \cos\left(\delta + \frac{\pi}{6}\right), \\
\bar{\sigma}_z \sqrt{\frac{3}{2}} &= S_1 \sin\left(\delta - \frac{\pi}{6}\right) - S_2 \cos\left(\delta - \frac{\pi}{6}\right).
\end{aligned} \tag{1.7.8}$$

Inverse relations are

$$\begin{aligned}
\frac{S_1}{\sqrt{2}} &= \bar{\sigma}_x \cos\left(\delta + \frac{\pi}{6}\right) - \bar{\sigma}_y \sin\delta, \\
\frac{S_2}{\sqrt{2}} &= \bar{\sigma}_x \sin\left(\delta + \frac{\pi}{6}\right) + \bar{\sigma}_y \cos\delta.
\end{aligned} \tag{1.7.9}$$

In particular, at $\delta = 0$ we have

$$\bar{\sigma}_x = \sqrt{\frac{2}{3}} S_1, \quad \bar{\sigma}_y = -\frac{S_1}{\sqrt{6}} + \frac{S_2}{\sqrt{2}}, \quad \bar{\sigma}_z = -\frac{S_1}{\sqrt{6}} - \frac{S_2}{\sqrt{2}}. \tag{1.7.10}$$

Inverse relations are

$$S_1 = \sqrt{\frac{3}{2}} \bar{\sigma}_x, \quad S_2 = \frac{\bar{\sigma}_x}{\sqrt{2}} + \sqrt{2} \bar{\sigma}_y. \tag{1.7.11}$$

Normal strain deviator components, $\bar{\varepsilon}_x$, $\bar{\varepsilon}_y$, and $\bar{\varepsilon}_z$, are also dependent on each other. Instead of $\bar{\varepsilon}_x, \bar{\varepsilon}_y, \ldots, \bar{\varepsilon}_{yz}$, five independent components $e_1, e_2, \ldots, e_5$ are introduced in a similar way:

$$\begin{aligned}
e_1 &= \sqrt{\frac{3}{2}} \bar{\varepsilon}_x, \quad e_2 = \frac{\bar{\varepsilon}_x}{\sqrt{2}} + \sqrt{2} \bar{\varepsilon}_y, \\
e_3 &= \sqrt{2} \varepsilon_{xz}, \quad e_4 = \sqrt{2} \varepsilon_{xy}, \quad e_5 = \sqrt{2} \varepsilon_{yz}.
\end{aligned} \tag{1.7.12}$$

Inverse relations are

$$\bar{\varepsilon}_x = \sqrt{\frac{2}{3}} e_1, \quad \bar{\varepsilon}_y = -\frac{e_1}{\sqrt{6}} + \frac{e_2}{\sqrt{2}}, \quad \bar{\varepsilon}_z = -\frac{e_1}{\sqrt{6}} - \frac{e_2}{\sqrt{2}}. \tag{1.7.13}$$

Now, the shear stress intensity and shear strain intensity take the following forms

$$\tau_0^2 = \frac{3}{2} S_i S_i = \frac{3}{2}\left(S_1^2 + S_2^2 + S_3^2 + S_4^2 + S_5^2\right),$$
$$\gamma_0^2 = \frac{2}{3} e_i e_i = \frac{2}{3}\left(e_1^2 + e_2^2 + e_3^2 + e_4^2 + e_5^2\right). \quad (1.7.14)$$

Therefore, in contrast to Eqs. (1.2.9) and (1.2.8), terms S_k^2 and e_k^2 appear "uniformly" in Eqs. (1.7.14).

The components $S_1, S_2, \ldots, S_5$ (as well $e_1, e_2, \ldots, e_5$) can be introduced in another way, for example [3],

$$S_1 = \frac{3}{2}(\overline{\sigma}_x + \overline{\sigma}_z), \quad S_2 = \frac{\sqrt{3}}{2}(-\overline{\sigma}_x + \overline{\sigma}_z),$$
$$S_3 = \sqrt{3}\tau_{xy}, \quad S_4 = \sqrt{3}\tau_{yz}, \quad S_5 = \sqrt{3}\tau_{xz}, \quad (1.7.15)$$

and, instead of Eq. (1.7.14), we have

$$\tau_0^2 = S_i S_i . \quad (1.7.16)$$

We will further use the standard relations (1.7.10)–(1.7.14) throughout.

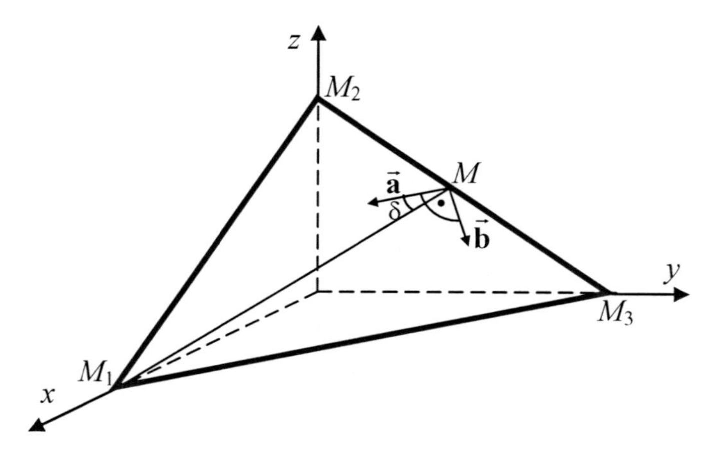

Fig. 1.7 Plane equally inclined to x-, y-, and z-axes in three-dimensional Cartesian coordinate system on which vectors $\vec{\mathbf{a}}$ and $\vec{\mathbf{b}}$ lie.

The five-dimensional space with Cartesian rectangular coordinate system, where components S_i from Eqs (1.7.3) and (1.7.11) and components e_i from Eq. (1.7.12) are laid off along coordinate axes, is called Ilyushin's space denoted as $\mathbf{R}^5$.

1.8 Isotropy Postulate

The vector whose coordinates in Ilyushin's space $\mathbf{R}^5$ are given by Eqs. (1.7.3) and (1.7.11) is called the stress vector $\vec{\mathbf{S}}$:

$$\vec{\mathbf{S}} = S_k \vec{\mathbf{g}}_k , \tag{1.8.1}$$

where $\vec{\mathbf{g}}_k$ are unit base vectors. The vector $\vec{\mathbf{S}}$ varies with position in a body. During loading, the endpoint of vector $\vec{\mathbf{S}}$ describes some space curve which is called the loading trajectory (path).

Similarly, the strain vector can be introduced:

$$\vec{\mathbf{e}} = e_k \vec{\mathbf{g}}_k ', \tag{1.8.2}$$

where $\vec{\mathbf{g}}_k '$ are unit base vectors oriented identically with base vectors $\vec{\mathbf{g}}_k$. It must be noted that the scale of orts $\vec{\mathbf{g}}_k '$ may differ from that of $\vec{\mathbf{g}}_k$. The hodograph of vector $\vec{\mathbf{e}}$ in $\mathbf{R}^5$ is called the strain trajectory (deformation trajectory). Vectors $\vec{\mathbf{S}}$ and $\vec{\mathbf{e}}$ and their hodographs are constructed for every point in a body.

Let us establish the relation between $\vec{\mathbf{S}}$ and $\vec{\mathbf{e}}$ for homogeneous and initially isotropy materials. Let e represent the arc length of strain path. An element of this arc is

$$de = \left| d\vec{\mathbf{e}} \right| , \quad de^2 = de_k de_k . \tag{1.8.3}$$

The above relations give

$$\dot{e} = \sqrt{\dot{e}_k \dot{e}_k} , \quad e = \int_0^t \sqrt{\dot{e}_k \dot{e}_k} \, dt , \tag{1.8.4}$$

where t is time or some other parameter monotonously growing with time. Equations (1.8.4) give a relation between e and t. Therefore, the equation of strain trajectory, a smooth curve parametrized by arc length, can be written as

$$\vec{\mathbf{e}} = \vec{\mathbf{e}}(e). \tag{1.8.5}$$

At each point on the curve (1.8.5), let us construct the Frenet-Serret frame consisting of five orthonormal base vectors $\vec{\mathbf{q}}_n$ $(n = 1,2,\ldots,5)$. Employing vectors $d^n\vec{\mathbf{e}}/de^n$ (these five vectors, in general, linearly independent but not mutually perpendicular) we obtain:

$$\vec{\mathbf{q}}_1 = \frac{d\vec{\mathbf{e}}}{de}, \tag{1.8.6}$$

where $\vec{\mathbf{q}}_1$ is a unit vector tangent to the strain path;

$$\vec{\mathbf{q}}_2 = \frac{1}{\kappa_1}\frac{d^2\vec{\mathbf{e}}}{de^2}, \quad \kappa_1 = \left|\frac{d^2\vec{\mathbf{e}}}{de^2}\right|. \tag{1.8.7}$$

Ort $\vec{\mathbf{q}}_2$ is directed along the main normal of curve (1.8.5), hence $\vec{\mathbf{q}}_1 \cdot \vec{\mathbf{q}}_2 = 0$. To find $\vec{\mathbf{q}}_3$, $\vec{\mathbf{q}}_4$ and $\vec{\mathbf{q}}_5$, we use the Frenet formulas,

$$\frac{d\vec{\mathbf{q}}_n}{de} = -\kappa_{n-1}\vec{\mathbf{q}}_{n-1} + \kappa_n\vec{\mathbf{q}}_{n+1}, \quad (n = 1,2,\ldots,5) \tag{1.8.8}$$

where $\kappa_1, \ldots \kappa_5$ ($\kappa_0 = \kappa_5 = 0$) are the parameters of the curvature and torsion of strain trajectory. It is known that vectors found from the above equality are orthonormal:

$$\vec{\mathbf{q}}_i \cdot \vec{\mathbf{q}}_j = \delta_{ij}, \tag{1.8.9}$$

where the dot between vectors means a scalar product.

Taking Eq. (1.8.6) into account, Eq. (1.8.8) at $n = 1$ leads to Eq. (1.8.7). In Eq. (1.8.8) at $n \geq 2$, both $\vec{\mathbf{q}}_3$, $\vec{\mathbf{q}}_4$, and $\vec{\mathbf{q}}_5$ and κ_2, κ_3, and κ_4 are unknown. Let us write Eq. (1.8.8) in the other form:

$$\kappa_n\vec{\mathbf{q}}_{n+1} = \kappa_{n-1}\vec{\mathbf{q}}_{n-1} + \frac{d\vec{\mathbf{q}}_n}{de}. \tag{1.8.10}$$

Raising to the square Eq. (1.8.10), we obtain

$$\kappa_n^2 = \kappa_{n-1}^2 + 2\kappa_{n-1}\vec{\mathbf{q}}_{n-1} \cdot \frac{d\vec{\mathbf{q}}_n}{de} + \left(\frac{d\vec{\mathbf{q}}_n}{de}\right)^2. \tag{1.8.11}$$

Multiplying the equality (1.8.10) by $\vec{\mathbf{q}}_{n-1}$, we get

$$\vec{\mathbf{q}}_{n-1} \cdot \frac{d\vec{\mathbf{q}}_n}{de} = -\kappa_{n-1}. \tag{1.8.12}$$

Substituting the above result into Eq. (1.8.11), we find

$$\kappa_n^2 = -\kappa_{n-1}^2 + \left(\frac{d\vec{\mathbf{q}}_n}{de} \right)^2. \tag{1.8.13}$$

The recurent use of Eqs (1.8.10) and (1.8.13) determines all unknown parameters. Indeed, since the unit vectors $\vec{\mathbf{q}}_1$ and $\vec{\mathbf{q}}_2$, and a curvature κ_1 are given by Eqs. (1.8.6) and (1.8.7), the parameter κ_2 can be calculated by Eq. (1.8.13) at $n = 2$, and then, the vector $\vec{\mathbf{q}}_3$ can be determined from Eq. (1.8.10), and so on.

On the base of Frenet formulas, Eq. (1.8.8), the derivative of any order of the vector $\vec{\mathbf{e}}$ with respect to *e* (for curves for which the derivative of any order of $\vec{\mathbf{e}}$ with respect to *e* exists) can be expressed through $\vec{\mathbf{q}}_1, \vec{\mathbf{q}}_2, \ldots, \vec{\mathbf{q}}_5$. Indeed, write the *r*-degree derivative of $\vec{\mathbf{e}}$ with respect to *e*:

$$\frac{d^r \vec{\mathbf{e}}}{de^r} = \alpha_k \vec{\mathbf{q}}_k,$$

where α_k is the component of vector $d^r \vec{\mathbf{e}} / de^r$. Then,

$$\frac{d^{r+1} \vec{\mathbf{e}}}{de^{r+1}} = \frac{d\alpha_k}{de} \vec{\mathbf{q}}_k + \alpha_k \frac{d\vec{\mathbf{q}}_k}{de}.$$

On the other hand, Eq. (1.8.8) shows that the derivative $d\vec{\mathbf{q}}_\kappa / de$ is expressed through $\vec{\mathbf{q}}_1, \vec{\mathbf{q}}_2, \ldots, \vec{\mathbf{q}}_5$, therefore, the relationship,

$$\frac{d^{r+1} \vec{\mathbf{e}}}{de^{r+1}} = \alpha_k' \vec{\mathbf{q}}_k,$$

determines the derivative of higher degree.

The above formula enables us to conclude the following: any linear differential operator (let us denote it as $L(\vec{\mathbf{e}})$) acting on $\vec{\mathbf{e}}$ with respect to the *e*, can be represented as

$$L(\vec{\mathbf{e}}) = \mathrm{X}_k \vec{\mathbf{q}}_k , \tag{1.8.14}$$

where the coefficient X_k depends on e and κ_k . Therefore, the operator $L(\vec{\mathbf{e}})$ is entirely determined by the internal geometry of deformation trajectory. In the other words, the operator $L(\vec{\mathbf{e}})$ is an invariant of the rotation and reflection of the strain trajectory in Ilyushin's space.

Due to the universality of the operator $L(\vec{\mathbf{e}})$, the statement that a stress vector is bound to the strain vector by the equation,

$$\vec{\mathbf{S}} = L(\vec{\mathbf{e}}) = \mathrm{X}_k \vec{\mathbf{q}}_k , \tag{1.8.15}$$

is postulated. Equation (1.8.15) expresses the dependence of the stress vector $\vec{\mathbf{S}}$ on deformation. This dependence satisfies the isotropy postulate [48]: *the magnitude and orientations of the vector* $\vec{\mathbf{S}}$ *relative to the strain trajectory depends only on the internal geometry of this trajectory.*

Let us introduce a new notion, the image of straining process defined as a strain trajectory and the stress vector constructed at each point of the trajectory. Then, the isotropy postulate is formulated as: *an image of straining process is an invariant of the rotation and reflection of strain curve(trajectory).*

Experimental works on many metals [73,186,187] has amply justified the validity of isotropy postulate.

Now, let us investigate a loading process expressed by stress vector $\vec{\mathbf{S}}(t)$. Let us denote the arc length of curve $\vec{\mathbf{S}} = \vec{\mathbf{S}}(t)$ as s and write down the equation of loading(stress) trajectory as $\vec{\mathbf{S}} = \vec{\mathbf{S}}(s)$. Further, we assign a natural coordinate system for the stress trajectory with orthonormal base vectors, $\vec{\mathbf{p}}_n$ ($n = 1,\ldots,5$), expressed through $d^n\vec{\mathbf{S}}/ds^n$ in a form similar to Eqs. (1.8.6)–(1.8.13). The loading trajectory and corresponding strain vector $\vec{\mathbf{e}}$ constructed at each point of this trajectory are called the image of loading process in $\mathbf{R}^5$. Then, the isotropy postulate is formulated as: *an image of the loading process is invariant of rotation and reflection, i.e., the orientation of strain vector relative to the loading trajectory depends on only internal geometry of this in space* $\mathbf{R}^5$. Therefore, the following relation is held

$$\vec{\mathbf{e}} = \mathrm{X}_n' \vec{\mathbf{p}}_n , \tag{1.8.16}$$

where X_n' are functions of hydrostatic pressure and parameters of loading trajectory, s and κ_n.

Let us note that Eqs. (1.8.15) and (1.8.16) can be represented not only through units vectors $\vec{\mathbf{q}}_n$ and $\vec{\mathbf{p}}_n$ but also through $d^n\vec{\mathbf{S}}/ds^n$.

Consider the case of proportional loading. It is assumed that the stress vector $\vec{\mathbf{S}}$ is directed along the trajectory of deformation (along the vector $\vec{\mathbf{e}}$)[3]. In this case, Eq. (1.8.15) results in the following:

$$\vec{\mathbf{S}} = \mathrm{X}_1\vec{\mathbf{q}}_1 = \mathrm{X}_1\frac{d\vec{\mathbf{e}}}{de}, \quad \mathrm{X}_m = 0 \text{ for } m = 2,3,\ldots 5. \tag{1.8.17}$$

Since the vector $d\vec{\mathbf{e}}/de$ in Eq. (1.8.17) is unit, the absolute value of vector $\vec{\mathbf{S}}$ is equal to X_1 and therefore

$$d\vec{\mathbf{e}} = \frac{\vec{\mathbf{S}}}{\left|\vec{\mathbf{S}}\right|}de. \tag{1.8.18}$$

Since the unit vector $\vec{\mathbf{S}}/\left|\vec{\mathbf{S}}\right|$ is constant during proportional loading, Eq. (1.8.18) is integrable:

$$\vec{\mathbf{e}} = \frac{e}{\left|\vec{\mathbf{S}}\right|}\vec{\mathbf{S}} \tag{1.8.19}$$

or

$$e_k = \frac{e}{\left|\vec{\mathbf{S}}\right|}S_k. \tag{1.8.20}$$

Equations (1.8.19) and (1.8.20), on the base of Eq. (1.7.14), yield

$$e = \sqrt{\frac{3}{2}}\gamma_0, \quad e_k = \frac{3}{2}\frac{\gamma_0}{\tau_0}S_k \tag{1.8.22}$$

that expresses the law of the proportionality between the stress and strain deviators. Indeed, applying Eqs (1.7.3), (1.7.11), (1.7.12), and (1.2.21), we arrive at the law of deviator proportionality in the form of Eq. (1.2.4).

On the base of thermodynamics of irreversible processes and isotropy postulate, A.A. Ilyushin has offered a general mathematical theory of plasticity [48]. He set up the system of relations between stress, deformation, time, and temperature

[3] This statement is assumed a priori whereas experiments give small but regular deviations from the consequence of the statement (see Sec. 2.14).

which are valid for any processes of deformation. But alas, theoretical thoughts end in the formula (1.8.15). The form of function X_k is not set for the case of an arbitrary combining loading.

Let us note that the replacement of $\overline{\sigma}_{ij}$ - $\overline{\varepsilon}_{ij}$ relations by the formulation of relations between the vectors $\vec{\mathbf{S}}$ and $\vec{\mathbf{e}}$ may rise doubts because the stress/strain deviator tensor is a wider notion than a five-dimensional vector $\vec{\mathbf{e}}$ (or $\vec{\mathbf{S}}$), a stress deviator tensor possesses two non-zero invariant while a stress vector has only one invariant, its length.

1.9 Loading Surface

It is well known that, an element of a body incurs pure elastic deformation until the stress vector $\vec{\mathbf{S}}$ reaches a yield surface. According to Guber's yield criteria, the yield surface is a sphere whose equation in Ilyushin's stress space $\mathbf{S}^5$ has the form:

$$\left|\vec{\mathbf{S}}\right|^2 = S_1^2 + S_2^2 + S_3^2 + S_4^2 + S_5^2 = \frac{2}{3}\sigma_S^2 . \tag{1.9.1}$$

If the length of the vector $\vec{\mathbf{S}}$ exceeds the distance from the origin in stress space to the yield surface, irreversible strains will arise. Consider the test in which the specimen is first loaded to some value beyond the initial yield point and then completely unloaded. Now, if this specimen is reloaded, further plastic deformation will be induced not at initial yield point (see Fig. 1.1). Indeed, after unloading from point M and subsequent loading, the irreversible deformation does not occur at $\sigma = \sigma_S$ but only at the stress of σ_M . Now, the elastic region is bounded by some new surface, loading surface. The loading surface changes in shape and size during loading, provided the acting stress induces plastic strain. The equation of loading surface can be written as

$$f\left(S_1,\ldots,S_5,e_1^S,e_2^S,\ldots,e_5^S,\chi_1,\chi_2,\ldots\right)=0, \tag{1.9.2}$$

where $\chi_1,\chi_2,\ldots$ are some parameters changing with the increment of plastic strain. It is known from experiments that a material can be deformed elastically (be unloaded) from an arbitrary state. It means that the loading surface must pass through the end-point of vector $\vec{\mathbf{S}}$ (the loading point) during plastic deforming. In order to induce an increment of $\vec{\mathbf{e}}^S$ at each moment of loading it is necessary to go beyond the current loading surface.

Consider a stress vector $\vec{\mathbf{S}}$ and corresponding loading surface $f=0$ shown by dotted line in Fig. 1.8a. If the increment of the stress vector, $d\vec{\mathbf{S}}$, moves outside the loading surface, a plastic strain increment will be produced. The new loading surface $f=0$ is shown by continuous line in Fig. 1.8a. As it is seen from Fig. 1.8a, due to additional loading by the vector $d\vec{\mathbf{S}}$, the material of body hardens e.g., in the direction of vector $\vec{\mathbf{n}}_1$ (current yield stress rises in this direction) but softens in the direction of vector $\vec{\mathbf{n}}_2$.

Negative values of function *f*,

$$f\left(S_1,\ldots,S_5,e_1^S,e_2^S,\ldots,e_5^S,\chi_1,\chi_2,\ldots\right)<0, \tag{1.9.3}$$

correspond to an elastic region.

As it is known, there is no unique idea about a transformation of the loading surface during plastic deformation. In terms of some theories, it is supposed that loading surface remains smooth at loading point. Other theories assume that loading surface has a corner at loading point.

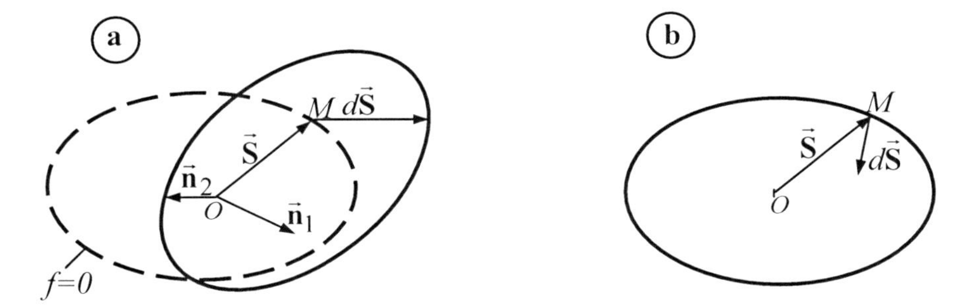

Fig. 1.8 Loading surface $f=0$ at loading (a) and unloading (b).

Consider the case when the loading surface $f=0$ is smooth at the loading point. Let us express the criterion of loading, unloading, and neutral loading through derivatives of the function *f*. If the endpoint of stress vector is on the surface $f=0$ and, due to increment $d\vec{\mathbf{S}}$, it moves inside the loading surface, and an unloading process takes place (Fig. 1.8b). In this case, the loading surface remains unchangeable and there are no increments in plastic strain:

$$de_k^S=0, \quad d\chi_k=0. \tag{1.9.4}$$

During unloading, Eqs. (1.9.2) and (1.9.3) give:

$$\frac{df}{dS_k} dS_k < 0. \tag{1.9.5}$$

If an increment $d\vec{\mathbf{S}}$ is directed along a tangent to the surface $f = 0$, i.e., the loading point remains on the surface $f = 0$, such loading is called a neutral loading. Then Eqs. (1.9.4) are valid and Eq. (1.9.2) yields

$$\frac{df}{dS_k} dS_k = 0. \tag{1.9.6}$$

If an infinitesimal increment $d\vec{\mathbf{S}}$ causes further plastic deformation, then we have the case of loading which is accompanied by the change of surface $f = 0$. Thus, the condition for the progress of plastic flow, or the criterion for loading, is

$$f\left(S_1 + dS_1, \ldots, e_1^S + de_1^S, \ldots, \chi_1 + d\chi_1, \ldots\right) = 0. \tag{1.9.7}$$

Expanding the function f by its Taylor series, with respect to increments dS_k, de_k^S, and $d\chi_k$, we obtain

$$\frac{\partial f}{\partial S_k} dS_k + \frac{\partial f}{\partial e_k^S} de_k^S + \frac{\partial f}{\partial \chi_i} d\chi_i = 0. \tag{1.9.8}$$

For arriving at the result (1.9.8), known as the *consistency condition* for a general work-hardening material, we have taken into consideration the relation (1.9.2). In addition, on the base of (1.9.5) and (1.9.6), we have

$$\frac{df}{dS_k} dS_k > 0. \tag{1.9.9}$$

Summarizing, if the loading point is on the surface $f = 0$, Eqs. (1.9.5), (1.9.6), and (1.9.9) express the criterion of unloading, neutral loading, and loading, respectively.

1.10 Drucker's Postulate

Drucker's postulate [26-29] is that criterion whose fulfillment is a minimal requirement for any theory of plasticity. Consider an initial state of stress with stress vector $\vec{\mathbf{S}}^*$ and the vector $\vec{\mathbf{S}}$ that varies in such way that its initial and final position in stress space coincide with that of the vector $\vec{\mathbf{S}}^*$, i.e., the vector $\vec{\mathbf{S}}$ describes a closed curve. The stress vector $\vec{\mathbf{S}}-\vec{\mathbf{S}}^*$ is referred to as the added stress vector.

Drucker's postulate reads that the work done by the added stresses on the changes in strain over a closed loading path is positive:

$$\oint\left(\vec{\mathbf{S}}-\vec{\mathbf{S}}^*\right)d\vec{\mathbf{e}} \geq 0 . \tag{1.10.1}$$

(Un)equality (1.10.1) must hold for any closed trajectories and for any vector $\vec{\mathbf{S}}^*$.

Since only small strains are considered, it is possible to decompose them on elastic (ε_{ij}^e) and plastic (ε_{ij}^S) parts (Eq. (1.5.35)):

$$e_k = e_k^e + e_k^S, \quad \vec{\mathbf{e}} = \vec{\mathbf{e}}^e + \vec{\mathbf{e}}^S . \tag{1.10.2}$$

As elastic deformations are fully reversible, all the elastic energy is recovered and the work performed by the added stresses on elastic strains is equal to zero for any closed loading trajectory. Therefore, the criterion (1.10.1) is equivalent to the following inequality

$$\oint\left(\vec{\mathbf{S}}-\vec{\mathbf{S}}^*\right)d\vec{\mathbf{e}}^S \geq 0 . \tag{1.10.3}$$

Two important consequences follow from this formula. Consider the closed loading trajectory(path), $M^*MM_1M^*$, shown in Fig. 1.9a (the direction of loading path is indicated by arrows). The vectors $\vec{\mathbf{S}}^*$ and $\vec{\mathbf{S}}$ correspond to the points M^* and M lying on the loading surface $f'=0$. In general, the vector $\vec{\mathbf{S}}^*$ (point M^*) can lie inside the surface $f'=0$. As the first segment of loading path, M^*M, is fully inside the surface $f'=0$, no plastic deformation occurs. If we add to the stress $\vec{\mathbf{S}}$ the increment $d\vec{\mathbf{S}}$, the stress vector, now, is at point M_1 lying outward the surface $f'=0$. As a consequence, the increment in plastic strain $d\vec{\mathbf{e}}^S$ is induced along the segment MM_1 and the loading surface evolves to the new form $f''=0$ that passes through point M_1. It is clear that the next portion of loading path, M_1M^*, symbolizes an unloading process under elastic law because it is oriented inward the current loading surface $f''=0$. Since the elastic energy is fully recovered, Eq. (1.10.3) can be reduced to the form, the first consequence of Drucker's postulate,

$$\left(\vec{\mathbf{S}}-\vec{\mathbf{S}}^*\right)d\vec{\mathbf{e}}^S \geq 0. \tag{1.10.4}$$

Let us consider another closed loading path (Fig. 1.9b). We take $\vec{\mathbf{S}}=\vec{\mathbf{S}}^*$ for the initial state. Then the stress vector $\vec{\mathbf{S}}^* + d\vec{\mathbf{S}}$ reaches point M_1 and finally the vector returns to initial point M^*. The second portion of such trajectory is an unloading process, therefore, Eq. (1.10.3) gives the second consequence of Drucker's postulate:

$$d\vec{\mathbf{S}}\cdot d\vec{\mathbf{e}}^S \geq 0. \tag{1.10.5}$$

As it was earlier noted, there is no single (common) view on the transformation of loading surface during plastic straining process – whether loading surface remains smooth or a conical point arises during loading? Unfortunately, Drucker's postulate is incapable of solving this problem. On the other hand, the first consequence expressed by Eq. (1.10.4) has important meaning in the theory of plasticity. If plastic strain coordinates are superimposed upon stress coordinates, Eq. (1.10.4) can be interpreted geometrically. The positive scalar product of the vector $\vec{\mathbf{S}}-\vec{\mathbf{S}}^*$ and the strain increment vector $d\vec{\mathbf{e}}^S$ require an acute angle between these two vectors. It must be noted that the endpoint of the vector $\vec{\mathbf{S}}^*$ can lie inside the loading surface as well as on it. If the loading surface is smooth, Drucker's postulate requires that it must be convex and the plastic strain-increment vector $d\vec{\mathbf{e}}^S$ must be directed along the normal to this surface, otherwise it is always possible to find vector such as $\vec{\mathbf{S}}^*$ where Eq. (1.10.4) is not valid, that is, inadmissible.

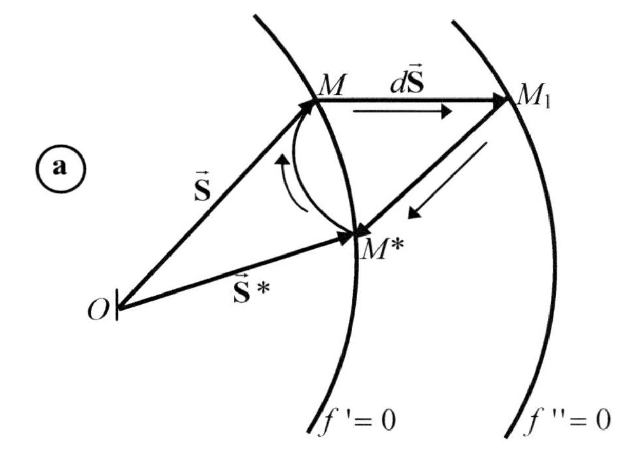

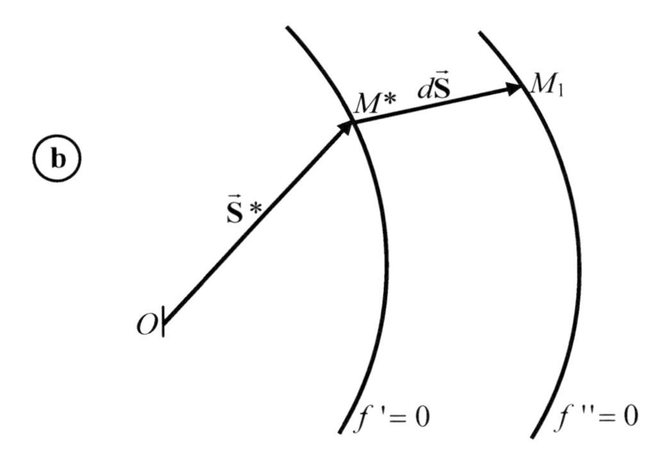

Fig. 1.9 Loading paths $M^*MM_1M^*$ (a) and $M^*M_1M^*$ (b) and Drucker's postulate.

Consider the situation when a corner arises at the loading point (point M in Fig. 1.10) on a loading surface $f = 0$. Let us construct the cone with top at point M whose generator is perpendicular to the arms of angle on the surface $f = 0$ at point M. Then, from the (un)equality (1.10.4), it follows that $d\vec{\mathbf{e}}^S$ must be inside this cone, shaded region in Fig. 1.10.

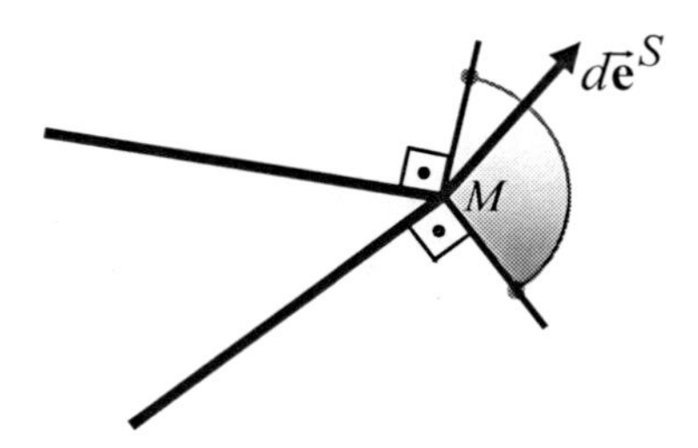

Fig. 1.10 Range of possible orientations of plastic-strain-increment-vector at a corner point on loading surface.

The second consequence of Drucker's, Eq. (1.10.5), for the case of pure torsion can be expressed as $d\tau_{xy} d\gamma_{xy}^S > 0$, hence, the plastic strain increment $d\gamma_{xy}^S$ and the stress increment $d\tau_{xy}$ must have the same sign. If $d\tau_{xy} < 0$, $d\gamma_{xy}^S > 0$, meaning that the plastic deformation will grow with the reduction of τ_{xy}. Such situation is realized for the materials with sharp yield limit on the stress-strain diagram. Therefore, the fulfillment of Drucker's postulates can be considered as the definition of the kind of stable materials.

1.11 Analysis of the Hencky–Nadai Theory

1. Numerous experiments [7,25,187] show that the deviator proportionality and the existence of universal diagram $\tau_0 - \gamma_0$ is fulfilled with sufficient accuracy for many materials at simple loading.

2. Hencky–Nadai relations are tensor-linear. The nonlinearity of stress-strain relations is provided by G_S which depends on the invariant τ_0 and thus varies during loading.

The most general linear relation between stress tensor and strain tensor can be written as

$$L_1(D_{\overline{\sigma}}) = L_2(D_{\overline{\varepsilon}}), \tag{1.11.1}$$

where $D_{\overline{\sigma}}$ and $D_{\overline{\varepsilon}}$ are the matrices of the stress/strain deviator tensor and L_1 and L_2 are linear operators.

The second Ilyushin's theorem (Sec. 1.5) ensures that any law constructed on the base of Eq. (1.11.1) coincides with the deviator proportionality, meaning that (1.11.1) expresses a general law within the category of tensor-linear laws at simple loading. This result considerably enhances the value of Hencky–Nadai relations.

Let us note that the Prager-Novozsilov relations, Eq. (1.6.12), cannot be written in the form of Eq. (1.11.1) because they are tensor-nonlinear.

3. Drucker's postulate ensures that the elementary work (dU^S) of stress deviators on plastic strains is positive. Therefore, the yield criterion can be assumed as $dU^S > 0$:

$$dU^S = \overline{\sigma}_{ij} d\overline{\varepsilon}_{ij}^S = S_k de_k^S . \tag{1.11.2}$$

The deviator proportionality in Eq. (1.2.4) can be rewritten as

$$e_k = \frac{S_k}{2G_S} . \tag{1.11.3}$$

Relations for plastic components, similar to Eq. (1.5.39), take the following form

$$e_k^S = \frac{S_k}{2}\left(\frac{1}{G_S} - \frac{1}{G}\right). \tag{1.11.4}$$

Therefore, Eqs. (1.11.3) and (1.11.4) give

$$de_k^S = \frac{dS_k}{2}\left(\frac{1}{G_S} - \frac{1}{G}\right) - \frac{S_k}{2G_S^2}\frac{dG_S}{d\tau_0} d\tau_0 . \tag{1.11.5}$$

Hence, the elementary work is

$$dU^S = \frac{S_k dS_k}{2}\left(\frac{1}{G_S} - \frac{1}{G}\right) - \frac{S_k S_k}{2G_S^2}\frac{dG_S}{d\tau_0} d\tau_0 . \tag{1.11.6}$$

In view of Eq. (1.7.14), we obtain

$$dU^S = \frac{1}{3}\left(\frac{1}{G_S} - \frac{1}{G} - \frac{\tau_0}{G_S^2}\frac{dG_S}{d\tau_0}\right)\tau_0 d\tau_0 . \tag{1.11.7}$$

As $G_S \le G$ and $dG_S / d\tau_0 \le 0$ (Fig. 1.3), the parenthesis in Eq. (1.11.7) is positive, therefore, the yield criterion $dU^S > 0$ is equivalent to the condition $d\tau_0 > 0$ in terms of the Hencky–Nadai theory.

The yield criterion is not arbitrary but associated with the stress-strain relation, Eq. (1.11.7) is derived from the stress-strain relation given by Eq. (1.11.4). Stress-strain relations differing from those in (1.11.4) lead to other formula for dU^S and other yield criterion.

4. Consider the loading close to a neutral. The deviator proportionality, Eq. (1.11.3), yields

$$de_k = \frac{dS_k}{2G_S} - \frac{S_k}{2G_S^2}\frac{dG_S}{d\tau_0}d\tau_0. \tag{1.11.8}$$

At neutral loading, $d\tau_0 = 0$, the above formula gives

$$de_k = \frac{dS_k}{2G_S}. \tag{1.11.9}$$

On the other hand, the neutral loading is a limiting case of unloading when Hooke's law is fulfilled, therefore

$$de_k = \frac{dS_k}{2G}, \tag{1.11.10}$$

where G is the elastic modulus.

Equations (1.11.9) and (1.11.10) are mutually contradictory because the modulus G_S can be largely smaller than G. It is obvious that the transition from plastic straining to unloading must occur continuously. But, as it follows from Eqs. (1.11.9) and (1.11.10), this transition has a non-continuous character. Therefore, it is impossible to consider the obtained result as acceptable.

Let us present one more example illustrating the inapplicability of the Hencky–Nadai relations at loading close to neutral. Consider two specimens loaded in tension beyond the yield stress, point M in Fig. 1.11. Further, let the loading path for the first specimen be the curve MM_1 and that for the second specimen be the curve MM_2. Both trajectories are close to the curve of neutral loading, $d\tau_0 = 0$, (dotted curve 3 in Fig. 1.11) and their state of stress at a point M_1 and M_2 is a pure shear ($\tau_{xy} > 0$). Finally, both specimens are unloaded, segments M_1O and M_2O. The difference is that the loading path MM_1 represents the curve for which $d\tau_0 > 0$, and the displacement along the path MM_2 leads to the decrease in shear stress intensity.

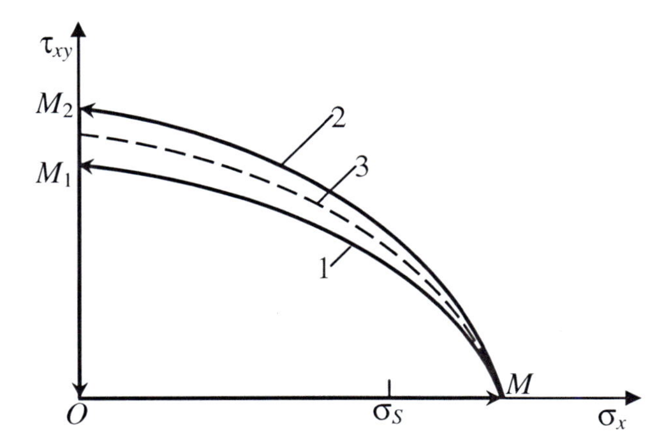

Fig. 1.11 Loading paths of two specimens.

Compare residual strains of these two specimens. Since we have $d\tau_0 < 0$ along the path MM_1O, there is no irreversible deformation on this segment. Therefore, the residual (plastic) strains of the first specimen are produced only by the uniaxial tensile stress at point M, i.e., $\varepsilon_x^S > 0$, $\varepsilon_y^S < 0$, $\gamma_{xy}^S = 0$. Along the loading path MM_2 of the second specimen, the shear stress intensity rises, $d\tau_0 > 0$. In terms of the Hencky–Nadai theory, a plastic strain does not depend on loading history, i.e., the value of plastic strain is fully determined by the current value of stress (proviso that $d\tau_0 > 0$) independently of the loading pass along which this stress has been reached. Therefore, the value of plastic strain due to loading along the path OMM_2 is the same as under torsion OM_2. This means that the Hencky–Nadai deformation theory prescribes the following plastic strains for the second specimen: $\varepsilon_x^S = \varepsilon_y^S = 0$, $\gamma_{xy}^S > 0$. The strains of the two specimens differ in finite magnitude while their trajectories are infinitesimally close to each other. This means that the principle of continuity (Prager V., Klyushnikov V.D., Shvajko Ju.M. [171]), approximately equal deformations must be induced by close loading paths, is violated.

5. Consider an orthogonal break of loading path. Further throughout, by the term "break of loading path" we mean that a continuous loading path with a discontinuous tangent is considered, i.e., the loading trajectory has the form of broken line (piecewise linear loading path). Let an element of a body be plastically loaded in uniaxial tension, $\sigma_x > \sigma_S$. Holding the value σ_x constant, the element undergoes infinitesimal torsion by $d\tau_{xy}$. Let us determine the increment

in shear strain $d\gamma_{xy}$ due to the action of $d\tau_{xy}$ and the shear modulus of plasticity G_S at additional loading.

It must be noted that the additional loading is neutral because, accurate up to infinitesimal negligible magnitude of higher order $(d\tau_{xy})^2$, we have $d\tau_0 = 0$. According to the deviator proportionality, on the base of Eqs. (1.7.3), (1.7.12), and (1.11.8), we have

$$d\gamma_{xy} = \frac{d\tau_{xy}}{G_S}. \tag{1.11.11}$$

where G_S is the shear modulus of plasticity at the beginning of additional-loading in torsion. Due to the infinitesimal magnitude of $d\tau_{xy}$, the value of G_S can be calculated from the Hencky–Nadai relations for uniaxial tension:

$$\varepsilon_x - \frac{\sigma}{3K_0} = \frac{\sigma_x - \sigma}{2G_S}, \quad \sigma = \frac{\sigma_x}{3}. \tag{1.11.12}$$

Using the notion of secant module E_S given by Eq. (1.2.30), Eqs. (1.2.14), (1.2.25), and (1.11.12) give the following expression for the shear modulus of plasticity:

$$G_S = \frac{G}{1 + 3G\left(\dfrac{1}{E_S} - \dfrac{1}{E}\right)}, \tag{1.11.13}$$

Calculations by Eq. (1.11.13) result in smaller values of G_S than those obtained by the Cicala formula (Sec. 2.5) and do not concur with experimental data, the values of G_S given by Eq. (1.11.13) are less than experimental G_S.

6. The rise of conic singularity on surface $f = 0$ at loading point follows from the deviator proportionality. To be convinced of validity of this statement, we use Eq. (1.11.5) written in vector form:

$$d\vec{\mathbf{e}}^S = \frac{1}{2}\left(\frac{1}{G_S} - \frac{1}{G}\right)d\vec{\mathbf{S}} - \frac{\vec{\mathbf{S}}}{2G_S^2}\frac{dG_S}{d\tau_0}d\tau_0. \tag{1.11.14}$$

This formula shows that, at fixed $\vec{\mathbf{S}}$, the change in the orientation of incremental vector $d\vec{\mathbf{S}}$ leads to the change in the direction of vector $d\vec{\mathbf{e}}^S$. This is in contradiction to the concept of a smooth surface $f = 0$, according to which the plastic-strain-vector increment must be directed along the normal to surface

$f = 0$ independently of the orientation of stress vector increment $d\vec{\mathbf{S}}$. Hence, according to Eq. (1.11.14), the loading surfaces can not be smooth.

1.12 Boundaries of the Applicability of the Hencky–Nadai Relations

Following investigations developed by Budiansky [15], let us determine the boundaries of acceptability of the Hencky–Nadai relations.

In the previous Section it is showed that the Hencky–Nadai relations prescribe the singular surface of loading $f = 0$ that widens the boundaries of applicability of deformation theory. Hencky–Nadai theory is valid not only at proportional (simple) loading but also at some deviation from it.

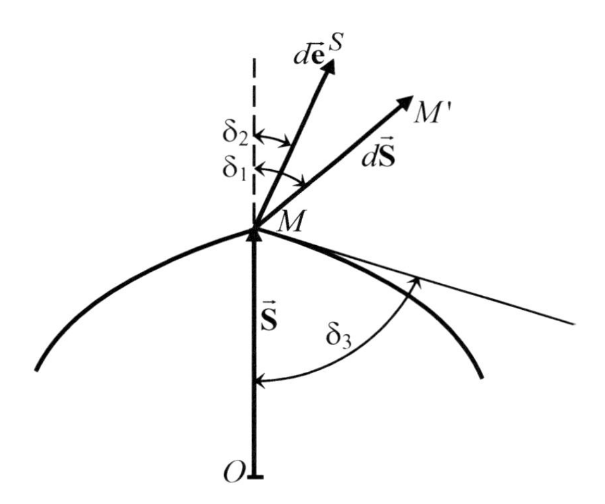

Fig. 1.12 Stress- and plastic strain-increment vectors.

Figure 1.12 shows the loading process which is simple up to point *M* and then the stress increment vector $d\vec{\mathbf{S}}$ and plastic strain increment vector $d\vec{\mathbf{e}}^S$ make an angle δ_1 and δ_2 with the stress vector $\vec{\mathbf{S}}$, respectively. Tangents to the surface $f = 0$ at point *M* create a cone whose generator makes an angle, δ_3, with the stress vector $\vec{\mathbf{S}}$. Proceeding from the consequence of Drucker's postulate that the vector $d\vec{\mathbf{e}}^S$ must be inside the cone constituted by normals to the surface $f = 0$ (Fig. 1.10), the following inequality is obtained

$$\delta_2 + \delta_3 \le \frac{\pi}{2}. \tag{1.12.1}$$

The vector $d\vec{\mathbf{e}}^S$ is determined by Eq. (1.11.14) which can be written in a more simple form. First, from Eq. (1.2.21), as follows

$$\frac{d}{d\tau_0}\left(\frac{1}{G_S}\right) = \frac{3}{\tau_0}\left(\frac{d\gamma_0}{d\tau_0} - \frac{\gamma_0}{\tau_0}\right). \tag{1.12.2}$$

Let us introduce the "tangent” modulus of universal diagram $\tau_0 - \gamma_0$ by formula

$$G_t = \frac{1}{3}\frac{d\tau_0}{d\gamma_0}. \tag{1.12.3}$$

Then,

$$\frac{d}{d\tau_0}\left(\frac{1}{G_S}\right) = \frac{1}{\tau_0}\left(\frac{1}{G_t} - \frac{1}{G_S}\right). \tag{1.12.4}$$

Thus Eq. (1.11.14) results in

$$d\vec{\mathbf{e}}^S = G_1 d\vec{\mathbf{S}} + G_2 \frac{\vec{\mathbf{S}}}{\tau_0} d\tau_0, \tag{1.12.5}$$

where

$$G_1 = \frac{1}{2}\left(\frac{1}{G_S} - \frac{1}{G}\right), \quad G_2 = \frac{1}{2}\left(\frac{1}{G_t} - \frac{1}{G_S}\right). \tag{1.12.6}$$

In order to eliminate τ_0 from Eq. (1.12.5), we express Eq. (1.7.14) in the form

$$\vec{\mathbf{S}} \cdot \vec{\mathbf{S}} = \left|\vec{\mathbf{S}}\right|^2 = \frac{2}{3}\tau_0^2. \tag{1.12.7}$$

From here, it follows

$$\vec{\mathbf{S}} \cdot d\vec{\mathbf{S}} = \frac{2}{3}\tau_0 d\tau_0. \tag{1.12.8}$$

The above scalar product is

$$\vec{\mathbf{S}} \cdot d\vec{\mathbf{S}} = \left|\vec{\mathbf{S}}\right| \cdot \left|d\vec{\mathbf{S}}\right| \cos\delta_1 \, . \tag{1.12.9}$$

On the base of Eqs. (1.12.7)–(1.12.9), Eq. (1.12.5) takes the form

$$d\vec{\mathbf{e}}^S = G_1 d\vec{\mathbf{S}} + G_2 \frac{\vec{\mathbf{S}}}{\left|\vec{\mathbf{S}}\right|} \left|d\vec{\mathbf{S}}\right| \cos\delta_1 \, . \tag{1.12.10}$$

Together with Eq. (1.12.9), the scalar product of Eq. (1.12.10) into itself gives

$$\frac{\left|d\vec{\mathbf{S}}\right|}{\left|d\vec{\mathbf{e}}^S\right|} = \frac{1}{\left[G_1^2 + \left(2G_1 G_2 + G_2^2\right)\cos^2\delta_1\right]^{1/2}} \, . \tag{1.12.11}$$

Multiplying Eq. (1.12.10) by unit vector $\vec{\mathbf{S}} / \left|\vec{\mathbf{S}}\right|$, we get

$$\left|d\vec{\mathbf{e}}^S\right| \cos\delta_2 = \left(G_1 + G_2\right)\cos\delta_1 \left|d\vec{\mathbf{S}}\right| . \tag{1.12.12}$$

Therefore, by means of Eqs. (1.12.11) and (1.12.12), we find:

$$\cos\delta_2 = \frac{\left(G_1 + G_2\right)\cos\delta_1}{\left[G_1^2 + \left(2G_1 G_2 + G_2^2\right)\cos^2\delta_1\right]^{1/2}} \, . \tag{1.12.13}$$

Let G_3 denote

$$G_3 = \frac{G_1 + G_2}{G_1} = \frac{\dfrac{1}{G_t} - \dfrac{1}{G}}{\dfrac{1}{G_S} - \dfrac{1}{G}} \, . \tag{1.12.14}$$

Thus, Eq. (1.12.13) results in the form

$$\cos\delta_2 = \frac{G_3 \cos\delta_1}{\left[1 + \left(G_3^2 - 1\right)\cos^2\delta_1\right]^{1/2}} \tag{1.12.15}$$

and the inequality (1.12.1) can be replaced by

$$\cos\delta_2 \geq \sin\delta_3$$

or, on the base of Eq. (1.12.15),

$$\frac{G_3^2 \cos^2\delta_1}{1+\left(G_3^2-1\right)\cos^2\delta_1} \geq \sin^2\delta_3 .$$

The final result is

$$\tan\delta_1 \leq \frac{G_3}{\tan\delta_3} . \tag{1.12.16}$$

Let us consider two cases of additional loading from point M, $\delta_1 \leq \delta_3$ (Fig. 1.13a) and $\delta_1 > \delta_3$ (Fig. 1.13b). For the case $\delta_1 \leq \delta_3$ the loading surface contains the surface $f = 0$ corresponding to the previous moment of loading (dotted line in Fig. 1.13a). Therefore, the return from point M_1 to M is not accompanied by irreversible strain. Unlike this, at $\delta_1 > \delta_3$, the vicinity of point M_1 does not contain the surface $f = 0$. Now the return from point M_1 to M is accompanied by plastic strain that is an absurd result. Therefore, if to use relations of an deformation type, i.e., to assume that the loading surface does not depend on the history of loading, we need to restrict the possible directions of additional loading by the condition, $\delta_1 \leq \delta_3$. Equating the angle δ_1 to δ_3 in Eq. (1.12.16), we obtain the maximal permissible value of angle δ_1, $\tan\delta_1 = \sqrt{G_3}$. So, according to Budiansky, the boundaries of acceptability of the Hencky–Nadai deformation can be expressed as

$$\tan\delta_1 \leq \sqrt{G_3} . \tag{1.12.17}$$

Let some boundary problem of plasticity be solved with the application of the Hencky–Nadai relations. It means that the loading path and the angle δ_1 between the vectors $\vec{\mathbf{S}}$ and $d\vec{\mathbf{S}}$ are determined at each point in a body. Furthermore, the modulus G_3 is given by Eq. (1.12.14). Let the angle δ_1 satisfy the inequality (1.12.17) at each point on the loading trajectory. It means that application of the Hencky–Nadai theory is substantiated from the point of view that Drucker's postulate is not violated.

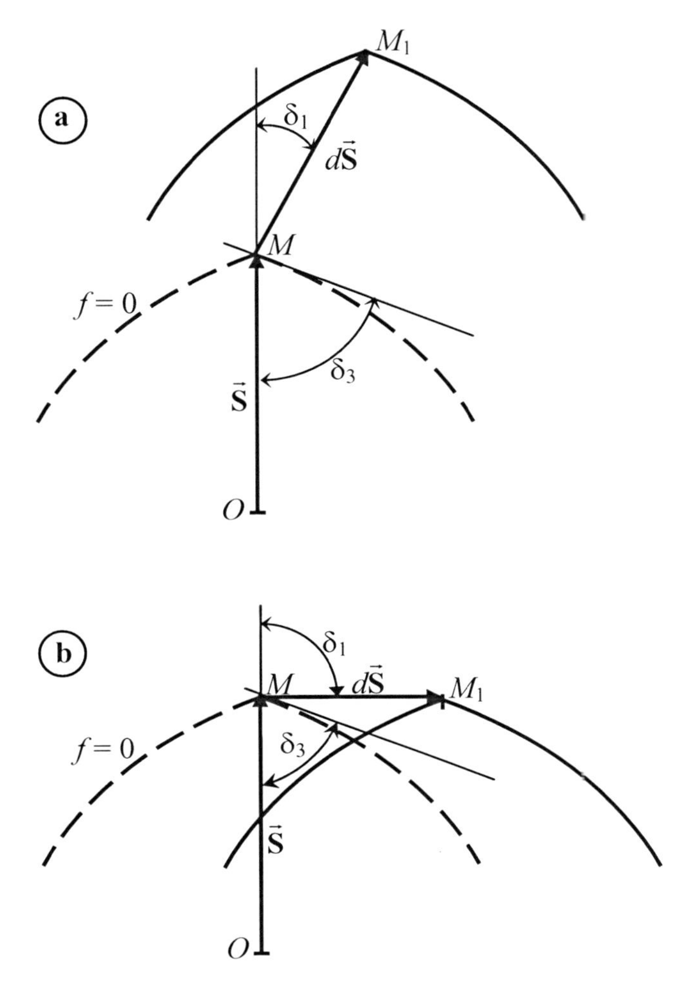

Fig. 1.13 Evolution of loading surface at slight (a) and large (b) deviation from proportional loading.

Let us verify the fulfillment of the criterion (1.12.17) at an orthogonal break of the loading path. Let a specimen, at fixed stress $\sigma_x > \sigma_S$, be additionally loaded by combined infinitesimal torsion, $d\tau_{xy} > 0$, and tension, $d\sigma_x > 0$. For this state of stress, on the base of Eqs. (1.2.2) and (1.7.11), we obtain

$$S_1 = \sqrt{\frac{2}{3}}\sigma_x,\ S_2 = \ldots = S_5 = 0,$$

$$dS_1 = \sqrt{\frac{2}{3}}d\sigma_x,\ dS_3 = \sqrt{2}d\tau_{xy}, \tag{1.12.18}$$

$$dS_2 = dS_4 = dS_5 = 0.$$

The angle of deviation from simple deformation, δ_1, is

$$\tan\delta_1 = \frac{dS_3}{dS_1} = \sqrt{3}\frac{d\tau_{xy}}{d\sigma_x}. \tag{1.12.19}$$

Therefore, the criterion (1.12.17) takes the form

$$\sqrt{3}\frac{d\tau_{xy}}{d\sigma_x} \le \sqrt{G_3}. \tag{1.12.20}$$

The modulus G_3 in the above equation is determined by Eq. (1.12.14) and can be found from the stress-strain diagram in uniaxial tension. We have

$$1 \le G_3 \le \frac{G_S}{G_t}, \tag{1.12.21}$$

where $G_3 = 1$ on the yield stress and $G_3 = G_S / G_t$ at big plastic strains which means that the right-hand side in Eq. (1.12.20) is a finite quantity. This requires that the left-hand side in Eq. (1.12.20) can also be finite. Therefore, according to the Hencky–Nadai criterion (1.12.17), the deformation theory cannot be applied to the determination of plastic strains in the vicinity of the orthogonal break ($d\sigma_x = 0$) of the loading path due to Eq. (1.12.20) does not hold. Hence, Eq. (1.11.13) is unacceptable and this is proved by experimental verifications.

1.13 Flow Plasticity Theories: Isotropic Hardening Rule

Since the plastic deformation depends on the history of loading, stress-strain relations cannot be expressed by finite equations but must be in a differential form [45,46]. In terms of flow plasticity theories, relations between the increments of stresses and strains are established. The change in elastic components $d\varepsilon_{ij}^e$ obeys Hooke's law. Therefore, the aim of flow plasticity theory consists in the determination of incremental plastic strain components $d\varepsilon_{ij}^{S}$ that, on the base of Eqs. (1.7.13), is equivalent to the determination of plastic strain vector $d\vec{\mathbf{e}}^{S}\left(de_1^S,\ldots,de_5^S\right)$. This vector must be determined so that the transition from plastic straining to unloading is continuous.

Let us assume the following assumptions.

1. A loading surface, $f=0$, is smooth that implies the existence of vector, $\vec{\mathbf{N}}$, normal to the surface. Formula for the components, N_k, of the unit normal vector $\vec{\mathbf{N}}$ has the form

$$N_k=\frac{\partial f}{\partial S_k}\Bigg/\left(\frac{\partial f}{\partial S_j}\frac{\partial f}{\partial S_j}\right)^{1/2},\quad k=1,\ldots,5 \tag{1.13.1}$$

2. Only those differential-linear theories are considered which prescribe linear relations between strain- and stress-increments. This means that the change in loading vector, $d\vec{\mathbf{S}}$, can be resolved into two components and the increment in $d\vec{\mathbf{e}}^{S}$ can be associated separately to each component. Therefore, the increment $d\vec{\mathbf{S}}$ can be decomposed as:

$$d\vec{\mathbf{S}}=d\vec{\mathbf{S}}_N+d\vec{\mathbf{S}}_\tau, \tag{1.13.2}$$

where the components $d\vec{\mathbf{S}}_N$ and $d\vec{\mathbf{S}}_\tau$ are directed along the normal and tangent to the loading surface, respectively. The component $d\vec{\mathbf{S}}_\tau$ does not result in an irreversible deformation due to the loading along the vector $d\vec{\mathbf{S}}_\tau$ is neutral. Hence, only the normal component $d\vec{\mathbf{S}}_N$ of vector $d\vec{\mathbf{S}}$ produces the irreversible deformation:

$$de_k^S=C_k{}'dS_N. \tag{1.13.3}$$

Since

$$dS_N=d\vec{\mathbf{S}}\cdot\vec{\mathbf{N}}=dS_iN_i, \tag{1.13.4}$$

from Eqs. (1.13.1), (1.13.3), and (1.13.4), we obtain

$$de_k^S=C_k{}'\frac{\partial f}{\partial S_i}dS_i, \tag{1.13.5}$$

and the denominator in Eq. (1.13.1) is included in the factor $C_k{}'$.

3. Drucker's postulate requires that the vector $d\vec{\mathbf{e}}^S$ be directed along the normal $\vec{\mathbf{N}}$ of the loading surface. Therefore, Eqs. (1.13.1) and (1.13.5) give

$$de_k^S = C\frac{\partial f}{\partial S_k}\frac{\partial f}{\partial S_i}dS_i, \quad C_k' = C\frac{\partial f}{\partial S_k}, \tag{1.13.6}$$

where C is the factor to be determined later.

Thus, the law of plastic straining, Eq. (1.13.6), is essentially associated with the loading surface. But, alas, flow plasticity theories do not offer theoretical reasoning from which it would be possible to set up the equation of loading surface. Only certain hypotheses (hardening rules) relative to the form of function f are supposed.

4. Isotropic hardening rule. The simplest suggestion is that the functions C and f depend only on the shear stress intensity τ_0, that is, Eq. (1.13.6) takes the form

$$de_k^S = C(\tau_0)\left(\frac{df}{d\tau_0}\right)^2 \frac{\partial \tau_0}{\partial S_k}\frac{\partial \tau_0}{\partial S_i}dS_i. \tag{1.13.7}$$

As the function $C(\tau_0)$ is not determined, the Equation (1.13.7) allows taking for the dependence $f(\tau_0)$ an arbitrary monotonously growing function of τ_0. Therefore, the equation of the loading surface (1.9.2) can be written (in the view of Eq. (1.7.14)) in the simpler forms

$$f = \tau_0 - \tau_{0m}, \quad \tau_{0m} \geq \sigma_S, \quad S_1^2 + S_2^2 + S_3^2 + S_4^2 + S_5^2 = \frac{2}{3}\tau_{0m}^2, \tag{1.13.8}$$

where τ_{0m} is the maximal value of τ_0 for all history of loading, provided that plastic flows take place in a body. If, for all history of loading, elastic deformations only arise in a body, the loading surface (1.13.8) is reduced to the yield surface for which $\tau_{0m} = \sigma_S$.

Counting the derivatives in Eqs. (1.7.14) and (1.13.8),

$$\frac{\partial \tau_0}{\partial S_k} = \frac{3}{2}\frac{S_k}{\tau_0}, \quad \frac{\partial \tau_0}{\partial S_i}dS_i = d\tau_0, \quad \frac{df}{d\tau_0} = 1, \tag{1.13.9}$$

Equation (1.13.7) leads to the Laning formula [60,61]

$$de_k^S = S_k C(\tau_0) d\tau_0, \tag{1.13.10}$$

where the function $C(\tau_0)$ contains the factor ($3/2\tau_0$). The above expression has the following vector- and tensor-form

$$\begin{aligned} d\vec{\mathbf{e}}^S &= \vec{\mathbf{S}} C(\tau_0) d\tau_0, \\ d\varepsilon_{ij}^S &= \bar{\sigma}_{ij} C(\tau_0) d\tau_0. \end{aligned} \tag{1.13.11}$$

It is the Eqs. (1.13.10) and (1.13.11) that determine the plastic strain increments. It is assumed that the endpoint of loading vector $\vec{\mathbf{S}}$ reaches the loading surface, $\tau_0 = \tau_{0m}$, and $d\tau_0 > 0$. If at least one of these conditions is not fulfilled, then $de_k^S = 0$.

5. Let us determine the plastic strain components, in terms of the flow theory with isotropic hardening rule, for the case of proportional loading. The integration in Eq (1.13.10) gives

$$e_k^S = \int_{\sigma_S}^{\tau_0} S_k C(\tau_0) d\tau_0, \quad \tau_{0m} \geq \sigma_S. \tag{1.13.12}$$

Under a simple loading, the ratio S_k/τ_0 is constant and therefore it can be factored outside the integral sign:

$$e_k^S = \frac{S_k}{\tau_0} \int_{\sigma_S}^{\tau_0} \tau_0 C(\tau_0) d\tau_0. \tag{1.13.13}$$

If to denote

$$\frac{2}{\tau_0} \int_{\sigma_S}^{\tau_0} \tau_0 C(\tau_0) d\tau_0 = \frac{1}{G_S} - \frac{1}{G}, \tag{1.13.14}$$

we arrive at Eq. (1.11.4) which means that Eq. (1.13.13) expresses the law of deviator proportionality.

Therefore, at proportional loading, from the flow plasticity theory with isotropic hardening rule, the Hencky–Nadai relations are obtained including the existence of universal relation between the intensities of shear stresses and shear strains.

6. So far, the function $C(\tau_0)$ remains indeterminate in the constitutive relations (1.13.10). It can be found from Eq. (1.13.14) as:

$$C(\tau_0)=\frac{1}{2\tau_0}\left(\frac{1}{G_S}-\frac{1}{G}\right)+\frac{1}{2}\frac{d}{d\tau_0}\left(\frac{1}{G_S}\right). \quad (1.13.15)$$

From the above formula and from Eq. (1.12.4), we obtain

$$C(\tau_0)=\frac{1}{2\tau_0}\left(\frac{1}{G_t}-\frac{1}{G}\right). \quad (1.13.16)$$

7. As shown in point 5, the deviator proportionality follows from the relations of the flow plasticity theory with smooth surface $f=0$. But it does not mean that the smoothness of loading surface is a necessary condition for arriving at the Hencky–Nadai relations. In Sec. 1.19, it will be shown that, at the proportional change in components S_k, the Hencky–Nadai relations can also be derived from the theory with a singular loading surface.

8. The formula (1.13.8) means that, in the stress space $\mathbf{R}^5$ the initial yield surface, sphere of radius $\sqrt{2/3}\sigma_S$ with center in the origin of coordinates, expands uniformly without distortion and translation as plastic flow occurs. The radius of this sphere is governed by the shear stress intensity τ_{0m}, which depends upon plastic strain history. The solid body, whose plastic properties are described in such way, is called the isotropic hardening body. For a specimen with a preloading τ_{0m}, during subsequent loading along an arbitrary loading path, the further plastic flows will be produced only under condition $\tau_0 \geq \tau_{0m}$. In particular, from this it follows that a loading in tension causes the yield stress to increase in a compression, that is, in contradiction with the Bauschinger effect. Thus, the flow plasticity theory with isotropic hardening rule is inapplicable for the case of alternating loading.

9. Following the work [51], let us verify the fulfillment of Drucker's postulate for a model with smooth loading surface depending on stress and strain components. In terms of such model, strain rates and stress rates are related to each other by the consistency condition which now has the form:

$$\frac{\partial f}{\partial S_k}\dot{S}_k+\frac{\partial f}{\partial e_k^S}\dot{e}_k^S=0. \quad (1.13.17)$$

Representing the associated flow law, Eq. (1.13.6), as

$$\dot{e}_k^S = C^0 \frac{\partial f}{\partial S_k}, \quad C^0 = C \frac{\partial f}{\partial S_i} \dot{S}_i, \tag{1.13.18}$$

the formula (1.13.17) takes the form

$$\dot{S}_i \dot{e}_i^S + \left(C^0\right)^2 \frac{\partial f}{\partial e_k^S} \frac{\partial f}{\partial S_k} = 0. \tag{1.13.19}$$

According to the second consequence of Drucker's postulate, the first term in Eq. (1.13.19) must be positive,

$$\dot{S}_i \dot{e}_i^S \geq 0, \tag{1.13.20}$$

therefore

$$\frac{\partial f}{\partial e_k^S} \frac{\partial f}{\partial S_k} \leq 0. \tag{1.13.21}$$

If a loading surface depends on an shear stress intensity τ_0 and plastic strain intensity γ_0^S, i.e., $f = f(\tau_0, \gamma_0)$, Eq. (1.13.21) can be written as

$$\frac{\partial f}{\partial \gamma_0^S} \frac{\partial f}{\partial \tau_0} S_i e_i^S \leq 0. \tag{1.13.22}$$

The inequality (1.13.21), in particular, (1.13.22) is a consequence of Drucker's postulate. Equations (1.13.21) or (1.13.22) must hold for any stress vector $\vec{\mathbf{S}}$, which reaches the loading surface $f = 0$.

Let us verify the inequality (1.13.22) for the body of linear- sotropic hardening material, for which the loading surface can be written as

$$f = c_1(\tau_0 - \sigma_S) - c_2 \gamma_0^S = 0, \quad c_1 = const > 0, \; c_2 = const > 0 \tag{1.13.23}$$

This is an equation of sphere,

$$S_1^2 + S_2^2 + S_3^2 + S_4^2 + S_5^2 - R^2 = 0, \; R = \sqrt{\frac{2}{3}} \left(\sigma_S + \frac{c_2}{c_1} \gamma_0^S \right), \tag{1.13.24}$$

whose radius increases with the rise of accumulated plastic strain γ_0^S. Since the terms τ_0 and γ_0^S appear in the equation of the loading surface (1.13.23) in a first degree, the specified surface describes linear-isotropic hardening. The derivatives from Eq. (1.13.23) are

$$\frac{\partial f}{\partial \tau_0} = c_1 > 0, \quad \frac{\partial f}{\partial \gamma_0^S} = -c_2 < 0, \tag{1.13.25}$$

i.e., from Eq. (1.13.22), we obtain

$$S_i e_i^S \geq 0 \tag{1.13.26}$$

or, in terms of the tensors of stress and plastic strain,

$$\sigma_{ij} \varepsilon_{ij}^S \geq 0. \tag{1.13.27}$$

Equation (1.13.26) or (1.13.27) is a consequence of Drucker's postulate for the model with the isotropic hardening rule and loading surface (1.13.23).

Consider the case when a proportional loading, represented by vector $\vec{\mathbf{S}}_0$, produces a plastic strain, vector $\vec{\mathbf{e}}^S$, as shown in Fig. 1.14a where dotted line corresponds to a yield limit. According to the isotropic hardening rule, a loading surface, corresponding to the vector $\vec{\mathbf{S}}_0$, is shown by solid line in Fig. 1.14a. Further, let an unloading and subsequent loading, with vector $\vec{\mathbf{S}}$, take place. It is evident that the inequality (1.13.26) is satisfied for an acute angle between the vectors $\vec{\mathbf{S}}$ and $\vec{\mathbf{e}}^S$. Otherwise, if the vector $\vec{\mathbf{S}}$ forms an obtuse angle (as shown in Fig. 1.14a) with the vector $\vec{\mathbf{e}}^S$, the inequality (1.13.26) is not fulfilled. The directions of the vector $\vec{\mathbf{S}}$ along which Drucker's postulate is violated constitute so-called zone of unstable straining; the border of this zone is plane $S_i e_i^S = 0$ which is perpendicular to the vector $\vec{\mathbf{e}}^S$ (dotted line in Fig. 1.14a).

Let us assume that the vector $\vec{\mathbf{S}}$ at subsequent loading is oriented as shown in Fig. 1.14a, i.e., Drucker's postulate is not fulfilled and, instead of Eq. (1.10.5), we have

$$d\vec{\mathbf{S}} \cdot d\vec{\mathbf{e}}^S < 0. \tag{1.13.28}$$

A plastic strain-increment -vector, by definition, is in the direction of the external normal to the loading surface. Therefore, according to Eq. (1.13.28) the increment $d\vec{\mathbf{S}}$ must have been pointed inward the loading surface. This is a paradoxical result, a change in plastic strain is induced by stress vector directed inward the surface $f = 0$. On the stress-strain diagram, this situation corresponds to an unreal segment MM_1 in Fig. 1.14.b.

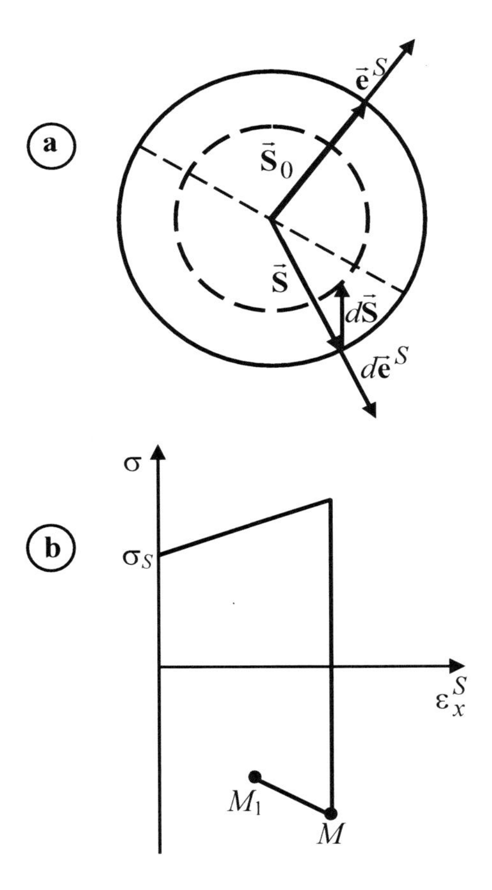

Fig. 1.14 Loading, unloading, and reloading. Loading surface at isotropy hardening rule (a); stress-strain diagram for the case of the violation of Drucker's postulate (b).

Thus, the model with isotropic hardening rule may lead to unstable processes of plastic straining. Such model, by Drucker's, does not belong to the model of stable materials.

1.14 Flow Theory for Elastic-Perfectly Plastic Materials

For ideal elasto-plastic materials, stress-strain relations are constructed under the following assumptions [39,118]: (a) strain components can be decomposed on elastic and plastic part, Eq. (1.5.35), (b) stress deviator tensor components and plastic strain deviator tensor components are proportional, Eq. (1.13.11), meaning that vectors $d\vec{\mathbf{e}}^S$ and $\vec{\mathbf{S}}$ are co-axial. Adopting these assumptions, we have:

$$d\varepsilon_{ij}^S = \left(\sigma_{ij} - \sigma\delta_{ij}\right)dW, \quad dW = W_1 dt, \tag{1.14.1}$$

where W_1 is the unknown function depending on the coordinates of points in a body; t is the time. The function W_1 is not characteristic function of material and must be determined (as well as $\sigma_x, \sigma_y, \ldots$ components) from the solution of boundary-value problem.

Dividing the left- and right-hand sides of Eq. (1.14.1) on dt and ignoring the elastic components of deformation, i.e., $\varepsilon_{ij} = \varepsilon_{ij}^S$, we obtain the following equation for strain rate components, $\dot{\varepsilon}_{ij}$,

$$\dot{\varepsilon}_{ij} = \left(\sigma_{ij} - \sigma\delta_{ij}\right)W_1. \tag{1.14.2}$$

This formula, together with a yield criterion $\tau_0 = \tau_S$, was proposed by von-Mises [91,92].

From Eq. (1.14.2) the following relations can be obtained

$$\begin{gathered} \dot{\varepsilon}_x - \dot{\varepsilon}_y = \left(\sigma_x - \sigma_y\right)W_1, \quad \dot{\varepsilon}_x - \dot{\varepsilon}_z = \left(\sigma_x - \sigma_z\right)W_1, \\ \dot{\varepsilon}_y - \dot{\varepsilon}_z = \left(\sigma_y - \sigma_z\right)W_1, \\ \dot{\gamma}_{xy} = 2\tau_{xy}W_1, \quad \dot{\gamma}_{xz} = 2\tau_{xz}W_1, \quad \dot{\gamma}_{yz} = 2\tau_{yz}W_1. \end{gathered} \tag{1.14.3}$$

Eliminating from Eq. (1.14.3) the factor W_1, we obtain

$$\frac{\sigma_x - \sigma_y}{\dot{\varepsilon}_x - \dot{\varepsilon}_y} = \frac{\sigma_y - \sigma_z}{\dot{\varepsilon}_y - \dot{\varepsilon}_z} = \frac{\sigma_z - \sigma_x}{\dot{\varepsilon}_z - \dot{\varepsilon}_x} = \frac{2\tau_{xy}}{\dot{\gamma}_{xy}} = \frac{2\tau_{yz}}{\dot{\gamma}_{yz}} = \frac{2\tau_{xz}}{\dot{\gamma}_{xz}}. \tag{1.14.4}$$

These relations, together with the Tresca yield criterion, were proposed by Saint-Venant [162] and Lévy [77]. The Saint-Venant-Lévy theory differs from the von-Mises theory only in the yield criterion.

Adding the increments of elastic strain components to the right-hand side in Eq. (1.14.1), we obtain the total strain tensor components,

$$\begin{aligned}
d\varepsilon_x &= \frac{1}{E}\left[d\sigma_x - \nu\left(d\sigma_y + d\sigma_z\right)\right] + (\sigma_x - \sigma)dW \\
d\varepsilon_y &= \frac{1}{E}\left[d\sigma_y - \nu\left(d\sigma_x + d\sigma_z\right)\right] + (\sigma_y - \sigma)dW \\
d\varepsilon_z &= \frac{1}{E}\left[d\sigma_z - \nu\left(d\sigma_x + d\sigma_y\right)\right] + (\sigma_z - \sigma) \\
d\gamma_{xy} &= \frac{d\tau_{xy}}{G} + 2\tau_{xy}dW \\
d\gamma_{xz} &= \frac{d\tau_{xz}}{G} + 2\tau_{xz}dW \\
d\gamma_{yz} &= \frac{d\tau_{yz}}{G} + 2\tau_{yz}dW
\end{aligned} \qquad (1.14.5)$$

Equations (1.14.5), together with the Guber yield criterion, were offered by Reuss [10,57,124] and Prandtl [120]. If the Guber criterion is fulfilled, a plastic deformation occurs. If $d\tau_0 < 0$, the plastic flow is not produced in a body and the unloading occurs that is governed by Hooke's law.

As an illustration of the application of Eq. (1.14.5), determine the stress acting in a thin-walled pipe of a perfectly plastic material under combined tension and torsion. We choose a cylindrical coordinate system with the z-axis along the axis of the pipe. Equation (1.14.5) gives

$$\begin{aligned}
d\varepsilon_z &= \frac{d\sigma_z}{E} + \frac{2}{3}\sigma_z dW \\
d\gamma_{\varphi z} &= \frac{d\tau_{\varphi z}}{G} + 2\tau_{\varphi z}dW .
\end{aligned} \qquad (1.14.6)$$

The Guber yield criterion for an ideally elasto-plastic material has the form as

$$\sigma_z^2 + 3\tau_{\varphi z}^2 = \sigma_S^2 , \qquad (1.14.7)$$

or, in a differential form,

$$\sigma_z d\sigma_z + 3\tau_{\varphi z} d\tau_{\varphi z} = 0 . \qquad (1.14.8)$$

The problem is that it is necessary to determine stresses σ_z and $\tau_{\varphi z}$ corresponding to given (fix) strain components ε_z and $\gamma_{\varphi z}$. The components

σ_z and $\tau_{\varphi z}$ can be determined from Eqs. (1.14.6) and (1.14.8) to be the system of three equations for the three unknowns σ_z, $\tau_{\varphi z}$, and W. Eliminating dW from Eqs. (1.14.6), we find

$$3\tau_{\varphi z} d\varepsilon_z - \sigma_z d\gamma_{\varphi z} = \frac{3\tau_{\varphi z} d\sigma_z}{E} - \frac{\sigma_z d\tau_{\varphi z}}{G}. \tag{1.14.9}$$

Eliminating $\tau_{\varphi z}$ from Eqs. (1.14.7)–(1.14.9), we obtain the equation for σ_z

$$\left(\frac{\sigma_S^2 - \sigma_z^2}{E} + \frac{\sigma_z^2}{3G}\right) d\sigma_z = \left(\sigma_S^2 - \sigma_z^2\right) d\varepsilon_z - \frac{\sigma_z \sqrt{\sigma_S^2 - \sigma_z^2}}{\sqrt{3}} d\gamma_{\varphi z}. \tag{1.14.10}$$

Similarly, the equation which contains one unknown function $\tau_{\varphi z}$ is obtained

$$\left(\frac{\sigma_S^2 - 3\tau_{\varphi z}^2}{G} + \frac{9\tau_{\varphi z}^2}{E}\right) d\tau_{\varphi z} = -3\tau_{\varphi z}\sqrt{\sigma_S^2 - 3\tau_{\varphi z}^2}\, d\varepsilon_z + \left(\sigma_S^2 - 3\tau_{\varphi z}^2\right) d\gamma_{\varphi z}. \tag{1.14.11}$$

It is the Eqs. (1.14.10) and (1.14.11) which determine the components σ_z and $\tau_{\varphi z}$. To simplify these relations, we assume that the material is incompressible, $\nu = 1/2$, $E = 3G$, then

$$\begin{gathered}
\frac{\sigma_S^2}{E} d\sigma_z = \left(\sigma_S^2 - \sigma_z^2\right) d\varepsilon_z - \frac{\sigma_z \sqrt{\sigma_S^2 - \sigma_z^2}}{\sqrt{3}} d\gamma_{\varphi z} \\
\frac{3\sigma_S^2}{E} d\tau_{\varphi z} = -3\tau_{\varphi z}\sqrt{\sigma_S^2 - 3\tau_{\varphi z}^2}\, d\varepsilon_z + \left(\sigma_S^2 - 3\tau_{\varphi z}^2\right) d\gamma_{\varphi z}
\end{gathered} \tag{1.14.12}$$

If the loading path is given as $\gamma_{\varphi z} = \gamma(\varepsilon_z)\sqrt{3}$, where γ is some known function, the Eq. (1.14.12) will be written as

$$\begin{gathered}
\frac{\sigma_S^2}{E} d\sigma_z = \sqrt{\sigma_S^2 - \sigma_z^2}\left[\sqrt{\sigma_S^2 - \sigma_z^2} - \sigma_z \gamma'(\varepsilon_z)\right] d\varepsilon_z, \\
\frac{\sigma_S^2}{E} d\tau_{\varphi z} = \sqrt{\sigma_S^2 - 3\tau_{\varphi z}^2}\left[-\tau_{\varphi z} + \frac{1}{\sqrt{3}}\sqrt{\sigma_S^2 - 3\tau_{\varphi z}^2}\,\gamma'(\varepsilon_z)\right] d\varepsilon_z.
\end{gathered} \tag{1.14.13}$$

Each of the above relations from (1.14.13) is the Ricatti differential equation. Let us consider two partial cases, two straining paths. Let us begin from the step loading. Assume that the pipe first incurs an uniaxial strain, ε_z, due to tension stress $\sigma_z = \sigma_S$. Then, under the condition of two-axial state of stress (the tensile and shear stress must be governed so that the shear stress intensity τ_0 remains unchangeable) only a shear strain, $\gamma_{\varphi z}$, develops. Such straining path is shown in Fig. 1.15a by broken line OM_1M_2. At tension, we have

$$\sigma_z = \sigma_S, \quad d\sigma_z = 0, \quad \tau_{\varphi z} = d\tau_{\varphi z} = 0, \quad d\gamma_{\varphi z} = 0, \tag{1.14.14}$$

and Eq. (1.14.13) is satisfied automatically.

On the second portion of straining, pure torsion, we have $d\varepsilon_z = 0$ and Eq. (1.14.12) takes the form

$$\begin{gathered}\frac{\sigma_S^2\sqrt{3}}{E}\int\limits_{\sigma_S}^{\sigma_z}\frac{d\sigma_z}{\sigma_z\sqrt{\sigma_S^2-\sigma_z^2}} = -\gamma_{\varphi z},\\ \frac{3\sigma_S^2}{E}\int\limits_{0}^{\tau_{\varphi z}}\frac{d\tau_{\varphi z}}{\sigma_S^2-3\tau_{\varphi z}^2} = -\gamma_{\varphi z}\end{gathered} \tag{1.14.15}$$

The integration in the above equations gives

$$\begin{gathered}\frac{\sigma_S\sqrt{3}}{E}\ln\frac{\sigma_S+\sqrt{\sigma_S^2-\sigma_z^2}}{\sigma_z} = \gamma_{\varphi z}\\ \frac{\sigma_S\sqrt{3}}{2E}\ln\frac{\sigma_S+\sqrt{3}\tau_{\varphi z}}{\sigma_S-\sqrt{3}\tau_{\varphi z}} = \gamma_{\varphi z}\end{gathered} \tag{1.14.16}$$

From here we have

$$\sigma_z = \frac{2x}{x^2+1}\sigma_S, \quad \sqrt{3}\tau_{\varphi z} = \frac{x^2-1}{x^2+1}\sigma_S, \quad x = \exp\left(\frac{E\gamma_{\varphi z}}{\sigma_S\sqrt{3}}\right). \tag{1.14.17}$$

Equations (1.14.17) determine the stress on the segment M_1M_2 (Fig. 1.15a). From Eqs. (1.14.17), it follows that $\sigma_z = \sigma_S$ and $\tau_{\varphi z} = 0$ at $\gamma_{\varphi z} = 0$ and the component $\tau_{\varphi z}$ increases and σ_z decreases with the growth of $\gamma_{\varphi z}$.

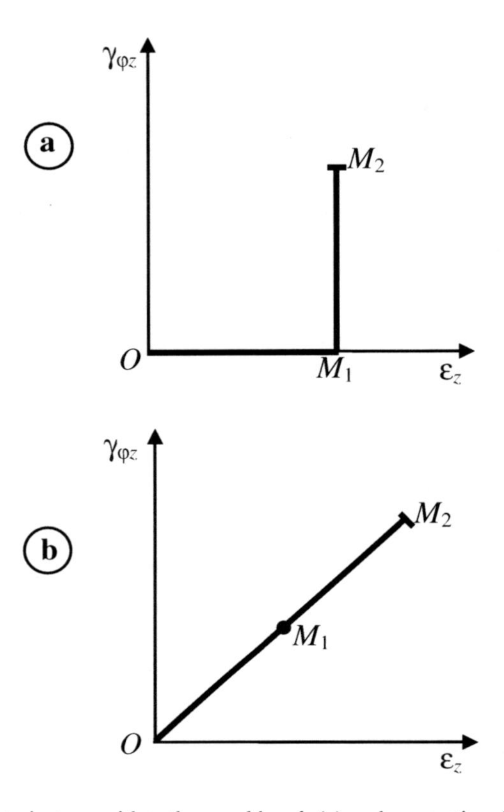

Fig. 1.15 Straining trajectory with orthogonal break (a) and proportional $\gamma_{\varphi z}$-ε_z relation (b).

Consider the second case when the straining path is a straight line, $\gamma_{\varphi z} = \sqrt{3}\varepsilon_z$, as shown in Fig. 1.15b (from point M_1 a plastic flow begins). Then, the constitutive system of equations, Eqs. (1.14.13), results in

$$\frac{\sigma_S^2}{E} d\sigma_z = \sqrt{\sigma_S^2 - \sigma_z^2}\left[\sqrt{\sigma_S^2 - \sigma_z^2} - \sigma_z\right] d\varepsilon_z$$
$$\frac{\sigma_S^2}{E} d\tau_{\varphi z} = \sqrt{\sigma_S^2 - 3\tau_{\varphi z}^2}\left[\frac{1}{\sqrt{3}}\sqrt{\sigma_S^2 - 3\tau_{\varphi z}^2} - \tau_{\varphi z}\right] d\varepsilon_z \qquad (1.14.18)$$

At the vicinity of point M_1, when $\tau_0 = \tau_S - 0$, Hooke's law can be applied:

$$\sigma_z = E\varepsilon_z, \quad \tau_{\varphi z} = G\gamma_{\varphi z} = \frac{E}{3}\gamma_{\varphi z} = \frac{E}{\sqrt{3}}\varepsilon_z. \qquad (1.14.19)$$

Substituting these components into the Guber criterion (1.14.7), strain component $\varepsilon_z = \sigma_S / (\sqrt{2}E)$ is obtained, which together with Eq. (1.14.19) gives the tensile and shear stress corresponding to this criterion:

$$\sigma_z = \frac{\sigma_S}{\sqrt{2}}, \quad \tau_{\varphi z} = \frac{\sigma_S}{\sqrt{6}}. \qquad (1.14.20)$$

Further, substituting σ_z and $\tau_{\varphi z}$ from Eq. (1.14.20) into the right-hand sides in the Eqs. (1.14.18), we obtain $d\sigma_z = d\tau_{\varphi z} = 0$. This means that Eqs. (1.14.20) are the solutions of the problem, the stresses remain unchangeable during straining along the segment M_1M_2.

The result obtained is valid for the case of an arbitrary proportional straining regardless of the magnitude of non-zero ε_{ij} components. The stresses determined from the yield criterion remain unchangeable during the further growth of the absolute value of components ε_{ij}.

Making comparison between Eq. (1.14.17) and Eq. (1.14.20), it is clear that, in terms of the flow plasticity theory, stresses essentially depend on straining path.

1.15 More Complicated Flow Plasticity Theories

Consider first the Handelman-Lin-Prager theory [37, 79-81]. This theory is based on the following assumptions:

(i) Stress deviator components and their increments determine the increments of plastic strains

(ii) The increments of plastic strains and stress deviator components are linearly related

(iii) The condition of continuity is valid.

(iv) Loading criterion has the form $d\tau_0 > 0$.

The assumptions (i) and (ii) lead to the following relations

$$d\varepsilon_{kl}^S = C_{klij} d\overline{\sigma}_{ij}, \qquad (1.15.1)$$

where C_{klij} do not depend on $d\overline{\sigma}_{ij}$. From the condition of continuity we have

$$C_{klij} d\overline{\sigma}_{ij} = 0 \quad \text{at} \quad d\tau_0 = 0 \qquad (1.15.2)$$

at neutral loading. On the other hand, the equality $d\tau_0 = 0$ is equivalent to the following condition

$$\overline{\sigma}_{ij} d\overline{\sigma}_{ij} = 0 . \tag{1.15.3}$$

Equations (1.15.2) and (1.15.3), being linear forms relatively to $d\overline{\sigma}_{ij}$, imply that

$$C_{klij} = C_{kl}{}'\overline{\sigma}_{ij} . \tag{1.15.4}$$

Hence

$$d\varepsilon_{ij}^{S} = C_{ij}{}'' d\tau_0 , \qquad C_{ij}{}'' = \frac{2}{3}\tau_0 C_{ij}{}' . \tag{1.15.5}$$

Prager has proposed that parameter $C_{ij}{}''$ has to be determined by the right-hand side in Eq. (1.6.12). Equations (1.6.12) and (1.15.5) determine the increments of plastic strain in terms of the Handelman-Lin-Prager theory. If we assume that $A_2 = 0$ and $A_1 = A_1(\tau_0)$, they will agree with the Laning relations.

Prager-Hodge theory [40,41,116,117,119]. In terms of this theory, the above assumptions (i)-(iii) are adopted, whereas the loading criterion, $d\tau_0 > 0$, is rejected. In addition, it is assumed that the arguments of function f are the components of vectors $\vec{\mathbf{S}}$, $\vec{\mathbf{e}}^{S}$ and parameters $\chi_i, \ldots$ whose increment are related to plastic strain increments as

$$d\chi_i = \chi_{ij}{}' de_j^S , \tag{1.15.6}$$

where $\chi_{ij}{}'$ is characteristic functions of material.

Assumption (ii) and (iii) lead to Eqs. (1.13.6) and (1.9.8) that remain valid. Now Eq. (1.9.8), on the base of Eq. (1.15.6), takes the form

$$\frac{\partial f}{\partial S_i} dS_i + \frac{\partial f}{\partial e_i^S} de_i^S + \frac{\partial f}{\partial \chi_i} \chi_{ij}{}' de_j^S = 0 . \tag{1.15.7}$$

By using Eq. (1.13.6) in Eq. (1.15.7) we get

$$-\frac{1}{C} = \frac{\partial f}{\partial e_i^S} \frac{\partial f}{\partial S_i} + \frac{\partial f}{\partial \chi_i} \frac{\partial f}{\partial S_j} \chi_{ij}{}' . \tag{1.15.8}$$

It is Eqs. (1.13.6) and (1.15.8) that determine plastic strain increments. The loading condition is

$$\frac{\partial f}{\partial S_i} dS_i > 0 . \tag{1.15.9}$$

If the inequality (1.15.9) is not held or the vector $\vec{S}$ does not reach the loading surface, then $de_k^S = 0$.

The obtained results are the top (summit) of the theory of plasticity with a regular loading surface $f = 0$. No restriction is imposed upon the loading surface $f = 0$. Eqs. (1.13.6) and (1.15.8) are valid in a general case when the loading surface depends not only on stress and plastic strain components but also on some parameters χ_i. According to Eq. (1.13.6), a plastic-strain-increment vector is equal to zero at neutral loading, i.e., the continuity of strains is satisfied at the transition from a plastic to elastic straining. But, as it was already specified, there are no theoretical thoughts giving the equation of loading surface and the form of functions χ_{ij}'.

Let us show that the Laning relations can be derived from Eqs. (1.13.6), (1.15.6), and (1.15.8). The equation of the loading surface for the case of isotropy hardening, Eq. (1.13.8), can be written as

$$f = \tau_0 - \chi_1, \quad \chi_1 = \tau_{0m}. \tag{1.15.10}$$

Hence, the function f depends on stress components and one parameter χ_1, which is the greatest value of τ_0 for the whole history of loading. Counting up derivatives $\partial f/\partial \chi_1 = -1$, $\partial f/\partial S_i = 3S_i/(2\tau_0)$, and $\partial f/\partial e_i^S = 0$ in Eq. (1.15.8), we obtain

$$C = \frac{2\tau_0}{3S_i\chi_{1i}'}. \tag{1.15.11}$$

If to assume

$$\chi_{1i}' = \frac{S_i}{\tau_0 C(\tau_0)}, \tag{1.15.12}$$

then Eqs. (1.15.11) and (1.15.12) show that $C = C(\tau_0)$ and from Eq. (1.13.6) we obtain the Laning relations (1.13.11); the factor $(3/2\tau_0)$ is included in $C(\tau_0)$ (as in Sec. 1.13).

1.16 Flow Plasticity Theory with Kinematic Hardening

Since the model with isotropy hardening rule under certain conditions is in an obvious contradiction with the Bauschinger effect, other theories must be formulated to describe alternating loading paths. To them belongs the Ishlinskij-Kadashevich-Novozsilov theory [49,54]. Similar to the A.Ju. Ishlinskij model [49], the kinematic hardening rule assumes that during plastic deformation, the loading surface translates as a rigid body in stress deviator space, maintaining the size shape, and orientation of the initial yield surface.

The basic statements of this theory are

1. The Guber yield criterion is adopted.
2. Stress vector components S_k are represented as the sum of two parts:

$$S_k = S_k{}' + S_k{}'', \tag{1.16.1}$$

where $S_k{}'$ is called the active stress components and $S_k{}''$ is residual micro-stress components.

3. The equation of loading surface is written as

$$f = (S_i - S_i{}'')(S_i - S_i{}'') - \frac{2}{3}\sigma_S^2 = 0, \tag{1.16.2}$$

i.e., this is a sphere of constant radius. The residual micro-stress components $S_k{}''$ govern the coordinates of the center of this sphere.

4. It is assumed that

$$S_k{}'' = c_0 e_k^S, \tag{1.16.3}$$

where c_0 is constant.

Equation (1.16.1)–(1.16.3) constitute the basic relations of the plastic flow theory with the kinematic hardening rule. The notions of active stress and residual micro stress can be explained by modeling of the behavior of material at uniaxial state of stress by a spring-mass system. Consider a spring whose one end is attached to a mass, which is located on a horizontal rough plane (Fig. 1.16a), and to another end, point M, a stretching force S_1 is applied. The displacement of the mass symbolizes the rise of a plastic strain in a body. The displacement of point M, the elongation of the spring, can be associated with the strain, e_1, arising in a body due to the action of the force S_1. While the force S_1 is less than a frictional force in the mass-plane contact, F_r, the spring is stretched under a linear law, which corresponds to a linear portion of the $S_1 - e_1$ diagram in Fig. (1.16.b). Once $S_1 > F_r$, the displacement of point M, together with the mass, takes place at the constant value of S_1, which corresponds to the horizontal portion of $S_1 - e_1$ diagram (Fig. 1.16b) of perfect elasto-plastic material.

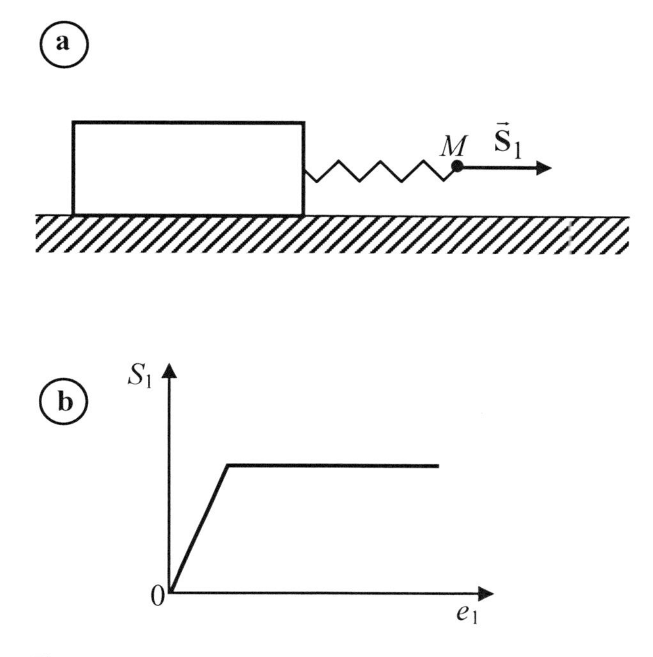

Fig. 1.16 Spring-analogy (a) for perfect elasto-plastic body (b).

It is possible to extend this model to the case of linear work-hardening material, which exhibits a linear $S_1 - e_1$ relation during plastic straining. Consider the case when two springs are attached to a mass (see Fig. 1.17). The mass remains immovable ($S_1''=0$) while $S_1 \leq F_r$ and the displacements of the end of the right spring, point *M*, corresponds to the initial elastic portion of the $S_1 - e_1$ diagram in Fig. 1.17b. Once $S_1 = F_r$, the mass is moved and the left spring works as well, $S_1''>0$. The force of the left spring S_1'' is governed by Eq. (1.16.3), which corresponds to the second straight-line branch of $S_1 - e_1$ diagram in Fig. 1.17b. After unloading ($S_1 = 0$), the force S_1'' does not vanish due to the assumption that $S_1'' \leq F_r$. If now to apply a compressive force, plastic compressive yielding begins at the value of compressive force smaller than F_r due to the additional action of the force S_1''. It is obvious that the model describes the ideal Bauschinger effect: the degree of hardening of material due to plastic tension causes an equal degree of softening relative to a subsequent compression. Stresses S_k'' are the macro-manifest of the latent internal micro-stresses, which arise in a body at plastic flow and do not disappear after unloading. From here the name of S_k'' is originated.

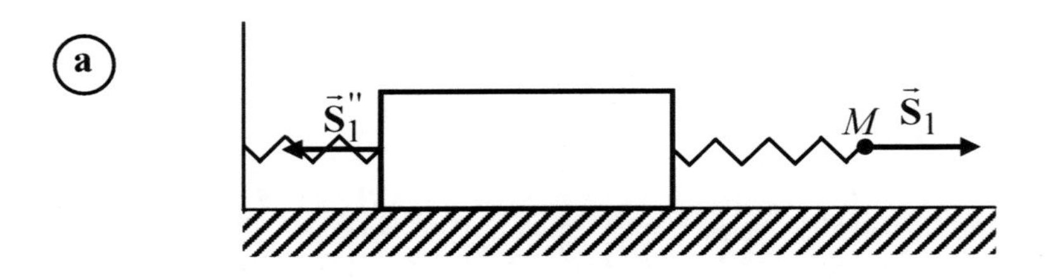

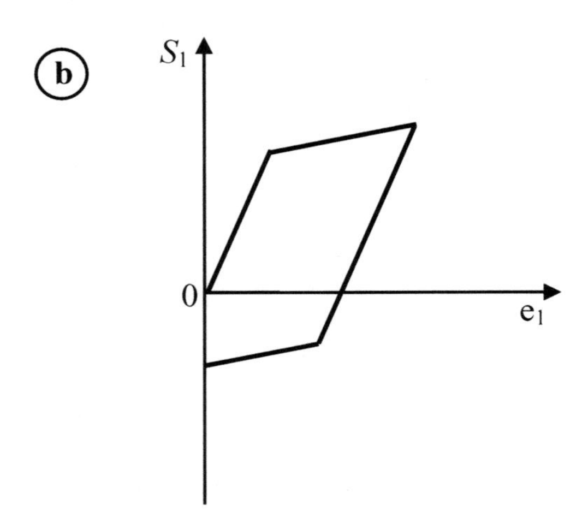

Fig. 1.17 Spring-analogy (a) for linear work-hardening body (b).

To determine plastic-strain-component increments, the equation of the loading surface (1.16.2), in view of Eq. (1.16.3), we write as

$$f = \left(S_i - c_0 e_i^S\right)\left(S_i - c_0 e_i^S\right) - \frac{2}{3}\sigma_S^2 = 0, \tag{1.16.4}$$

i.e., the function f depends on the stress- and plastic strain-vector components but is independent of variable parameters $\chi_1, \chi_2, \ldots$ (see Eq. (1.9.2)).

Let us use the Prager-Hodge formulae. Calculating derivatives in Eq. (1.16.4)

$$\frac{\partial f}{\partial S_i} = 2\left(S_i - c_0 e_i^S\right), \quad \frac{\partial f}{\partial \chi_i} = 0, \quad \frac{\partial f}{\partial e_i^S} = -2c_0\left(S_i - c_0 e_i^S\right) \tag{1.16.5}$$

and substituting them into Eqs. (1.13.6) and (1.15.8), we obtain

$$de_k^S = \frac{3}{2c_0\sigma_S^2}\left(S_k - c_0 e_k^S\right)\left(S_i - c_0 e_i^S\right)dS_i. \tag{1.16.6}$$

The plastic loading condition has the form

$$\left(S_i - c_0 e_i^S\right) dS_i > 0. \tag{1.16.7}$$

If this condition is not satisfied or the stress vector $\vec{\mathbf{S}}$ does not reach the surface $f = 0$, then $de_k^S = 0$. Therefore, Eq. (1.16.6), together with the condition of Eq. (1.16.7), determines the increments of plastic strain components.

In a general case of loading, the expression (1.16.6) is the system of differential equations for plastic strain components. In solving this system, the kinematic flow rule (1.16.4) must be taken into account. For example, for the case of pure shear $\tau_{xy} > 0$ when only S_3 and e_3^S are nonzero, Eq. (1.16.6) takes the form

$$de_3^S = \frac{3}{2 c_0 \sigma_S^2} \left(S_3 - c_0 e_3^S\right)^2 dS_3. \tag{1.16.8}$$

On the other hand, Eq. (1.16.4) gives

$$\left(S_3 - c_0 e_3^S\right)^2 = \frac{2}{3} \sigma_S^2. \tag{1.16.9}$$

Therefore,

$$de_3^S = \frac{dS_3}{c_0} \tag{1.16.10}$$

and

$$e_3^S = \frac{1}{c_0} \left(S_3 - \sqrt{\frac{2}{3}} \sigma_S \right). \tag{1.16.11}$$

The same result can be obtained from Eq. (1.16.9).

Let us note that according to Eq. (1.16.6) the increment-vector $d\vec{\mathbf{e}}^S$ is co-axial with vector $\vec{\mathbf{S}} - c_0 \vec{\mathbf{e}}^S$. As consequence, the model describes so-called Feigin effect that consists in the following. Consider the specimen that is first loaded in combined tension and torsion then unloaded and subsequently reloaded in uniaxial tension. In this case a plastic "untwisting" of the specimen takes place, i.e., the decrease of residual shear deformation [32].

Let us investigate the fulfillment of Drucker's postulate in terms of the translational flow theory with the loading surface in the form of (1.16.4). For this purpose we use the inequality (1.13.21) written in the following form

$$c_0\left(S_i - c_0 e_i^S\right)\left(S_i - c_0 e_i^S\right) \geq 0. \tag{1.16.12}$$

Equations (1.16.12) and (1.16.4) show that the kinematic model, at $c_0 > 0$, satisfies Drucker's postulate for arbitrary loading paths.

Kadashevich's and Novozsilov's works [54,55] generalize relations obtained in this section.

1.17 Lode-Nadai Variable

The Lode-Nadai variable [84,85] is often used at experimental verifications of the laws of plasticity. Let $\sigma_1, \sigma_2, \sigma_3$ be principal values of stress tensor,

$$\sigma_1 \geq \sigma_2 \geq \sigma_3. \tag{1.17.1}$$

The Lode-Nadai variable μ_σ, relative to stresses, is defined as

$$\mu_\sigma = \frac{2\sigma_2 - (\sigma_1 + \sigma_3)}{\sigma_1 - \sigma_3}. \tag{1.17.2}$$

The variable μ_σ characterizes the ratio between the principal stresses, in particular the position of point σ_2 relative to values of σ_1 and σ_3 laid off along a common axis. Therefore, we can say that μ_σ is the characteristic of the type of stress-state. The variable μ_σ does not depend on a hydrostatic pressure and is constant at proportional loading.

The variable μ_σ ranges from -1 to 1 so that we have $\mu_\sigma = -1$ at uniaxial tension ($\sigma_1 > 0, \sigma_2 = \sigma_3 = 0$), $\mu_\sigma = 0$ at pure shear ($\sigma_1 > 0, \sigma_2 = 0, \sigma_3 = -\sigma_1$), and $\mu_\sigma = 1$ at uniaxial compression ($\sigma_1 = \sigma_2 = 0, \sigma_3 < 0$).

The Lode-Nadai variable μ_ε, relative to strains, is similarly introduced

$$\mu_\varepsilon = \frac{2\varepsilon_2 - (\varepsilon_1 + \varepsilon_3)}{\varepsilon_1 - \varepsilon_3}, \tag{1.17.3}$$

where ε_1, ε_2, and ε_3 are principal strains. Similarly, the following quantities, expressed in increments, can also be introduced,

$$\mu_{\Delta\sigma} = \frac{2\Delta\sigma_2 - (\Delta\sigma_1 + \Delta\sigma_3)}{\Delta\sigma_1 - \Delta\sigma_3}, \tag{1.17.4}$$

$$\mu_{\Delta\varepsilon^S} = \frac{2\Delta\varepsilon_2^S - \left(\Delta\varepsilon_1^S + \Delta\varepsilon_3^S\right)}{\Delta\varepsilon_1^S - \Delta\varepsilon_3^S}. \tag{1.17.5}$$

If Eq. (1.17.1) holds for stresses σ_1, σ_2, and σ_3, this does not mean that it will also hold for their increments. Therefore, in contrast to μ_σ, the variable $\mu_{\Delta\sigma}$ can vary in the range $-\infty \le \mu_{\Delta\sigma} \le \infty$.

If the Hencky–Nadai relations are valid, Eqs. (1.17.2) and (1.17.3) give

$$\mu_\sigma = \mu_\varepsilon. \tag{1.17.6}$$

For the case when the relations of flow plasticity theory, Eq. (1.13.10), are held, we get

$$\mu_\sigma = \mu_{\Delta\varepsilon^S}. \tag{1.17.7}$$

At the verification of the concurrency of theoretical results with experimental data, Eqs (1.17.6) or (1.17.7) will be used.

1.18 Experimental Check of the Laws of Plasticity at Combined Loading

This section treats the results of experiments in two-axial tension [7,184,185]. On their base, we will make some conclusions about the applicability of theories considered earlier to two-segment loading paths. Our main goal is to verify the existence of proportional relation between: (i) stress and strain deviators; and (ii) stress-deviators and plastic strains-deviator increments.

The experiments were performed on thin-walled pipes of steel 12XH3A in combined tension and internal pressure. The experimental technique, measurement procedure, homogeneity- and initial-isotropy-check, and data processing are given in [179].

Let us denote the components of normal stresses, which act in the axial, tangential, and radial directions of pipe, as σ_z, σ_φ, and σ_r. It is clear that no shear stress acts in the elements of the pipe, i.e., the components σ_z, σ_φ, and σ_r are principal stresses ($\sigma_r = \sigma_3 \approx 0$) and ε_z, ε_φ, and ε_r are principal deformations.

In contrast to the stress σ_r, the radial strain ε_r was not zero in experiments however its direct measurements gave unreliable results. Therefore, in contrast to ε_z and ε_φ, which were measured by means of the change in the length and diameter of tube, the strain ε_r was calculated from Eq. (1.2.23), i.e., from the condition of an elastic volume change,

$$\varepsilon_r = -\varepsilon_z - \varepsilon_\varphi + \frac{\sigma_z + \sigma_\varphi}{3K_0} \tag{1.18.1}$$

while plastic strains do not change the volume of body:

$$\varepsilon_r^S = -\varepsilon_z^S - \varepsilon_\varphi^S. \tag{1.18.2}$$

It is obvious that the axes of principal stresses and strains coincide between themselves and their directions do not change during loading.

The results of the experiments are presented in terms of Ilyushin's stress/strain-deviator space, where plastic strain coordinates are superimposed upon stress coordinates. It is suffice to consider two-dimensional subspace, $S_1(e_1)-S_2(e_2)$ coordinate plane, due to that $S_3 = S_4 = S_5 = 0$, $e_3 = e_4 = e_5 = 0$. If to replace the abbreviates of *x*-, *y*-, and *z*-axis by *z*, φ, and *r*, respectively, Eqs. (1.7.11) and (1.7.12) give

$$S_1 = \sqrt{\frac{3}{2}}\bar{\sigma}_z = \sqrt{\frac{2}{3}}\left(\sigma_z - \frac{\sigma_\varphi}{2}\right), \quad S_2 = \frac{\bar{\sigma}_z}{\sqrt{2}} + \sqrt{2}\bar{\sigma}_\varphi = \frac{\sigma_\varphi}{\sqrt{2}}, \tag{1.18.3}$$

$$e_1 = \sqrt{\frac{2}{3}}\left[\varepsilon_z - \frac{1}{2}(\varepsilon_\varphi + \varepsilon_r)\right], \quad e_2 = \frac{1}{\sqrt{2}}(\varepsilon_\varphi - \varepsilon_r). \tag{1.18.4}$$

In addition, on the base of Eq. (1.7.14) we have

$$|\vec{\mathbf{S}}| = \sqrt{S_1^2 + S_2^2} = \sqrt{\frac{2}{3}}\tau_0, \quad |\vec{\mathbf{e}}| = \sqrt{e_1^2 + e_2^2} = \sqrt{\frac{3}{2}}\gamma_0. \tag{1.18.5}$$

The increments of strain components from Eq. (1.18.4) are

$$de_1^S = \sqrt{\frac{2}{3}}\left[d\varepsilon_z^S - \frac{1}{2}\left(d\varepsilon_\varphi^S + d\varepsilon_r^S\right)\right], \quad de_2^S = \frac{1}{\sqrt{2}}\left(d\varepsilon_\varphi^S - d\varepsilon_r^S\right). \tag{1.18.6}$$

The experiment was performed so that a loading path in $S_1(e_1)-S_2(e_2)$ coordinate plane is a broken line. The rate of loading (0,08-0,3 MPa/sec) was measured during the second portion of loading (additional loading) and it was measured by the rate of the change of component σ_z if $\Delta\sigma_z > \Delta\sigma_\varphi$ and σ_φ if $\Delta\sigma_\varphi > \Delta\sigma_z$. As shown in [179], such loading rates do not result in temporary effects during plastic straining.

Figures 1.18a–1.22a show the results of experiments for two-segment loading paths. Dashed lines represent loading paths and solid lines depict straining trajectories. Point M, at which a loading paths incur a brake, and points marked by 1,2... correspond to point M' and points 1', 2',... on a straining paths. There are a stress vector $\vec{\mathbf{S}}$ (dotted lines) and strain vector $\vec{\mathbf{e}}$ (solid lines) at some points of the straining curve, they show their orientation only. For each loading process corresponding $\mu_{\Delta\varepsilon}s(\mu_\sigma)$ and $\mu_\varepsilon(\mu_\sigma)$ relations are plotted in Fig. 1.18b–1.22b.

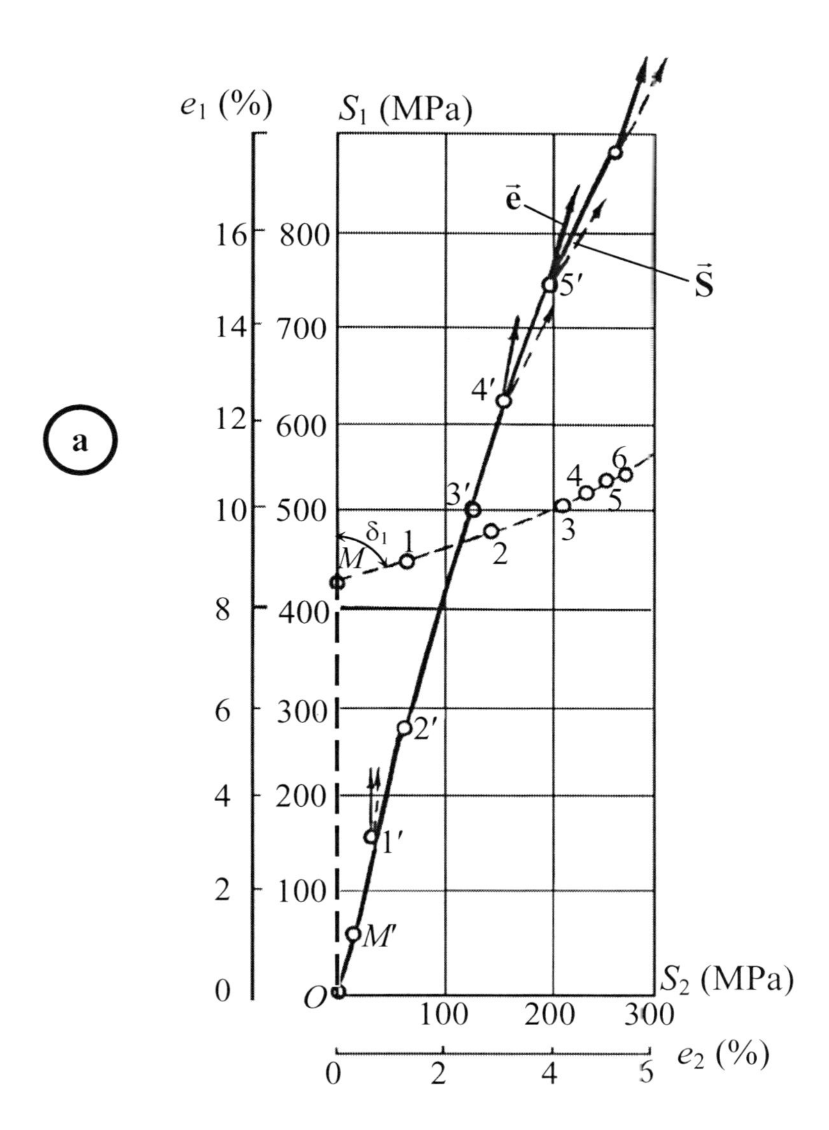

Fig. 1.18a Two-segment loading path (dotted line $0M123\ldots$) with break-angle $\delta_1 = 60^\circ$. Continuo line corresponds to the straining path.

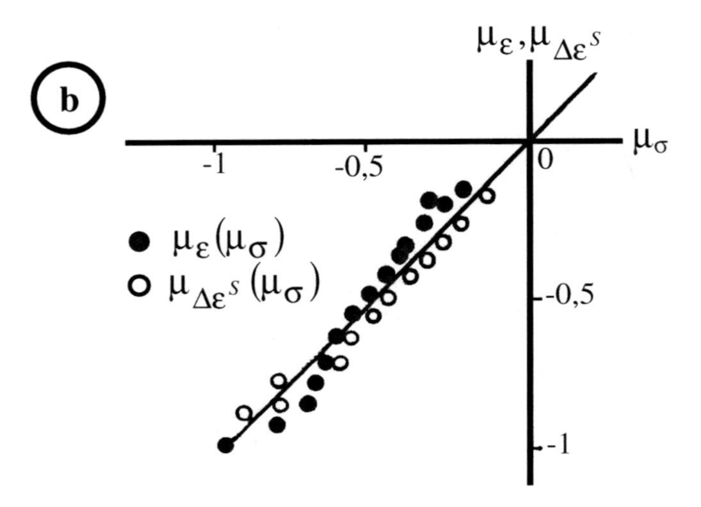

Fig. 1.18b Plot of relations between the Lode-Nadai parameters μ_ε, $\mu_{\Delta\varepsilon^S}$, and μ_σ.

Let us consider the following loading paths.

1. The tube is first subjected to uniaxial tension (segment *OM*, $S_1 > 0$ and $S_2 = 0$ in Fig. 1.18a) and then loaded in two-axial tension so that there is the break of loading path (at point *M*) at an angle δ_1 of 60° with the segment *OM*. The break of loading path of angle $\delta_1 = 60^\circ$ requires that $\Delta\sigma_z = \Delta\sigma_\varphi$ during additional loading (this follows from the relation $\Delta S_2/\Delta S_1 = \sqrt{3}$ and Eq. (1.18.3)), which implies that the Lode-Nadai variable $\mu_{\Delta\sigma} = 1$. Beyond point *M*, the vectors $\vec{\mathbf{S}}$ and $\vec{\mathbf{e}}$ do not coincide in direction but the maximal divergence angle does not exceed 7° that corresponds to the fulfilling of equality $\mu_\varepsilon = \mu_\sigma$ within 5–7%. Corresponding graphs $\mu_\varepsilon(\mu_\sigma)$ and $\mu_{\Delta\varepsilon^S}(\mu_\sigma)$ are plotted in Fig. 1.18b. The divergence angle, Δw, between vectors $\vec{\mathbf{S}}$ and $\vec{\mathbf{e}}$ is calculated as

$$\Delta w = \arctan\frac{e_1}{e_2} - \arctan\frac{S_1}{S_2}. \tag{1.18.7}$$

2. The tube is loaded so that a segment *OM* makes an angle of 60° with S_2-axis ($\sigma_z = 2\sigma_\varphi$) and the second portion of loading is horizontal to this axis

(Fig. 1.19a). It must be noted that the segment *OM* ($\sigma_z = 2\sigma_0$, $\sigma_\varphi = \sigma_0$, $\sigma_r = 0$) corresponds to the case of pure shear. Indeed, if to add a hydrostatic pressure, $\sigma_z = \sigma_\varphi = \sigma_r = -\sigma_0$, to these components, we will obtain the case of pure shear ($\sigma_z = \sigma_0$, $\sigma_\varphi = 0$, $\sigma_r = -\sigma_0$). The straight line, horizontal to S_2-axis, implies that the Lode-Nadai variable $\mu_{\Delta\sigma} = 3$. The directions of vectors $\vec{\mathbf{S}}$ and $\vec{\mathbf{e}}$ do not coincide beyond point *M*. The greatest angle of divergence has the value of 10° and, at the end of the second segment, it decreases to the value of 5°. The proportional relation between the stress and strain deviators is violated, $\mu_\varepsilon \neq \mu_\sigma$ (see Fig. 1.9b). At the same time, the proportional relation between the stress deviators and the increments of strain, $\mu_{\Delta\varepsilon^S} = \mu_\sigma$, is valid during the whole extra-loading.

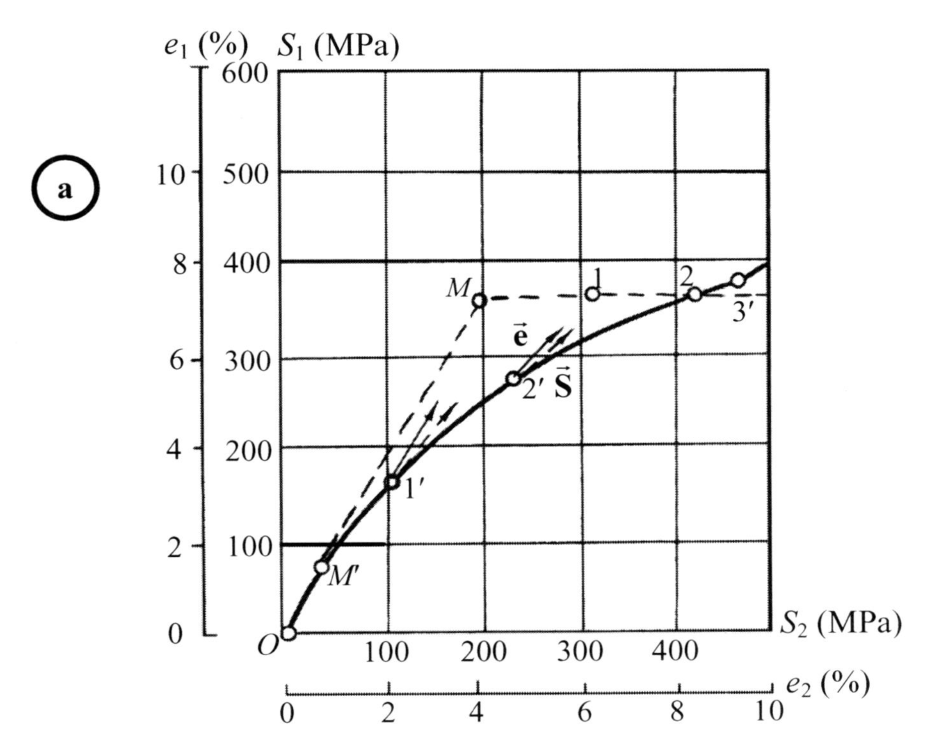

Fig. 1.19a Two-segment loading trajectory (dotted line *0M*12...) and corresponding straining trajectory (solid line).

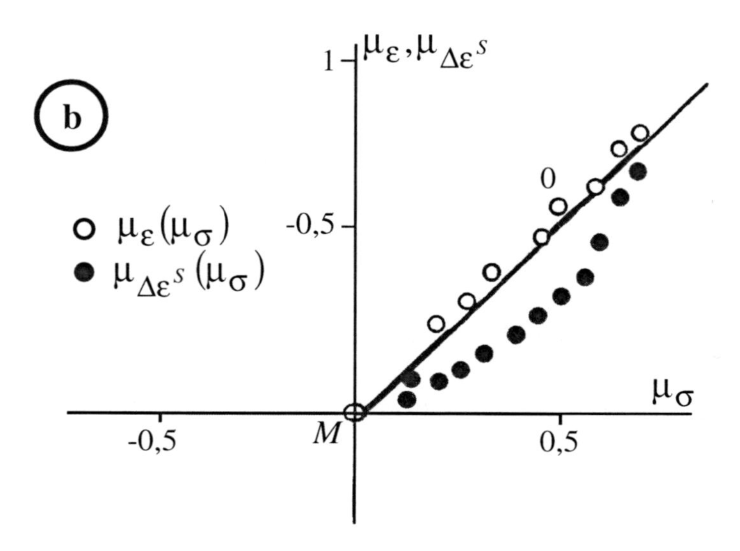

Fig. 1.19b Plot of relations between the Lode-Nadai parameters μ_ε, $\mu_{\Delta\varepsilon^S}$, and μ_σ.

3. The results of experiments at an orthogonal break of loading path are shown in Fig. 1.20a. The tube is first subjected to uniaxial tension along the segment *OM* and the additional loading is carried out so that the second portion of loading path is a horizontal straight line. Similar to the above experiments, the vectors $\vec{\mathbf{S}}$ and $\vec{\mathbf{e}}$ are not co-directed beyond the braking of loading trajectory. The angle between them increases first with moving away from point *M* (reaches its maximal value of 18°) and then decreases. A considerable deviation from the equality $\mu_\varepsilon = \mu_\sigma$ (Fig. 1.20b) is observed.

From the comparison of experiments given in Fig. 1.18–1.20 it follows that the (in)applicability of deformation theory depends not only on the value of angle δ_1, i.e., on the amount of the break of loading path, but also on the value of variable $\mu_{\Delta\sigma}$ as well. In the case of an orthogonal breaking of loading path, flow plasticity theories are valid only along some part of additional loading, beyond point M_1 (Fig. 1.20a), where $\mu_{\Delta\varepsilon^S} = \mu_\sigma$.

In the experiments shown in Fig. 1.18–1.20, the tangential stress intensity τ_0 increases during loading. $\tau_0 - \gamma_0$ curves plotted on the base of these experiments turned out to be close to $\tau_0 - \gamma_0$ curves obtained at simple loading, i.e., there exists an universal dependence between τ_0 and γ_0. However, it does not mean (Fig. 1.19b – 1.20b) that the proportionality between the stress and strain deviators always hold.

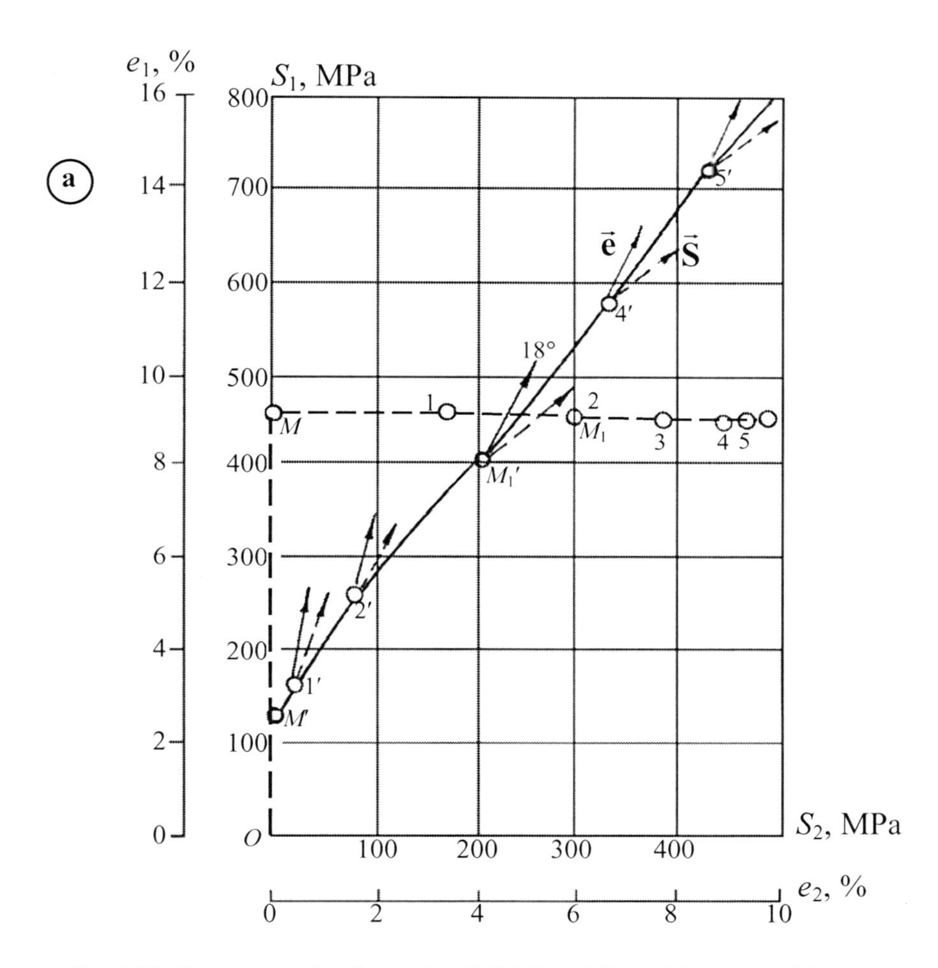

Fig. 1.20a Two-segment loading path $0M1234$ (dotted line) with orthogonal break.

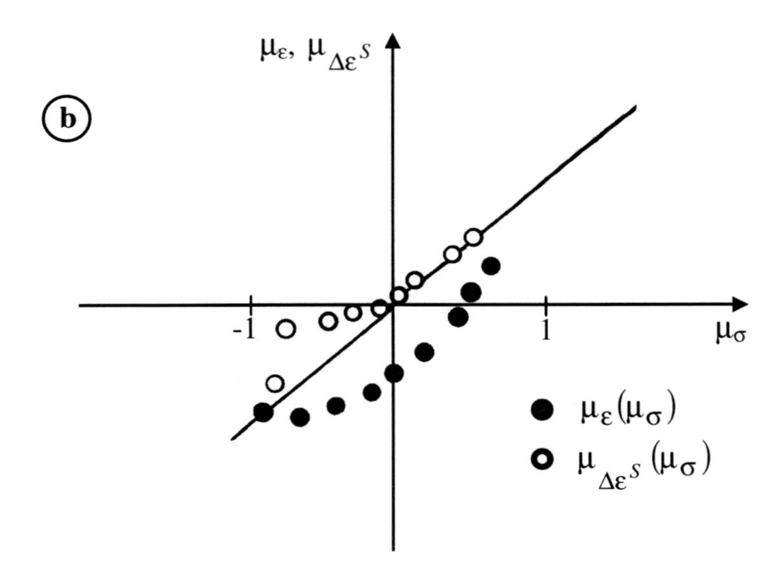

Fig. 1.20b Plot of relations between the Lode-Nadai parameters μ_ε, $\mu_{\Delta\varepsilon^S}$, and μ_σ.

4. The experiments given in Fig. 1.21 and 1.22 are of a great importance. The tube is first subjected to uniaxial tension and then is loaded so that the second portion of loading path makes an angle with the segment *OM* greater than 90°, $\delta_1 = 109°$ in Fig. 1.21a and $\delta_1 = 139°$ in Fig. 1.22a. The disorientation between the vectors $\vec{\mathbf{S}}$ and $\vec{\mathbf{e}}$ increases first, along segment $M'M_1'$, and then decreases. The maximal angle between the vectors $\vec{\mathbf{S}}$ and $\vec{\mathbf{e}}$ is 25° (Fig. 1.21a) and 43° (Fig. 1.22a). A large deviation from the equality $\mu_\varepsilon = \mu_\sigma$ (Fig. 1.21b, 1.22b) is observed. The equation $\mu_{\Delta\varepsilon^S} = \mu_\sigma$ holds only beyond point M_1. These results do not differ qualitatively from the case of the orthogonal breaking of loading path, however, they are of great value due to the following. In vicinity of point *M*, the modulus $|\vec{\mathbf{S}}|$ and the shear stress intensity τ_0, decreases. In terms of the flow theory with isotropic hardening, such change of the vector $\vec{\mathbf{S}}$ means that an additional loading is directed inside a loading surface $f = 0$ and, as a consequence, must be accompanied only by elastic strain. However, the performed experiments indicate that the increment of plastic strain takes place in the vicinity of point *M*, meaning that the flow theory with isotropic hardening rule is incapable of modeling the plastic properties of a material at sharp breaking of loading path.

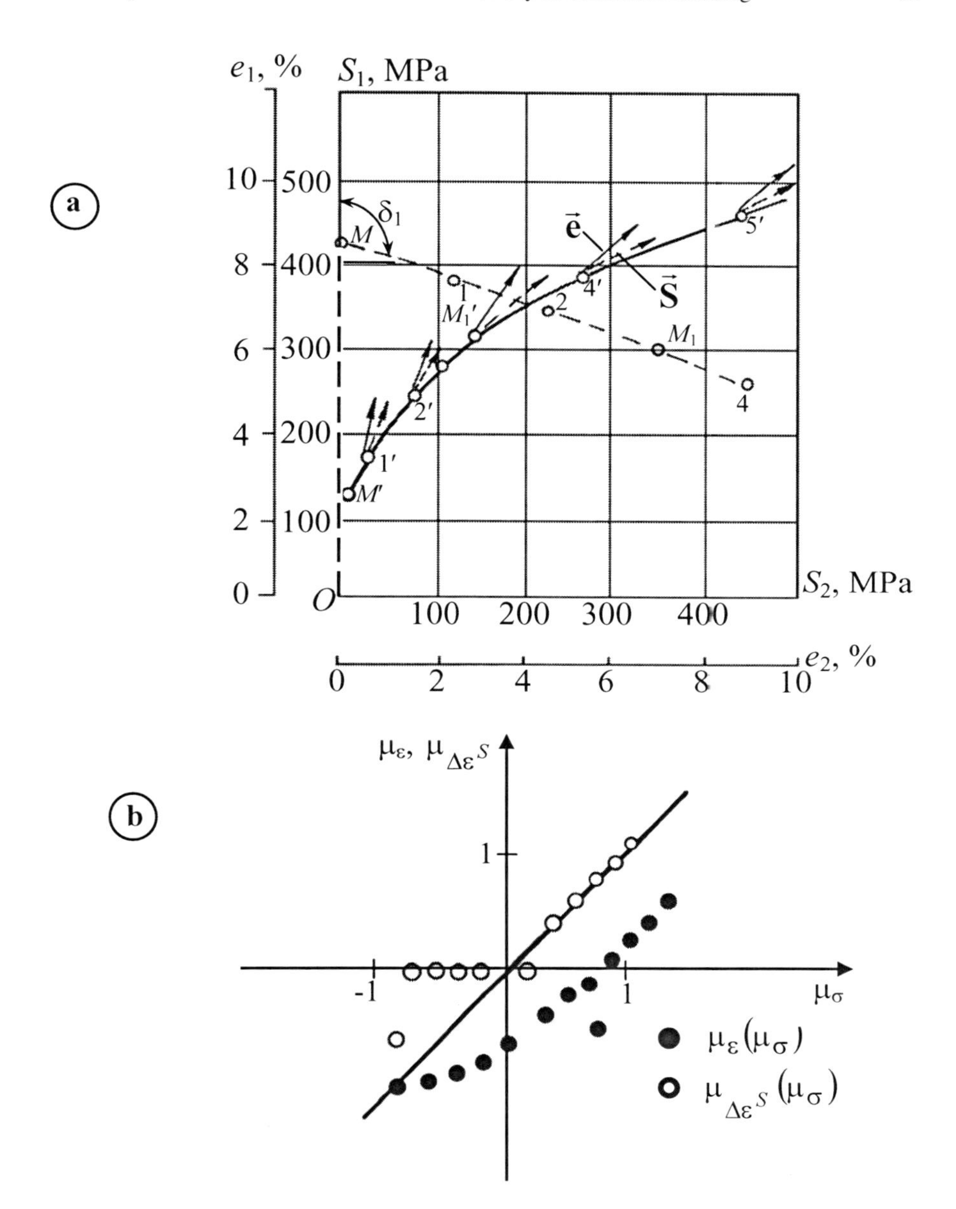

Fig. 1.21 (a) Break of loading trajectory (dotted line $0M123\ldots$) at angle $\delta_1 > 90°$ and corresponding straining trajectory (continuous line). (b) Plot of relations between the Lode-Nadai parameters μ_ε, $\mu_{\Delta\varepsilon^S}$, and μ_σ.

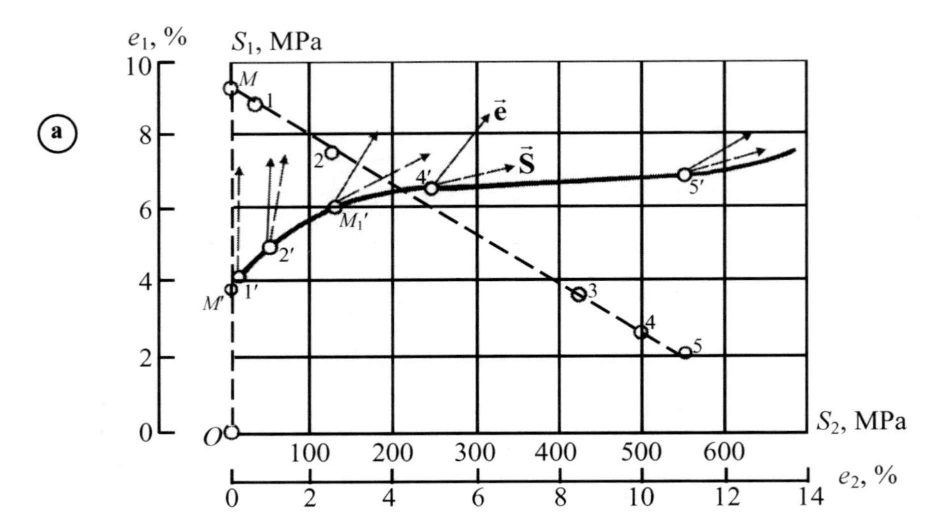

Fig. 1.22a. Break of loading path (dotted line $0M123\ldots$) of angle 139° and corresponding straining trajectory (solid line $0M'1'2'\ldots$).

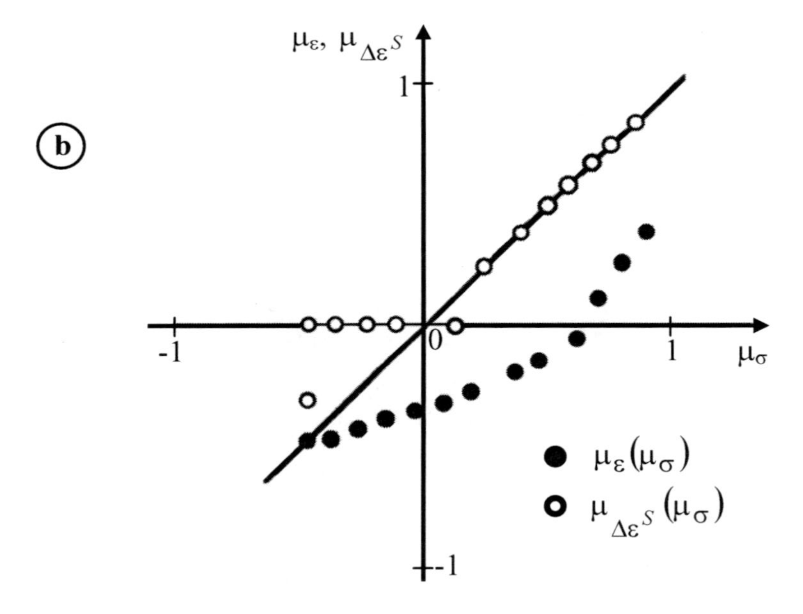

Fig. 1.22b Plot of relations between the Lode-Nadai parameters μ_ε, $\mu_{\Delta\varepsilon^S}$, and μ_σ.

This result can be extended to any smooth loading surface $f=0$ (independent of that how it relates to plastic strain components and other parameters). Indeed, consider any smooth loading surface passing through point M as shown in Fig. 1.23 (isotropic hardening rule) and Fig. 1.24 (kinematic hardening rule). It is obvious that the plane tangential to this surface at point M is perpendicular to S_1-axis. Therefore, the increment of stress vector $d\vec{\mathbf{S}}$, which makes an angle with S_1-axis greater than 90°, is directed inside an elastic region, i.e., produces no plastic strain. This is in a direct contradiction with the results of the above experiments.

The situation is entirely other in the case when there is a conic singularity at the loading point of surface $f=0$ (Fig. 1.25). Now the increment $d\vec{\mathbf{S}}$ is directed outward the loading surface $f=0$, which is constructed for a current value of vector $\vec{\mathbf{S}}$, inducing a plastic strain. We may conclude that the concept of the arising of the corner-point on loading surface is probably a unique possibility for the modeling of experiments from Fig. 1.21 and 1.22.

Let us note that a shear stress intensity τ_0 decreases beyond point M while a strain intensity γ_0 grows, i.e., the diagram $\tau_0-\gamma_0$ differs from that at a proportional loading. In this case a universal relation $\tau_0-\gamma_0$ does not exist.

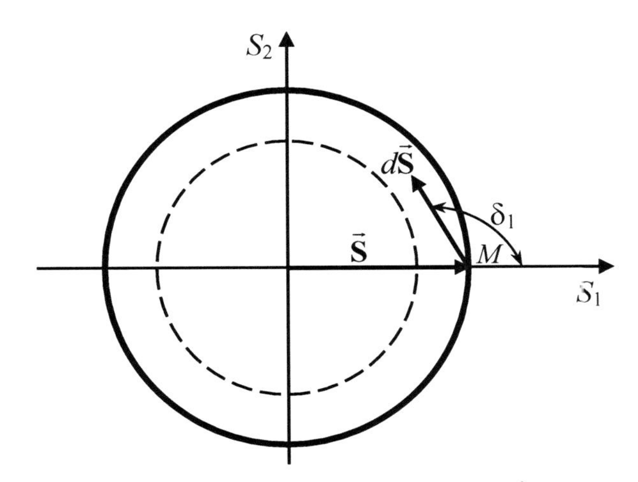

Fig. 1.23 Breaking of loading path at angle $\delta_1 > 90°$ at isotropic hardening rule (dotted line – initial yield surface).

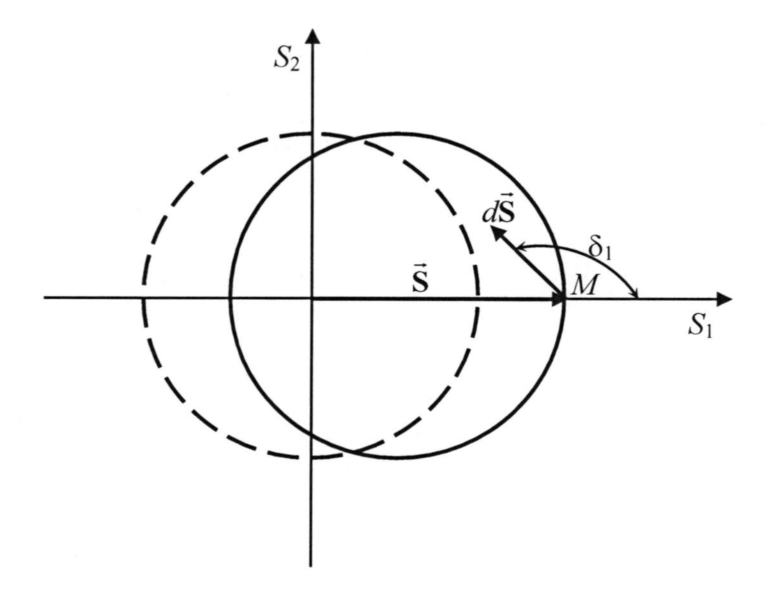

Fig. 1.24 Break of loading path at $\delta_1 > 90°$ at kinematic hardening rule (dotted line – initial yield surface).

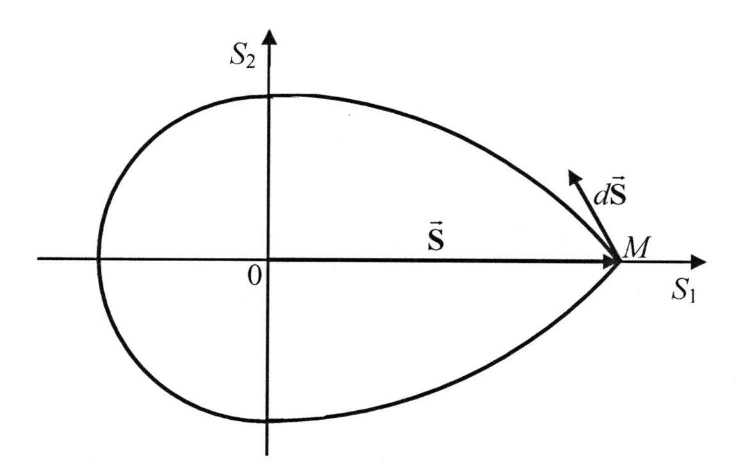

Fig. 1.25 Loading surface with conic singularity at loading point *M*.

1.19 Flow Plasticity Theory with a Singular Loading Surface

W. Koiter has offered the variant of flow plasticity theory [64,65] with a singular loading surface (i.e., a corner arises at a loading point). The loading surface is constructed by the set of smooth surfaces $f_i = 0$ ($i = 1,2,\ldots$; in general $i \to \infty$). The intersections of these surfaces create corner-points on loading surface. A plastic strain increment is calculated as the sum of increments "produced" by the transformations of surfaces $f_i = 0$ that are reached by stress vector $\vec{\mathbf{S}}$ [64]. It is assumed that the plastic strain increment vector is normal to the surface $f_i = 0$. Therefore, let us start with Eq. (1.13.6) generalized in the following way[4]

$$de_k^S = C_i \frac{\partial f_i}{\partial S_k} \frac{\partial f_i}{\partial S_j} dS_j . \tag{1.19.1}$$

If the stress vector $\vec{\mathbf{S}}$ reaches the surface $f_i = 0$ and

$$\frac{\partial f_i}{\partial S_j} dS_j > 0 , \tag{1.19.2}$$

then $C_i > 0$ in Eq. (1.19.1). If at least one of these two conditions is not satisfied, then $C_i = 0$.

It is possible to characterize the considered flow plasticity theories [60,63] as the reconciliation between the principles of regular and singular surface in plastic straining. It has basic importance in construction of the general theory of plasticity. But, as well as in case of one function of loading, it is not known how to determine f_i and theoretical reasons end in Eqs. (1.19.1) and (1.19.2) in the work [64].

I. Sanders offered [165] to take planes to be the functions $f_i = 0$. His approach of the determination of plastic strains consists in the following

(i) In the Ilyushin five-dimensional space of stress deviators, we draw a yield surface.

(ii) Through the each point of the yield surface, we draw the tangent plane, $f_i = 0$. Therefore, the yield surface is regarded as an inner-envelope of the tangent planes (see Fig. 1.26a).

---

[4] In Eqs. (1.19.1) and (1.19.5). index i appears three times, meaning the usual summation over index i

$$de_k^S = C_1 \frac{\partial f_1}{dS_k} \frac{\partial f_1}{dS_j} dS_j + C_2 \frac{\partial f_2}{dS_k} \frac{\partial f_2}{dS_j} dS_j + \ldots$$

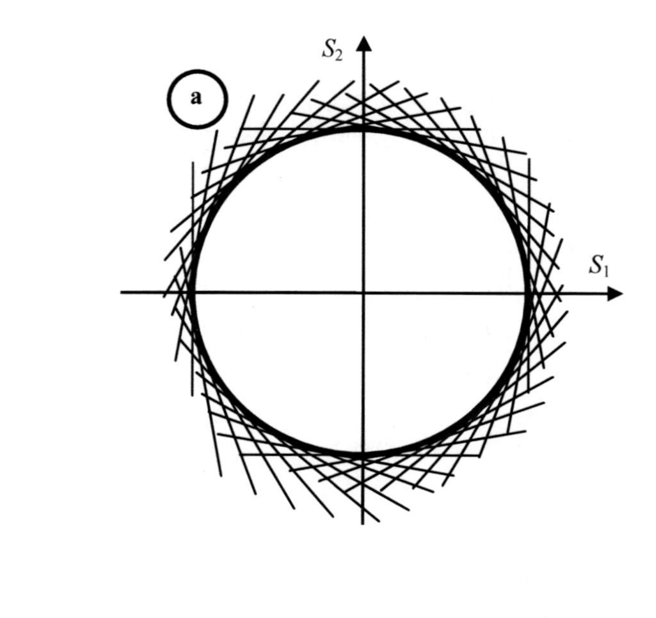

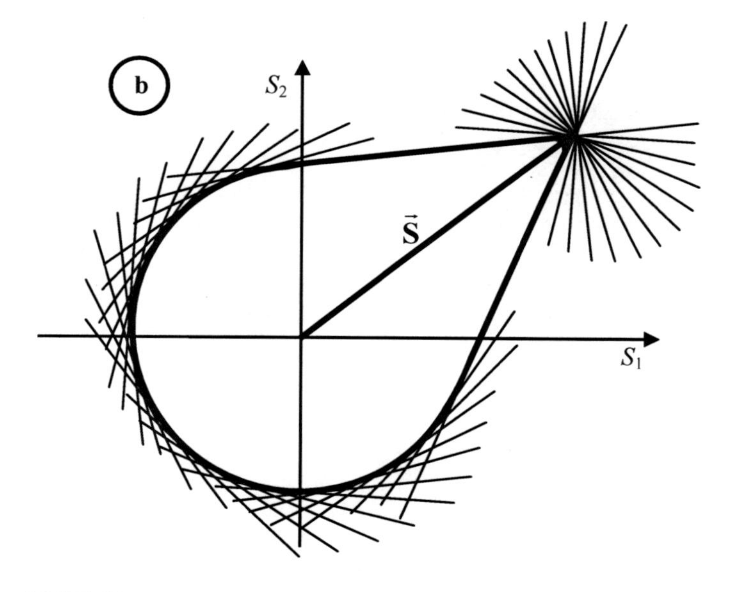

Fig. 1.26 Yield (a) and loading (b) surface as an inner envelope of tangent planes.

(iii) The tangent planes are assumed to be capable of displacements. In the process of loading, a stress vector, $\vec{\mathbf{S}}$, shifts tangent planes. The motion of the planes located at the end-point of the vector $\vec{\mathbf{S}}$ is translational. The form of loading surface, being the inner envelope of tangent planes, is fully determined by the current positions of the tangent planes (see Fig. 1.26b). It is clear that a corner-point arises on the loading surface at the loading point, i.e., at the endpoint of the vector $\vec{\mathbf{S}}$. The loading surface consists of two parts: (a) the cone, with its top at the endpoint of the vector $\vec{\mathbf{S}}$, whose generator is constituted by the boundary tangent planes that took part in displacements; (b) the initial yield surface constituted by planes with their unchangeable distances to the origin of coordinates.

(iv) It is assumed that the displacement of each plane on the endpoint of vector $\vec{\mathbf{S}}$ symbolizes the increment of micro plastic strain whose vector is normal to the plane and depends on the amount of its displacement. The total/macro strain is calculated as the sum of elementary strain components "generated" by the displacements of planes. The planes not reached by the vector $\vec{\mathbf{S}}$ remain immovable.

The equation of plane in five-dimensional stress space is

$$f_i = N_{ij} S_j - H_i = 0, \qquad (1.19.3)$$

where i is the index of plane (the number of plane in the case of the limited quantity of planes), H_i is the distance from the origin of coordinates to the plane, and N_{ij} are directing cosines of the unit vector $\vec{\mathbf{N}}_i$ normal to the plane:

$$N_{i1}^2 + N_{i2}^2 + N_{i3}^2 + N_{i4}^2 + N_{i5}^2 = 1. \qquad (1.19.4)$$

With Eq. (1.19.3), the equation for the increment of plastic strain, Eq. (1.19.1), becomes

$$de_k^S = C_i N_{ik} N_{ij} dS_j. \qquad (1.19.5)$$

Let us consider the case when $S_3 = S_4 = S_5 = 0$ and only those planes are considered that are tangential to the yield surface in the $S_1 S_2$-plane. Then, instead of considering a five-dimensional loading surface and its tangent planes, it is possible to investigate their traces (projections) in the subspace $S_1 S_2$.

The dashed line in Fig. 1.27a shows the yield surface (circle), which obeys to the Guber yield criterion (due to the smoothness of the loading surface, it is obvious that we can draw an infinite number of lines tangential to this circle); the

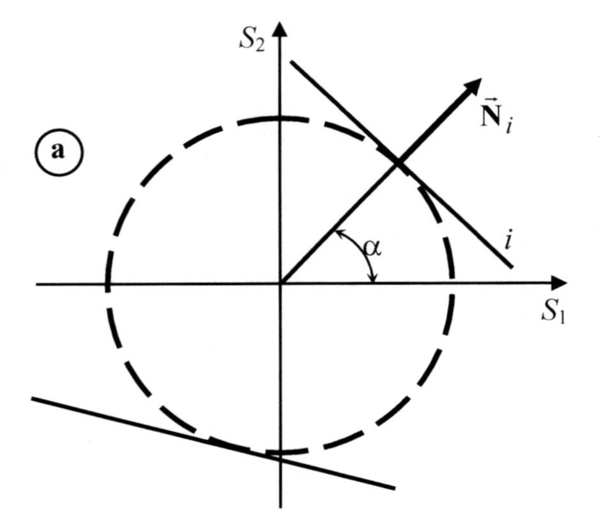

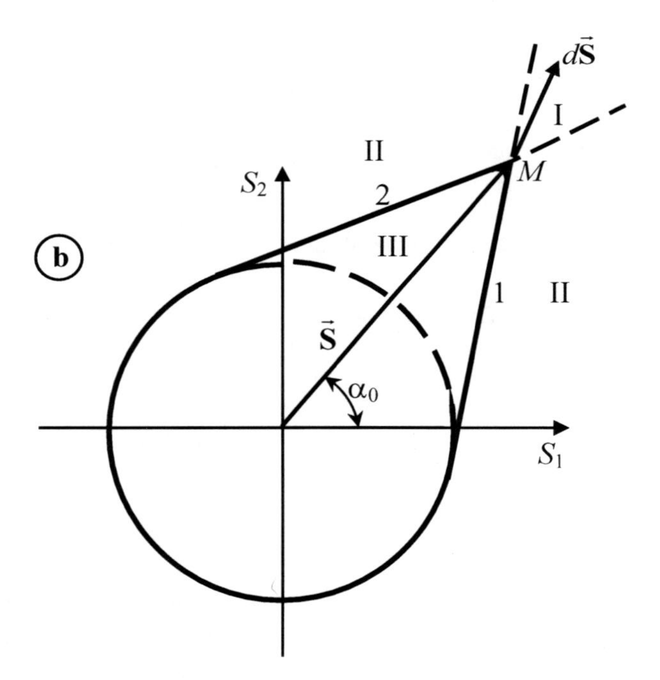

Fig. 1.27 The Guber yield surface and vector $\vec{\mathbf{N}}_i$ normal to i-tangent plane(line) in S_1S_2 stress plane (a); loading surface due to loading by vector $\vec{\mathbf{S}}$ (b).

vector $\vec{\mathbf{N}}_i$ is normal to the tangent plane(line) marked by i. The vector $\vec{\mathbf{S}}$ shifts a part of these lines up to point M during loading. The rest of planes remain immovable. Lines 1 and 2 are the boundaries between the planes that are and are not shifted by the vector $\vec{\mathbf{S}}$, respectively. The loading surface(line), as an inner envelope, is shown in Fig. 1.26b. It has a corner at the loading point, point M.

In the case when a yield surface is bounded by a finite number of planes, e.g., lines M_1M_2, M_2M_3, M_3M_4, and M_4M_1 in Fig. 1.28 (the loading surface is a rectangle), the stress vector $\vec{\mathbf{S}}$ shifts only the plane(line) M_3M_4 while the rest remains immovable. The loading surface maintains the form of the rectangle, but its lateral faces incur the change in their length.

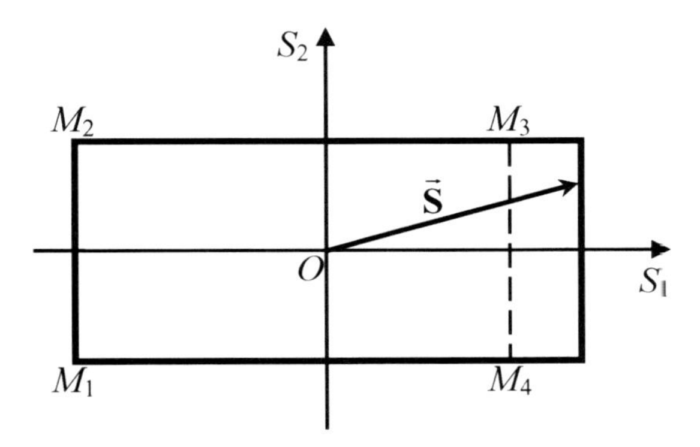

Fig. 1.28 Loading surface (line) for the case when initial elastic area restricted by $M_1M_2M_3M_4$ rectangle.

The summation in Eq. (1.19.5) can be replaced by integration for the case of an infinite number of tangent planes (regular yield surface in Fig. 1.27a):

$$\Delta e_k^S = \int_{\alpha_1}^{\alpha_2} C(\alpha) N_k(\alpha) N_j(\alpha) \Delta S_j d\alpha , \tag{1.19.6}$$

$$N_1 = \cos\alpha , \quad N_2 = \sin\alpha . \tag{1.19.7}$$

Limits of integration, angles α_1 and α_2, set the normals to the boundary lines that are on the endpoint of the vector $\vec{\mathbf{S}}$. To distinguish the differential of the variable of integration from the increment of stress and strain vector, we enter the designation ΔS_j and Δe_k^S (instead of dS_j and de_k^S).

For the case of proportional loading, assuming that $C=\sqrt{3}/\left(\sqrt{2}r\sigma_S\right)$, $r=const$, Eq. (1.19.6) degenerates to a finite form:

$$e_k^S=\frac{1}{r}\int_{\alpha_1}^{\alpha_2}\left(\frac{\sqrt{3}N_jS_j}{\sqrt{2}\sigma_S}-1\right)N_k d\alpha . \tag{1.19.8}$$

In arriving at the result (1.19.8) we have taken a constant of integration so that parenthesis under the integral is equal to zero on the yield limit. During proportional loading, let the vector $\vec{\mathbf{S}}$ make an angle α_0 with S_1-axis. Then

$$S_1=\left|\vec{\mathbf{S}}\right|\cos\alpha_0,\quad S_2=\left|\vec{\mathbf{S}}\right|\sin\alpha_0,\quad N_jS_j=\left|\vec{\mathbf{S}}\right|\cos\left(\alpha-\alpha_0\right). \tag{1.19.9}$$

In view of Eq. (1.7.14), we obtain

$$e_k^S=\frac{1}{r}\int_{\alpha_1}^{\alpha_2}\left[\frac{\tau_0}{\sigma_S}\cos\left(\alpha-\alpha_0\right)-1\right]N_k d\alpha . \tag{1.19.10}$$

The limits of integration α_1 and α_2 are calculated under condition that the square parenthesis is equal to zero:

$$\alpha_1=\alpha_0-\arccos\frac{\sigma_S}{\tau_0},\quad \alpha_2=\alpha_0+\arccos\frac{\sigma_S}{\tau_0}. \tag{1.19.11}$$

By integrating in Eq. (1.19.10), on the base of Eq. (1.19.11), we obtain

$$\begin{aligned} e_1^S&=\frac{1}{r}\left(\frac{\tau_0}{\sigma_S}\arccos\frac{\sigma_S}{\tau_0}-\sqrt{1-\frac{\sigma_S^2}{\tau_0^2}}\right)\cos\alpha_0\\ e_2^S&=\frac{1}{r}\left(\frac{\tau_0}{\sigma_S}\arccos\frac{\sigma_S}{\tau_0}-\sqrt{1-\frac{\sigma_S^2}{\tau_0^2}}\right)\sin\alpha_0 \end{aligned} \tag{1.19.12}$$

Eliminating the angle α_0 from the formulae above, together with Eq. (1.19.9), we arrive at Eq. (1.11.4) in which

$$\frac{1}{G_S}-\frac{1}{G}=\frac{\sqrt{6}}{r\sigma_S}\left(\arccos\frac{\sigma_S}{\tau_0}-\frac{\sigma_S}{\tau_0}\sqrt{1-\frac{\sigma_S^2}{\tau_0^2}}\right). \tag{1.19.13}$$

Equations (1.11.4) and (1.19.13) express the deviator proportionality.

In the Section 1.13 (point 7) has been specified which smoothens the loading surface, and is not the necessary condition of the deviator proportionality, that is, justified by Eqs. (1.11.4) and (1.19.13).

1.20 Conclusions

1. Experiments with two-segment loading paths presented in Sec. 1.18 show that the deviator proportionality and the existence of the universal diagram $\tau_0 - \gamma_0$ are satisfied with sufficient accuracy at insignificant deviation from simple loading. The situation is possible when the deviator proportionality is violated (Fig. 1.19b, 1.20b) while there exists the universal diagram $\tau_0 - \gamma_0$. If the shear stress intensity τ_0 decreases during additional loading (Fig. 1.21, 1.22), the universal diagram $\tau_0 - \gamma_0$ does not exist.

2. Flow plasticity theories have wider boundaries of applicability than the Hencky–Nadai relations. Thus, for example, according to the experiments shown in Fig. 1.19, we have $\mu_\varepsilon \neq \mu_\sigma$ beyond point *M*, however the proportionality between stress deviators and plastic-strain-increment deviators is valid along the whole loading path. In addition, in contrast to deformation theories, a plastic strain changes continuously in terms of flow plasticity theories at the transition from loading to unloading.

3. The concept of conic point appears to be the only hypothesis to explain the experiments with two-segment trajectories at the reduction of shear stress intensity during additional loading.

4. In terms of Koiter's theory, a conic singularity arises on loading surface at a loading point. The theories with regular and singular surface $f = 0$ considerably differ from each other. Some basic differences are given below.

5. Suppose that the loading be proportional up to point *M* (see Fig. 1.27b). Further, the increment of stress vector $d\vec{\mathbf{S}}$, additional loading, is directed inward the area I is limited by the boundary planes(lines) marked by 1 and 2. Then all planes, which moved on the endpoint of vector $\vec{\mathbf{S}}$ to point *M*, would continue their displacements under the action of the additional vector $d\vec{\mathbf{S}}$. Therefore, the increments of plastic strains can be calculated by Eqs. (1.19.6) and (1.19.11). These increments can be obtained by the differentiation of Eq. (1.11.4) as well, i.e., from the deformation theory. Thus, even if the extra-loading vector $d\vec{\mathbf{S}}$ is not directed along the vector $\vec{\mathbf{S}}$, however, it falls within the region I, and the deviator proportionality is satisfied.

The obtained result differs to a great extent from the conclusion in Section 1.13 (point 5). If the surface $f = 0$ is regular, Eq. (1.11.4) is obtained only at proportional loading. In the case of singular surface $f = 0$ the boundaries of the applicability of deformation theory are wider.

6. Let us consider the case when the vector $d\vec{\mathbf{S}}$ is directed into the area II (Fig. 1.27b). The part of tangent planes, which moved up to point M, remains immovable at the action of $d\vec{\mathbf{S}}$ because it does not shifts them. Such case is called the incomplete loading or partial unloading. Now one of the limits of integration in Eq. (1.19.6) is not determined by Eq. (1.19.11) any more. The plastic strain increments cannot be obtained by the differentiation of Eq. (1.11.4), i.e., the equations of the deformation theory (1.11.4) do not hold.

If the vector is directed into area III (Fig. 1.27b), all planes are immovable, unloading occurs, and the increments of plastic strains are zero.

Therefore, we have an essential difference between the results obtained for regular and singular loading surfaces. If the loading surface is regular, we have plastic loading or elastic unloading depending on the orientation of additional vector $d\vec{\mathbf{S}}$. If the loading surface $f = 0$ is singular, there are three variants: complete loading, partial loading (partial unloading), and unloading.

7. Let us emphasize what follows from Drucker's postulate (Section 1.10). Namely, if a loading surface, $f = 0$, is regular, an incremental plastic strain vector, $d\vec{\mathbf{e}}^S$, is directed along the normal to the surface. If a surface $f = 0$ is singular, there is no such rigid restriction imposed on the orientation of vector $d\vec{\mathbf{e}}^S$. $d\vec{\mathbf{e}}^S$ must be directed inside the cone as shown in Fig. 1.10.

8. The Sanders formulae (1.19.12) are obtained from the assumption that plastic deformation is "developed" by the movements of tangent lines in S_1S_2-plane. However this assumption is wrong. In terms of any flow theory, the plastic strain is accumulated due to the change of loading surface in five-dimensional deviatoric space independent of the quantity of nonzero stress vector components. The Sanders theory prescribes a singular surface; however it results in extremely bulky calculations at non-proportional loading, which are practically not feasible. In more detail, this question will be discussed in Chapter 3.

Making conclusion to the whole chapter, it is possible to state the following. The problem of establishing of constitutive stress-strain relations can be considered as solved for the case of simple loading. This relation is the deviator proportionality whose most general form was proposed by Prager.

On the other hand, at loading different from proportional, the question of the analytical description of irreversible strains still exists today. The Sanders flow theory with singular surface results in extremely bulky calculations at non-proportional loading, which are practically not feasible.

The further investigations of new theories of plasticity are necessary.

Chapter 2
The Concept of Slip

The theories of plasticity, considered in the Chapter 1, are based on the hypotheses of formal character and the mechanism of the deformation of a body is not taken into account. In contrast to them, the theory termed as the concept of slip allows for the polycrystalline conglomerate that consists of crystallites (grains) whose irreversible deformations are the slips of their parts relative to one another. This chapter treats the Batdorf-Budiansky concept of slip and its several updating. Since the calculations of strains are bulky in terms of the concept, only the formulation of problem and received results are presented (intermediate equations are omitted frequently).

2.1 One- and Two-Level Models of Deformation

In formulating the constitutive relationships in the continuous mechanics, the state of stress at a point in a body is assumed to be defined by the state of some small representative macroregion, which, at the same time, should be large enough to represent the average properties of a material. The real structure of the macroregion is ignored at a pure phenomenological approach, the parameters averaged in macroregion, in particular, such as a macrostress and macrostrain, are only considered. The models of such kind are called one-level models.

Two-level models define a macroregion as some set of the interconnected microparticles whose strain-stress states are determined by micro-stresses and -deformations. Therefore, at least two levels of characteristic sizes are supposed to exist: a *macrolevel*, which is determined by the size of a representative macroregion and *microlevel* whose characteristic size is determined by the size of microparticle.

However, in terms of most two-level models, the size of micro- and macro-element is not fixed, and, consequently, the concept of micro- and macro-particle is purely symbolic and manifests only the fact of the existence of two structural levels at the formulation of constitutive equations.

A microparticle is assumed to be a continuous medium with homogeneous stress-strain state, which is determined by the local law of straining. It is obvious that formulating of the local law is simpler, than that for a whole macroregion. Constitutive relations, written for a microparticle, can with sufficient accuracy display the real laws of deformation. This provides an advantage of a two-level model above one-level one.

At the same time, the extreme difficulty of the physical processes of deforming of real bodies complicates the formulation of local laws on a microlevel. In addition, the question of the interaction between microparticles is extremely complicated. As a consequence, simplified assumptions and hypotheses are used in microstructural models. Therefore, the theories based on two-level models are, as a matter of fact, phenomenological. The basic statements of such theories are the following:

(i) The macroregion is assumed to be consisted of a finite or infinitive (in limit) number of microparticles; in each of them a homogeneous state of stress arises
(ii) The law for the calculation of micro-particle strains on local level is set up
(iii) The relations between micro-stresses and -strains are set up
(iv) The rule of averaging, according to which the macro-stresses and -strains are determined through the field of micro-stresses and -strains, is established.

The Batdorf-Budiansky theory is based on the two-level approach for the determination of irreversible strains. In terms of this theory, crystal grains play the role of a microparticle. Therefore, some basic knowledge of the plastic deformation of crystal grains is presented below.

2.2 Some Basic Knowledge in the Physics of Metals

As it is known, the polycrystalline aggregate consists of interconnected crystal grains with characteristic arrangements of atoms within them. Furthermore, the basic mechanism of the plastic deformation of grain is that its part slips relative to each other. The slip occurs in the planes of close packing of atoms.

Consider, as an example, the face centered cubic structure (FCC). FCC structure (see Fig. 2.1a.) has atoms located at each of the corners and the centers of all the cubic faces. Each of the corner atoms is the corner of another cube so the corner atoms are shared among eight unit cells. Additionally, each of its six face centered atoms is shared with an adjacent unit cell. In FCC, the planes of close packing of atoms are equally inclined to the crystal axes (one of them showed in Fig. 2.1b). There are six atoms in these planes falling at the area $l^2\sqrt{3}/2$ of triangle $M_1M_3M_5$, where l is the interatomic distance. In the diagonal plane $M_1M_2M_4M_5$ there are six atoms as well, however they fall at the area $l^2\sqrt{2}$ of rectangular $M_1M_2M_4M_5$, i.e., larger than $l^2\sqrt{3}/2$. There are five atoms on lateral faces and on the base of cube falling at area equal l^2. As $5:l^2 < 6:\left(l^2\sqrt{3}/2\right)$, the lateral faces and the bases of cube are not the planes of close packing either. The slips can occur only in the directions of close packing of atoms in the each slip plane as shown in Fig. 2.1b by arrows. The slip plane and the direction of possible slip in this plane constitute the so-called slip system. Thus, FCC prescribes the four planes, in each of which there are three slip directions, in total there are 12 slip systems.

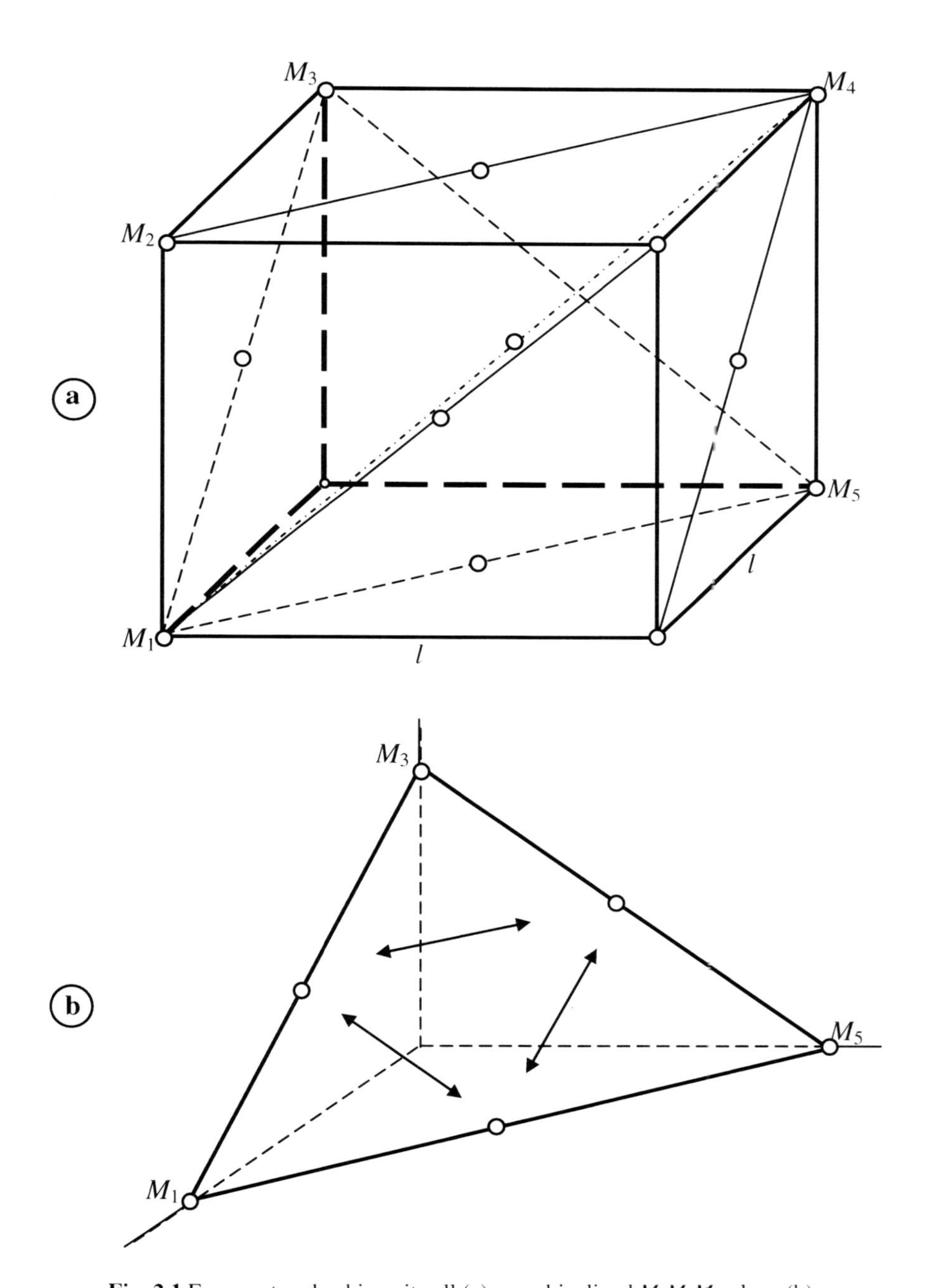

Fig. 2.1 Face centered cubic unit cell (a); equal inclined $M_1M_3M_5$-plane (b).

Let us estimate the necessary magnitude of shear stress, which act in a grain, to bring about the shift of an atomic plane relative to a neighboring parallel plane, meaning that all atoms located in the plane incur a simultaneous slip. This shear stress, τ_T, defined for a perfect crystal, is called the theoretical slip strength.

Consider a simplified model of crystal lattice [44,121], shown in Fig. 2.2. Here l_1 is an interatomic distance (atomic spacing) in the slip plane and l_2 is the distance between slip planes. If to shift the upper layer of atoms relative to the bottom one by an amount u, then each atom is subjected to the restoring force P. This is possible if $u < l_1/2$. At $u = l_1/2$, the force P is equal to zero, but the equilibrium is unstable. At $u > l_1/2$, the force P is negative because the upper row of atoms tries to take such new position of equilibrium when the upper layer of atoms is displaced at distance of l_1 relative to the bottom one. Thus, as a first approximation, we can assume, by Frenkel, that

$$P = P_0 \sin\frac{2\pi u}{l_1}, \tag{2.2.1}$$

where the value of P_0 will be evaluated below. To obtain the shear stress τ (Fig. 2.2) acting in a grain due to the force P, it is necessary to add all forces P applied to the atoms of a given plane and to divide the resultant force by the area occupied by the atoms:

$$\tau = \frac{P}{l_1^2}, \qquad \tau_T = \frac{P_0}{l_1^2}. \tag{2.2.2}$$

Thus,

$$\tau = \tau_T \sin\frac{2\pi u}{l_1}. \tag{2.2.3}$$

At small displacements of atoms, the above equation can be represented in the form

$$\tau = \frac{2\pi u}{l_1}\tau_T \tag{2.2.4}$$

and the displacement u can be expressed as $u = \gamma l_2$, where γ is the shear strain (Fig 2.2). Hence,

$$\tau = \frac{2\pi\gamma l_2}{l_1}\tau_T. \tag{2.2.5}$$

The stress τ and the strain γ are connected by Hooke's law, $\tau = G\gamma$, and Eq. (2.2.5) gives

$$\tau_T = \frac{l_1}{2\pi l_2} G . \tag{2.2.6}$$

Equation (2.2.6) determines the order of the theoretical slip strength. As the distances l_1 and l_2 are of the same order of magnitude, consequently $\tau_T \approx G/6$. This value is hundred times as large as the real value of the slip strength for real crystals obtained experimentally.

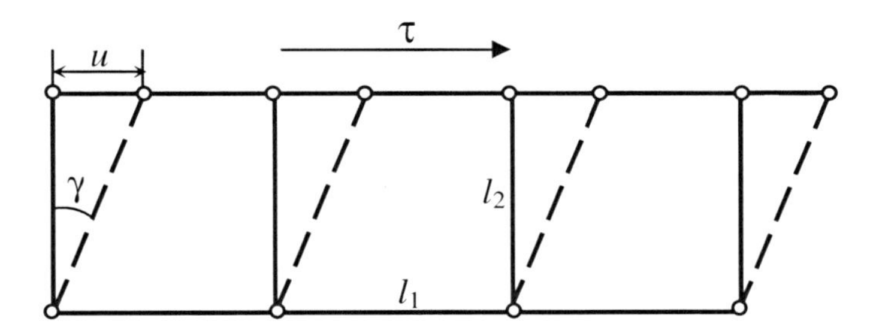

Fig. 2.2 Simplified lattice model.

The considerable difference between theoretical and experimental slip strength in crystals can be explained by that the crystal structure is not perfect; there are defects in crystals (vacancies, edge and crew dislocations, grain boundaries, etc.). The presence of structural imperfections is a general property of bodies existed in nature. The elastic deformations of crystal lattice, thermal fluctuations of atoms, or other insignificant defects are not included into the concept of irregularities. The presence of defects makes a real crystal more malleable to the irreversible deformation in comparison with a perfect crystal.

Dislocations play the main role in explanation of the mechanism of the plastic deformation of crystallite. Let us consider the schematic model of primitive cubic lattice (Fig. 2.3a) whose projection is shown in Fig. 2.3b. Imagine that the crystal has been slitted between the adjacent crystal planes of atoms from point *M* upward and the extra-layer (half-plane) of atoms have been inserted into the slit. The crystal lattice distorted in this way is shown in Fig. 2.3c. A dislocation is said to be created at point *M*; the dislocation line is perpendicular to the plane of Fig. 2.3c and 2.3d. The dislocation is that of unit capacity if the extra half inserted into the body consists of one layer of atoms. Dislocations generate stresses: a compression experienced by the atoms near the "extra" plane and tension experienced by those atoms near the "missing" plane; these stresses decrease with distance from the core of the dislocation.

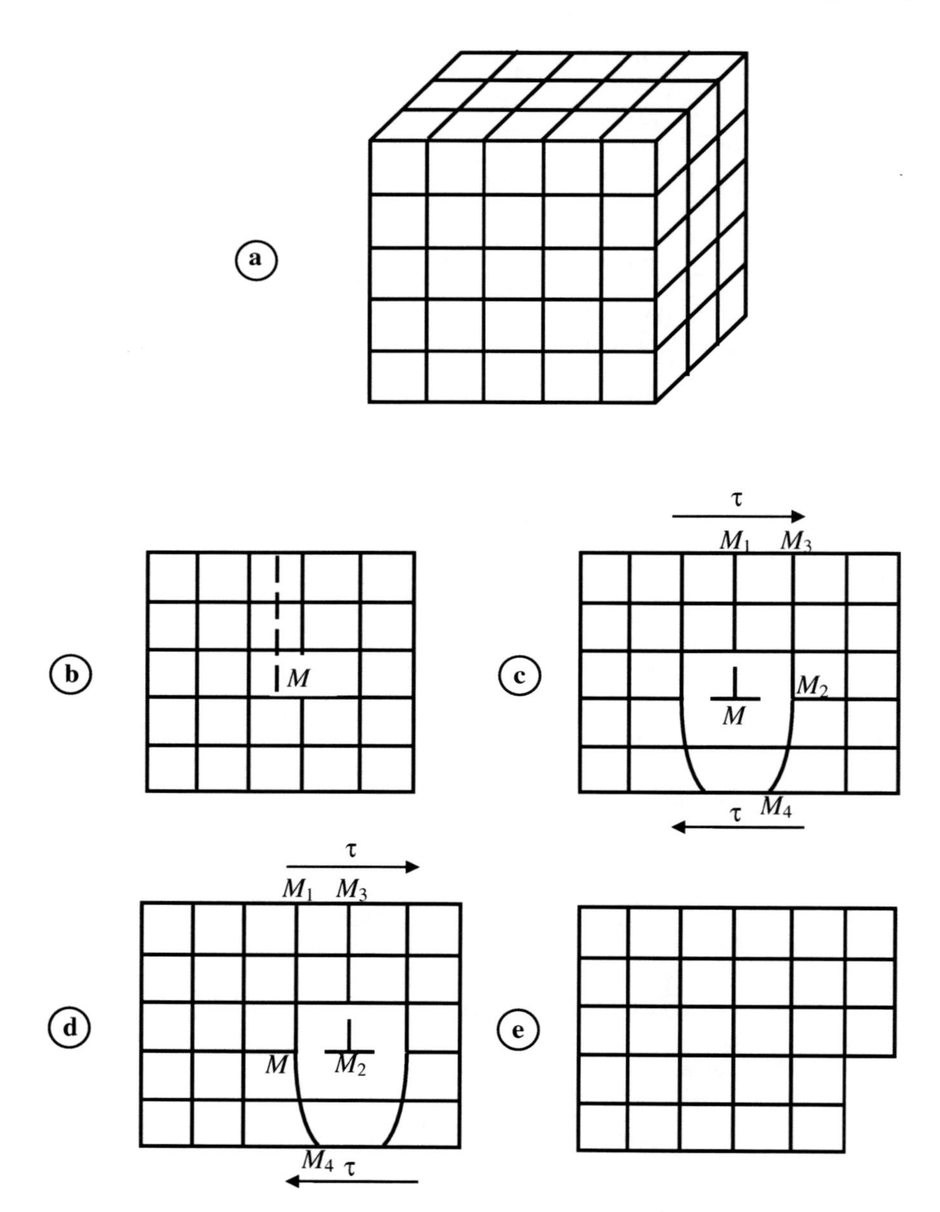

Fig. 2.3 Evolution of crystal lattice with the propagation of edge dislocation.

Once a critical shear stress is reached, the dislocation starts moving and deformation is no longer elastic, but plastic. This movement consists of the following. Under the action of shear stress, atomic plane MM_1 supersedes (to the right) half-plane M_2M_3. When half planes MM_1 and M_2M_4 are brought together, they constitute the complete atomic plane M_1MM_4. Now, the half-plane M_2M_3 is an atomic extra-plane, i.e., the dislocation has displaced from point M to point M_2 (Fig. 2.3d), etc. After the dislocation has passed through a crystal and left it, the lattice is completely restored, and no traces of the dislocation are left in the lattice. Parts of the crystal are now shifted in the plane of the movement of the dislocation (Fig. 2.3.e).

The slip within an crystallite caused by the dislocation motions explains why the plastic slip is possible at much lower stresses than in a perfect crystal. Indeed, at definition of τ_T, it is assumed that two atomic planes simultaneously slip relative to each other. However the fact is that there is no simultaneous slip of the whole atomic plane during the movement of dislocation. The atomic line (dislocation core), projected on point M in Fig. 2.3, covers approximately only half an interatomic distance. Then adjacent atomic line (point M_2) covers the same distance, i.e., the relay-race of movement is passed from one atomic line to another. Obviously, the displacement of single atomic line occurs at much lower stress than at simultaneous displacement of the whole atomic plane thereby accounting for the difference between a theoretical and real slip strength.

As it is stated above, after a dislocation has passed through a crystal and left it, the lattice is completely restored. As may appear at first sight, the crystal becomes perfect after certain amount of plastic deformation and incapable of generating further plastic deformation. Actually, this is not observed, the dislocation density considerably grows with the increase in plastic deformation. There exists the mechanism of the generation of new dislocations during plastic deformation.

The schema of plastic form change in separate crystal grain (monocrystal) is the following (Fig. 2.4). Once the shear stress reaches a critical value in some slip system, simultaneous slips by a large amount (about of 1000 atomic spacing) occur in appropriate planes. With increase in loading, another close slip planes join with them, creating a "pack" of slip. The distance between active slip planes in the pack has the degree of 100 atomic spacing.

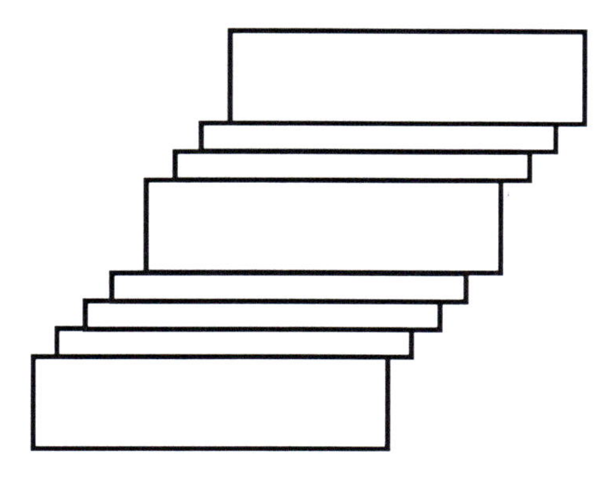

Fig. 2.4 Plastically deformed crystal.

2.3 The Batdorf-Budiansky Theory of Slip

Let us consider a polycrystalline solid, which consists of a great number of crystal grains, under the conditions of macro homogeneous stress state. A plastic deformation of a grain is assumed to be the result of the slips of its parts relative to one another. Figure 2.5 shows the grain in which the slips occur along parallel planes. Let us introduce a system of coordinate, *l*-*n*, with base unit vectors $\vec{\mathbf{n}}$ and $\vec{\mathbf{l}}$, $\vec{\mathbf{n}} \cdot \vec{\mathbf{l}} = 0$. The *l*- axis is co-directed with the direction of slips, while *n*- axis is perpendicular to the planes of slips. The parallel planes and direction of slip, given by the vectors $\vec{\mathbf{n}}$ and $\vec{\mathbf{l}}$, constitute a slip system. The plastic deformation developed in a grain is denoted as γ_{nl}^{g}.

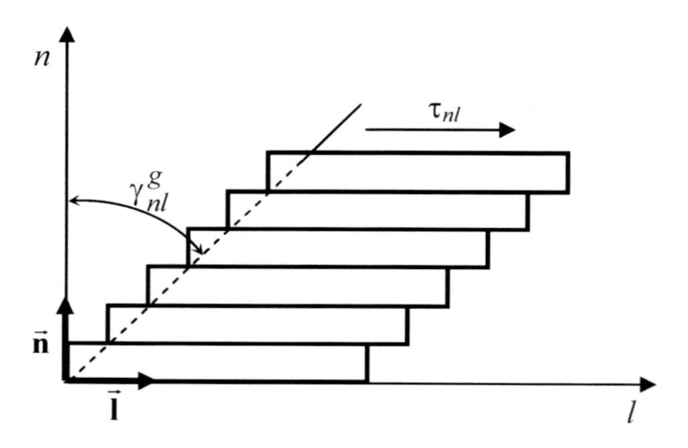

Fig. 2.5 Plastic shear strain γ_{nl}^{g} of crystal grain.

Budiansky has proposed the following simplified assumptions [8,9] relative to the mechanical properties of crystals and their interaction during plastic deforming:

1) There is a single slip system in each grain
2) The shear strain in any grain (crystallite) is developed independently of that in other grains
3) The state of stress in all grains is identical with that at a point in a polycrystalline body
4) Crystallites are identical by size and form.

In a real grain, slips occur in several slip systems simultaneously within the grain, each of which influences another. In addition, slips in adjacent grains with small disorientations exert an influence on each other. The assumptions 1) and 2) remove the question relative to the interactions of the slip systems. Therefore, a shear strain γ_{nl}^{g} in a given crystal depends only on the history of loading with respect to n- and l-axes. Further, we assume that the shear strain γ_{nl}^{g} is produced only by the shear stress τ_{nl} and it can be determined as

$$\gamma_{nl}^{g} = F_1(\tau_{nl}), \tag{2.3.1}$$

where F_1 is the characteristic function of a material. Equation (2.3.1) expresses the local law of straining of microparticles (point (iii) in Sec. 2.1).

Crystal grains interact with each other, they are anisotropic, and their orientations have a casual character. Therefore, the determination of the state of stress that arises in grains is extremely a difficult problem. The assumption 3) eliminates these difficulties by stating that microstresses in particles coincide with the macrostresses in macroregion. Therefore, the shear stress acting in the microparticle can be expressed through stress-tensor components at a point of polycrystalline body by known transformations in continuous mechanics:

$$\tau_{nl} = \sigma_{ij} l_i n_j, \quad (i, j = x, y, z) \tag{2.3.2}$$

where l_i and n_j are directing cosines of l- and n-axes relative to fixed coordinate system, x, y, and z.

Real crystals differ one from another in form and size. Therefore the crystals have a different contribution in the total strain of a body. The assumption (4) ignores the size and forms of crystals and allows to characterize the grain by only vectors $\vec{\mathbf{n}}$ and $\vec{\mathbf{l}}$, i.e., only by its orientation and the directions of slips.

Let us determine the total plastic strain at a point in a polycrystalline body caused by strains developed in grains. For this purpose, we introduce the orthogonal coordinate system with x-, y-, and z-axes and construct a hemisphere ($z > 0$) of unit radius with center in the origin of coordinates (Fig. 2.6). Then, imaginarily, we carry the planes of slip of all grains over to the hemisphere so that each of planes touches the hemisphere. The orientation of the slip system is given by the unit vector $\vec{\mathbf{n}}$, which is determined by the spherical coordinates, angles α and β (Fig. 2.6).

Since the material is assumed to be initially isotropic and the orientation of grains is of a random nature, there are no privileged directions in a body. Furthermore, as there are, in limiting case, infinitely many crystal grains in a body, the touch points are distributed homogeneously and they cover continuously the whole hemisphere.

To set up a direction of slip, let us introduce at every point of the hemisphere (every plane of slip) an orthogonal two-dimensional coordinate system with ξ_1- and ξ_2-axes, ξ_2-axis is tangential to the meridian of the hemisphere. The direction of slip, the direction of unit vector $\vec{\mathbf{l}}$ (Fig. 2.6b), in ξ_1-ξ_2 plane is determined by an angle ω measured from ξ_1-axis. Angle ω can take values from 0 up to 2π.

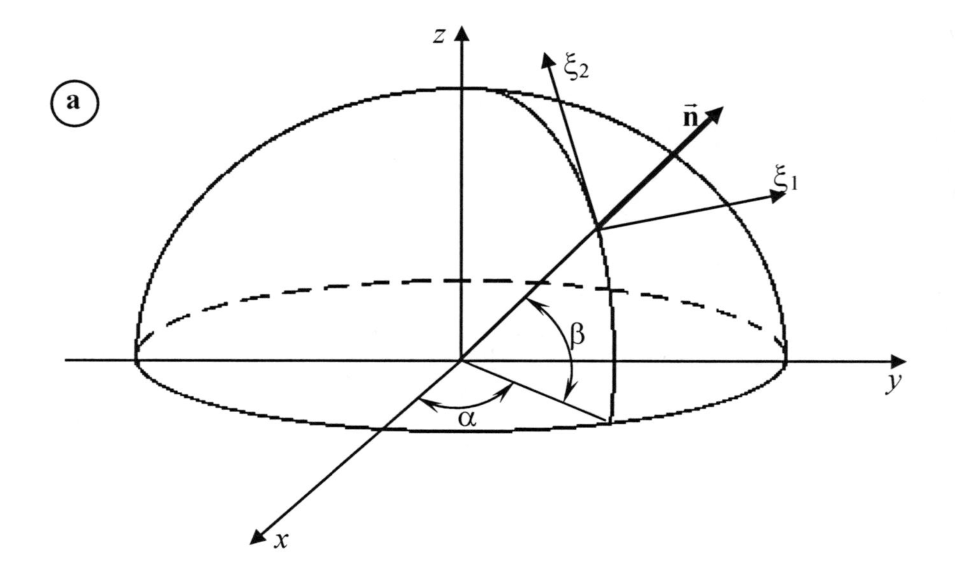

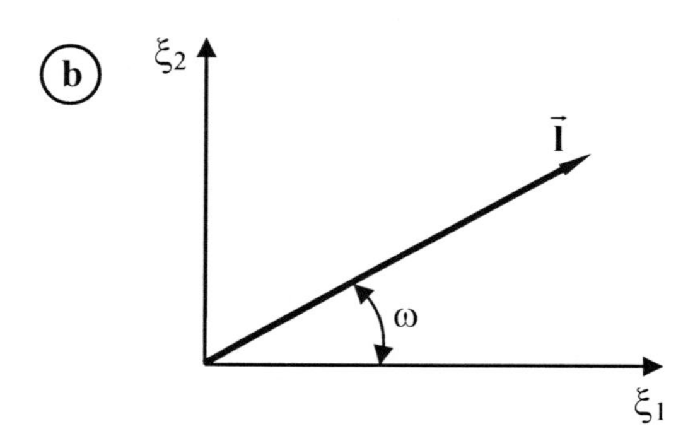

Fig. 2.6 Spherical coordinates of slip plane on the hemisphere of unit radius, angles α and β (a); a slip direction in the plane is given by angle ω in ξ_1-ξ_2 coordinate system (b).

Hence, a slip system at a point of a body is fully determined by three parameters, angles α, β, and ω; angles α and β give the orientation of the slip system, i.e., the orientation of the vector $\vec{\mathbf{n}}$, with respect to the introduced *xyz*-coordinate system at the point of body and angle ω gives the slip direction (the direction of the vector $\vec{\mathbf{l}}$) in this slip system. Since, as follows from Fig. 2.6, the ranges of the angles α, β, and ω are $[0,2\pi]$, $[0,\pi/2]$, and $[0,2\pi]$, respectively, all possible slip systems are taken into account in terms of these angles.

Let us determine an elementary macro-plastic-shear-strain, $d\gamma_{nl}^{S}$, which resulted from slips, γ_{nl}^{g}, produced in slip systems with normals $\vec{\mathbf{n}}$ falling within area $d\Omega$ of hemisphere and in directions $\vec{\mathbf{l}}$ lying in the range of $d\omega$. Batdorf and Budiansky [8] have offered the following hypothetical relation between $d\gamma_{nl}^{S}$ and γ_{nl}^{g} in a such way

$$d\gamma_{nl}^{S} = \frac{\gamma_{nl}^{g}}{4\pi^2} d\Omega d\omega, \tag{2.3.3}$$

where $4\pi^2$ is obtained as the product of the surface of hemisphere of unit radius, 2π, and the round angle of ω, 2π. Therefore, Eq. (2.3.3) means that the deformation of polycrystalline equals to the product of the grain strains, γ_{nl}^{g}, by the relative number of grains, $d\Omega d\omega / \left(4\pi^2\right)$, where the strains occur.

Since n-axis is perpendicular to the plane of slip and l-axis gives a slip direction, the unique distinct from zero strain component relative to l- and n-axes is $d\gamma_{nl}^{S}$. Thus the plastic strain components relative to x-, y-, and z-axes are determined by the transformation of second rank tensor components:

$$\begin{aligned} d\gamma_{ij}^{S} &= \left(l_i n_j + l_j n_i\right) d\gamma_{nl}^{S} \qquad i \neq j \\ d\varepsilon_{i}^{S} &= l_i n_i d\gamma_{nl}^{S} \qquad (i, j = x, y, z) \;\; \not\Sigma \end{aligned} \tag{2.3.4}$$

Here and below the mark $\not\Sigma$ means that there is no summation over repeated indexes.

The total deformation at a point in a body is obtained by the summation (three-fold integration) of slips induced in all grains. Equations (2.3.1), (2.3.3) and (2.3.4) give:

$$\begin{aligned} \gamma_{ij}^{S} &= \iint_{\Omega} d\Omega \int_{\omega_1}^{\omega_2} \left(l_i n_j + l_j n_i\right) F_1\left(\tau_{nl}\right) d\omega, \\ \varepsilon_{i}^{S} &= \iint_{\Omega} d\Omega \int_{\omega_1}^{\omega_2} l_i n_i F_1\left(\tau_{nl}\right) d\omega, \qquad \not\Sigma \end{aligned} \tag{2.3.5}$$

Here Ω is the surface of hemisphere where slips occur and ω_1 and ω_2 are the boundary directions of slip in the planes tangential to the hemisphere in the region Ω.

The immaterial factor $1/(4\pi^2)$ is included into the function $F_1(\tau_{nl})$. The function $F_1(\tau_{nl})$ is presented as polynomial

$$F_1(\tau_{nl}) = b_i\left(\frac{\tau_{nl}}{\tau_s} - 1\right)^i \tag{2.3.6}$$

with the summation over index i which ranges from 1 to 5.

Let us designate by τ_{nl}^0 the greatest value of tangential stress τ_{nl} in the considered n-l system for the whole history of the load. The slip criterion in n-l system has the form

$$\tau_{nl} > \tau_S, \quad \tau_{nl} = \tau_{nl}^0, \qquad d\tau_{nl} > 0. \tag{2.3.7}$$

If even one of these three conditions is not fulfilled, the slip in n-l system at a given moment of loading does not occur. The conditions (2.3.7) determine the surface of integration on the hemisphere Ω and the limits ω_1 and ω_2 in the plane of slip. It is obvious that when the Tresca yield criterion is satisfied, the first slip occurs in the plane where a maximal shear stress acts in the direction of its action.

In Eq. (2.3.5) the direction cosines l_x, l_y,..., n_z of l- and n- axes are still not determined. Projecting the unit vectors $\vec{\mathbf{n}}$ and $\vec{\mathbf{l}}$ on x-, y-, and z-*axes*, we obtain

$$\begin{gathered} l_x = -\sin\alpha\cos\omega - \cos\alpha\sin\beta\sin\omega, \\ l_y = \cos\alpha\cos\omega - \sin\alpha\sin\beta\sin\omega \\ l_z = \cos\beta\sin\omega, \quad n_x = \cos\alpha\cos\beta, \quad n_y = \sin\alpha\cos\beta, \\ n_z = \sin\beta. \end{gathered} \tag{2.3.8}$$

Thus, in terms of the Batdorf-Budiansky theory, the calculation of plastic strain components is reduced to the calculation of the integrals (2.3.5) whose integrands are determined by Eqs. (2.3.6) and (2.3.8) and the limits of integration can be determined from the conditions (2.3.7). Equations (2.3.5)–(2.3.8) complete the construction of two-level model.

At a combined loading, the situations, when slips in certain n-l systems first take place but then cease, are possible. Therefore, instead of strains it is worthwhile to determine their increments by formulas similar to (2.3.5):

$$\begin{gathered} \Delta\gamma_{ij}^S = \iint\limits_{\Omega} d\Omega \int\limits_{\omega_1}^{\omega_2} \left(l_i n_j + l_j n_i\right) \Delta F_1 d\omega \qquad (i \neq j) \\ \Delta\varepsilon_i^S = \iint\limits_{\Omega} d\Omega \int\limits_{\omega_1}^{\omega_2} l_i n_i \Delta F_1(\tau_{nl}) d\omega \quad \Sigma \end{gathered} \tag{2.3.9}$$

The values of Ω, ω_1, and ω_2 in Eq. (2.3.9) are considerably different from those in Eq. (2.3.5). In Eq. (2.3.5), the region Ω and the values of ω_1 and ω_2 determine the region of slips and their directions for the whole history of straining, whereas in Eq. (2.3.9) they give the limits of integration only for a given moment of loading.

Total plastic strain components are

$$\gamma_{ij}^S = \int \Delta\gamma_{ij}^S , \quad \varepsilon_i^S = \int \Delta\varepsilon_i^S . \tag{2.3.10}$$

which obviously are equivalent to the following relation expressed through the strain rates:

$$\gamma_{ij}^S = \int_0^t \dot{\gamma}_{ij}^S dt , \quad \varepsilon_i^S = \int_0^t \dot{\varepsilon}_i^S dt . \tag{2.3.11}$$

To obtain the strain rates, one must to replace the increment ΔF_1 by the derivative $\dot{F}_1$ in the right-hand side in Eq. (2.3.9); ΔF_1 and $\dot{F}_1$ are determined from Eq. (2.3.6) by differentiating function F_1.

2.4 Uniaxial Strain Pure Shear

A) First consider the case of uniaxial tension: $\sigma_z > 0$, $\sigma_x = \ldots \tau_{yz} = 0$. Then, Eqs. (2.3.2) and (2.3.8) give

$$\tau_{nl} = l_z n_z \sigma_z = \frac{\sigma_z}{2} \sin 2\beta \sin \omega, \tag{2.4.1}$$

which is the shear stress acting in the plane of slip determined by angles α and β in the direction of ω. As seen from Eq. (2.4.1), the stress τ_{nl} does not depend on the angle α, i.e., τ_{nl} is constant at $\omega = const$ for the all points of the fixed parallel ($\beta = const$) of hemisphere. Let the greatest value of τ_{nl} in a plane, i.e., the maximum of the function τ_{nl} in variable ω at fixed values of angles α and β, be called the total shear stress τ_n. From Eq. (2.4.1) it follows that the maximum is reached at $\omega = \pi/2$,

$$\tau_n = \frac{\sigma_z}{2} \sin 2\beta , \tag{2.4.2}$$

i.e., at any point of the hemisphere, the total shear stress is directed along ξ_2-axis, i.e., along a tangent to the meridian of the hemisphere.

Equation (2.3.5) gives the plastic strain component

$$\varepsilon_z^S = \frac{1}{2}\iiint\limits_V \sin 2\beta \sin \omega F_1(\tau_{nl}) dV, \tag{2.4.3}$$

$$dV = d\Omega d\omega, \qquad d\Omega = \cos\beta d\alpha d\beta,$$

where Ω is the area of slips on the hemisphere; V is "volume" where irreversible strains are developed. This "volume" is determined from the condition $\tau_{nl} \geq \tau_S$, i.e.,

$$\frac{\sigma_z}{2}\sin 2\beta \sin\omega \geq \tau_S. \tag{2.4.4}$$

From the limits of integration in Eq. (2.4.3), which can be determined from the above inequality, we designate through $\alpha_{1,2}$, $\beta_{1,2}$, and $\omega_{1,2}$. Since angle α does not appear in neither integrand in Eq. (2.4.3) nor in Eq. (2.4.4), then $\alpha_1 = 0$, $\alpha_2 = 2\pi$. The region of slips on the hemisphere Ω is a strip symmetric above z-axis and limited by angles β_1 and β_2. The slips occur in directions $\omega_1 \leq \omega \leq \omega_2$ at internal points of the strip. We have $\omega_1 = \omega_2 = \pi/2$ at boundary points of strip ($\beta = \beta_1$ or $\beta = \beta_2$). Equation (2.4.4) at $\omega = \pi/2$ gives the range of angle β and its boundary values:

$$\beta_1 \leq \beta \leq \beta_2, \qquad \sin 2\beta_{1,2} = \frac{2\tau_S}{\sigma_z}. \tag{2.4.5}$$

This formula can be also obtained by equating the complete shear stress from (2.4.2) to τ_S. The bound on variable ω depends on angle β and results from Eq. (2.4.4):

$$\omega_1 \leq \omega \leq \omega_2, \qquad \sin\omega_{1,2} = \frac{2\tau_s}{\sigma_z \sin 2\beta}. \tag{2.4.6}$$

The first plastic slips occur at $\sigma_z = 2\tau_S = \sigma_S$ at all points of the parallel of the hemisphere $\beta = \pi/4$ in the direction $\omega = \pi/2$ tangent to the meridian.

With Eqs. (2.4.5) and (2.4.6)' Eq. (2.4.3), which determines the plastic strain component ε_z^S for uniaxial tension, can be written as

$$\varepsilon_z^S = \pi \int_{\beta_1}^{\beta_2} \sin 2\beta \cos\beta d\beta \int_{\omega_1}^{\omega_2} F_1(\tau_{nl}) \sin\omega d\omega . \tag{2.4.7}$$

The component ε_z^S has been found by numerical integration [8]. If $i = 1$ in polynomial (2.3.6), the ε_z^S is expressed by elliptic integrals [142,143].

B) Let us proceed to the case of pure shear, $\tau_{xy} > 0$, $\sigma_x = \ldots = \tau_{yz} = 0$. In this case the shear stress τ_{nl}, on the base of Eq. (2.3.2), is

$$\tau_{nl} = (l_x n_y + l_y n_x)\tau_{xy} . \tag{2.4.8}$$

From Eq. (2.3.8) we obtain

$$l_x n_y + l_y n_x = (\cos 2\alpha \cos\omega - \sin 2\alpha \sin\beta \sin\omega)\cos\beta . \tag{2.4.9}$$

Let's designate through τ_1 and τ_2 the value of τ_{nl} at $\omega = 0$ and $\omega = \pi/2$, respectively, which can be expressed through Eqs (2.4.8) and (2.4.9) in the form

$$\tau_1 = \cos 2\alpha \cos\beta \tau_{xy} , \qquad \tau_2 = -\frac{1}{2}\sin 2\alpha \sin 2\beta \tau_{xy} . \tag{2.4.10}$$

The total shear stress is determined by the following formula

$$\tau_n = \sqrt{\tau_1^2 + \tau_2^2} . \tag{2.4.11}$$

Thus,

$$\tau_n = \cos\beta \sqrt{1 - \sin^2 2\alpha \cos^2 \beta}\, \tau_{xy} . \tag{2.4.12}$$

The above formula can also be obtained by finding the maximum of ω from the right-hand side in Eq.(2.4.9).

The plastic shear strain component from Eq. (2.3.5) takes the form

$$\gamma_{xy}^S = \iiint_V (l_x n_y + l_y n_x) F_1(\tau_{nl}) dV . \tag{2.4.13}$$

The area of integration V is obtained from the condition $\tau_{nl} \geq \tau_S$, i.e.,

$$l_x n_y + l_y n_x \geq \frac{\tau_S}{\tau_{xy}} . \tag{2.4.14}$$

From Eq. (2.4.14) it follows that the first plastic slips, $\tau_{xy} = \tau_S$, occur at the following four points of the hemisphere and in the following directions:

$$\left.\begin{array}{ll} 1.\,\alpha = 0, \omega = 0 & 2.\,\alpha = \pi/2,\ \omega = \pi \\ 3.\,\alpha = \pi,\ \omega = 0 & 4.\,\alpha = 3\pi/2,\ \omega = \pi \end{array}\right\}, \quad \beta = 0, \tag{2.4.15}$$

at which τ_{nl} reaches the maximal value, τ_S. The range of further slips expands on the hemisphere Ω around the specified values of α and β with increase in τ_{xy}. The equation of region Ω can be obtained equating the total shear stress from Eq. (2.4.12) to τ_S, i.e.,

$$\cos\beta\sqrt{1-\sin^2 2\alpha\cos^2\beta}\,\tau_{xy} = \tau_S\,. \tag{2.4.16}$$

The limits of integration over ω are

$$\sin\omega_{1,2} = \left(-\tau_S\sin 2\alpha\sin\beta \pm \left|\cos 2\alpha\right|\sqrt{\tau_n^2 - \tau_S^2}\right)\frac{\tau_{xy}\cos\beta}{\tau_n^2}. \tag{2.4.17}$$

In the internal points of Ω, where $\tau_n > \tau_S$, the slips occur in the directions $\omega_1 < \omega < \omega_2$. At the boundary points of Ω, where $\tau_n = \tau_S$, we have $\omega_1 = \omega_2$.

Thus, the determination of shear strains in pure shear is reduced to the calculation of the integrals in Eq. (2.4.13) between the limits $\omega_{1,2}$ from Eq. (2.4.17) and above the domain Ω set by Eq. (2.4.16). Similar to the case of uniaxial tension, the plastic strain component γ_{xy}^S is found by numerical integration in the work [8].

2.5 The Cicala Formula

Let us return once again to the determination of additional-loading modulus G_S at an orthogonal break of loading trajectory. At Section 1.11 (point 5), the modulus G_S was found in terms of the Hencky-Nadai theory. Now let us determine G_S in terms of the concept of slip.

Consider the element of a body that is first subjected to uniaxial tension, $\sigma_z > \sigma_S$, and then, holding the stretching stress ($\Delta\sigma_z = 0$), an infinitesimal shear stress $\Delta\tau_{xz}$ is applied. Let us determine the plastic strain component, $\Delta\gamma_{xz}^{S}$, induced by the additional stress $\Delta\tau_{xz}$ if function $F_1(\tau_{nl})$ is linear,

$$F_1(\tau_{nl}) = b_1\left(\frac{\tau_{nl}}{\tau_S} - 1\right). \tag{2.5.1}$$

Prior to the breaking of the loading path, i.e., at tension, the planes of slips and the directions of slips within these planes are given by the following range of angles α, β, and ω:

$$0 \le \alpha \le 2\pi,\ \beta_1 \le \beta \le \beta_2, \quad \omega_1 \le \omega \le \omega_2, \tag{2.5.2}$$

where the boundary angles $\beta_{1,2}$ and $\omega_{1,2}$ are determined by Eqs. (2.4.5) and (2.4.6). Outside the region (2.5.2), we have $\tau_{nl} < \tau_S$.

If the element of the body is subjected to an infinitesimal torsion $\Delta\tau_{xz}$ at the constant value of σ_z, the shear stress relative to n- and l-axes incurs an increment,

$$\Delta\tau_{nl} = (l_x n_z + l_z n_x)\Delta\tau_{xz}. \tag{2.5.3}$$

where the term in brackets, according to Eq. (2.3.8), takes the form

$$l_x n_z + l_z n_x = \cos\alpha\cos 2\beta\sin\omega - \sin\alpha\sin\beta\cos\omega. \tag{2.5.4}$$

The increment $\Delta\tau_{nl}$ produces additional slips; the plastic strain component $\Delta\gamma_{xz}^{S}$ is determined by Eq. (2.3.9) as

$$\Delta\gamma_{xz}^{S} = \iint_{\Omega'} \cos\beta\, d\alpha\, d\beta \int_{\omega_1'}^{\omega_2'} (l_x n_z + l_z n_x)\Delta F_1\, d\omega, \tag{2.5.5}$$

where Ω', ω_1', and ω_2' constitute the region of integration at additional loading to be determined; the increment of function $F_1(\tau_{nl})$ according to Eq. (2.5.1) can be expressed as

$$\Delta F_1 = \frac{b_1}{\tau_S}\Delta\tau_{nl}. \tag{2.5.6}$$

Substituting ΔF_1 into the integrand in Eq. (2.5.5) gives

$$\Delta\gamma^S_{xz} = \frac{b_1\Delta\tau_{xz}}{\tau_S}\int_0^{2\pi} d\alpha \int_{\beta_1'}^{\beta_2'} \cos\beta d\beta \int_{\omega_1'}^{\omega_2'} (l_x n_z + l_z n_x)^2 d\omega . \tag{2.5.7}$$

There are no additional slips outside the domain (2.5.2) due to outside the area (2.5.2) the inequality $\tau_{nl} < \tau_S$ is holds because the change in shear stress $\Delta\tau_{nl}$ is infinitesimal. The additional slips occur in that part of area (2.5.2) where $\Delta\tau_{nl} > 0$ that enables to determine angles β_1', β_2', ω_1', ω_2' in Eq. (2.5.7). In the monograph [142], the integrals from Eqs. (2.5.7) and (2.4.7) have been computed:

$$\Delta\gamma^S_{xz} = \frac{3\varepsilon^S_z}{2\sigma_z}\Delta\tau_{xz}, \tag{2.5.8}$$

where ε^S_z is the plastic strain component, the relative elongation of a sample, due to tension before the break of loading path. Plastic strain components we obtain subtracting elastic components from the total strain:

$$\Delta\gamma^S_{xz} = \Delta\gamma_{xz} - \frac{\Delta\tau_{xz}}{G}, \quad \varepsilon^S_z = \varepsilon_z - \frac{\sigma_z}{E}. \tag{2.5.9}$$

Inserting results of the above formulae into Eq. (2.5.8), we get

$$\Delta\gamma^S_{xz} - \frac{\Delta\tau_{xz}}{G} = \frac{3}{2\sigma_z}\left(\varepsilon_z - \frac{\sigma_z}{E}\right)\Delta\tau_{xz} . \tag{2.5.10}$$

Let us introduce, similar to section §1.11 (point 5), an extra-loading modulus, $G_S = \Delta\tau_{xz}/\Delta\gamma_{xz}$, and a secant modulus at uniaxial tension, $E_S = \sigma_z/\varepsilon_z$. Then from Eq. (2.5.10) we obtain the relation

$$G_S = \frac{G}{1+\frac{3}{2}G\left(\frac{1}{E_S} - \frac{1}{E}\right)} \tag{2.5.11}$$

termed as the Cicala formula [19,20,68].

2.6 The Cicala Surface

The concept of slip works without the notion of loading surface. In contrast to flow theories, where a surface $f = 0$ is constructed according to a certain hardening rule, the constitutive equations (2.3.5) and (2.3.9) are not associated with any surface $f = 0$. The loading surface can be constructed by making use of the slip criterion (2.3.7).

Let us construct the loading surface for the following regime of loading. The first stage of loading is an uniaxial tension with tensile stress $\sigma_z^0 > \sigma_S$; then, the complete unloading of an element of a body is carried out. Further, the element is subjected to a combined (biaxial) proportional loading under tensile and shear stresses ($\sigma_z > 0$, $\tau_{xz} > 0$). Let us determine values of σ_z and τ_{xz} which first produce plastic deformations after the previous uniaxial loading and unloading. The locus of the obtained values of σ_z and τ_{xz} gives the loading (subsequent yield) surface in σ_z - τ_{xz} stress plane.

Let us designate the shear stress corresponding to the maximal stretching stress σ_z^0 in initial tension as τ_{nl}^0. Eq. (2.4.1) gives

$$\tau_{nl}^0 = \frac{\sigma_z^0}{2} \sin 2\beta \sin \omega . \tag{2.6.1}$$

The stresses τ_{nl}^0 produce plastic slips in the region determined by Eqs. (2.4.5) and (2.4.6) at $\sigma_z = \sigma_z^0$. It is clear that there is no slip in any slip system at unloading. The shear stresses, which arise in the slip systems due to the proportional combined reloading, are determined by Eq. (2.3.2)

$$\begin{gathered} \tau_{nl} = \frac{\sigma_z}{2}\left[\sin 2\beta \sin \omega + k\left(l_x n_z + l_z n_x\right)\right], \\ k = \frac{2\tau_{xz}}{\sigma_z} = const . \end{gathered} \tag{2.6.2}$$

Further, we introduce an auxiliary function,

$$\tau = \begin{cases} \tau_{nl}^0 & \text{inside the region (2.5.2)} \\ \tau_S & \text{outside the region (2.5.2)} \end{cases} \tag{2.6.3}$$

i.e., the function τ is equal to the shear stresses in *n-l* coordinates at the maximal tensile stress σ_z^0 and τ is constant outside the region (2.5.2).

Let us consider further a four-dimension space of variables α, β, ω and τ where the function τ(α,β,ω) exhibits certain hypersurface. The shear stress from Eq. (2.6.2) represents a hypersurface as well. Equation (2.6.2) gives that τ_{nl} grows with increase in components σ_z and τ_{xz}, whereas Eqs. (2.6.1) and (2.6.3) lead to the unchangeable value of τ. According to the condition for the onset of slip, Eq. (2.3.7), plastic slips occur in a given *n-l* system if specified hypersurfaces touch each other at certain point. Indeed, at the touch point, the stress τ_{nl} reaches the maximal value for all history of loading in the given *n-l* system. In work [142], it is shown that the touch point is on the border of area (2.5.2). The determination of coordinates of the specified touch points is reduced to the finding of maximum τ_{nl} in variables α, β, and ω, provided that these variables are interconnected by relationship

$$\tau_{nl}^{0} = \tau_S . \tag{2.6.4}$$

Let us introduce an auxiliary function,

$$\tau_* = \tau_{nl} - \chi\left(\tau_{nl}^{0} - \tau_S\right), \tag{2.6.5}$$

where χ is the Lagrange factor to be determined. We find the coordinates of conditional extremum from the equations

$$\frac{\partial \tau_*}{\partial v} = 0 \qquad (v = \alpha\,, \beta, \omega, \chi) \tag{2.6.6}$$

As τ_{nl}^{0} does not depend on α, the first of these equations takes the form

$$\frac{\partial \tau_{nl}}{\partial \alpha} = 0. \tag{2.6.7}$$

The second and third equation in Eq. (2.6.6) is

$$\frac{\partial \tau_{nl}}{\partial \beta} - \chi \frac{\partial \tau_{nl}^{0}}{\partial \beta} = 0, \quad \frac{\partial \tau_{nl}}{\partial \omega} - \chi \frac{\partial \tau_{nl}^{0}}{\partial \omega} = 0 \tag{2.6.8}$$

and the latter will be reduced to Eq. (2.6.4). By eliminating the multiplier χ from Eq. (2.6.8) we obtain

$$\frac{\partial \tau_{nl}}{\partial \beta}\frac{\partial \tau_{nl}^{0}}{\partial \omega} - \frac{\partial \tau_{nl}}{\partial \omega}\frac{\partial \tau_{nl}^{0}}{\partial \beta} = 0 . \tag{2.6.9}$$

Equations (2.6.4), (2.6.7), and (2.6.9), together with the equality $\tau_{nl} = \tau_S$, constitute the system of four equations for four variables α, β, ω, and σ_z. The solution of this system is

$$\alpha = 0, \ \sin 2\beta = \frac{2\tau_S}{\sigma_z^0}, \ \omega = \frac{\pi}{2}, \ \sigma_z = \frac{2\tau_S \sigma_z^0}{2\tau_S + k\sqrt{\left(\sigma_z^0\right)^2 - 4\tau_S^2}}. \tag{2.6.10}$$

Upon substitution of coefficient k from Eq. (2.6.2) into Eq. (2.6.10), we have

$$\tau_S \sigma_z + \tau_{xz}\sqrt{\left(\sigma_z^0\right)^2 - 4\tau_S^2} = \tau_S \sigma_z^0. \tag{2.6.11}$$

Equation (2.6.11) is the equation of straight lines the pass through point $\sigma_z = \sigma_z^0$, $\tau_{xz} = 0$ and touch to the yield surface

$$\sigma_z^2 + 4\tau_{xz}^2 = 4\tau_S^2. \tag{2.6.12}$$

Figure 2.7 shows the loading surface for uniaxial tension in σ_z-$2\tau_{xz}$ coordinate plane, which was originally developed by Cicala [19,20]. Its important property is the presence of corner singularity at point $\sigma_z = \sigma_z^0$, $\tau_{xz} = 0$.

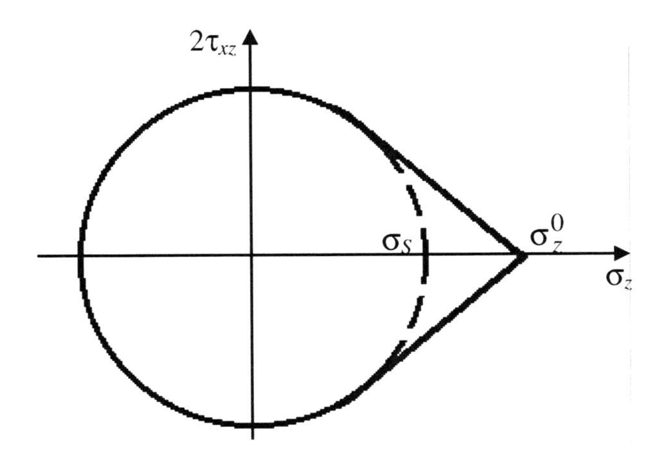

Fig. 2.7 Cicala's surface.

Figure 2.8 [2,142] shows the loading surface for an element in a body first loaded in uniaxial tension σ_z^0 and then subjected to additional torsion of τ_{xz}^0 under unchangeable tensile stress.

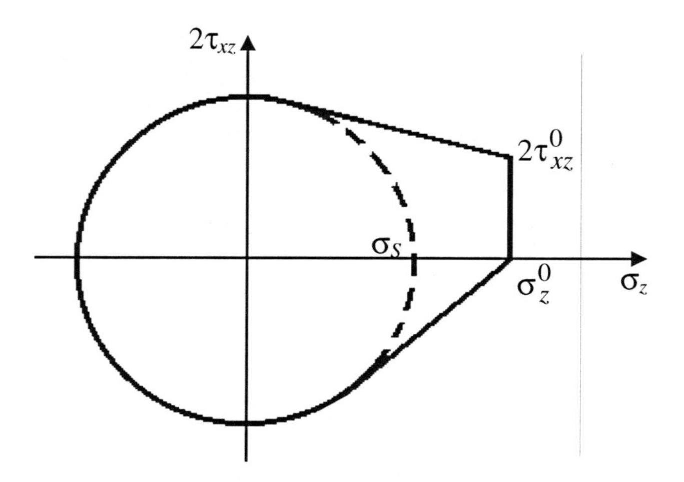

Fig. 2.8 Loading surface under the additional loading in torsion.

Figure 2.9a,b [2,142] illustrates loading surfaces for a biaxial proportional stress state. Their distinguishing features are:

a. The presence of singularity at a point of loading, point M
b. The possibility of vanishing of corner point at M_1 (Fig. 2.9b)

The results shown in Fig. 2.7–2.9 enable us to formulate a technique for the construction of loading surface for an arbitrary loading path [60,63]. A loading surface can be obtained as an infinitive set of fibers pulled on the yield surface and the endpoint of stress vector $\vec{\mathbf{S}}$.

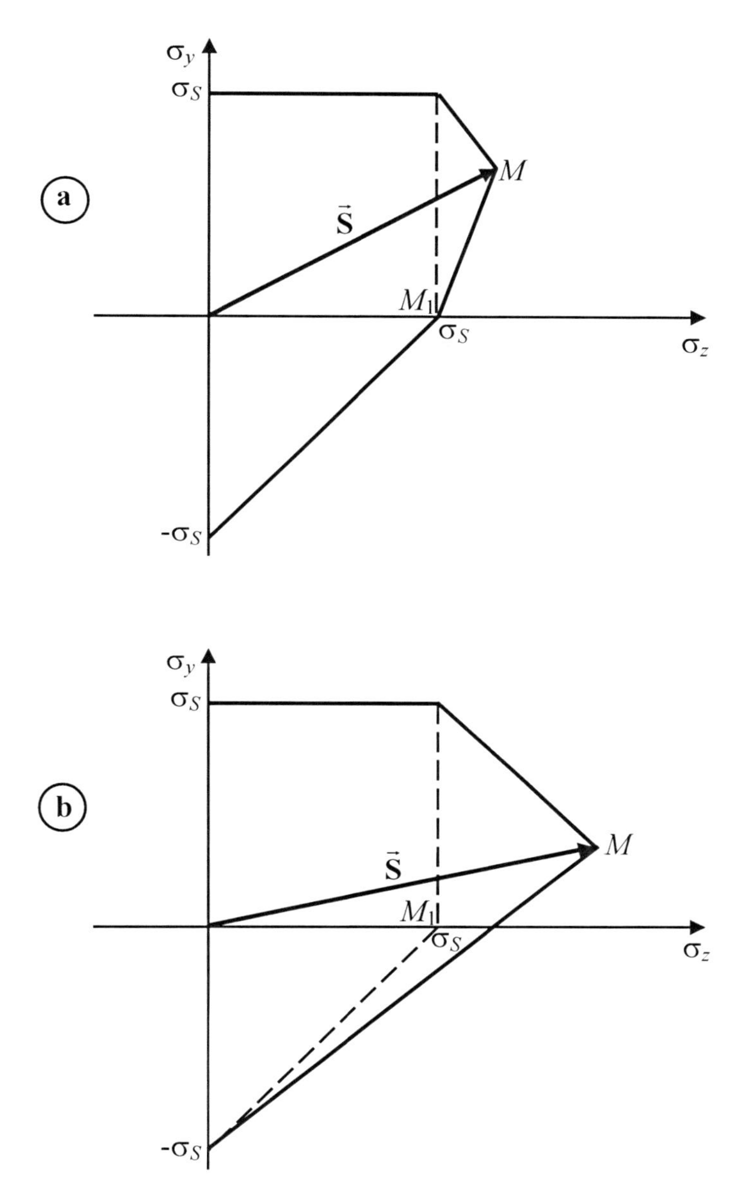

Fig. 2.9 Loading under two-axial tension for the case of small (a) and great (b) exceeding of yield limit.

2.7 Analysis of the Budiansky Theory

1. Let an element of a body be first stretched in the direction of fixed z-axis by tensile stress $\sigma_z > \sigma_S$. Further, the element is subjected to a uniaxial stretching as well, however in directions differing from the direction of the z-axis. Such loading can be obtained in a thin-walled pipe loaded by an axial force, twisting moment, and internal pressure. In cylindrical coordinates (z-axis is directed along the pipe axis), an element of the pipe is subjected to the action of stresses σ_z, $\sigma_\varphi \equiv \sigma_x$, and $\tau_{\varphi z} \equiv \tau_{xz}$. The principal stresses are

$$\sigma_{1,2} = \frac{\sigma_x + \sigma_z}{2} \pm \sqrt{\left(\frac{\sigma_x - \sigma_z}{2}\right)^2 + \tau_{xz}^2}, \quad \sigma_r \equiv \sigma_y \approx 0. \tag{2.7.1}$$

To ensure a uniaxial straining we require that $\sigma_2 = 0$.Then from Eq. (2.7.1) we obtain

$$\sigma_1 = \sigma_x + \sigma_z, \quad \sigma_x \sigma_z = \tau_{xz}^2. \tag{2.7.2}$$

An angle, δ_0, which makes the direction of tension with z-axis in xz-plane, can be calculated as

$$\tan 2\delta_0 = \frac{2\tau_{xz}}{\sigma_z - \sigma_x} \quad \left(\delta_0 < \frac{\pi}{4}\right). \tag{2.7.3}$$

Equations (2.7.2) and (2.7.3) give the value of stresses referred to x- and z-axes,

$$\sigma_x = \sigma_1 \sin^2 \delta_0, \; \sigma_z = \sigma_1 \cos^2 \delta_0, \text{ and } \tau_{xz} = \frac{\sigma_1}{2} \sin 2\delta_0, \tag{2.7.4}$$

which induce an uniaxial tension (σ_1) in an element of a body in the direction making an angle of δ_0 with z-axis. By making use of Eq.(2.3.2), the shear stress τ_{nl} corresponding to the stress components from Eq. (2.7.4) are

$$\tau_{nl} = l_z n_z \sigma_z + l_x n_x \sigma_x + (l_x n_z + l_z n_x)\tau_{xz}. \tag{2.7.5}$$

In particular, at $\alpha = 0$ and $\omega = \pi/2$, on the base Eq. (2.3.8), we obtain

$$\tau_{nl} = \frac{\sigma_z - \sigma_x}{2} \sin 2\beta + \tau_{xz} \cos 2\beta. \tag{2.7.6}$$

By substituting the components σ_x, σ_z, and τ_{xz} from Eq. (2.7.4) into Eq. (2.7.6), we get

$$\tau_{nl} = \frac{\sigma_1}{2} \sin 2(\beta + \delta_0). \tag{2.7.7}$$

It is easy to see that there is exists a set of the values of the angle β where

$$\sin 2(\beta + \delta_0) > \sin 2\beta \tag{2.7.8}$$

for small values of δ_0. The term in the right-hand side in the above inequality appears in the relation for τ_{nl} corresponding to the case of uniaxial tension in the direction of z-axis (see Eq. (2.4.2)). Therefore, Eqs. (2.7.7) and (2.7.8) give that, even at a slight decreasing in σ_1 due to the rotation of tensile direction, the shear stress τ_{nl} grows. This means that specified rotation results in a plastic deformation, which is justified by experimental works [8]. This phenomenon cannot be described neither in terms of the Hencky-Nadai theory nor in the terms the flow theory with isotropic hardening, since the decrease of σ_1 implies that $d\tau_0 < 0$; the Ishlinsky-Kadashevich-Novozsylov theory [54,55] is not suitable for a relative small angles δ_0. The concept of slip can describe the given phenomenon, which is the result for allowing the real mechanisms of deformation in a polycrystalline body.

2. Another positive feature of the concept of slip to be emphasized is the forming of a conic singularity on a loading surface. As it has been already noted, the concept of corner point is, probably, the unique possibility to explain the experiments in Section 1.18.

Many works are devoted to the experimental definition of loading surfaces, $f = 0$ (their detail review is given in work [108]). The analysis of these writings shows that experimenters give different results with respect to the arising of a singularity (corner-point) on a loading surface: one of them justifies this; the others refute [108]. In various publications concerning with a surface $f = 0$, their authors use different definitions for the moment of the onset of plastic deformation; temporary effects influence upon the obtained results. Therefore, there is certain discrepancy in comparing of the results obtained by different researchers. In addition, the Linn [80] and the Klyushnikov papers [62] show the following. The loading surface constructed for the yield strength defined at a residual strain tending to zero exhibits a corner point, while a finite value of the residual strain (for example 0.2%) results in the disappearing of angular point. For this reason it is impossible experimentally either to prove or to reject the existence of a conic singularity on a loading surface. At the same time, the calculations by the Cicala formula (2.5.11) result in too small values of plastic extra-modulus G_S in comparison with the research data [176].

3. One of the basic hypothesis of the concept of slip is the existence of universal function $F_1(\tau_{nl})$ independently of the state of stress. Yosimuras' investigations [52] are devoted for testing this hypothesis. Two states of stress, uniaxial tension and pure shear, are considered for the following stresses

$$\frac{\sigma_z}{2\tau_S} = \frac{\tau_{xy}}{\tau_S} = 1{,}1; 1{,}25; 1{,}4; 1{,}6; 1{,}8\,. \tag{2.7.9}$$

Equations (2.4.7) and (2.4.13) are used for the calculation of strains in uniaxial tension and pure shear, respectively; in both cases function $F_1(\tau_{nl})$ has the form of Eq. (2.3.6). By inserting experimental strains into the left-hand sides of Eqs. (2.4.7) and (2.4.13), two linear 5x5-systems of equations in coefficients b_1, b_2,..., b_5 for ratios (2.7.9) are obtained. The solutions of these two systems for b_1, b_2,..., b_5 , two groups of b_1, b_2,..., b_5 for tension and shear, lead to considerably different results. Consequently, the values of $F_1(\tau_{nl})$ calculated on the base of Eq. (2.3.6) for the obtained two groups of coefficient b_2,..., b_5 are considerably different as well. This means that the function $F_1(\tau_{nl})$ is not a characteristic function of a material as it strongly depends on the state of stress.

Another work [127] also shows that the function $F_1(\tau_{nl})$ is not a characteristic function of a material. Once again the experiments performed by Yosimura were considered and their results were recalculated. The following has been revealed. If to find the coefficients b_i from the experiment in tension and then use them in Eq. (2.4.13) for the calculating of strain for the case of pure shear, then Eq. (2.4.13) will give the strain which is one-half of the strain obtained experimentally. If to take the values of coefficients b_i taken from the experiment in torsion, the strain calculated by Eq. (2.4.7) is approximately twice as experimental.

4. In work [8], the deformation properties of a material in the direction opposite to current slips is not formulated. Consequently, the concept of slip can not to describe the Bauschinger effect. A significant mathematical complexity in the calculation of strains is not in favor of model either.

Thus, the Budiansky theory can describe the experiment (point 1) which cannot be described in terms of other theories of plasticity, demonstrating the advantage of the concept of slip above other theories of an irreversible deformation.

At the same time, the concept of slip shows large disagreements with experiments, the value of plastic extra-modulus at the point of the orthogonal break of loading path is less than experimental. The greatest disparity with the theory showed the investigations given in works [6,52,127]. If to take the function $F_1(\tau_{nl})$ to be a characteristic function of material, the theory is incapable of simultaneously describing experimental diagrams in tension and torsion. As a consequence, in the scientific literature the conclusions had been done [123,127] that the Budiansky theory was not applicable in the modeling of plastic properties of a material. In work [123], pessimistic conclusions are stated relative to a potential progress in the theories of plasticity based on slip mechanisms.

2.8 Generalization of Yield Criterion

The discrepancy between experiments and the predictions of the Budiansky theory has generated a number of works with improvements of the concept of slip. The most effective one has been proposed by Leonov in his writings [74,75], where a yield criterion is presented in the following form:

$$\tau_m = f_1(\tau_0), \quad \frac{df_1}{d\tau_0} \le 0, \qquad (2.8.1)$$

where τ_m is a maximal shear stress at a given moment of loading; f_1 is a characteristic function of a material depending on the shear stress intensity τ_0. The maximal tangential stress τ_m grows during loading, and an element of a body is deformed elastically till $\tau_m < f_1(\tau_0)$. The plastic deformation arises when τ_m becomes equal to function $f_1(\tau_0)$.

To explain Eq. (2.8.1), we consider a polycrystalline body, which consists of a number of crystal grains, under a macro homogenous state of stress. At an earlier discussion, we have specified that the plastic deformation of a grain is the result of the shift of its parts relative to each other. The yield criterion for a grain obtained experimentally by Schmidt reads [170]: if in a system of a possible slip of a crystal a shear stress reaches the value of τ_S (constant of material), an irreversible deformation arise in the crystallite (grain).

Let us suppose that the crystal grains of the considered polycrystalline body obey to the Schmidt law. Let us assume so far that the grains do not hinder each other in their deformations. The yield criterion for the body is the Tresca criterion. Indeed, once the maximal value of shear stress τ_m reaches the value of τ_S, there is a group of grains in whose slip systems the shear stress τ_m acts. The latter statement follows from a random nature of the orientation of grains and a large number (in limit, an infinitive number of grains) of grains. As the grains are deformed without restriction, a plastic deformation arises in the specified micro particles.

Now we refuse the assumption that grains exhibit no interaction relative to the development of plastic deformation, i.e., a real polycrystalline body, with crystals obeying the Schmidt yield criterion, is considered. Let forces applying to a body produce a shear stress in a slip system (grain), which reaches the value of τ_S. In this grain, the yield criterion is formally held. But actually it can be not a case if adjacent grains are under an elastic condition and limit the plastic slips. Therefore, the condition that a shear stress in a grain exceeds a yield stress is not sufficient for the initiation of plastic deformation on a macroscopic scale. To produce a macro plastic deformation, the Schmidt condition must be satisfied in the relative great number of grains. The quantity of the grains depends on the set of planes where high-value shear stresses act. The set depends on the type of stress-state.

Let us consider two bodies under different states of stress, which produce equal maximal shear stresses. Since the number of slip systems, where large shear stresses act, is different in the bodies, the yield limit is reached for not identical quantity of grains; therefore, irreversible macro deformations arise at different values of τ_m. Thus, the Schmidt law does not lead to the Tresca yield criterion.

To formulate the condition for the onset of macro plastic deformation, let us once more turn to the hemisphere of unit radius (Fig. 2.6) whose points characterize slip planes. We wish to find the average value of total stresses τ_n^2 above all slip planes, i.e., we need to calculate the surface integral of τ_n^2 over the hemisphere surface:

$$J = \frac{1}{2\pi}\iint_{\Omega}\tau_n^2 d\Omega = \frac{1}{2\pi}\int_0^{2\pi} d\alpha \int_0^{\pi/2} \tau_n^2 \cos\beta d\beta, \tag{2.8.2}$$

where 2π in the denominator is equal to the surface of the hemisphere of unit radius. For example, consider the following two states of stress, uniaxial tension and pure shear. For the case of uniaxial tension, $\sigma_z > 0$, Eqs. (2.4.2) and (2.8.2) give

$$J = \frac{8}{15}\tau_m^2, \quad \tau_m = \frac{\sigma_z}{2},$$

and Eqs. (2.4.12) and (2.8.2), for the case of pure shear, $\tau_{xy} > 0$, lead to the following

$$J = \frac{2}{5}\tau_m^2, \quad \tau_m = \tau_{xy},$$

where τ_m is the maximal shear stress. As follows from the above two relations, the ratio J/τ_m^2 is greater at uniaxial tension, 8/15, than that at pure shear, 2/5. The statement, the greater the number of high-stress-slip-planes, the greater the value of J/τ_m^2, for the considered states of stress is explained in the following way. Under uniaxial tension, the maximal stress τ_m acts in the set of slip planes within the ranges $0 \le \alpha \le 2\pi$ and $\beta = \pi/4$, while, under pure shear, τ_m acts only on four planes (2.4.15).

The ratio J/τ_m^2 is taken for the characteristic of the number of slip systems experiencing high shear stresses. Since, as shown in [103],

$$J = \nu_0\tau_0^2 \quad (\nu_0 = const)$$

for an arbitrary state of stress, the ratio τ_0^2/τ_m^2 will express the number of high-stress-slip-systems. The greater this number, the lesser the τ_m induces first plastic slips. Therefore, we propose a new yield criterion expressed in the following form

$$\tau_m = f_2\left(\frac{\tau_0^2}{\tau_m^2}\right), \tag{2.8.3}$$

where f_2 is a descending function. By solving this equation for τ_m, we arrive at the criterion (2.8.1). If the function f_1 in Eq. (2.8.1) is assumed to be linear, then the yield criterion (2.8.1) takes the form

$$\tau_m = f_3 - f_4\tau_0, \tag{2.8.4}$$

where f_3 and f_4 are constants of material. For uniaxial tension, this relation becomes

$$\frac{\sigma_S}{2} = f_3 - f_4\sigma_S, \tag{2.8.5}$$

and for a pure shear

$$\tau_S = f_3 - \sqrt{3}f_4\tau_S. \tag{2.8.6}$$

The solution of Eqs. (2.8.5) and (2.8.6) for f_3 and f_4 has the form

$$f_3 = \frac{\left(2-\sqrt{3}\right)\tau_S\sigma_S}{2\left(\sigma_S - \sqrt{3}\tau_S\right)}, \quad f_4 = \frac{2\tau_S - \sigma_S}{2\left(\sigma_S - \sqrt{3}\tau_S\right)}. \tag{2.8.7}$$

All existing theories of slip are based on the Tresca yield criterion that is a partial case of Eq. (2.8.4):

$$\tau_m = f_3 = \tau_S = \frac{\sigma_S}{2}, \quad f_4 = 0. \tag{2.8.8}$$

Another limiting case, $f_3 \to \infty$ and $f_4 \to \infty$, corresponds to the Guber yield criterion,

$$\tau_0 = \frac{f_3}{f_4} = \sqrt{3}\tau_S = \sigma_S. \tag{2.8.9}$$

If constants f_3 and f_4 satisfy inequalities

$$\tau_S < f_3 < \infty \quad \text{and} \quad 0 < f_4 < \infty, \tag{2.8.10}$$

then the yield surface (2.8.4) lies between the Tresca and Guber surfaces in a stress space.

Although the yield criterion (2.8.1) possesses the physical significance, at the same time, it follows from formal derivations. As solids are assumed to be initially isotropic, the yield condition should depend on the three invariants of stress tensor. Since considered bodies do not depend on hydrostatic pressure, the found criterion depends on two invariants of stress deviator tensor. These invariants are τ_m and τ_0; this serves for the substantiation of Eq. (2.8.1).

As it is known, the Tresca and Guber criteria, which define the conditions for the onset of irreversible deformations, lead to the results differing from each other insignificantly. Therefore, there can be doubts relative to the expediency of the generalized yield criterion (2.8.1); this will be discussed in Section 2.11.

2.9 The Leonov Theory of Slip

One of the basic notions of the Leonov theory is the slip intensity. In the definition of this notion, a start point is the hemisphere of unit radius introduced in Section 2.3. Through every point on the hemisphere the tangent plane passes oriented by a normal vector $\vec{\mathbf{n}}$ (Fig. 2.6); the plane symbolizes a plane of slip in a grain. Within each of the planes, the direction of slip is set up by the direction of vector $\vec{\mathbf{l}}$.

Let the plastic shifts in slip systems, which are given by normal vectors $\vec{\mathbf{n}}$ within a solid angle $d\Omega$ and vectors $\vec{\mathbf{l}}$ from a range $d\omega$, result in a total plastic shift, $d\gamma_{nl}^S$. Then, a slip intensity in the direction $\vec{\mathbf{l}}$ in the plane of normal vector $\vec{\mathbf{n}}$, φ_{nl}, is defined as

$$\varphi_{nl} = \frac{d\gamma_{nl}^S}{d\omega d\Omega}. \tag{2.9.1}$$

Let us rewrite the latter formula to the form,

$$d\gamma_{nl}^S = \varphi_{nl} d\omega d\Omega, \tag{2.9.2}$$

and compare it with the Budiansky formula (2.3.3). Equations (2.3.3) and (2.9.2), being identical by form, differ in their significations. To find out this difference, we turn to the physics of the phenomenon of plastic deformation partially discussed in Section 2.2.

As it is known, the plastic deformation of crystal bodies can be induced by various ways, the slips of the parts of crystal relative to each other is not a unique mechanism. Between them a twinning must be named. Crystal twinning occurs when the deformation of crystal structure is such that parts of crystals share some of the same crystal lattice points in a symmetrical manner. It occurs if the tangential stress reaches a certain critical value and, the same way as slips, develops in certain crystallographic planes and directions in them. Furthermore, let us note that the role of twinning in the deformation of crystals grows with the decrease of temperature and (or) increase of loading rate.

So far, we have considered the cases when the plastic deformations of polycrystalline aggregate have resulted from deformations generated exclusively within crystal grains, they are the shifts of the parts of crystals relative to each other caused by the movements of dislocations or twinning. However, there is qualitatively another form of the deformation of body when grains remain non-deformed. Let us consider, first of all, the slips at the grain boundaries. Fig. 2.10a and 2.10b show schematically the group of crystal grains prior an after the slip at the grain boundary, respectively. It is clear that the movements of grains result in the deformation of polycrystalline body. However, in addition, the slip at the grain boundaries results in the generating of pores in a body (shaded areas in Fig 2.10b). These pores are liquidated by a mass transfer caused by diffusional processes.

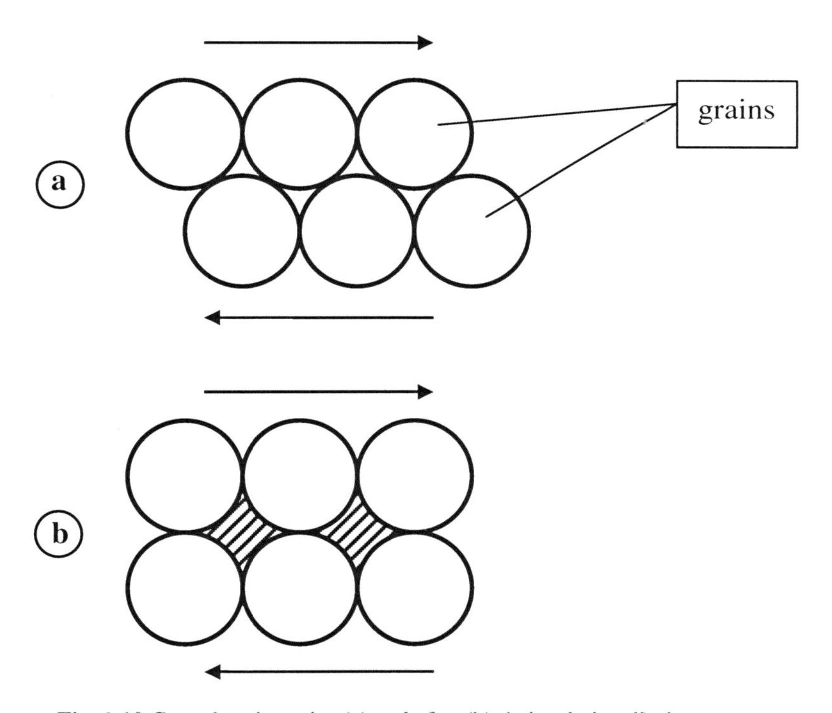

Fig. 2.10 Crystal grains prior (a) and after (b) their relative displacements.

By a microparticle (Section 2.1) we now mean a group of crystal grains (not necessarily only one crystal grain). Then the function φ_{nl} does not govern a slip of a unique crystal but plays the role of slip characteristic above a continuum region containing a group of crystal grains.

According to Eq. (2.9.1), the notion of φ_{nl} is defined in the same way as, for example, the notion of the intensity of distributed loading, $q = dp/dx$, in the

Strength of Materials, where dp is the amount of resultant force applied over a given distance increment dx of a beam, or as the notion of density, $\rho = dm/dV$, where dm is the mass of a body of volume dV. Such interpreting of intensity φ_{nl} leads to that the amount of the planes and directions of slip in crystals now do not matter. Thus, the first restriction adopted in the Batdorf-Budiansky theory has been removed (Section 2.3). Further, the intensity φ_{nl} is indifferent to the causes of the deformation of crystal (movements of dislocations, twinning, slips at the grain boundaries, the diffusion of atoms or something else). In addition, the function φ_{nl} can describe the deformation of another (non-shear) nature as well. Indeed, the deformation of phase transformation is described by means of formulas similar to Eq. (2.3.9); however it is not of a shear character [83]. In general, by G. Tailor, if there is no volume change in a body, any elementary irreversible deformation can be represented as the result of slips in five slip systems.

Equation (2.9.2) determines the strain caused by slips over the area of hemisphere $d\Omega$ in the directions of $d\omega$. The total plastic strain components caused by all slips are determined as

$$\gamma_{ij}^{S} = \iint_{\Omega} d\Omega \int_{\omega_1}^{\omega_2} \left(l_i n_j + l_j n_i\right)\varphi_{nl}\, d\omega, \quad (i \neq j,\ i, j = x, y, z)$$
$$\varepsilon_i^{S} = \iint_{\Omega} d\Omega \int_{\omega_1}^{\omega_2} l_i n_i \varphi_{nl}\, d\omega. \quad (i = x, y, z;\ \Sigma) \tag{2.9.3}$$

where Ω, ω_1 and ω_2 designate the same as in Eq. (2.3.5). In an incremental form, we have

$$\dot{\gamma}_{ij}^{S} = \iint_{\Omega} d\Omega \int_{\omega_1}^{\omega_2} \left(l_i n_j + l_j n_i\right)\dot{\varphi}_{nl}\, d\omega, \quad (i \neq j,\ i, j = x, y, z)$$
$$\dot{\varepsilon}_i^{S} = \iint_{\Omega} d\Omega \int_{\omega_1}^{\omega_2} l_i n_i \dot{\varphi}_{nl}\, d\omega, \quad (i = x, y, z;\ \Sigma) \tag{2.9.4}$$

It is obvious that Eqs. (2.9.3) and (2.9.4) are obtained from Eqs. (2.3.5) and (2.3.9) by replacing of the function $F_1(\tau_{nl})$ by the intensity φ_{nl}. Similar to Section 2.3, the limits of integration in Eqs. (2.9.3) and (2.9.4) can be different.

Another basic notion of the Leonov theory, plastic shear strength, is presented in the discussion to follow.

2.10 Shear Strength

The value of stress acting in a given slip system that produces a plastic slip will be term as a shear strength.

An originally isotropic body behaves in non-isotropic manner during plastic straining, exhibits different resistances to plastic slips in different directions. As known, a slip in a single slip system in a crystal grain hardens other slip systems. In addition, the slips can occur in several directions simultaneously. Thus, the hardening law in slip systems is complicated enough even for a single crystal. Furthermore, a plastic deformation of single grain influences the condition of the onset of plastic deformation in nearby grains. The latter is caused by the inter-grain forces that vary with plastic deformation, i.e., the stress-strain state varies in crystals during plastic deforming. Since the orientation of grains is of a random nature, the interacting forces acting between grains have a random pattern as well. Therefore, the establishment of the hardening law, i.e., the change in the shear strength, is an extremely complicated problem of the physics of solids and the mathematical statistics. In terms of the M. Leonov theory, this difficulty is managed by phenomenological introduction of the shear strength for a perfectly homogeneous model.

It is assumed that a slip occurring in a slip system (we recall that a slip system is oriented by the vector $\vec{\mathbf{n}}$ (n-axis) at a point of the hemisphere (Fig 2.6) in the direction given by the vector $\vec{\mathbf{l}}$ (l-axis)) changes the shear strength in all planes and directions. Together with n- and l-axes, let us introduce n_0- and l_0-axes indicating the slip system where the shear strength's to be determined. n_0- and l_0-axes, being of the same meaning as n- and l-axes, are set up by angles α_0, β_0, and ω_0 which are counted off in the same fashion as angles α, β, and ω in Fig. 2.6. A plastic shear strength, $S_{n_0 l_0}$, referred to as n_0- and l_0-axes is defined as [74,75]:

$$S_{n_0 l_0} = f_1(\tau_0)\left(1 + R_{n_0 l_0}(\varphi_{nl})\right),$$

$$R_{n_0 l_0}(\varphi_{nl}) = r_1 \varphi_{n_0 l_0} + r_2 \int_{\omega_1}^{\omega_2} \varphi_{n_0 l} \cos(\omega_0 - \omega) d\omega + r_3 \gamma^S_{n_0 l_0}. \tag{2.10.1}$$

$\gamma^S_{n_0 l_0}$ is the total shear plastic strain, with respect to n_0- and l_0-axes, produced by all slip systems experiencing plastic slips; ω_1 and ω_2 are the boundary values of slip directions in a given plane (n_0 - axis); r_1, r_2, and r_3 are constants of material. For that slip system where plastic straining occurs, we have

$$\tau_{n_0 l_0} = S_{n_0 l_0}. \tag{2.10.2}$$

For the case if slips do not occur:

$$\tau_{n_0 l_0} < S_{n_0 l_0}. \tag{2.10.3}$$

Equations (2.9.3) and (2.10.1) – (2.10.3) constitute the core of the Leonov theory.
Let us comment the terms in Eq. (2.10.1).

1. The shear strength should be given so that the yield criterion (2.8.1) follows from Eqs. (2.10.1) and (2.10.2). Prior to a first plastic slip, we have

$$R_{n_0 l_0} \equiv 0, \quad S_{n_0 l_0} = f_1(\tau_0). \tag{2.10.4}$$

In an elastic region, in all slip systems the inequality (2.10.3) is satisfied: $\tau_{n_0 l_0} < f_1(\tau_0)$. Since $f_1(\tau_0)$ does not depend on angles α_0, β_0, and ω_0 during increase in load, Eq. (2.10.2) holds, first of all, in the plane of the maximal tangent stress in the direction of its action, that is Eqs. (2.10.1)–(2.10.3) lead to the yield criterion in the form of Eq. (2.8.1). According to Eq. (2.10.4), the function $f_1(\tau_0)$ is an initial shear strength. Since the initial shear strength depends on τ_0, it is natural that we expect the shear strength $S_{n_0 l_0}$ to be dependent on τ_0 during further plastic slips. This fact is expressed in Eq. (2.10.1) through the presence of the factor $f_1(\tau_0)$.

2. The operator $R_{n_0 l_0}$ in Eq. (2.10.1) is given so that a plastic slip in a slip system affects shear strengths in all other systems. This is ensured by the second and third terms in Eq. (2.10.1). Indeed, consider a l_0-n_0 slip system, $\varphi_{n_0 l_0} \geq 0$ if this slip system produces plastic slips. The integral in Eq. (2.10.1), being an integral measure of slips for fixed n_0-axis and varying directions, allows for the change in the shear strength in the n_0-l_0 slip system due to the slips in other "active" n_0-l slip systems. In a similar fashion, the quantity $\gamma^S_{n_0 l_0}$ is an integral measure of plastic slips developing in all slip systems at a point in a body, both slip planes and slip directions vary. Thus $\gamma^S_{n_0 l_0}$ provides that the total plastic strain affects the shear strength in the considered fixed slip system.

It must be noted that $R_{n_0 l_0} \geq 0$ for the slip system that does not take part in the developing of plastic strain, $\varphi_{n_0 l_0} = 0$, as well. Indeed, if in other directions or slip systems plastic slips occur, then $R_{n_0 l_0} \geq 0$ due to the integral and $\gamma^S_{n_0 l_0}$ from Eq. (2.10.1) are nonzero.

As a consequence, the second restriction of the Batdorf-Budiansky theory (see Section 2.3), according to which the plastic slip in each grain takes place irrespective of the plastic slips in the rest of grains, is removed.

3. The general form of the operator determining the degree of hardening only in one slip plane is:

$$R_{n_0 l_0} = \int_{\omega_1}^{\omega_2} \varphi_{n_0 l} F_2(\omega_0 - \omega) d\omega . \tag{2.10.5}$$

The function F_2 determines the dependence between a local slip and hardening in one and the same plane n_0. The function F_2 must be pair and periodic with period of 2π. Furthermore, to describe the Bauschinger effect we require that $F_2(\omega_0 - \omega + \pi) = -F_2(\omega_0 - \omega)$. The simplest function satisfying these requirements is $\cos(\omega_0 - \omega)$. Then from Eq. (2.10.5) the second term of Eq. (2.10.1) results.

If to take into account the influence of slip in one system upon the hardening in all planes and directions, then $R_{n_0 l_0}$ takes the form:

$$R_{n_0 l_0} = \iint_{\Omega} d\Omega \int_{\omega_1}^{\omega_2} F_3(\alpha, \beta, \omega, \alpha_0, \beta_0, \omega_0) \varphi_{nl} d\omega \tag{2.10.6}$$

The function F_3 determines the dependence between the local slip in one n-l system and the hardening in another n_0-l_0 system. Shear strain component referred n_0- and l_0-axes is determined by similar formula:

$$\gamma^S_{n_0 l_0} = \iint_{\Omega} d\Omega \int_{\omega_1}^{\omega_2} F_4(\alpha, \beta, \omega, \alpha_0, \beta_0, \omega_0) \varphi_{nl} d\omega . \tag{2.10.7}$$

By making use the second-rank-stress-tensor transformation equation, it is easy to obtain the explicit expression for F_4 that has been done in Sec. 2.11. However we do not need it at the current discussion. If we assume that $F_4 = F_3$, then from Eq. (2.10.6) the third term in Eq. (2.10.1) follows.

4. The operator $R_{n_0 l_0}$ is linear as small plastic strains are only considered.

Summarizing, the shear strength $S_{n_0 l_0}$ introduced in the form of Eq. (2.10.1) expresses the deformational anisotropy of material. Further we will show that Eq. (2.10.1) leads to the agreement between experimental and calculated strains.

2.11 Rehabilitation of the Concept of Slip

The M. Leonov theory considerably rehabilitates the concept of slip, which follows from the results of this Section. Let us show by means of direct calculations the following. If slips occur at all points of the hemisphere (Fig. 2.6) and at each point in all directions where $\tau_{nl} > 0$, and if the slip intensity is proportional to the shear stress component,

$$\varphi_{nl} = K\tau_{nl}, \tag{2.11.1}$$

then the shear strain with respect to any n_0- and l_0-axes is

$$\gamma^S_{n_0 l_0} = \frac{2\pi^2}{5} K\tau_{n_0 l_0}. \tag{2.11.2}$$

In terms of considered theory, in the same way as in the Batdorf-Budiansky theory, the state of stress is assumed to be identical in all microparticles and coincides with the state of stress of whole polycrystalline aggregate. Therefore, Eq. (2.3.2) determining the tangential stress τ_{nl} remains correct.

According to the condition of the problem, slips occur in directions $\omega_* \le \omega \le \omega_* + \pi$ where $\tau_{nl} > 0$; here ω_* is the angle where $\tau_{nl} = 0$.

To be convinced of Eq. (2.11.2), let us evaluate the integrals in Eq. (2.10.7). By the equations of second-rank tensor transformation at a transition from nl- to other $n_0 l_0$-axes, we find the integration function F_4:

$$F_4 = F_{ll_0} F_{nn_0} + F_{ln_0} F_{l_0 n}, \tag{2.11.3}$$

where

$$F_{ll_0} = l_i l_{0i}, \quad F_{nn_0} = n_i n_{0i}, \quad F_{ln_0} = l_i n_{0i}, \quad F_{l_0 n} = l_{0i} n_i \quad (i = x, y, z) \tag{2.11.4}$$

and $l_x, \dots, n_z$ are determined by Eqs. (2.3.8); $l_{x_0}, \dots, n_{z_0}$ are the direction cosines of l_0- and n_0-axes also determinated by Eqs. (2.3.8) where angles α, β and ω are replaced by α_0, β_0 and ω_0. Substituting F_4 from Eqs. (2.11.3) and (2.11.4) into the integrand in Eq. (2.10.7) and taking the integral between the limits $0 < \alpha < 2\pi$, $0 < \beta < \pi/2$, and $0 < \omega < \pi$, we will obtain

$$\gamma^S_{n_0 l_0} = \frac{2\pi^2}{5} K \big[l_{0x} n_{0x} \sigma_x + l_{0y} n_{0y} \sigma_y + l_{0z} n_{0z} \sigma_z + \\ + (l_{0x} n_{0y} + l_{0y} n_{0x})\tau_{xy} + (l_{0y} n_{0z} + l_{0z} n_{0y})\tau_{yz} + (l_{0z} n_{0x} + l_{0x} n_{0z})\tau_{xz} \big] \tag{2.11.5}$$

or

$$\gamma^S_{n_0 l_0} = \frac{2\pi^2}{5} K l_{0i} n_{0j} \sigma_{ij}, \quad (i, j = x, y, z). \tag{2.11.6}$$

However, according to Eq. (2.3.2), Eqs. (2.11.2) and (2.11.6) are identical, which is the proof of the correctness of Eq. (2.11.2). The obtained Eq. (2.11.2) leads to that if the slip intensity is

$$\varphi_{nl} = K_1 \tau_{nl} + K_2 \gamma_{nl}^S, \tag{2.11.7}$$

then

$$\gamma_{n_0 l_0}^S = \frac{2\pi^2}{5}\left(K_1 \tau_{n_0 l_0} + K_2 \gamma_{n_0 l_0}^S\right), \tag{2.11.8}$$

i.e.,

$$\gamma_{n_0 l_0}^S = \frac{2\pi^2 K_1}{5 - 2\pi^2 K_1} \tau_{n_0 l_0}. \tag{2.11.9}$$

Let us turn, further, to the constitutive relations of the Leonov theory (2.10.1) – (2.10.3). Assuming, so far, that constant r_2 is equal to zero, we obtain

$$S_{n_0 l_0} = f_1(\tau_0)\left(1 + r_1 \varphi_{n_0 l_0} + r_3 \gamma_{n_0 l_0}^S\right). \tag{2.11.10}$$

By equating $S_{n_0 l_0}$ to $\tau_{n_0 l_0}$, we obtain the equation for slip intensity in the following form

$$r_1 \varphi_{nl} = \frac{\tau_{nl}}{f_1(\tau_0)} - r_3 \gamma_{nl}^S - 1. \tag{2.11.11}$$

In contrast to the Budiansky formula (2.3.6), where the shear stress τ_{nl} appears in a nonlinear fashion, τ_{nl} stands as first-order term in Eq. (2.11.11).

We have $\tau_m = f_1(\tau_0)$ on a yield limit and the increase in forces applied to a body causes the growth of the maximal shear stress τ_m, while the function $f_1(\tau_0)$ decreases (see Eq. (2.8.1)). In addition, the shear-plastic-strain component γ_{nl}^S grows as well. Therefore, at enough large stress, the unit on the right-hand side in Eq. (2.11.11) may be neglected in comparison to the sum of its first two terms (meaning that $r_3 < 0$). Then we arrive at Eq. (2.11.7), where

$$K_1 = \frac{1}{r_1 f_1(\tau_0)}, \quad K_2 = -\frac{r_3}{r_1} > 0, \tag{2.11.12}$$

from which Eq. (2.11.9) follows, i.e.,

$$\gamma^S_{n_0 l_0} = \frac{K_3 \tau_{n_0 l_0}}{f_1(\tau_0)}, \quad K_3 = \frac{2\pi^2}{\left(5 - 2\pi^2 r_3\right) r_1}. \tag{2.11.13}$$

Adding the elastic shear strain $\tau_{n_0 l_0}/G$ to the plastic component from Eq. (2.11.13), we obtain the total elasto-plastic strain:

$$\gamma_{n_0 l_0} = \frac{\tau_{n_0 l_0}}{G_S}, \tag{2.11.14}$$

where

$$\frac{1}{G_S} = \frac{K_3}{f_1(\tau_0)} + \frac{1}{G}. \tag{2.11.15}$$

Equation (2.11.14) expresses the deviator proportionality. Indeed, since the n_0- and l_0-axes are arbitrary, they can be marked by x, y, z, i.e., Eq. (2.11.14) leads to the shear components in Eq. (1.2.5). Further, if to insert the normal strain components from Eq. (1.2.5) at $\varepsilon = 0$ into the second-rank-strain-tensor transformation equation,

$$\gamma_{nl} = \left(l_i n_j + l_j n_i\right)\varepsilon_{ij},$$

we arrive, on the base of Eq. (2.3.2), at the result (2.11.14)

Thus, if the function $f_1(\tau_0)$ and constant r_3 are such that the unit in Eq. (2.11.11) can be neglected in comparison to the other terms, the deviator proportionality follows from the Leonov model and a shear modulus is dependent on shear stress intensity.

The obtained result conciderably rehabilitates the concept of slip. The basic shortcoming of the Budiansky theory is (point 3 of Sec. 2.7) that it is incapable of analytic modeling (with the unique set of model constants) of experimental diagrams in tension and pure shear. Since the Leonov theory predicts the deviator proportionality and the dependence of the shear modulus G_S on shear stress intensity, i.e., the existence of the universal diagram $\tau_0 \sim \gamma_0$, the specified shortcoming of the Budiansky theory is absent in terms of the Leonov model. This is a consequence of linear relation between the slip intensity and the shear stress, Eq. (2.11.11).

Further, let us turn to the work [142], where the following has been shown. If the linear function $\varphi_{nl}(\tau_{nl})$ describes simultaneously experimental diagrams in uniaxial tension and pure shear, (by "simultaneously" we mean that shear and stretching strains are determined by the only set of model constants), then, at a power function, $\varphi_{nl}\left(\tau^p_{nl}\right)$ ($p > 1$), the calculated values of strain in torsion turned out to be largely smaller and in tension considerably larger in comparison to experimental data.

From the all above it follows that the shortcomings of the Budiansky theory originate from Eq. (2.3.6), according to which the slip intensity relates to the shear stress τ_{nl} in a nonlinear manner.

Let us return to the function $f_1(\tau_0)$, which appears in the expressions for yield criterion (2.8.1) and shear strength (2.10.1). The function $f_1(\tau_0)$ can be determined from the yield criterion for a small diapason of τ_0. This follows from that the yield criterion (2.8.1), which lies between the Tresca and Guber criteria in stress space, gives the close values of τ_0 inducing plastic deformation for different states of stress. The function $f_1(\tau_0)$, for the values of τ_0 greater than those on the yield limit, can be defined through Eq. (2.11.15) as

$$f_1(\tau_0) = \frac{K_3 G G_S}{G - G_S} \qquad (G_S < G). \tag{2.11.16}$$

The function $f_1(\tau_0)$ has been entered into the constitutive relationships of the Leonov theory not so much for the generalization of the yield criterion as it enables us to obtain stress-strain diagrams of required form. Indeed, if to calculate $f_1(\tau_0)$ by Eq. (2.11.16) on the base of experimental values of G_S, the stress-strain curve constructed by Eqs. (2.11.16), (2.11.11), and (2.9.3) is identical to the experimental diagram.

There is a certain analogy between Eq. (2.11.11) and the deviator proportionality (1.2.4). In both relationships, quantities to be founded, φ_{nl} and ε_{ij}, depend linearly on driving forces, τ_{nl} and σ_{ij}; the nonlinearity is regulated by the functions of the stress-tensor invariant τ_0, $f_1(\tau_0)$, and $G_S(\tau_0)$.

2.12 Peculiarities of Plastic Straining under Proportional Loading

Let us investigate the Leonov theory with respect to the basic rules of the Hencky-Nadai theory, the deviator proportionality and the existence of the universal diagram $\tau_0 \sim \gamma_0$.

We carry out the analysis of the deviator proportionality in terms of the Lode-Nadai coefficients. At uniaxial tension (e.g. in the direction of z-axis) we have $\sigma_1 = \sigma_z$, and $\sigma_2 = \sigma_3 = 0$, therefore $\mu_\sigma = -1$. The components of relative lengthening are

$$\varepsilon_z = \frac{\sigma_z}{E} + \varepsilon_z^S, \quad \varepsilon_x = \varepsilon_y = -\frac{\nu\sigma_z}{E} - \frac{\varepsilon_z^S}{2}. \tag{2.12.1}$$

The principal strains are $\varepsilon_1 = \varepsilon_z$, and $\varepsilon_2 = \varepsilon_3 = \varepsilon_x$ and Eqs. (1.17.3) and (2.12.1) give $\mu_\varepsilon = -1$. Thus, for the case of uniaxial tension we have $\mu_\sigma = \mu_\varepsilon$ and this equality holds true regardless of ε_z^S, i.e., independently of the model of plasticity in terms of which strain components are determined. The situation is the same for uniaxial pressure and pure shear.

Let us verify the correctness of equality $\mu_\sigma = \mu_\varepsilon$ at two-axial tension. If a stretching develops in the direction of x- and *z*-axes and $\sigma_x = k\sigma_z > 0$, $0 < k < 1$, then $\sigma_1 = \sigma_z$, $\sigma_2 = \sigma_x$, and $\sigma_3 = 0$, so that

$$\mu_\sigma = 2k - 1. \tag{2.12.2}$$

Now we wish to determine the Lode-Nadai coefficient μ_ε. For this purpose let us determine approximately plastic strain components. Shear stresses acting in slip systems are

$$\tau_{nl} = l_x n_x \sigma_x + l_z n_z \sigma_z. \tag{2.12.3}$$

This stress is maximal at

$$\alpha = \frac{\pi}{2}, \quad \beta = \frac{\pi}{4}, \quad \omega = \frac{\pi}{2}. \tag{2.12.4}$$

As shown in the previous Section, if the stress exceeds considerably the yield limit, the deviator proportionality is fulfilled approximately, therefore $\mu_\sigma \approx \mu_\varepsilon$. The greatest deviation from this equality is expected at the vicinity of the yield limit. Therefore, let us estimate the relative elongations in the vicinity of the yield limit using Eq. (2.9.3) for the values of α, β, and ω given by Eq. (2.12.4). Eq. (2.3.8), together with Eq. (2.12.4), gives $l_x = 0$, $l_y n_y = -l_z n_z$ and Eq. (2.9.3) results in $\varepsilon_x^S = 0$, $\varepsilon_y^S = -\varepsilon_z^S$. Supplementing these values by elastic strains, we obtain

$$\begin{aligned} \varepsilon_1 = \varepsilon_z &= \frac{(1 - k\nu)\sigma_z}{E} + \varepsilon_z^S, \\ \varepsilon_2 = \varepsilon_x &= \frac{(k - \nu)\sigma_z}{E}, \\ \varepsilon_3 = \varepsilon_y &= -\frac{\nu(1 + k)\sigma_z}{E} - \varepsilon_z^S. \end{aligned} \tag{2.12.5}$$

Therefore, from Eqs. (1.17.3) and (2.12.5) we obtain the Lode-Nadai coefficients for the two-axial state of stress:

$$\mu_\varepsilon = \frac{(2k-1)(1+\nu)\sigma_z}{(1+\nu)\sigma_z + 2E\varepsilon_z^S}. \tag{2.12.6}$$

Let us compare the results obtained by Eqs. (2.12.2) and (2.12.6), i.e., μ_σ and μ_ε. For $0 < k < 1/2$ we have $\mu_\sigma < \mu_\varepsilon < 0$; $\mu_\sigma = \mu_\varepsilon = 0$ at $k = 1/2$ and if $1/2 < k < 1$, then $0 < \mu_\varepsilon < \mu_\sigma$. Therefore, in a general case,

$$|\mu_\varepsilon| \le |\mu_\sigma|. \tag{2.12.7}$$

It must be noted that in the range of the coefficients k, $-1 < k < 1$, all possible forms of plane stress states are covered. We also note that Eq. (2.12.6) is incorrect at tension or pressure. It has been obtained from the condition that the shear stress is maximal in the slip system (2.12.4), whereas the maximal shear stress τ_m at tension and pressure acts in planes $0 < \alpha < 2\pi$, $\beta = \pi/4$.

To answer the question on the existence of the uniform diagram $\tau_0 \sim \gamma_0$, we need to compare stress-strain diagrams in uniaxial tension and pure shear. For the case of uniaxial tension we have $\varepsilon_x^S = \varepsilon_y^S = -\varepsilon_z^S/2$ therefore the plastic shear strain components take the form

$$\gamma_{nl}^S = \frac{3}{2}\varepsilon_z^S \sin 2\beta \sin\omega. \tag{2.12.8}$$

Thus, the slip intensity (2.11.11) becomes

$$r_1\varphi_{nl} = k_4 \sin 2\beta \sin\omega - 1, \quad k_4 = \frac{\sigma_z}{2f_1(\tau_0)} - \frac{3}{2}r_3\varepsilon_z^S. \tag{2.12.9}$$

The plastic strain ε_z^S is determined by Eq. (2.9.3) and between the limits of integrals (2.4.5) and (2.4.6) the stress $\sigma_z/2\tau_S$ must be replaced by coefficient k_4.

Let us proceed to a pure shear, $\tau_{xy} > 0$. In this case

$$\gamma_{nl}^S = (l_x n_y + l_y n_x)\gamma_{xy}^S. \tag{2.12.10}$$

Thus the slip intensity (2.11.11) will be written as

$$r_1\varphi_{nl} = k_5(l_x n_y + l_y n_x) - 1, \quad k_5 = \frac{\tau_{xy}}{f_1(\tau_0)} - r_3\gamma_{xy}^S. \tag{2.12.11}$$

The plastic strain component is determined by Eq. (2.9.3) with the limits of integration of Eq. (2.4.16) and (2.4.17) where coefficient k_5 must be inserted instead of ratio τ_{xy}/τ_S. Now the shear strain intensity has the form

$$\gamma_0 = \frac{\sigma_z}{3G} + \varepsilon_z^S, \quad \tau_0 = \sigma_z \tag{2.12.12}$$

and, according to Eq. (1.2.15), at pure shear we obtain

$$\gamma_0 = \frac{1}{\sqrt{3}}\left(\frac{\tau_{xy}}{G} + \gamma_{xy}^S\right), \quad \tau_0 = \sqrt{3}\tau_{xy}. \tag{2.12.13}$$

The diagrams $\tau_0 \sim \gamma_0$ constructed on the base of Eqs. (2.9.3), (2.12.12), and (2.12.13) are plotted in Fig. 2.11 (the curves marked by 1 and 2 correspond to uniaxial tension and pure shear, respectively). The initial portions of plastic deformations do not coincide with each other, however these curves become closer with the growth in τ_0. The diagrams $\tau_0 \sim \gamma_0$ at any other states of stress are located between the curves shown in Fig. 2.11.

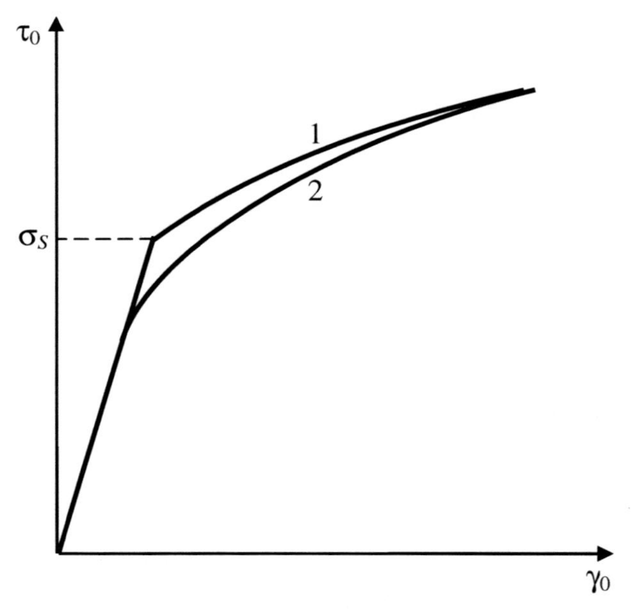

Fig. 2.11 Shear stress intensity vs. shear strain intensity diagram in uniaxial tension (1) and pure shear (2).

2.13 Drucker's Postulate and the Concept of Slip

The fulfillment of Drucker's postulate is the obligatory requirement for any theory of plasticity including the concept of slip. As it is known (Section 1.10) the consequence from the postulate is the perpendicularity of plastic-strain-increment vector to a yield/loading surface (in its regular points). Let us verify the perpendicularity of vector $d\vec{\mathbf{e}}^S$ to a yield surface under two-axial tension.

Similarly to the previous paragraph, we have the tension in the direction of *x*- and *z*- axes and $\sigma_x = k\sigma_z$, $0 < k < 1$. Analogical to the Section 1.17, and switching the role of *x and z* and *y and x*, Eq. (1.7.11) gives the following stress vector components:

$$S_1 = \sqrt{\frac{3}{2}}\overline{\sigma}_z, \quad S_2 = \frac{\overline{\sigma}_z}{\sqrt{2}} + \sqrt{2}\overline{\sigma}_x, \tag{2.13.1}$$

where $\overline{\sigma}_x$ and $\overline{\sigma}_z$ are the stress deviator vector components, so that

$$S_1 = \sqrt{\frac{2}{3}}\left(\sigma_z - \frac{\sigma_x}{2}\right), \quad S_2 = \frac{\sigma_x}{\sqrt{2}}, \quad S_3 = S_4 = S_5 = 0. \tag{2.13.2}$$

Plastic strain increment components are established in a similar way:

$$\begin{gathered} de_1^S = \sqrt{\frac{2}{3}}\left[d\varepsilon_z^S - \frac{1}{2}\left(d\varepsilon_x^S + d\varepsilon_y^S\right)\right], \\ de_2^S = \frac{1}{\sqrt{2}}\left(d\varepsilon_x^S - d\varepsilon_y^S\right), \\ de_3^S = de_4^S = de_5^S = 0. \end{gathered} \tag{2.13.3}$$

As already shown in the previous section, the plastic strain components caused by first plastic slips under two-axial tension are $d\varepsilon_x^S = 0$ and $d\varepsilon_y^S = -d\varepsilon_z^S$, therefore

$$de_1^S = \sqrt{\frac{3}{2}}d\varepsilon_z^S, \quad de_2^S = \frac{d\varepsilon_z^S}{\sqrt{2}}. \tag{2.13.4}$$

The plastic-strain-increment vector is

$$d\vec{\mathbf{e}} = \frac{1}{\sqrt{2}}\left(\sqrt{3}\vec{\mathbf{g}}_1 + \vec{\mathbf{g}}_2\right)d\varepsilon_z^S . \tag{2.13.5}$$

Let us turn to the Tresca hexagon (Fig. 2.12a) binding the region of elastic deformations at two-axial stress state. The Tresca yield criterion constructed according to Eq. (2.13.2) in variables S_1 and S_2 is shown in Fig. 2.12b, points M_1', M_2',... in Fig. 2.12b correspond to points M_1, M_2,... in Fig. 2.12a. At $\sigma_z > \sigma_x > 0$ a stress vector $\vec{\mathbf{S}}$ reaches the Tresca hexagon at some point M, which lies on segment M_1M_2, or at point M' on $M_1'M_2'$ in S_1-S_2 plane. The equation of line $M_1'M_2'$ is

$$\frac{1}{\sqrt{2}}\left(\sqrt{3}S_1 + S_2\right) = \sigma_S . \tag{2.13.6}$$

From the analysis of Eqs. (2.13.5) and (2.13.6) it is clear that the vector $d\vec{\mathbf{e}}$ is perpendicular to the straight $M_1'M_2'$.

Since the Budiansky theory is based on the Tresca yield criterion, it agrees with the fundamental consequence from Drucker's postulate, a plastic-strain-increment vector is normal to the yield surface. Similar to the Budiansky slip theory, in terms of the Leonov theory it is assumed that an irreversible strain is caused by plastic slips. The first of them occurs in the plane and in the direction of the action of maximal shear stress. Therefore, Eq. (2.13.5) remains correct for the Leonov model as well. At the same time, the condition for the onset of plastic deformation is not the Tresca criterion but a more general criterion, Eq. (2.8.1). The part of the yield surface (line) corresponding to Eq. (2.8.1) is shown by the dashed curve in Fig. 2.12b. As this surface is displaced (rotated) relative to Tresca surface, the perpendicularity of the vector (2.13.5) to the line of plasticity (2.8.1) is violated.

Furthermore, a corner point arises on a loading surface with the growth of stress vector. According to the consequence from Drucker's postulate, a plastic-strain-increment vector must lie inside the shaded cone in Fig. 1.10. If this vector is not orthogonal to the yield surface, then at enough small increase in loading the vector $d\vec{\mathbf{e}}^S$ will not get inside the specified cone. It means that the consequence from Drucker's postulate is not fulfilled not only on the yield surface but also on some finite plastic straining as well.

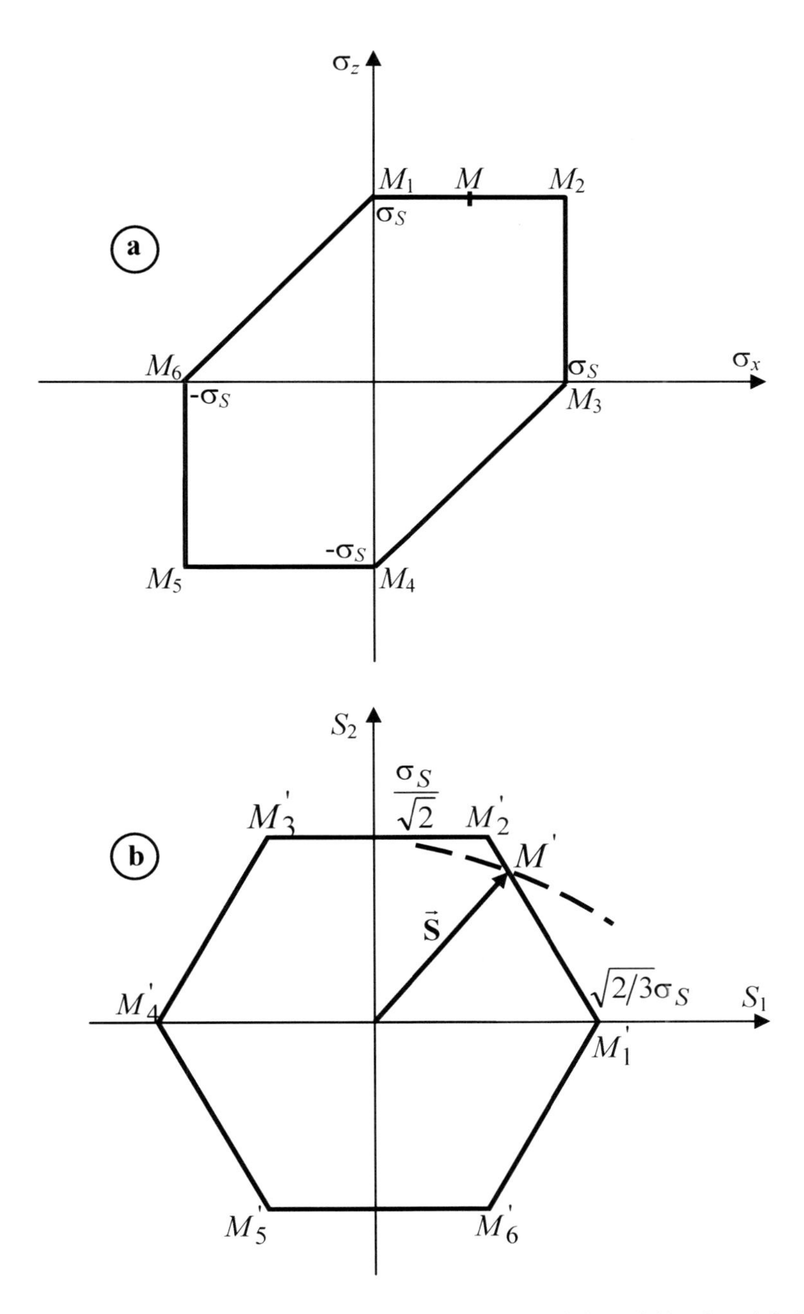

Fig. 2.12 Tresca's hexagon on σ_x-σ_z plane (a) and in Ilyushin's variables S_1 and S_2 (b).

2.14 Analysis of the Leonov Theory

1. This theory rehabilitates the concept of slip to a considerable extent. After the publication of works [123,127], it seemed that the concept of slip is inapplicable to the description of plastic properties of solids because it is incapable of describing experimental diagrams, e.g. in uniaxial tension and pure shear, by the unified constitutive equations. The introducing of linear relation between slip intensity and shear stress removes this shortcoming.

2. The Leonov theory results only in approximated fulfillment of the deviator proportionality, i.e., some deviation from the equality $\mu_\varepsilon = \mu_\sigma$ and from the existence of universal diagram $\tau_0 \sim \gamma_0$ is observed.

Let us consider the agreement of the above results with experiments. As shown in the point 1 of Section 1.11, the deviator proportionality is fulfilled with certain accuracy at simple loading for many materials. However, more accurate experiments give an insignificant but systematical deviations from the equality $\mu_\varepsilon = \mu_\sigma$. Experimental points fall not on the bisector of the first and third quadrants on $\mu_\varepsilon - \mu_\sigma$ plane but on the curve shown in Fig 2.13. Experiments performed by Lode [84] and Davis [24,25] (as well as many other experimental investigations) on thin-walled tubes of many materials (iron, copper, nickel, and steel) under combined tension and internal pressure have amply justified Eq. (2.12.7) obtained in terms of the Leonov theory prescribing an inequality between the Lode parameters μ_ε and μ_σ.

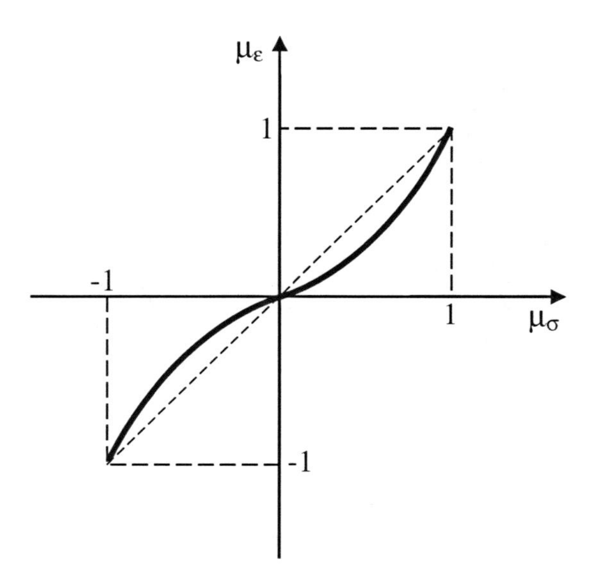

Fig. 2.13 Experimental relation between Lode-Nadai parameters.

Further, experimental data (we note the experiments carried out by N.M. Mitrochin and Yu. Jagn [95]) show considerable regular deviations from which the $\tau_0 \sim \gamma_0$ relation is independent on the state of stress for many materials. The experimental curve $\tau_0 \sim \gamma_0$ constructed in a pure shear locates below experimental curves $\tau_0 \sim \gamma_0$ obtained at tension or pressure. These curves calculated in terms of the Leonov theory behave in a similar fashion (Fig. 2.11).

3. The theory predicts an arising of conic singularity on a loading surface during plastic straining. The fact that function $f_1(\tau_0)$ appears in the expression for shear strength results in the increase in the top-angle of cone at loading point that must be treated as positives of the model. The rounding of the corner top caused by the function $f_1(\tau_0)$ will be shown in Section 2.20.

4. The Leonov theory leads to contradictions with fundamental consequences from Drucker's postulate, namely, this theory gives that a plastic-strain-increment vector, $d\vec{\mathbf{e}}^S$, is not orthogonal to a yield surface; the vector $d\vec{\mathbf{e}}^S$ does not always lie within the cone whose generator is normal to the loading surface.

The disagreement of the theory with the consequences of Drucker's postulate results from the fact that the function $f_1(\tau_0)$ appears in the expression for shear strength, in particular, in the yield criterion. The law of irreversible straining and the yield criterion are not associated. The mechanism of deformation is the same as in the Budiansky theory but the yield criterion is another. Let us note that specified contradiction was stated by M. Leonov.

5. In terms of the considered theory, the evaluation of integrals is much simpler than in the Budiansky model as the intensity φ_{nl} relates linearly to τ_{nl} and γ_{nl}^S. On the other hand, the integration does not give the formulae for strain components themselves but in the system of equations for the strains due to the found components $\varepsilon_x^S, \varepsilon_y^S \ldots$ appear in the right-hand side of Eq. (2.11.11) as well.

2.15 Plastic Strain under Combined Loading

To provide simpler stress-strain relations, let us restrict ourselves to particular form of shear strength, $S_{n_0 l_0}$. We assume that coefficients r_1 and r_3 in Eq. (2.10.1) are equal to zero:

$$S_{n_0 l_0} = f_1(\tau_0)\left[1 + r_2 \int\limits_{\omega_1}^{\omega_2} \varphi_{n_0 l} \cos(\omega_0 - \omega) d\omega\right], \tag{2.15.1}$$

meaning that a slip in one slip plane affects $S_{n_0 l_0}$ in all directions within the plane but does not influence upon $S_{n_0 l_0}$ in all other planes.

Let us show that the shear strength $S_{n_0 l_0}$ given by Eq. (2.15.1) results in a slip that under any loading occurs only in one direction in each slip plane. To show this we split a loading path into q small elements (steps). Let us designate the increment of $\varphi_{n_0 l}$ on any j-step $(1 \le j \le q)$ through $\Delta\varphi_j$ and the limits of slip on the slip planes through ω_{1j} and ω_{2j}. Further, we write the formula for shear strength so that the term $\Delta\varphi_q$ appears separately from all other analogical terms corresponding to previous $1-q$ steps, i.e., $S_{n_0 l_0}$ on q-element has the form

$$S_{n_0 l_0} = f_1(\tau_0)\left[1 + r_2 \int\limits_{\omega_{1q}}^{\omega_{2q}} \Delta\varphi_q \cos(\omega_0 - \omega)d\omega + r_2 \sum_{j=1}^{q-1} \int\limits_{\omega_{1j}}^{\omega_{2j}} \Delta\varphi_j \cos(\omega_0 - \omega)d\omega\right], \qquad (2.15.2)$$

where $\Delta\varphi_q$ is the slip intensity increment on the final q-step of loading, $\Delta\varphi_j$ ($j = 1 \ldots q-1$)is the slip intensity increment on the current j-step; for simplicity, indexes n and l have been missed.

Further, we write Eq. (2.3.2) for shear stress in the following form

$$\tau_{nl} = P_1 \cos\omega + P_2 \sin\omega, \qquad (2.15.3)$$

where P_1 and P_2 are expressed through stress components as

$$P_1 = \left[\frac{1}{2}(\sigma_y - \sigma_x)\sin 2\alpha + \tau_{xy}\cos 2\alpha\right]\cos\beta + (\tau_{yz}\cos\alpha - \tau_{xz}\sin\alpha)\sin\beta,$$
$$P_2 = \frac{1}{2}(\sigma_z - \sigma_x \cos^2\alpha - \sigma_y \sin^2\alpha - \tau_{xy}\sin 2\alpha)\sin 2\beta + (\tau_{yz}\sin 2\alpha + \tau_{xz}\cos\alpha)\cos 2\beta. \qquad (2.15.4)$$

These formulae are obtained by inserting the direction cosines from Eq. (2.3.8) into Eqs. (2.3.2). Equating the shear strength (2.15.2) to the shear stress (2.15.3), we obtain

$$r_2 \int\limits_{\omega_{1q}}^{\omega_{2q}} \Delta\varphi_q \cos(\omega_0 - \omega)d\omega = P_3 \cos\omega_0 + P_4 \sin\omega_0 - 1, \quad \omega_{1q} \le \omega_0 \le \omega_{2q}, \qquad (2.15.5)$$

where

$$P_3 = \frac{P_1}{f_1(\tau_0)} - r_2 \sum_{j=1}^{q-1} \int\limits_{\omega_{1j}}^{\omega_{2j}} \Delta\varphi_j \cos\omega d\omega, \quad P_4 = \frac{P_2}{f_1(\tau_0)} - r_2 \sum_{j=1}^{q-1} \int\limits_{\omega_{1j}}^{\omega_{2j}} \Delta\varphi_j \sin\omega d\omega. \qquad (2.15.6)$$

In order to determine the function $\Delta\varphi_q$ from the obtained integral equation (2.15.5), we rewrite Eq. (2.15.5) as

$$P_5 \cos\omega_0 + P_6 \sin\omega_0 = P_3 \cos\omega_0 + P_4 \sin\omega_0 - 1, \tag{2.15.7}$$

where

$$P_5 = r_2 \int_{\omega_{1q}}^{\omega_{2q}} \Delta\varphi_q \cos\omega d\omega, \quad P_6 = r_2 \int_{\omega_{1q}}^{\omega_{2q}} \Delta\varphi_q \sin\omega d\omega. \tag{2.15.8}$$

Since parameters $P_1, \ldots, P_6$ do not depend on angle ω_0, Eq. (2.15.7) holds only at the single value of angle ω_0. Consequently, the inequality $\omega_{1q} \le \omega_0 \le \omega_{2q}$ from Eq. (2.15.5) degenerates into the following equality:

$$\omega_0 = \omega_{1q} = \omega_{2q}. \tag{2.15.9}$$

The Equation (2.15.5) has no solution in the class of continuous functions. This follows from that the right-hand side in Eq. (2.15.5) is a finite non-zero value. The condition that the left-hand side in Eq. (2.15.5) is non-zero at $\omega_{1q} = \omega_{2q}$ requires that the solution be in the form

$$\Delta\varphi_q = \Delta\varphi_{0q}\delta(\omega - \omega_q), \tag{2.15.10}$$

where δ is the Dirac δ-function and $\Delta\varphi_{0q}$ is a new unknown function of angles α and β, ω_q is a new unknown angle. Thus, if the shear strength $S_{n_0 l_0}$ is determined by Eq. (2.15.1), the plastic slip in slip planes occurs only in one direction although with infinite large intensity. Substituting $\Delta\varphi_q$ from Eq. (2.15.10) into Eq. (2.15.5), on the base of Eq. (2.15.9) and the property of δ-function, we obtain the increment $\Delta\varphi_{0q}$ in the following form

$$r_2\Delta\varphi_{0q} = P_3 \cos\omega_q + P_4 \sin\omega_q - 1. \tag{2.15.11}$$

Inserting solutions similar to (2.15.10),

$$\Delta\varphi_j = \Delta\varphi_{0j}\delta(\omega - \omega_j), \tag{2.15.12}$$

into Eq. (2.15.6) we find

$$P_3 = \frac{P_1}{f_1(\tau_0)} - r_2 \sum_{j=1}^{q-1} \Delta\varphi_{0j} \cos\omega_j, \quad P_4 = \frac{P_2}{f_1(\tau_0)} - r_2 \sum_{j=1}^{q-1} \Delta\varphi_{0j} \sin\omega_j. \tag{2.15.13}$$

Equations (2.15.11) and (2.15.13) determine the slip intensity on any step of loading. Slip direction ω_q is unknown in these equations. To find the angle ω_q we note that the shear strength (2.15.1) and the shear stress (2.15.3) are smooth functions of variable ω_0. Therefore, for the slip direction $\omega_0 = \omega_q$, together with equality $S_{n_0 l_0} = \tau_{n_0 l_0}$, the formula

$$\frac{\partial S_{n_0 l_0}}{\partial \omega_0} = \frac{\partial \tau_{n_0 l_0}}{\partial \omega_0} \tag{2.15.14}$$

holds. To satisfy the above equation, it suffices to differentiate the left- and right-hand side in Eq. (2.15.5) with respect to ω_0 at $\omega_0 = \omega = \omega_q$, which results in the following

$$P_3 \sin \omega_q = P_4 \cos \omega_q . \tag{2.15.15}$$

Equations (2.15.10), (2.15.11), and (2.15.15) give the slip intensity on any step of combined loading.

Let us proceed to the calculations of strains determined by Eq. (2.9.3). Now, the integrals from Eq. (2.9.3) become two-folded as Eq. (2.15.12) contains δ-function:

$$\left.\begin{aligned} \Delta\varepsilon_{iq}^{S} &= \iint_{\Omega_q} \tilde{l}_i n_i \Delta\varphi_{0q} d\Omega \qquad (i = x, y, z) \\ \Delta\gamma_{ijq}^{S} &= \iint_{\Omega_q} \left(\tilde{l}_i n_j + \tilde{l}_j n_i\right) \Delta\varphi_{0q} d\Omega \qquad (i \neq j) \end{aligned}\right\} \Sigma = 0 \tag{2.15.16}$$

where $\tilde{l}_x, \tilde{l}_y$, and $\tilde{l}_z$ are direction cosines determined by Eq. (2.3.8) at $\omega = \omega_q$.

The relationships similar to (2.15.16) were proposed by A.K. Malmeister and G.A. Teters [89,177], slip direction in a given plane was hypothetically assumed to be co-directed with total shear stress. In contrast to this, the Leonov theory prescribes that a slip occurs in the direction of $\omega = \omega_q$, not necessarily in the direction of total shear stress.

2.16 Two Problems

The first problem is an uniaxial strain, $\sigma_z > 0$. The equality between the shear strength (2.15.1) and the shear stress (2.15.3) results in the following integral equation:

$$r_2 \int_{\omega_1}^{\omega_2} \varphi_{n_0 l} \cos(\omega_0 - \omega) d\omega = P_7 \sin \omega_0 - 1, \tag{2.16.1}$$

$$P_7 = \frac{\sigma_z \sin 2\beta_0}{2 f_1(\tau_0)}, \quad \tau_0 = \sigma_z, \tag{2.16.2}$$

which is similar to Eq. (2.15.5). According to the results of the previous Section, Eq. (2.16.1) holds only for one direction, $\omega_1 = \omega_2 = \omega_0 = \omega_q$, and its solution is

$$\varphi_{nl} = \varphi_0 \delta(\omega - \omega_q). \tag{2.16.3}$$

The φ_0 and ω_q are new unknown quantities. Substituting φ_{nl} from Eq. (2.16.3) into Eq. (2.16.1), we obtain the formula for φ_0:

$$r_2 \varphi_0 = P_7 \sin \omega_q - 1 \tag{2.16.4}$$

which is analogous to Eq. (2.15.11). Differentiating Eq. (2.16.1) with respect to ω_0 and letting $\omega_0 = \omega = \omega_q$, we obtain $\omega_q = \pi/2$ and Eqs. (2.16.4) and (2.16.3) take the form

$$r_2 \varphi_0 = P_7 - 1, \quad r_2 \varphi_{nl} = (P_7 - 1)\delta(\omega - \pi/2), \tag{2.16.5}$$

respectively. In this case Eq. (2.15.16), which determines relative elongations, takes the form

$$\varepsilon_z^S = \frac{2\pi}{r_2} \int_{\beta_1}^{\beta_2} \left(\frac{\sin 2\beta}{a} - 1 \right) \cos^2 \beta \sin \beta d\beta, \quad a = \frac{2 f_1(\tau_0)}{\sigma_z}. \tag{2.16.6}$$

The limits β_1 and β_2, which are determined by Eq. (2.4.5) where τ_S is replaced by $f_1(\tau_0)$, are

$$\sin 2\beta_{1,2} = a. \tag{2.16.7}$$

As a result of integration we find

$$\varepsilon_z^S = \frac{2\pi}{15 r_2} (1 - a)^{3/2} \left(\frac{4}{a} + 1 \right). \tag{2.16.8}$$

Now we proceed to the second problem and wish to obtain the analogue of the Cicala formula, Eq. (2.5.11), for the case of shear strength taken in the form of Eq. (2.15.1). The formulation of the problem is the same as in Section 2.5: an element in a body is first stretched under uniaxial tension and then, at constant tensile stress σ_z, is subjected to the action of small shear stress, $\Delta\tau_{xz}$.

At the moment before the orthogonal break of loading trajectory, i.e., under tension, slips occur in the planes determined by Eq. (2.5.2) with boundary angles $\beta_{1,2}$ determined by Eq. (2.16.7). In any of these planes, the slips occur only in one direction, $\omega_1 = \omega_2 = \omega = \pi/2$. Under the action of additional stress $\Delta\tau_{nl}$, shear stresses relative to *n*- and *l*- axes obtain increments (2.5.3) and (2.5.4) resulting in additional slips. They develop in that part of hemisphere (2.5.2) where $\Delta\tau_{nl} > 0$ at $\omega = \pi/2$:

$$\Delta\tau_{nl} = \cos\alpha\cos 2\beta\cdot\Delta\tau_{xz} > 0. \tag{2.16.9}$$

It is easy to see that the range of additional slips is the same as shown in Fig. 2.14 (shaded area). The shear strain increment caused by the additional slips is determined by Eq. (2.15.16):

$$\Delta\gamma_{xz}^S = \iint\limits_{\Omega}\cos\alpha\cos\beta\cos 2\beta\Delta\varphi_0 d\alpha d\beta. \tag{2.16.10}$$

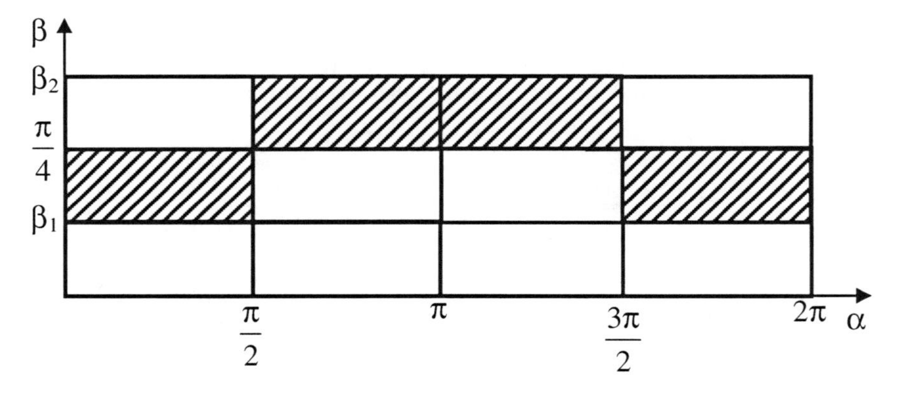

Fig. 2.14 The regions of additional slips on hemisphere induced by additional shear stress

In order to find the function $\Delta\varphi_0$ we use Eq. (2.15.11) splitting the loading path (trajectory) into two steps (the first corresponds to the tension and the second to torsion). On the second step, Eq. (2.15.11) gives

$$r_2\Delta\varphi_0 = P_4 - 1 \text{ at } \omega = \pi/2, \tag{2.16.11}$$

where parameter P_4 can be determined by Eq. (2.15.13) at $q = 2$:

$$P_4 = \frac{P_2}{f_1(\tau_0)} - r_2\varphi_0 \tag{2.16.12}$$

and according to equality (2.15.4), we have

$$P_2 = \frac{\sigma_z}{2}\sin 2\beta + \Delta\tau_{xz}\cos\alpha\cos 2\beta\,. \tag{2.16.13}$$

From Eqs. (2.16.2), (2.16.5), and (2.16.11) - (2.16.13) we find the additional slip intensity:

$$r_2\Delta\varphi_0 = \frac{\Delta\tau_{xz}}{f_1(\tau_0)}\cos\alpha\cos 2\beta\,. \tag{2.16.14}$$

The above equation and Eq. (2.16.10) lead to the following formula for the shear strain increment caused by the additional slips:

$$\Delta\gamma_{xz}^S = \frac{\Delta\tau_{xz}}{r_2 f_1(\tau_0)}\iint_\Omega \cos^2\alpha\cos^2 2\beta\cos\beta\, d\alpha\, d\beta \tag{2.16.15}$$

and, by integrating, we obtain

$$\Delta\gamma_{xz}^S = \frac{2\sqrt{2}\pi}{15 r_2 f_1(\tau_0)}\sin^3\Theta_1\left(5 - 3\sin^2\Theta_1\right)\Delta\tau_{xz}\,, \tag{2.16.16}$$

$$\Theta_1 = \frac{\beta_2 - \beta_1}{2}, \quad \cos 2\Theta_1 = a\,. \tag{2.16.17}$$

Relative elongation ε_z^S (2.16.8) expressed through angle Θ_1 takes the form:

$$\varepsilon_z^S = \frac{2\sqrt{2}\pi}{15 r_2 f_1(\tau_0)}\sin^3\Theta_1\left(5 - 2\sin^2\Theta_1\right)\sigma_z\,. \tag{2.16.18}$$

Dividing $\Delta\gamma_{xz}^S$ on ε_z^S, we obtain

$$\frac{\Delta\gamma_{xz}^S}{\varepsilon_z^S} = \kappa\frac{\Delta\tau_{xz}}{\sigma_z}\,, \tag{2.16.19}$$

$$\kappa = \frac{5 - 3\sin^2\Theta_1}{5 - 2\sin^2\Theta_1}\,. \tag{2.16.20}$$

Doing the same with Eq. (2.16.19) as at the deriving of the Cicala formula (Section 2.5), we have

$$G_S = \frac{G}{1 + \kappa G\left(\dfrac{1}{E_c} - \dfrac{1}{E}\right)}. \tag{2.16.21}$$

This formula determines the shear modulus at the orthogonal break of loading path at the representation of shear strength by Eq. (2.15.1). Parameter κ depends on the tensile stress σ_z insignificantly, $7/8 \le \kappa \le 1$, and its lower limit corresponds to the case if slips occur at every point on the hemisphere ($\beta_1 = 0$, $\beta_2 = \pi/2$, $\Theta_1 = \pi/4$), the upper limit corresponds to the value of σ_z equal to the yield limit ($\beta_1 = \beta_2 = \pi/4$, $\Theta_1 = 0$).

Comparing Eq. (2.16.21) with the Cicala formula (2.5.11), one can see that the value of G_S from Eq. (2.16.21) is larger than that from (2.5.11), i.e., Eq. (2.16.21) shows a better agreement with experiments than the Cicala formula. The latter can be treated as arguments in favor of the representation of shear strength in the form of Eq. (2.15.1).

2.17 Rearranging in Differential Equations

The procedures of Section 2.15 can be represented in the form of differential equations whose deriving is the goal of this section. Let us start with Eqs. (2.15.11) and (2.15.15) into which we insert the expressions for P_3 and P_4 from Eq. (2.15.13). As a result of elementary transformations we obtain

$$\begin{aligned} r_2 \Delta\varphi_{0q} \cos\omega_q &= \frac{P_1}{f_1(\tau_0)} - r_2 \sum_{j=1}^{q-1} \Delta\varphi_{0j} \cos\omega_j - \cos\omega_q, \\ r_2 \Delta\varphi_{0q} \sin\omega_q &= \frac{P_2}{f_1(\tau_0)} - r_2 \sum_{j=1}^{q-1} \Delta\varphi_{0j} \sin\omega_j - \sin\omega_q \end{aligned} \tag{2.17.1}$$

and

$$\begin{aligned} r_2 \sum_{j=1}^{q} \Delta\varphi_{0j} \cos\omega_j &= \frac{P_1}{f_1(\tau_0)} - \cos\omega_q, \\ r_2 \sum_{j=1}^{q} \Delta\varphi_{0j} \sin\omega_j &= \frac{P_2}{f_1(\tau_0)} - \sin\omega_q. \end{aligned} \tag{2.17.2}$$

Further, assuming that the number of segments, q, tends to infinity and the magnitude of step tends to zero, from Eq. (2.17.2) we obtain

$$r_2 \int_0^t \dot{\varphi}_0 \cos\omega dt = \frac{P_1}{f_1(\tau_0)} - \cos\omega,$$
$$r_2 \int_0^t \dot{\varphi}_0 \sin\omega dt = \frac{P_2}{f_1(\tau_0)} - \sin\omega, \tag{2.17.3}$$

where t is time or some time-dependent, monotonously growing parameter. Differentiating the above equation with respect to t, we find

$$r_2 \dot{\varphi}_0 \cos\omega = \frac{d}{dt}\left[\frac{P_1}{f_1(\tau_0)}\right] + \dot{\omega}\sin\omega,$$
$$r_2 \dot{\varphi}_0 \sin\omega = \frac{d}{dt}\left[\frac{P_2}{f_1(\tau_0)}\right] - \dot{\omega}\cos\omega. \tag{2.17.4}$$

This system of equations determines the slip intensity rate and its direction which is set up by angle ω in any slip plane and at any instant of loading. The parameters P_1 and P_2 are determined by Eqs. (2.15.4) and they depend on the stresses acting on an element in a body. It is easy to eliminate φ_0 from the system (2.17.4) resulting in the following equation for ω:

$$\dot{\omega} = -\frac{d}{dt}\left[\frac{P_1}{f_1(\tau_0)}\right]\sin\omega + \frac{d}{dt}\left[\frac{P_2}{f_1(\tau_0)}\right]\cos\omega \tag{2.17.5}$$

Once the quantity $\dot{\omega}$ is known, by solving the above equation, the function φ_0 can be determined either from the one of equations (2.17.4) or from more simple relation following from these equations:

$$r_2 \dot{\varphi}_0 = \frac{d}{dt}\left[\frac{P_1}{f_1(\tau_0)}\right]\cos\omega + \frac{d}{dt}\left[\frac{P_2}{f_1(\tau_0)}\right]\sin\omega. \tag{2.17.6}$$

Let us notice that the obtained equations determine φ_0 and ω on a given slip plane, i.e., coordinates of the plane, α and β, are constants. From Eq. (2.17.5) it follows that at proportional loading a slip occurs in the direction of the action of total shear stress. Indeed, at proportional loading the following formula is valid:

$$\frac{dP_2}{dP_1} = \frac{P_2}{P_1} = \tan\Psi, \tag{2.17.7}$$

where Ψ is an angle determining the direction of the action of total shear stress. The angle Ψ takes different values in different points of hemisphere, but in the fixed point it does not change with increase of loading. With Eq. (2.17.7), the variables in Eq. (2.17.5) are readily separable:

$$\frac{d\omega}{\sin(\omega - \Phi)} = -\frac{1}{\cos\Psi} d\left[\frac{P_1}{f_1(\tau_0)}\right]. \tag{2.17.8}$$

The integral of this equation is

$$\frac{1-\cos(\omega-\Psi)}{1+\cos(\omega-\Psi)} = C\exp\left[-\frac{2P_1}{f_1(\tau_0)\cos\Psi}\right]. \tag{2.17.9}$$

However, the initial condition $\omega = \Psi$ implies that $C = 0$, therefore $\omega \equiv \Psi$ ensures that the slip is co-directed with the total shear stress. If $\omega = \Psi = \text{const}$, Eq. (2.17.6) becomes as

$$r_2\varphi_0 = \frac{P_1}{f_1(\tau_0)}\cos\Psi + \frac{P_2}{f_1(\tau_0)}\sin\Psi - 1. \tag{2.17.10}$$

As a result, together with Eq. (2.17.7), it follows that

$$r_2\varphi_0 = \frac{1}{f_1(\tau_0)}\sqrt{P_1^2 + P_2^2} - 1 \tag{2.17.11}$$

during proportional loading.

At the same time, the slip direction under combined loading is not in the direction of total shear stress. Let us show this on the example of two-segment loading path (Fig. 2.15) assuming, for simplicity, that $f = \tau_S = \text{const}$. Let at $P_1 = P_1^0$ the break of loading path take place and, furthermore, on the second segment we have

$$\frac{dP_2}{dP_1} = \tan\Psi_2, \tag{2.17.12}$$

where Ψ_2 is a constant value for the fixed point of hemisphere. Equation (2.17.5) can be written as

$$\tau_S \frac{d\omega}{dP_1} = -\sin\omega + \tan\Psi_2 \cos\omega. \tag{2.17.13}$$

The integral of this equation, for boundary condition $\omega = \Psi_1$ at $P_1 = P_1^0$, is

$$\tan\frac{\omega - \Psi_2}{2} = \tan\frac{\Psi_1 - \Psi_2}{2}\exp\left(-\frac{P_1 - P_1^0}{\tau_S \cos\Psi_2}\right), \tag{2.17.14}$$

where angle ω determines the slip direction on the second segment of loading path while the angle $\omega_1 = \Psi_1$ gives that on the first segment. The direction of total shear stress on the second segment, Ψ_3, can be determined by integrating Eq. (2.17.12). In view of Fig. 2.15, we obtain

$$P_2 = P_2^0 + \left(P_1 - P_1^0\right)\tan\Psi_2, \quad P_1 \geq P_1^0. \tag{2.17.15}$$

Therefore, the direction of complete tangential stress, Ψ_3, is determined as

$$\tan\Psi_3 = \frac{P_2}{P_1} = \frac{P_2^0 - P_1^0 \tan\Psi_2}{P_1} + \tan\Psi_2. \tag{2.17.16}$$

It is seen from Eqs. (2.17.14) and (2.17.16) that $\omega \neq \Psi_2$ and $\omega \neq \Psi_3$, the direction of slip differs from that of total shear stress. Only for a load being at a sufficient "distance" from the break-point of loading path ($P_1 \to \infty$), we have $\omega = \Psi_2 = \Psi_3$.

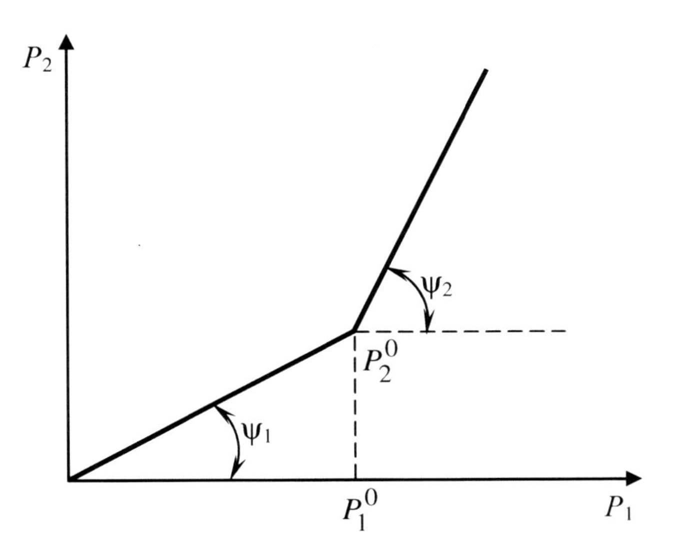

Fig. 2.15 Two-segment loading path at an arbitrary state of stress

2.18 Computer-Aided Calculations

Our goal now is to work out the algorithm for computer-aided calculations of the strains considered in Section 2.15. Similar to this Section, we split a loading path into q small segments and, according to Eq. (2.15.15), a direction of slip on the q-segment is determined as

$$\sin\omega_q = \frac{P_4}{\sqrt{P_3^2 + P_4^2}}, \quad \cos\omega_q = \frac{P_3}{\sqrt{P_3^2 + P_4^2}}. \tag{2.18.1}$$

By inserting these values of angles into Eq. (2.15.11), we obtain the increment in $\Delta\varphi_{0q}$:

$$r_2\Delta\varphi_{0q} = \sqrt{P_3^2 + P_4^2} - 1. \tag{2.18.2}$$

Equations (2.15.4), (2.15.13), (2.18.1), and (2.18.2) are recurrent formulas determining the slip direction and the increment in its intensity on any step of loading. Indeed, according to Eq. (2.15.4), P_1 and P_2 are known on any step at a given evolution of stress state. Firstly, we assume that $q = 1$. Then, on the base of Eq. (2.15.13), P_3 and P_4 are known as the sums in Eq. (2.15.13) become equal zero at $q = 1$ and from Eq. (2.18.1) and (2.18.2) $\sin\omega_1$, $\cos\omega_1$, and $\Delta\varphi_{01}$ are calculated. At $q = 2$, the parameters P_3 and P_4 are known as well because the sums in Eq. (2.15.13) contain terms $\Delta\varphi_{01}\sin\omega_1$ and $\Delta\varphi_{01}\cos\omega_1$ which is already found on the previous stage, therefore from Eqs. (2.18.1) and (2.18.2) we calculate $\sin\omega_2$, $\cos\omega_2$, and $\Delta\varphi_{02}\ldots$.

The calculation of strains in Eq. (2.15.16) can be easily carried out by the following algorithm. For this purpose, let us break the Budiansky hemisphere on a number of small "squares" and consider one of them, marked by k. Assigning successively $q = 1,2\ldots$, Eqs. (2.18.1) and (2.18.2) give all values of $\sin\omega_q$, $\cos\omega_q$, and $\Delta\varphi_{0q}$ at considered "point" k. If $\Delta\varphi_{0q} \le 0$ at a given point, a slip does not occur on q-step, i.e., $\Delta\varphi_{0q} = 0$. On the base of the found ω_q ($q = 1,2,\ldots$) we can determine direction cosines, $\tilde{l}_{xq}, \tilde{l}_{yq}, \tilde{l}_{zq}$, from Eq. (2.3.8). Making use of Eq. (2.15.16), plastic strains caused by slip on "square" k are

$$\Delta\varepsilon^S_{xqk} = \left(\tilde{l}_{xq} n_x \Delta\varphi_{0q}\right)_k \Delta\Omega_k, \quad \Delta\gamma^S_{xyqk} = \left[\left(\tilde{l}_{xq} n_y + \tilde{l}_{yq} n_x\right)\Delta\varphi_{0q}\right]_k \Delta\Omega_k, \quad \Sigma \tag{2.18.3}$$

where Ω_k is the area of k-square. Carrying out the same calculations for all squares and summating them, we obtain

$$\Delta\varepsilon_{xq}^{S} = \sum_k \Delta\varepsilon_{xqk}^{S}, \quad \Delta\gamma_{xyq}^{S} = \sum_k \Delta\gamma_{xyqk}^{S}. \tag{2.18.4}$$

Finally, the relations for plastic strains for an arbitrary loading path take the form

$$\varepsilon_{x}^{S} = \sum_q \Delta\varepsilon_{xq}^{S}, \qquad \gamma_{xy}^{S} = \sum_q \Delta\gamma_{xyq}^{S}, \tag{2.18.5}$$

2.19 The Concept of Slip and the Isotropy Postulate

The Ilyushin isotropy postulate was formulated for the Guber continuous medium. Let us determine the prediction of the theory of slip with respect to this postulate for materials whose yield criterion locates between the Guber and Tresca criterion. Using the step-method considered in previous Section, the plastic strain components have been calculated in the works [142,146,147] at a three-segment loading trajectory.

Consider the following loading paths in the two-dimensional subspace of the Ilyushin five-dimensional space along two three-segment trajectories, broken lines I (A-1'-2'-3'-4'-5'-6') and II (A-1"-2"-3"-4"-5"-6") (Fig. 2.16), which are the mirror images of each other in respect to the bisector III of the quadrantal angle (at point A yielding begins). Along the segments of these trajectories only one stress component varies, σ_z or τ_{xz}; stress- and strain-vector components are related to the stress- and strain-tensor components by Eq. (1.7.3) and (1.7.11):

$$S_1 = \sqrt{\frac{2}{3}}\sigma_z, \quad S_3 = \sqrt{2}\tau_{xz}, \quad S_2 = S_4 = S_5 = 0, \tag{2.19.1}$$

$$e_1^S = \sqrt{\frac{3}{2}}\varepsilon_z^S, \quad e_3^S = \frac{\gamma_{xz}^S}{\sqrt{2}}, \quad e_2^S = e_4^S = e_5^S = 0. \tag{2.19.2}$$

Now we wish to calculate plastic strain vector components, e_1^S and e_3^S, corresponding to I and II loading paths for the following material constants σ_S, τ_S, and r. For the yield criterion close to the Guber criterion $\sigma_S = 240\,\text{MPa}$, $\tau_S = 132\,\text{MPa}$, $r_2 = 224$, $f_3 = 373\,\text{MPa}$, $f_4 = 2.2$ (we refer these constants as to constants of a-group). For the intermediate case between the Guber and Tresca criteria $\sigma_S = 240\,\text{MPa}$, $\tau_S = 126\,\text{MPa}$, $r_2 = 47.6$,

$f_3 = 186\,\text{MPa}$, $f_4 = 0.58$ (the constants of *b*-group). For the Tresca continuum $\sigma_S = 240\,\text{MPa}$, $\tau_S = 120\,\text{MPa}$, $r_2 = 18.5$; $f_3 = 120\,\text{MPa}$, $f_4 = 0$ (the constants of *c*-group).

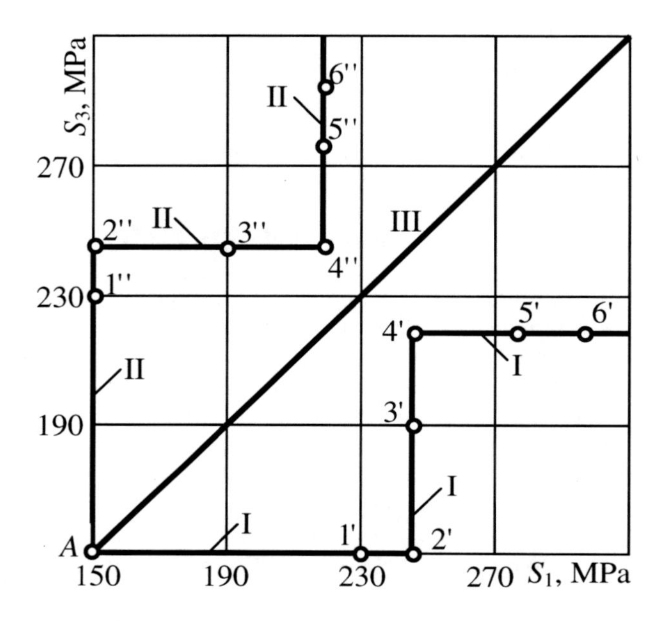

Fig. 2.16 Two three-segment loading trajectories

The values of e_1^S and e_3^S calculated by the procedure of Sections 2.15 and 2.18 for I and II loading path (Fig. 2.16) are shown in Fig. 2.17*a,b,c* in solid lines (0-1'-2'-3'-4'-5'-6') and (0-1"-2"-3"-4"-5"-6") for *a*-, *b*-, and *c*-group of constants, respectively (0-point corresponds to *A* point in Fig. 2.16). It must be noted that points in Fig. 2.16 have been selected arbitrarily, while the corresponding points in Fig. 2.17 are obtained by the calculations of strains.

The mirror image of I loading path in respect to the bisector of the quadrantal angle results in the following relations

$$S_1'' = S_3', \quad S_3'' = S_1' \tag{2.19.3}$$

If Eq. (2.19.3) holds, the isotropy postulate requires that the strain vector components be related in the same way:

$$e_1^{S\,''} = e_3^{S\,'}, \quad e_3^{S\,''} = e_1^{S\,'}. \tag{2.19.4}$$

Straining path constructed by Eq. (2.19.4), the reflection of trajectory 0-1'-2'-3'-4'-5'-6', is shown in Fig. 2.17 by dashed lines. This Figure shows the deviation between the solid and dashed lines for all groups of the constants. Therefore, we observe the deviation from the Ilyushin isotropy postulate, the closer the yield criterion to the Guber criterion (Fig. 2.17a), the less specified the deviation [146,147].

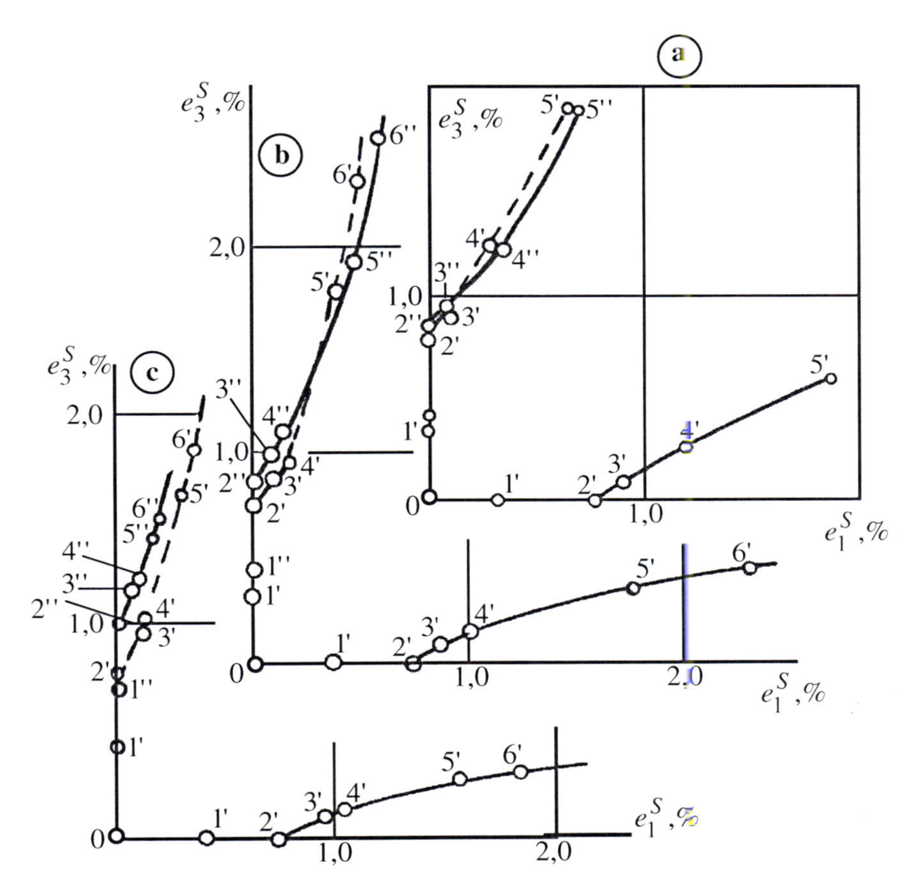

Fig. 2.17 Strain trajectories and their mirror images.

2.20 Loading Surface

Let us construct an analogue of the Cicala surface considered in Section 2.6 for the case when the shear strength is given in the form of Eq. (2.15.1). The formulation of problem is the same as in Section 2.5, an element in a body is loaded in tension to some value σ_z^0 and then unloaded. Then, the element is proportionally reloaded under combined tension and torsion ($\sigma_z > 0$, $\tau_{xz} > 0$). Let us construct a loading surface (subsequent yield surface) under reloading, i.e., determine the first value of σ_z and τ_{xz} which produce plastic deformations.

Slip intensity in tension is determined on the base of Eqs. (2.16.2) and (2.16.5) as

$$\varphi_{nl} = \frac{1}{r_2}\left[\frac{\sin 2\beta}{a} - 1\right]\delta\left(\omega - \frac{\pi}{2}\right), \quad a = \frac{2f_1\left(\sigma_z^0\right)}{\sigma_z^0}. \tag{2.20.1}$$

These slips occur in the planes given by Eq. (2.5.2) and the boundary values of angles β and ω are

$$\sin 2\beta_{1,2} = a, \qquad \omega_1 = \omega_2 = \pi/2. \tag{2.20.2}$$

Outside the range of Eq. (2.5.2) $\varphi_{nl} = 0$. For the case of uniaxial tension Eqs. (2.15.1) and (2.20.1) give the following expression for shear strength:

$$S_{n_0 l_0} = f_1(\tau_0)\left\{1 + \left[\frac{\sin 2\beta_0}{a} - 1\right]\sin\omega_0\right\}, \tag{2.20.3}$$

which holds for the range of Eq. (2.5.2) with boundary values of Eq. (2.20.2). In the direction of $\omega_0 = \pi/2$, the above equation becomes

$$S_{n_0 l_0} = \frac{f_1(\tau_0)}{a}\sin 2\beta_0 \tag{2.20.4}$$

and on the border of the range of slips, $\beta_0 = \beta_{1,2}$, we have

$$S_{n_0 l_0} = f_1(\tau_0). \tag{2.20.5}$$

Under the combined proportional reloading, a shear stress, τ_{nl}, is determined by Eq. (2.6.2) which at $\alpha = 0$ and $\omega = \pi/2$ can be written as

$$\tau_{nl} = \frac{\sigma_z}{2}(\sin 2\beta + k\cos 2\beta), \quad k = \frac{2\tau_{xz}}{\sigma_z} = \text{const}. \tag{2.20.6}$$

Under combined loading, the shear stress τ_{nl} grows (at $0 < \beta < \pi/4$) and reaches the value of shear strength (see Section 2.6) at the boundary points of region (2.5.2), i.e., becomes equal to $S_{n_0 l_0}$ given by Eq. (2.20.5):

$$\frac{\sigma_z}{2}\left(\sin 2\beta_1 + k\cos 2\beta_1\right) = f_1(\tau_0), \tag{2.20.7}$$

where the shear stress intensity τ_0 is

$$\tau_0 = \sqrt{\sigma_z^2 + 3\tau_{xz}^2}\,. \tag{2.20.8}$$

Substituting $\sin 2\beta_1$ and $\cos 2\beta_1$ from Eq. (2.20.2) into Eq. (2.20.7) and assuming the function $f_1(\tau_0)$ in the linear form of (2.8.4), Eq. (2.20.7) can be rewritten as

$$\sigma_z\left(f_3 - f_4\sigma_z^0\right) + \tau_{xz}\sqrt{\left(\sigma_z^0\right)^2 - 4\left(f_3 - f_4\sigma_z^0\right)^2} = \sigma_z^0\left(f_3 - f_4\sqrt{\sigma_z^2 + 3\tau_{xz}^2}\right). \tag{2.20.9}$$

This is the equation of the part of loading surface (line) after uniaxial tension. It passes through the loading point, ($\sigma_z = \sigma_z^0$, $\tau_{xz} = 0$) and touches the yield surface, $\tau_m = f_1(\tau_0)$, which has the form

$$\sqrt{\sigma_z^2 + 4\tau_{xz}^2} = 2f_3 - 2f_4\sqrt{\sigma_z^2 + 3\tau_{xz}^2}\,. \tag{2.20.10}$$

For the case of the Tresca material ($\sigma_S = 2\tau_S$, $f_3 = \tau_S$, $f_4 = 0$), Eq. (2.20.9) describes the Cicala surface, meaning that the equation of the loading surface does not depend on the form shear strength.

The whole loading surface (line) is shown in Fig. 2.18. Its characteristic feature is the appearance of singularity at the loading point. Let us determine the slope of tangent to the curve (2.20.9) at point M. It is obvious that

$$\tan\delta_3 = 2\left|\frac{d\tau_{xz}}{d\sigma_z}\right|_{\tau_{xz}=0} \tag{2.20.11}$$

Therefore,

$$\tan\delta_3 = \frac{2f_3}{\sqrt{\left(\sigma_z^0\right)^2 - 4\left(f_3 - f_4\sigma_z^0\right)^2}}\,. \tag{2.20.12}$$

In particular, for the Tresca material (the Cicala surface), Eq. (2.20.12) gives

$$\tan\delta_{3T} = \frac{\sigma_S}{\sqrt{\left(\sigma_z^0\right)^2 - \sigma_S^2}}\,. \tag{2.20.13}$$

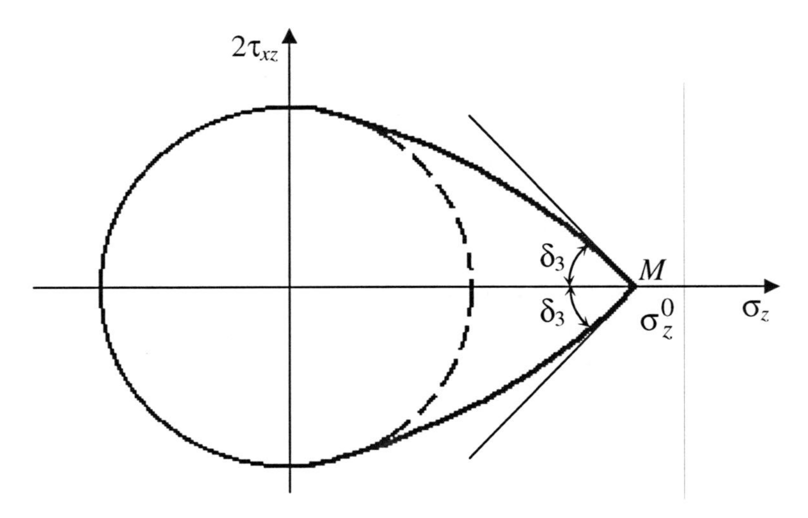

Fig. 2.18 Loading surface corresponding to shear strength (2.15.1)

Let us compare angles δ_3 and δ_{3T} at the fixed value of σ_z^0. If $\sigma_z^0 = 1{,}5\sigma_S$, then Eq. (2.20.13) implies that $\tan\delta_{3T} = 0{,}91$. Further, for materials close to the Guber continuum, $\sigma_S = 1{,}8\tau_S$, the coefficients f_3 and f_4 determined by Eq. (2.8.7) take the values $f_3 \approx 1{,}5\sigma_S$ and $f_4 = 1$ and, according to Eq. (2.20.12), $\tan\delta_3 = 2$. Hence, $\delta_3 > \delta_{3T}$, meaning that the surface(curve) (2.20.9) shows a better agreement with experiments than the surface(curve) (2.6.11), on the experimental surfaces there is no such sharp corner as on the Cicala surface.

2.21 The Bauschinger Effect

The applicability of the theories of plasticity is tested mainly by the comparison of their predictions with experimental results obtained under two-segment loading. However, such checks are insufficient. It is necessary to compare the results of theory with experimental data under alternating cyclic loading. Strains produced at alternating loading exhibit a number of features, which are not displayed at a monotonous loading.

First of all, we note that the Budiansky theory does not describe an alternating loading, the deformation properties of a material in the direction opposite to the current slips are not specified. To do this, we designate through $-l_0$ a direction opposite to l_0 and require that the operator $R_{l_0 n_0}$ be asymmetric:

$$R_{-l_0 n_0} = -R_{l_0 n_0}. \tag{2.21.1}$$

Then, in the base of Eq. (2.10.1), we need to require that

$$\varphi_{-ln} = -\varphi_{ln}. \tag{2.21.2}$$

This formula is used only for the calculation of $R_{l_0 n_0}$, not for the determination of strains, due to that Eq. (2.9.3), together with Eq. (2.21.2), would lead to the double value of strain components. Let us note that, independently of Eq. (2.21.2), the second and third items of the operator satisfy the condition (2.21.1).

Consider the following loading trajectory: proportional loading that produces plastic strains – unloading – loading in the direction opposite to initial loading. We wish to determine the condition for the onset of plastic deformation at the opposite loading. Let φ_{nl}^0, τ_m^0, and τ_0^0 designate the slip intensity, maximal shear stress, and shear stress intensity, respectively, corresponding to the beginning of unloading. The maximum of shear strength, S_m, is

$$S_m = f_1\left(\tau_0^0\right)\left(1 + R_m\right), \tag{2.21.3}$$

where R_m is the maximum of function $R_{l_0 n_0}\varphi_{nl}^0$ with respect to angles α_0, β_0, and ω_0. At proportional loading, the function $R_{l_0 n_0}\varphi_{nl}^0$ takes its maximal value in that slip plane and direction where the maximal shear stress τ_m^0 acts. The condition that plastic slips occur in the pointed out slip system under the action of τ_m^0 can be written as

$$f_1\left(\tau_0^0\right)\left(1 + R_m\right) = \tau_m^0. \tag{2.21.4}$$

Further, the shear strength, as a function of α_0, β_0, and ω_0, is minimal in the plane of action of initial maximal shear stress but in the direction opposite to τ_m^0. Then, according to Eq. (2.21.1), we obtain

$$S_{\min} = f_1\left(\tau_0\right)\left(1 - R_m\right). \tag{2.21.5}$$

At the loading opposite to the initial one, the maximal shear stress acts in that system where the shear resistance has a minimum. Therefore, now, the condition for the onset of plastic deformation is

$$f_1\left(\tau_0\right)\left(1 - R_m\right) = \tau_m. \tag{2.21.6}$$

By eliminating R_m from Eqs. (2.21.4) and (2.21.6), the yield criterion at the change of the sign of applied forces takes the form

$$\frac{\tau_m}{f_1(\tau_0)} + \frac{\tau_m^0}{f_1\left(\tau_0^0\right)} = 2. \tag{2.21.7}$$

As $\tau_m^0 > f_1\left(\tau_0^0\right)$, $\tau_m < f_1(\tau_0)$ and, comparing this inequality with the yield criterion of Eq. (2.8.1), it follows that an initial loading of one sign reduces the resistance of the material with respect to the onset of subsequent plastic deformation of the opposite sign. This is termed as the Bauschinger effect.

If $f_1 = \tau_S = const$, Eq. (2.21.7) can be written as

$$\tau_m^0 - \tau_S = \tau_S - \tau_m, \tag{2.21.8}$$

which expresses the ideal Bauschinger effect, the increase in the subsequent yield limit due to loading in one direction results in the equal decrease in the yield limit in the direction opposite to the initial loading.

It can be shown that Eq. (2.21.7) becomes $\tau_m^0 - \tau_S < \tau_S - \tau_m$ for the yield criterion (2.8.1) at $f'(\tau_0) < 0$.

2.22 Alternating Loading in Tension-Compression

Consider the case when a specimen is first loaded in tension, $\sigma_z = \sigma_z^0$, and then unloaded and loaded in pressure. Let us determine the strain of the specimen under the reloading, i.e., pressure, ε_z. If l_{01} denotes the slip direction under the initial loading, we get

$$\tau_{n_0 l_{01}} = f_1\left(\tau_0^0\right)\left(1 + R_{n_0 l_{01}} \varphi_{nl}^0\right), \tag{2.22.1}$$

where φ_{nl}^0 is the slip intensity under initial tension: Let us designate through $-l_{01}$ the direction opposite to l_{01}. Then Eqs. (2.21.1) and (2.22.1) give

$$R_{n_0 l_0} \varphi_{nl}^0 = 1 - \frac{\left|\tau_{n_0 l_0}^0\right|}{f_1\left(\tau_0^0\right)}, \qquad l_0 = -l_{01} \tag{2.22.2}$$

and the shear strength under pressure acquires the form

$$S_{n_0 l_0} = f_1(\tau_0)\left(1 + R_{n_0 l_0} \varphi_{nl}^0 + R_{n_0 l_0} \varphi_{-ln}^c\right), \tag{2.22.3}$$

where $-l$ is the slip direction at $\sigma_z < 0$, $\varphi^c_{-\ln} > 0$. If a slip occurs in direction $-l_{01}$ under pressure, i.e.,

$$S_{n_0 l_0} = \tau_{n_0 l_0} \quad (l_0 = -l_{01}), \tag{2.22.4}$$

Eq. (2.22.3) leads to the following

$$R_{n_0 l_0} \varphi^c_{-ln} = \frac{\tau_{n_0 l_0}}{f_1(\tau_0)} + \frac{\left|\tau^0_{n_0 l_0}\right|}{f_1\left(\tau^0_0\right)} - 2. \tag{2.22.5}$$

If the operator $R_{n_0 l_0}$ is assumed to be in the form of Eq. (2.15.1), Eq. (2.22.5) is an integral equation for φ^c_{-ln0}:

$$r_2 \int\limits_{\omega_1}^{\omega_2} \varphi^c_{-ln0} \cos(\omega_0 - \omega) d\omega = \frac{|\sigma_z|}{2 f_1(\tau_0)} \sin 2\beta_0 |\sin \omega_0| + \frac{\sigma^0_z}{2 f_1\left(\tau^0_0\right)} \sin 2\beta_0 |\sin \omega_0| - 2,$$

$$\omega_1 \le \omega_0 \le \omega_2, \quad \pi < \omega_{1,2} < 2\pi, \quad \sigma_z < 0. \tag{2.22.6}$$

The obtained relationship is similar to Eq. (2.16.1) and is solved in the same way as Eq. (2.16.1). The solution of Eq. (2.22.6) gives, through Eq. (2.15.16), the plastic strain component due to pressure

$$\varepsilon^{Sc}_z = \frac{4\pi}{15 r_2} (1 - a_c)^{3/2} \left(\frac{4}{a_c} + 1\right), \tag{2.22.7}$$

$$a_c^{-1} = \frac{1}{4} \left[\frac{|\sigma_z|}{f_1(\tau_0)} + \frac{\sigma^0_z}{f_1\left(\tau^0_0\right)}\right]. \tag{2.22.8}$$

The total relative elongation caused by the initial tension and the following pressure is

$$\varepsilon^S_z = \varepsilon^{S0}_z - \varepsilon^{Sc}_z, \tag{2.22.9}$$

where the component ε^{S0}_z corresponds to the stress σ^0_z determined by Eq. (2.16.8), where a is given by the second equality of (2.20.1).

The theoretical stress-strain diagram under the considered loading regime is shown in Fig 2.19. This diagram agrees with experimental fact that the plastic straining in pressure (in the vicinity of point M_4) experiences a greater hardening than in tension (along segment M_1M_2). The latter follows from that the function $f_1(\tau_0)$ decreases in growth of τ_0 (takes greater values in the vicinity of point

M_4 than at point M_1). According to Eq. (2.22.5), the larger the function $f_1(\tau_0)$, the smaller the plastic strain. The dashed line in Fig. 2.19 shows the compressive portion of the stress-strain diagram constructed at $f_1 = \tau_S = \text{const}$; obviously, it does not fit the experimental curves.

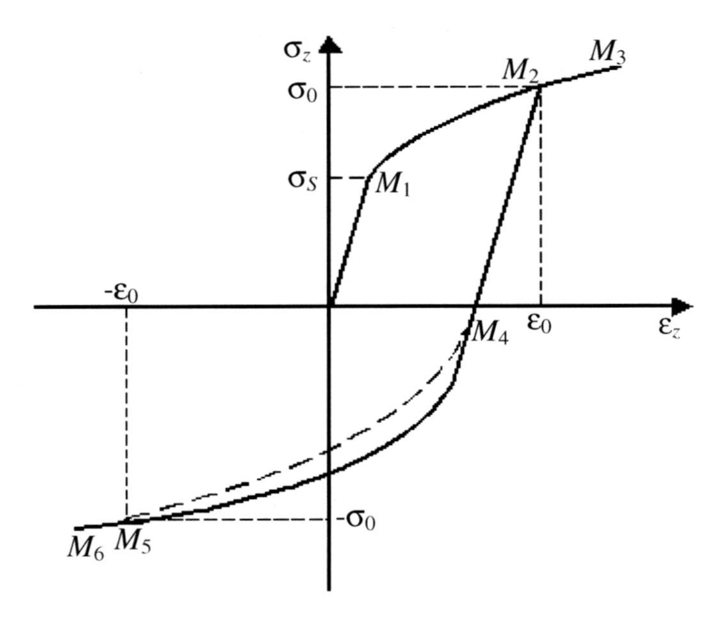

Fig. 2.19 Stress-strain diagram in alternate loading

For the case of symmetrical alternating loading, $\sigma_z = -\sigma_z^0$, Eqs. (2.20.1) and (2.22.8) imply that $a_c = a$ and Eqs. (2.16.8) and (2.22.7) give $\varepsilon_z^{Sc} = 2\varepsilon_z^{S0}$, i.e.,

$$\varepsilon_z^S = -\varepsilon_z^{S0} \quad \text{at} \quad \sigma_z = -\sigma_z^0. \tag{2.22.10}$$

The materials whose diagrams satisfy to the condition (2.20.10) are referred to as cyclically ideal materials. Thus, the shear strength from Eq. (2.15.1) describes cyclically ideal materials.

Let us consider the stress-strain diagram in pressure beyond point M_5 (Fig. 2.19). If $|\sigma_z| > \sigma_z^0$, segment M_5M_6 has the same form as M_2M_3. The material "forgets" the previous stretching and deforms as if at a simple pressure [56,153,154].

2.23 Loading Surface at Alternating Loading in Tension-Compression

Let the element in a body, previously loaded in tension by stress σ_z^0 and then unloaded, be subject to a combined proportional pressure $\sigma_z < 0$ and torsion $\tau_{xz} > 0$. Our goal is to determine the yield locus under such loading. The loading surface under initial tension, $\sigma_z^0 > 0$, has been constructed in Sec. 2.6 ($f_1 = \tau_S = \mathrm{const}$) and shown in Fig. 2.7.

Under initial tension the slips occurred in planes (2.5.2) in direction $\omega = \pi/2$. In the opposite direction, $\omega = -\pi/2$, the operator $R_{n_0 l_0}$ is determined by Eq. (2.22.3) so that the shear strength has the form

$$S_{n_0 l_0} = \tau_S \left(2 - \frac{\sigma_z^0}{2\tau_S} \sin 2\beta_0 \right). \tag{2.23.1}$$

The shear stress at combined torsion and pressure is determined by Eq. (2.6.2), which at $\omega = -\pi/2$ takes the form:

$$\tau_{n_0 l_0} = -\frac{\sigma_z}{2} \left(\sin 2\beta_0 + k \cos 2\beta_0 \right). \tag{2.23.2}$$

In the considered element, the plastic shears occur, if the curve (2.23.1) touches the line (2.23.2). This can be expressed as

$$\frac{\partial S_{n_0 l_0}}{\partial \beta_0} = \frac{\partial \tau_{n_0 l_0}}{\partial \beta_0}, \quad S_{n_0 l_0} = \tau_{n_0 l_0}. \tag{2.23.3}$$

From these equations we find that

$$\sigma_z = \sigma_z^0 - 2\sigma_S \sin 2\beta_0, \qquad \tau_{xz} = -\sigma_S \cos 2\beta_0. \tag{2.23.4}$$

Eliminating from the above formulae angle β_0, we obtain the equation of the loading surface (line) due to the combined pressure-torsion action:

$$\left(\sigma_z - \sigma_z^0 \right)^2 + 4\tau_{xz}^2 = 4\sigma_S^2. \tag{2.23.5}$$

This is an equation of the circle of radius $2\sigma_S$ with its center at point $\sigma_z = \sigma_z^0$, $\tau_{xz} = 0$, arc $M_1 M_2$ in Fig. 2.20a; the dotted line is the yield circle. Consider the points of intersection, M_1 and M_2, between the yield and loading surfaces. As seen from Fig. 2.20, the lines tangential to the loading surface on the left and right of these points are different, i.e., new angular points arise.

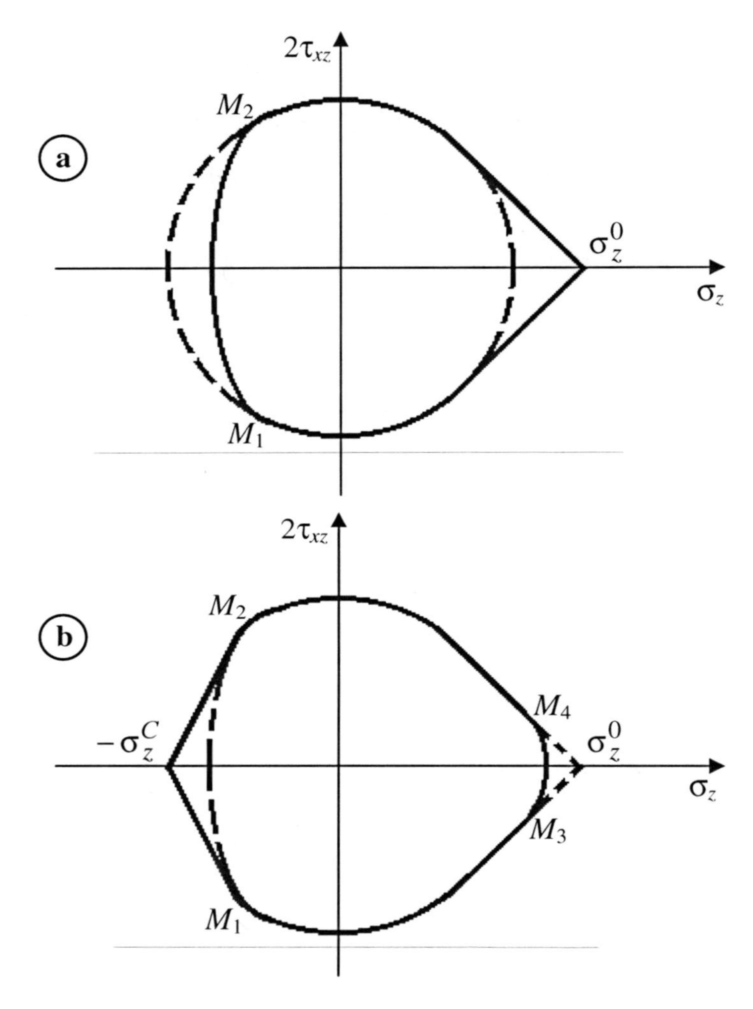

Fig. 2.20 Loading surface in uniaxial tension (a) and subsequent compression (b)

Further, assuming that after tension $\sigma_z = \sigma_z^0$ the pressure $\sigma_z = -\sigma_z^C < 0$ is applied and missing intermediate formulae, the portions of loading surface acquire the form:

$$
\begin{gathered}
-2\tau_S\sigma_z + \tau_{xz}\sqrt{\left(\sigma_z^0 - \sigma_z^c\right)^2 - 16\tau_S^2} = 2\tau_S\sigma_z^c, \\
\left(\sigma_z + \sigma_z^c\right)^2 + 4\tau_{xz}^2 = 16\tau_S^2.
\end{gathered}
\tag{2.23.6}
$$

The first relation is the equations of a straight line passing through point $\sigma_z = -\sigma_z^c$, $\tau_{xz} = 0$ and touching to the circle (2.23.5). The second formula is the equation of circle, M_3M_4. Figure 2.20b (the dotted line corresponds to the solid one from Fig. 2.20a) shows that the singularity at point $\sigma_z = \sigma_z^0$, $\tau_{xz} = 0$, which have arisen under the initial tension, disappears, while there are new corner points M_3 and M_4.

2.24 Conclusions Relative to Shear Strength

1. The use of shear strength in the form of Eq. (2.15.1) leads to the determination of plastic strain that is reduced to the calculation of double integral (2.15.16) instead of three-fold integral in Eqs. (2.3.5) or (2.9.3). This is a considerable simplification of the problem.

As specified earlier, the formulae similar to (2.15.16) were offered by A.K. Malmeister, the initial formulation prescribes that the slip direction in any slip plane is co-directed with the direction of total shear stress τ_n. However, later, the model of kinematic hardening proposed by Kadashevich-Novozsilov-Prager-Cigler was employed in the following way [89,177]. A yield surface, a circle of radius τ_S, is introduced in each slip system (not in the stress space as in terms of flow theories). If the vector τ_n reaches the circle, it translates it a rigid body. The translation occurs in the direction of the vector normal to the circle at the loading point (where the endpoint of vector τ_n touches the yield circle). A slip in a given plane is assumed to occur in the direction of the circle translation, i.e., the direction of slip and τ_n is not identical for curvilinear loading paths, giving the considerable improvement of the theory.

2. Regarding the orthogonal break of loading trajectory, the additional shear modulus (2.16.21) obtained on the base of shear strength (2.15.1) is greater than the Cicala modulus (2.5.11) meaning that Eq. (2.16.21) shows a better agreement with experiments than the Cicala formula giving too small values of G_S.

3. Let us turn, again, to the role of function $f_1(\tau_0)$ appearing in the yield criterion of Eq. (2.8.1) and the expression for shear strength, Eqs. (2.10.1) or (2.15.1). It is the presence of this function which provides the considerable rehabilitation of the slip concept. The investigation of the loading surface under alternating tension-pressure also justifies the thesis on the necessity of the presence of descending function $f_1(\tau_0)$ in Eqs. (2.10.1) and (2.15.1). The loading

surface constructed together with $f_1(\tau_0)$ is better coordinated with experimental one, which has no such sharp corner as on the Cicala surface. Further, theoretical alternate tension-pressure diagrams are much closer to experimental diagrams if f_1 is taken to be the descending function of τ_0 than at $f_1 = const$. At the same time, the factor $f_1(\tau_0)$ in Eqs. (2.10.1) or (2.15.1) leads to the results that are in a contradiction to the consequence from Drucker's postulate.

4. The strains under alternating loading have a number of features, strain-stress diagrams vary from cycle to cycle. A material can become harder, softer, or remains stable. In addition, the stabilization of hardening/softening processes or the transition from one process to another may be observed. It is obvious that the slip theory with shear strength (2.10.1) and (2.15.1) is not applicable to describe the specified features of alternate loading.

2.25 Leonov-Shvajko Model

This Section concerns with a plastic plane strain in terms of the Leonov-Shvajko model [76].

The plane strain arises if three strain tensor components possessing identical index, for example z, are equal to zero,

$$\varepsilon_z^S = \gamma_{xz}^S = \gamma_{yz}^S = 0, \tag{2.25.1}$$

Letting

$$\tau_{xz} = \tau_{yz} = 0 \text{ and } \sigma_z = \frac{1}{2}\left(\sigma_x + \sigma_y\right), \tag{2.25.2}$$

the relationships of the Hencky-Nadai theory, Eq. (1.5.39), and the flow plasticity theory, Eq. (1.13.11), lead to the result of Eq. (2.25.1).

The main assumption of the Leonov-Shvajko model [76] consists in that slips can occur only in the set of planes parallel to z-axis, $\beta = 0$ (shaded in Fig. 2.21); in each slip planes a single slip direction is assumed to be possible, $\omega = 0$ or $\omega = \pi$, i.e parallel to xy – plane. Under the adopted restrictions Eq. (2.25.1) holds automatically. In order to induce these slips, the maximal shear stress has to act in the one of the specified planes. Therefore, stress component σ_z need to be principal and to fall within two other principal stresses, σ_1 and σ_2:

$$\sigma_2 < \sigma_z < \sigma_1, \tag{2.25.3}$$

$$\sigma_{1,2} = \frac{\sigma_x + \sigma_y}{2} \pm \frac{1}{2}\sqrt{\left(\sigma_x - \sigma_y\right)^2 + 4\tau_{xy}^2}. \tag{2.25.4}$$

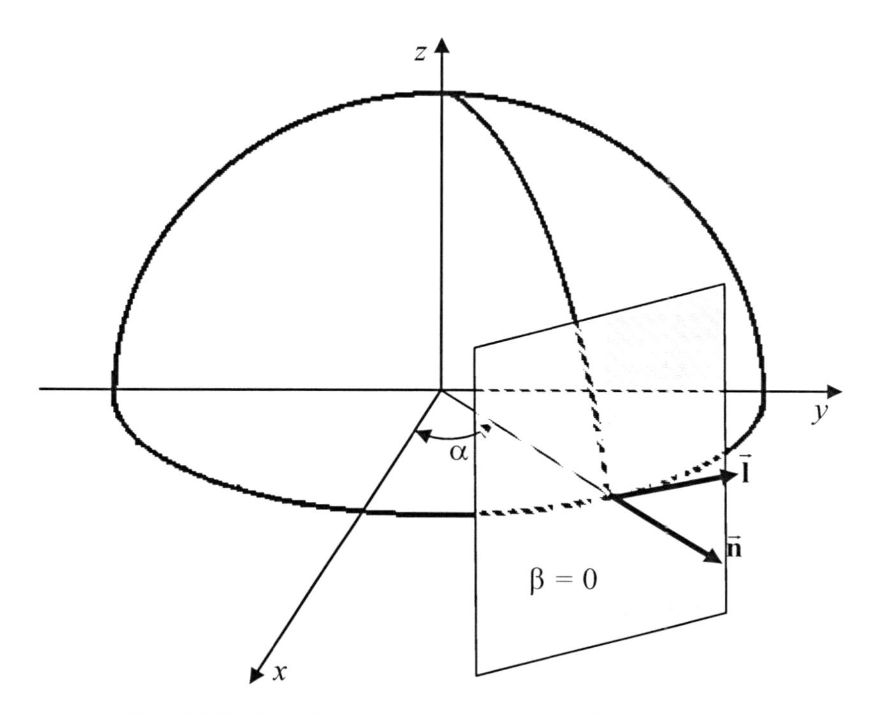

Fig. 2.21 Vertical slip plane considered in the plain-plastic model.

From Eqs. (2.25.3) and (2.25.4) we obtain the restrictions imposed on the component σ_z:

$$\left|\sigma_z - \frac{\sigma_x + \sigma_y}{2}\right| < \frac{1}{2}\sqrt{(\sigma_x - \sigma_y)^2 + 4\tau_{xy}^2}\,. \tag{2.25.5}$$

In this case, the maximal shear stress has the form

$$\tau_m = \frac{1}{2}\sqrt{(\sigma_x - \sigma_y)^2 + 4\tau_{xy}^2}\,. \tag{2.25.6}$$

The total shear stress acting in the slip system $\beta = 0$ and $\omega = 0$, on the base of Eqs. (2.3.2) and (2.3.8), is

$$\tau_n = \frac{\sigma_y - \sigma_x}{2}\sin 2\alpha + \tau_{xy}\cos 2\alpha\,. \tag{2.25.7}$$

Since stress σ_z does not appear in Eq. (2.25.7), we take σ_z in the form of Eq. (2.25.2) demanding only that the condition (2.25.5) be satisfied. Then, the shear stress intensity (1.2.9) takes the form

$$\tau_0 = \frac{\sqrt{3}}{2}\sqrt{\left(\sigma_x - \sigma_y\right)^2 + 4\tau_{xy}^2} = \sqrt{3}\tau_m. \tag{2.25.8}$$

The shear strength is assumed as follows [76]:

$$S_{n_0} = \tau_S + \int\limits_{\alpha_1}^{\alpha_2} \varphi_n(\alpha)\tilde{F}_1(\alpha_0 - \alpha)d\alpha, \tag{2.25.9}$$

where the function $\tilde{F}_1(\alpha_0 - \alpha)$ is proposed as

$$\tilde{F}_1(\alpha_0 - \alpha) = r_1 \ln \frac{k}{|\alpha_0 - \alpha|}, \qquad (r_1, k = \text{const}). \tag{2.25.10}$$

The representation of shear strength S_{n_0} in the form of Eq. (2.25.9) allows for the interaction of different slip systems, meaning that even if in some plane $\varphi_n = 0$, the shear strength varies, $S_{n_0} \neq \tau_S$. This follows from that the integral (2.25.9) is taken between the limits where $\varphi_n > 0$.

The condition for the onset of slip in a given plane is $S_{n_0} = \tau_{n_0}$. If there is no slip at a given instant of loading, we have $S_{n_0} > \tau_{n_0}$. Equating the shear strength (2.25.9) to the shear stress (2.25.7), we get the following integral equation for φ_n [76]

$$\int\limits_{\alpha_1}^{\alpha_2} \tilde{F}_1(\alpha_0 - \alpha)\varphi_n(\alpha)d\alpha = \frac{\sigma_y - \sigma_x}{2}\sin 2\alpha + \tau_{xy}\cos 2\alpha - \tau_S, \tag{2.25.11}$$

where the boundaries of integration, α_1 and α_2, can be found from the condition $S_{n_0} > \tau_{n_0}$.

According to Eqs. (2.3.8) and (2.9.3), the plastic strain components can be written as

$$\varepsilon_y^S = -\varepsilon_x^S = \frac{1}{2}\int\limits_{\alpha_1}^{\alpha_2} \varphi_n(\alpha)\sin 2\alpha d\alpha, \quad \gamma_{xy}^S = \int\limits_{\alpha_1}^{\alpha_2} \varphi_n(\alpha)\cos 2\alpha d\alpha. \tag{2.25.12}$$

Equations (2.25.9), (2.25.11), and (2.25.12) are the basic relationships of the Leonov-Shvajko model, so called plastic plane theory of slip. From Eqs. (2.25.6) and (2.25.8) it follows that if the maximal shear stress is constant, the shear stress intensity is constant as well, i.e., the Tresca continuum coincides with the Guber continuum. Therefore, on the base of the isotropy postulate, the obtained relationships can be generalized on any plane not parallel to xy-plane. The fact that the model obeys the Guber criterion leads to the fulfillment of the deviator proportionality and the existence of the universal diagram $\tau_0 \sim \gamma_0$. Thus, the basic shortcoming of the Batdorf-Budiansky theory, point 3 in Sec. 2,7, is absent. Furthermore, the module G_S at an orthogonal break of loading path is better coordinated [156] with experimental data than that obtained from the Cicala formula.

Let us discuss the condition that slips can occur only at $\beta = 0$ and $\omega = 0$ or $\omega = \pi$ in terms of the Budiansky slip concept. Consider the case of pure shear, $\tau_{xy} > 0$. By making use Eqs. (2.3.2) and (2.3.8) at $\beta = 0$ and $\omega = 0$, the resolved shear stress is

$$\tau_{nl} = \tau_{xy} \cos 2\alpha . \tag{2.25.13}$$

On the other hand, the Budiansky theory imposes no restrictions on the orientation of slip planes. A resolved shear stress at any point of hemisphere (Fig. 2.6 a,b) and in any direction is determined by Eqs. (2.3.2) and (2.3.8), which at, e.g., $\alpha = \omega = 0$ becomes

$$\tau_{nl} = \tau_{xy} \cos \beta . \tag{2.25.14}$$

If $\beta < 2\alpha$, shear stress (2.25.13) is smaller than that from (2.25.14), meaning that plastic slips occur on the inclined planes ($\beta > 0$). Since these planes are ignored in terms of the Leonov-Shvajko model, the slip intensity φ_n should be considered as an integral measure of φ_{nl} generated by slips occurring at $\beta > 0$.

The presented theory is considerably developed in M.J. Shvajko's writings [171,172].

2.26 Comparison of Two Models

For the case of plane deformation simpler formulas can be obtained if to neglect the interaction of slip systems. In this case, the function $\widetilde{F}_1$ in Eq. (2.25.10) degenerates in the Dirac δ-function

$$\widetilde{F}_1(\alpha - \alpha_0) = r\delta(\alpha - \alpha_0), \quad r = const . \tag{2.26.1}$$

Then Eq. (2.25.9) becomes

$$S_{n_0} = \tau_S + r\varphi_{n_0} . \tag{2.26.2}$$

Hence,

$$r\varphi_n = \tau_n - \tau_S\,. \tag{2.26.3}$$

Shear stress τ_n is maximal at $\alpha = \alpha_m$ and

$$\tan 2\alpha_m = \frac{\sigma_y - \sigma_x}{2\tau_{xy}}. \tag{2.26.4}$$

Let us express stress τ_n through the angle α' counted off from the direction given by α_m:

$$\alpha = \alpha_m + \alpha'. \tag{2.26.5}$$

Then

$$r\varphi_n = \tau_m \cos 2\alpha' - \tau_S\,. \tag{2.26.6}$$

In terms of variable α, the strain components (2.25.12) can be written as

$$\begin{gathered} \varepsilon_y^S = -\varepsilon_x^S = \frac{\sigma_y - \sigma_x}{4r\tau_m} \int_{-\alpha_1}^{\alpha_1} (\tau_m \cos 2\alpha' - \tau_S) \cos 2\alpha' \, d\alpha', \\ \gamma_{xy}^S = \frac{\tau_{xy}}{r\tau_m} \int_{-\alpha_1}^{\alpha_1} (\tau_m \cos 2\alpha' - \tau_S) \cos 2\alpha' \, d\alpha', \end{gathered} \tag{2.26.7}$$

where α_1 is the root of equation

$$\cos 2\alpha_1 = \frac{\tau_S}{\tau_m}. \tag{2.26.8}$$

Integrating in Eq. (2.26.7) gives

$$\varepsilon_y^S = -\varepsilon_x^S = \left(\frac{1}{G_S} - \frac{1}{G} \right) \frac{\sigma_y - \sigma_x}{4}, \quad \gamma_{xy}^S = \left(\frac{1}{G_S} - \frac{1}{G} \right) \tau_{xy}\,. \tag{2.26.9}$$

$$\frac{1}{G_S} - \frac{1}{G} = \frac{1}{2r} \left(\arccos \frac{\tau_S}{\tau_m} - \frac{\tau_S}{\tau_m} \sqrt{1 - \frac{\tau_S^2}{\tau_m^2}} \right). \tag{2.26.10}$$

In view of Eq. (2.25.2), stress deviator tensor components, $\overline{\sigma}_x$ and $\overline{\sigma}_y$, are

$$\overline{\sigma}_x = \frac{\sigma_x - \sigma_y}{2}, \quad \overline{\sigma}_y = \frac{\sigma_y - \sigma_x}{2}, \quad \overline{\sigma}_z = 0 \tag{2.26.11}$$

and Eq. (2.26.9) gives

$$\varepsilon_x^S = \frac{1}{2}\left(\frac{1}{G_S} - \frac{1}{G}\right)\overline{\sigma}_x, \quad \varepsilon_y^S = \frac{1}{2}\left(\frac{1}{G_S} - \frac{1}{G}\right)\overline{\sigma}_y,$$
$$\gamma_{xy}^S = \frac{1}{2}\left(\frac{1}{G_S} - \frac{1}{G}\right)\tau_{xy}, \quad \varepsilon_z^S = 0. \tag{2.26.12}$$

Let us compare the obtained formulae to the results obtained in terms of the Sanders theory for the case of plane stress state. In both variants the deviator proportionality is satisfied. The Sanders theory gives the shear modulus by Eq. (1.19.13), while the concept of slip leads to Eq. (2.26.10). By using Eq. (2.25.8) and the equality for the Guber medium $\sigma_S = \sqrt{3}\tau_S$, we obtain

$$\frac{\tau_m}{\tau_S} = \frac{\tau_0}{\sigma_S}. \tag{2.26.13}$$

Therefore, the specified shear modules are identically equal. It is worth to note that in Eqs. (1.19.13) and (2.26.10) the constant r takes different values; r has the unit of stress in Eq. (2.26.10), while $[r]=1$ in Eq. (1.19.13). Consequently, two essentially different approaches to the determination of plastic strain – the Sanders theory for the case of plane stress state and the plane plastic theory of slip proviso that the interaction between slip systems are ignored – result in the identical results.

2.27 Discrete Scheme of Slips

The Zsigalkin-Shemyakin-Christianovich theory [7,184,185] belongs to the concept of slip as well. Those loadings are considered for which the principal directions of stress- and strain-tensor coincide and do not change their orientation. The following is supposed: a plastic deformation can develop only in the slip systems co-directed with the principal shear stresses. The deformation anisotropy arising during plastic straining and the impact of slip in one slip system on the shear resistance in other slip systems are taken into account.

Let principal stresses σ_1, σ_2, and σ_3 be numbered so that inequalities (1.17.1) hold. We designate principal shear stresses through $\tau_{13} = \tau_m$, τ_{12} and τ_{23},

$$2\tau_{13} = \sigma_1 - \sigma_3, \quad 2\tau_{23} = \sigma_2 - \sigma_3, \quad 2\tau_{12} = \sigma_1 - \sigma_2. \tag{2.27.1}$$

Let a material be in a plastic state, $\tau_{13} > \tau_S$. After a small increment in the principal stresses ($\Delta\sigma_1$, $\Delta\sigma_2$, $\Delta\sigma_3$), without the orientation of principal stresses being changed, the increments of the principal shear stresses are

$$2\Delta\tau_{13} = \Delta\sigma_1 - \Delta\sigma_3, \ 2\Delta\tau_{23} = \Delta\sigma_2 - \Delta\sigma_3, \ 2\Delta\tau_{12} = \Delta\sigma_1 - \Delta\sigma_2. \tag{2.27.2}$$

Supposing that an element of the material in plastic state is orthotropic, the response of material on the increments $\Delta\sigma_1$, $\Delta\sigma_2$, $\Delta\sigma_3$, strain increments, can be written as follows

$$\begin{aligned}
\Delta\varepsilon_1 &= \frac{\Delta\sigma_1}{E_1} - \frac{\nu_{21}}{E_2}\Delta\sigma_2 - \frac{\nu_{31}}{E_3}\Delta\sigma_3, \\
\Delta\varepsilon_2 &= -\frac{\nu_{12}}{E_1}\Delta\sigma_1 + \frac{\Delta\sigma_2}{E_2} - \frac{\nu_{32}}{E_3}\Delta\sigma_3, \\
\Delta\varepsilon_3 &= -\frac{\nu_{13}}{E_1}\Delta\sigma_1 - \frac{\nu_{23}}{E_2}\Delta\sigma_2 + \frac{\Delta\sigma_3}{E_3},
\end{aligned} \tag{2.27.3}$$

where E_i and ν_{ij} are the coefficients characterizing orthotropy, they are equal to the instant Young modulus and Poisson's ratio. By making use of denotation

$$\begin{aligned}
&\Delta\sigma_1 = \Delta\sigma + \frac{2}{3}(\Delta\tau_{13} + \Delta\tau_{12}), \ \Delta\sigma_2 = \Delta\sigma + \frac{2}{3}(\Delta\tau_{23} - \Delta\tau_{12}), \\
&\Delta\sigma_3 = \Delta\sigma - \frac{2}{3}(\Delta\tau_{13} + \Delta\tau_{23}), \quad \Delta\sigma_1 + \Delta\sigma_2 + \Delta\sigma_3 = 3\Delta\sigma
\end{aligned} \tag{2.27.4}$$

and inserting the obtained values of principal stress increments into Eq. (2.27.3), we obtain the volume change,

$$\Delta\varepsilon_1 + \Delta\varepsilon_2 + \Delta\varepsilon_3 = \kappa_0\Delta\sigma + \kappa_{13}\Delta\tau_{13} + \kappa_{12}\Delta\tau_{12} + \kappa_{23}\Delta\tau_{23}, \tag{2.27.5}$$

where κ_0 and κ_{ij} depend, obviously, on E_i and ν_{ij}. It is assumed that the shear stress increment $\Delta\tau_{ij}$ do not cause the volume change, therefore

$$\kappa_{12} = \kappa_{13} = \kappa_{23} = 0. \tag{2.27.6}$$

Further, on the base of Eqs. (2.27.3) and (2.27.4), the expressions for slip increments take the form:

$$\Delta\gamma_{13} = \Delta\varepsilon_1 - \Delta\varepsilon_3 = \kappa_{13}'\Delta\sigma + A_{13},$$
$$\Delta\gamma_{12} = \Delta\varepsilon_1 - \Delta\varepsilon_2 = \kappa_{12}'\Delta\sigma + A_{12}, \quad (2.27.7)$$
$$\Delta\gamma_{23} = \Delta\varepsilon_2 - \Delta\varepsilon_3 = \kappa_{23}'\Delta\sigma + A_{23},$$

where κ_{ij}' is a function of E_i, ν_{ij}; A_{ij} and does not depend on $\Delta\sigma$. As slip increments do not depend on the hydrostatic stress,

$$\kappa_{12}' = \kappa_{13}' = \kappa_{23}' = 0. \quad (2.27.8)$$

The strain increments (2.27.3) depend on nine variable parameters E_1, E_2,...ν_{32}. They satisfy Eqs. (2.27.6) and (2.27.8), of which five are independent. Therefore, instead of the nine specified parameters, it is possible to introduce four independent parameters and Eq. (2.27.7), together with Eq. (2.27.3), acquires the form

$$9\Delta\varepsilon_1 = \left(\frac{3}{G_{13}} + \frac{1}{G_{12}} - \frac{1}{G_{23}} + \frac{1}{G'}\right)\Delta\tau_{13} + \left(\frac{1}{G_{13}} + \frac{3}{G_{12}} - \frac{1}{G_{23}} - \frac{1}{G'}\right)\Delta\tau_{12} + \left(\frac{2}{G_{13}} - \frac{2}{G_{12}} - \frac{2}{G'}\right)\Delta\tau_{23}$$
$$9\Delta\varepsilon_2 = \left(-\frac{2}{G_{12}} + \frac{2}{G_{23}} - \frac{2}{G'}\right)\Delta\tau_{13} + \left(\frac{1}{G_{13}} - \frac{3}{G_{12}} - \frac{1}{G_{23}} - \frac{1}{G'}\right)\Delta\tau_{12} + \left(-\frac{1}{G_{13}} + \frac{1}{G_{12}} + \frac{3}{G_{23}} - \frac{1}{G'}\right)\Delta\tau_{23} \quad (2.27.9)$$
$$9\Delta\varepsilon_3 = \left(-\frac{3}{G_{13}} + \frac{1}{G_{12}} - \frac{1}{G_{13}} + \frac{1}{G'}\right)\Delta\tau_{13} + \left(-\frac{2}{G_{13}} + \frac{2}{G_{23}} + \frac{1}{G'}\right)\Delta\tau_{12} + \left(-\frac{1}{G_{13}} + \frac{1}{G_{12}} - \frac{3}{G_{23}} - \frac{1}{G'}\right)\Delta\tau_{23}$$

To find out the nature of modules G_{ij}, we calculate components $\Delta\gamma_{ij}$:

$$\Delta\gamma_{13} = \frac{\Delta\tau_{13}}{G_{13}} + \frac{\Delta\sigma_2'}{3}\left(\frac{1}{G_{23}} - \frac{1}{G_{12}} + \frac{1}{G'}\right),$$
$$\Delta\gamma_{12} = \frac{\Delta\tau_{12}}{G_{12}} + \frac{3\Delta\tau_{13} + \Delta\sigma_2'}{3}\left(\frac{1}{G_{13}} - \frac{1}{G_{23}} + \frac{1}{G'}\right),$$
$$\Delta\gamma_{23} = \frac{\Delta\tau_{23}}{G_{23}} + \frac{3\Delta\tau_{13} - \Delta\sigma_2'}{3}\left(\frac{1}{G_{13}} - \frac{1}{G_{12}} - \frac{1}{G'}\right), \quad (2.27.10)$$
$$\Delta\sigma_2' = \Delta\sigma_2 - \frac{1}{3}(\Delta\sigma_1 + \Delta\sigma_3).$$

The above equations show that the G_{13}, G_{12}, and G_{23} are shear modules in the direction of the principal shear stresses $\Delta\tau_{13}$, $\Delta\tau_{12}$, and $\Delta\tau_{23}$ and can be determined under the extra-loading $\Delta\sigma_2' = 0$, $3\Delta\tau_{13} + \Delta\sigma_2' = 0$, and $3\Delta\tau_{13} - \Delta\sigma_2' = 0$, respectively; parameter G' is a function of modules G_{ij}.

The constitutive relations of the considered theory, Eqs (2.27.9), have been tested under numerous experiment and shown to have good agreements with them. At the same time, this model is applicable only for the loading when the principal directions of stress- and strain-tensor are identical and do not change their directions. This requirement exposes a considerable restriction on loading trajectories. In addition, from the point of view of the Batdorf-Budiansky theory, it is not clear why slips occur only in discrete slip systems, where maximal shear stresses act.

Let us summarize the main results of the chapter 2.

The analytical results obtained in terms of the Budiansky slip theory are coordinated with some experiments, which the plastic flow theories are incapable of describing. At the same time, the Budiansky concept of slip is not justified by other very important experiments that have resulted in pessimistic conclusions [123] relative to the progress of the theory of plasticity based on the slip-mechanisms of deformation. The researches of the second chapter (Sec. 2.8-2.25, 2.27) dispel this pessimism and rehabilitate the concept of slip. Not the concept of slip is poor, but some assumptions of the Budiansky theory are not substantiated.

Further searches of new theories of plasticity are needed.

Chapter 3
Synthetic Theory of Plasticity

3.1 Introductory Remarks

A new theory, ***the synthetic theory of plasticity***, is presented in this chapter. It combines the Sanders flow plasticity theory and the Batdorf-Budiansky concept of slip.

We demand that the synthetic theory should satisfy the following requirements:

(i) As it was repeatedly noted, the simplest theory of plasticity, the Hencky-Nadai deformation theory, shows satisfactory agreements with experiments in proportional (simple) loading for many materials. Therefore, it is logical to require that stress-strain relationships of the synthetic theory reduce to the deviator proportionality at simple loading.

(ii) The arising of corner point on loading surface; it appears to be the only way to model plastic strain increments in orthogonal additional loading.

(iii) The two-level determination of plastic strain; as shown earlier two-level models are more effective than one-level ones (see Sec. 2.7 points 1 and 2, and Sec. 2.14 points 1-3).

It must be noted that the synthetic theory is not applicable to an arbitrary state of stress, the maximal number of non-zero stress tensor components is four: all three normal and single shear stress tensor components (other two shear stress components are equal to zero).

The bases of synthetic theory have been elaborated in the framework of the J. Andrusik candidate dissertation under K. Rusinko's supervision.

3.2 Partial Cases of the Tresca Yield Surface

Since the Batdorf–Budiansky concept of slip is based on the Tresca yield criterion, let us consider in more detail the construction of the Tresca yield surface. Consider the plane stress state ($\sigma_x \neq 0$, $\sigma_z \neq 0$, $\tau_{xz} \neq 0$) and the case when $\sigma_x \neq 0$, $\sigma_z \neq 0$, $\tau_{xy} \neq 0$. According to the Tresca yield criterion, yielding begins when the shear stress reaches a critical value, $\sigma_S/2$.:

$$\sigma_{max} - \sigma_{min} = \sigma_S , \tag{3.2.1}$$

where σ_{max} and σ_{min} are the maximal and minimal principal stresses acting in an element of a body; they are determined by well-known cubic equation. For the plane stress system the principal stresses are determined by Eq. (2.7.1). Consider the following three variants

a) $\sigma_1 > \sigma_2 > 0$, $\sigma_3 = 0$, $\sigma_{max} = \sigma_1$, $\sigma_{min} = 0$. Then from Eq. (3.2.1) it follows that $\sigma_1 = \sigma_S$, i.e.,

$$\tau_{xz}^2 = \sigma_x \sigma_z - \sigma_S (\sigma_x + \sigma_z) + \sigma_S^2 . \tag{3.2.2}$$

This is the equation of cone whose apex is at the point with coordinates $\sigma_x = \sigma_z = \sigma_S$ and $\tau_{xz} = 0$, and its axis coincides with bisector $\sigma_x = \sigma_z$ of $\tau_{xz} = 0$-plane.

b) $\sigma_1 > 0$, $\sigma_2 < 0$, $\sigma_3 = 0$, $\sigma_{max} = \sigma_1$, $\sigma_{min} = \sigma_2$. Now, the yield condition, $\sigma_1 - \sigma_2 = \sigma_S$, is

$$(\sigma_x - \sigma_z)^2 + 4\tau_{xz}^2 = \sigma_S^2 . \tag{3.2.3}$$

This is the equation of elliptic cylinder whose axis is the bisector $\sigma_x = \sigma_z$ of $\tau_{xz} = 0$-plane; the semi-major and semi-minor axes of the base-ellipse are $\sigma_S / \sqrt{2}$ and $\sigma_S / 2$, respectively.

c) $\sigma_2 < \sigma_1 < 0$, $\sigma_3 = 0$, $\sigma_{max} = 0$, $\sigma_{min} = \sigma_2$. Then $-\sigma_2 = \sigma_S$, i.e.,

$$\tau_{xz}^2 = \sigma_x \sigma_z + \sigma_S (\sigma_x + \sigma_z) + \sigma_S^2 \tag{3.2.4}$$

giving a cone with apex at the point $\sigma_x = \sigma_z = -\sigma_S$ and $\tau_{xz} = 0$, the cone axis is the same as in (3.2.2).

The yield surface constructed in σ_x-σ_z-τ_{xz} stress space is shown in Fig. 3.1, portions 1, 2, and 3 are constructed on the base of Eqs. (3.2.4), (3.2.3), and (3.2.2), respectively. The surface consists of the elliptic cylinder 2, bounded by the cones 1 and 3. The intersection of the surface and $\sigma_x = 0$-plane gives the line given by Eq. (2.6.12). The projection of the surface on $\tau_{xz} = 0$-plane gives the Tresca hexagon.

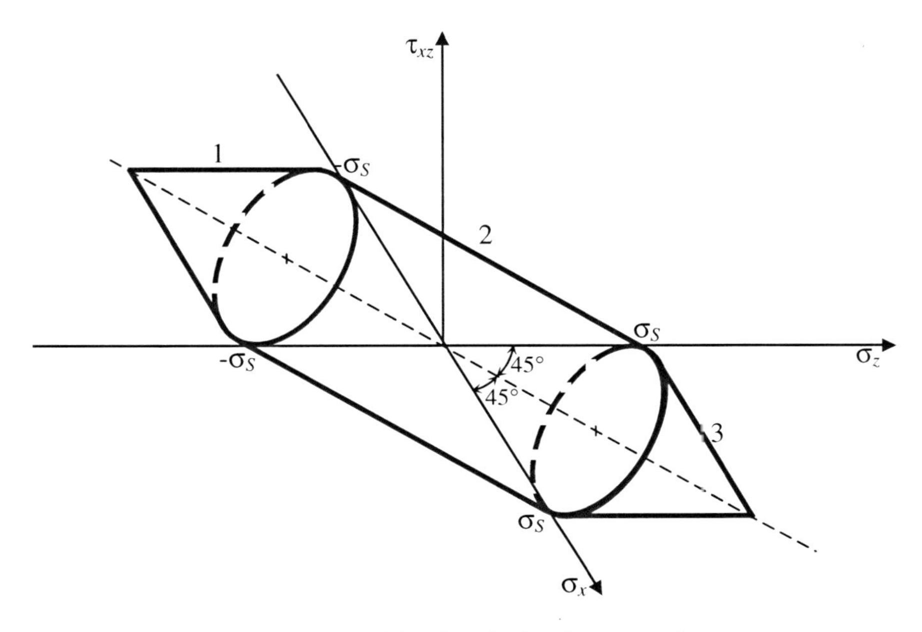

Fig. 3.1 Tresca's yield surface for the plane state of stress.

Now we proceed to the construction of the Tresca surface under the loading given by three stress tensor components, $\sigma_x \neq 0$, $\sigma_z \neq 0$, $\tau_{xy} \neq 0$. According to Eq. (2.25.4), the principal stresses are

$$\sigma_1 = \frac{\sigma_x}{2} + \frac{1}{2}\sqrt{\sigma_x^2 + 4\tau_{xy}^2} \geq 0,\ \sigma_2 = \frac{\sigma_x}{2} - \frac{1}{2}\sqrt{\sigma_x^2 + 4\tau_{xy}^2} \leq 0,\ \text{and}\ \sigma_3 = \sigma_z. \tag{3.2.5}$$

Here, three variants are possible as well:

a). $\sigma_{\max} = \sigma_z$, $\sigma_{\min} = \sigma_2$. The yield criterion has the form $\sigma_z - \sigma_2 = \sigma_S$, i.e.,

$$\sigma_z - \frac{\sigma_x}{2} + \frac{1}{2}\sqrt{\sigma_x^2 + 4\tau_{xy}^2} = \sigma_S. \tag{3.2.6}$$

This is the equation of cone with apex at the point $\sigma_x = \tau_{xy} = 0$ and $\sigma_z = \sigma_S$, its axis lies on $\tau_{xy} = 0$-plane making the angle of 22.5° with σ_z-axis.

b). $\sigma_{\max} = \sigma_1$, $\sigma_{\min} = \sigma_2$. Therefore, $\sigma_1 - \sigma_2 = \sigma_S$, i.e.,

$$\sigma_x^2 + 4\tau_{xy}^2 = \sigma_S^2. \tag{3.2.7}$$

This is the equation of elliptic cylinder whose axis coincides with σ_z-axis.

c). $\sigma_{\max} = \sigma_1$, $\sigma_{\min} = \sigma_z$. Then $\sigma_1 - \sigma_z = \sigma_S$, i.e.,

$$-\sigma_z + \frac{\sigma_x}{2} + \frac{1}{2}\sqrt{\sigma_x^2 + 4\tau_{xy}^2} = \sigma_S. \tag{3.2.8}$$

We obtain the cone with apex at the point $\sigma_x = 0$, $\tau_{xy} = 0$, and $\sigma_z = -\sigma_S$.

The Tresca yield surface constructed on the base of Eqs. (3.2.6)–(3.2.8) is shown in Fig. 3.2. Portions 1, 2, and 3 are determined by Eqs. (3.2.8), (3.2.7), and (3.2.6), respectively. The surface consists of elliptic cylinder 2 and two cones, 1 and 3. The intersection of the surface and $\tau_{xy} = 0$-plane gives the Tresca hexagon.

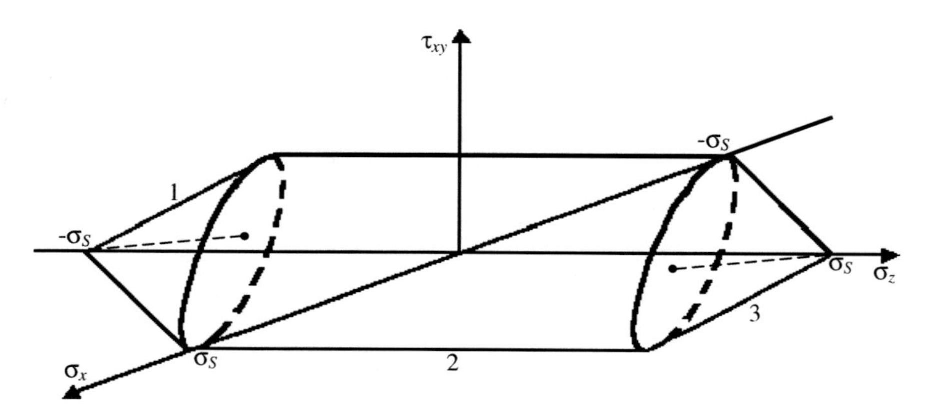

Fig. 3.2 Tresca's yield surface for the partial case of three-axial state of stress

Consider the case when the normal stresses are distinct from zero, whereas the shear stresses are $\tau_{xy} = \tau_{yz} = \tau_{xz} = 0$. In contrast to the above cases, the Tresca surface is an infinite surface in three-dimensional σ_x-σ_y-σ_z stress space, which encloses the straight line $\sigma_x = \sigma_y = \sigma_z$. Along this line, according to Eq. (2.3.2), $\tau_{nl} = 0$ for all slip planes, therefore, the condition (3.2.1) does not hold true.

3.3 Tresca Yield Surface in Five-Dimensional Stress Deviator Space

The construction of the surface (3.2.1) for an arbitrary stress state in six-dimensional stress space is very difficult, until now not solved, problem. The difficulty of the construction consists in that the yield surface consists of many parts and it is difficult to imagine their location in six- or five-dimensional space. As follows from the previous Section, the yield surface must acquire the form illustrated in Figs. 3.1 and 3.2 in σ_x - σ_z - τ_{xz} and σ_x - σ_z - τ_{xy} stress subspaces, respectively.

Let us construct not the yield surface itself but planes tangential to this surface. For this purpose, we turn to the hemisphere of unit radius discussed in Sec. 2.3 and illustrated in Fig. 2.6 and to Eqs. (2.3.2) and (2.3.8) giving the value of resolved shear stress τ_{nl} acting in a slip system. The orientation of slip system, the plane where τ_{nl} acts, is given by spherical angles α and β; the direction of the action of τ_{nl} on this plane is given by angle ω. By using Eqs. (1.7.3) and (1.7.10), Eq. (2.3.2) can be expressed in terms of stress vector components:

$$\tau_{nl} = \frac{1}{\sqrt{2}}\left[l_x n_x \sqrt{3} S_1 + \left(l_y n_y - l_z n_z\right) S_2 + \left(l_x n_z + l_z n_x\right) S_3 + \left(l_x n_y + l_y n_x\right) S_4 + \left(l_y n_z + l_z n_y\right) S_5\right]. \quad (3.3.1)$$

Since τ_{nl} is a smooth function of angles α, β, and ω which appear in the direction cosines $l_x, \ldots, n_z$, the Tresca yield criterion, yielding begins when the shearing stress reaches a critical(maximal) value, is equivalent to the following equation

$$\frac{\partial \tau_{nl}}{\partial \alpha} = \frac{\partial \tau_{nl}}{\partial \beta} = \frac{\partial \tau_{nl}}{\partial \omega} = 0, \quad \tau_{nl} = \tau_S, \quad (3.3.2)$$

and the latter formula, on the base of Eq. (3.3.1), takes the form

$$l_x n_x \sqrt{3} S_1 + \left(l_y n_y - l_z n_z\right) S_2 + \left(l_x n_z + l_z n_x\right) S_3 + \left(l_x n_y + l_y n_x\right) S_4 + \left(l_y n_z + l_z n_y\right) S_5 - \sqrt{2}\tau_S = 0. \quad (3.3.3)$$

Since $S_1, \ldots, S_5$ appear linearly in the above formula, Eq. (3.3.3) is an equation of plane in stress deviator stress, provided that the values of $l_x, \ldots, n_z$ are fixed. It is obvious that different directions cosines $l_x, \ldots, n_z$ give different planes.

The key to the situation is that Eqs. (3.3.2) express the fact that the ***Tresca surface is the inner envelope of planes (3.3.3)***. In other words, ***planes from Eq. (3.3.3) are tangential to the Tresca yield surface***.

Let us specify two possible variants of the using of Eq. (3.3.2).

A) If the slip system, where first plastic slip to be occurred, is given, i.e., the values of $l_x, \ldots, n_z$ are fixed, Eqs. (3.3.2) become the system of equations for components $S_1, \ldots, S_5$ which induce the first yielding in this slip system. The founded values of $S_1, \ldots, S_5$ determine the point where plane (3.3.3) touches the Tresca surface.

B) Consider a loading along some trajectory, e.g., along a direct line when stress vector components are

$$S_k = S_k^0 \eta, \tag{3.3.4}$$

where S_k^0 determines the direction of the stress vector $\vec{\mathbf{S}}$ and parameter η varies from 0 to such value that $\vec{\mathbf{S}}$ reaches the Tresca surface. The question to be decided is *in what slip system and in what direction first yielding will occur?* At such formulation, Eqs. (3.3.2) are used to find coefficient η and angles α, β, and ω. Once η is known, the coordinates where the plane (3.3.3) touches the yield surface, S_k, is also known.

Hence, it is worthwhile to repeat that, as follows from Eq. (3.3.2), the Tresca surface is the inner envelope of planes (3.3.3). However, the following doubts can arise, whether the set of planes given by (3.3.3) contains in itself all the planes tangential to the Tresca surface. These doubts are caused by the following. In five-dimensional space, planes passing through some point can be defined by four parameters, less by unit than the dimension of space, while planes (3.3.3) are defined by only three angles α, β, and ω. This can be explained by that the system of equations (3.3.2) is indeterminate (four equations for five components $S_1,\ldots,S_5$), meaning that the locus of the tangency of plane (3.3.3) to the surface (3.2.1) is a line, which plays the part of fourth parameter. For example, Fig. 3.1 and 3.2 illustrate the cases when a plane touches a surface along the generator of cylinder or cone.

Further, let us designate through $\vec{\mathbf{N}}(N_1,\ldots,N_5)$ the vector normal to plane (3.3.3):

$$\vec{\mathbf{N}} = N_i \vec{\mathbf{g}}_i$$

$$N_1 = \sqrt{3} l_x n_x,\ N_2 = l_y n_y - l_z n_z,\ N_3 = l_x n_z + l_z n_x, \tag{3.3.5}$$

$$N_4 = l_x n_y + l_y n_x,\ N_5 = l_y n_z + l_z n_y$$

where $\vec{\mathbf{g}}_i$ is (as in Chapter 1) unit vectors directed along S_i-axes in Ilyushin's space. Together with Eq. (2.3.8), it is easy to show that the vector $\vec{\mathbf{N}}$ is unit,

$$\left|\vec{\mathbf{N}}\right|^2 = N_i N_i = 1, \tag{3.3.6}$$

meaning that any plane from Eq. (3.3.3) is distant of $\sqrt{2}\tau_S$ from the origin of coordinates. It must be emphasized that this does not mean at all that the Tresca surface is a sphere, even in a plane stress-state it acquires the form differing from sphere (Fig. 3.1).

3.4 Koiter's Result

Following Sanders [165], we assume that a stress vector shifts planes tangential to the yield surface during loading. The displacements of the planes on the endpoint of the stress vector are translational, i.e without the change in the orientations of planes. The movements of planes symbolize the occurring of plastic deformation. An elementary plastic strain vector is assumed to be normal to tangent plane and dependent on the distance covered by the plane. The macro-deformation is determined as the sum of elementary strains "produced" by planes shifted by stress vector, i.e., the scheme described in Sec. 1.19 is fully accepted.

Let us determine plastic strain vector components in terms of the above formulation. Doing as in the work [3,4,140], we consider the increments in normal vector $\vec{\mathbf{N}}$ due to the change in angles α, β, and ω. These increments are:

$$d\vec{\mathbf{N}}_\alpha = \frac{\partial \vec{\mathbf{N}}}{\partial \alpha} d\alpha, \quad d\vec{\mathbf{N}}_\beta = \frac{\partial \vec{\mathbf{N}}}{\partial \beta} d\beta, \quad d\vec{\mathbf{N}}_\omega = \frac{\partial \vec{\mathbf{N}}}{\partial \omega} d\omega. \tag{3.4.1}$$

$$d\vec{\mathbf{N}}_\alpha = \frac{\partial N_i}{\partial \alpha} \vec{\mathbf{g}}_i d\alpha, \quad d\vec{\mathbf{N}}_\beta = \frac{\partial N_i}{\partial \beta} \vec{\mathbf{g}}_i d\beta, \quad d\vec{\mathbf{N}}_\omega = \frac{\partial N_i}{\partial \omega} \vec{\mathbf{g}}_i d\omega. \tag{3.4.2}$$

The volume of parallelepiped constructed on vectors $d\vec{\mathbf{N}}_\alpha$, $d\vec{\mathbf{N}}_\beta$, and $d\vec{\mathbf{N}}_\omega$ can be determined by the Gram determinant:

$$(dV)^2 = \begin{vmatrix} d\vec{\mathbf{N}}_\alpha \cdot d\vec{\mathbf{N}}_\alpha & d\vec{\mathbf{N}}_\alpha \cdot d\vec{\mathbf{N}}_\beta & d\vec{\mathbf{N}}_\alpha \cdot d\vec{\mathbf{N}}_\omega \\ d\vec{\mathbf{N}}_\beta \cdot d\vec{\mathbf{N}}_\alpha & d\vec{\mathbf{N}}_\beta \cdot d\vec{\mathbf{N}}_\beta & d\vec{\mathbf{N}}_\beta \cdot d\vec{\mathbf{N}}_\omega \\ d\vec{\mathbf{N}}_\omega \cdot d\vec{\mathbf{N}}_\alpha & d\vec{\mathbf{N}}_\omega \cdot d\vec{\mathbf{N}}_\beta & d\vec{\mathbf{N}}_\omega \cdot d\vec{\mathbf{N}}_\omega \end{vmatrix}, \tag{3.4.3}$$

which, after bulky calculations [3,4,140], is

$$dV = \kappa \cos\beta d\alpha d\beta d\omega, \tag{3.4.4}$$

where κ is immaterial factor.

According to the above-stated initial provisions, the plastic strain induced by the movement of single plane, $\vec{\mathbf{e}}_0^S$, is presented as

$$\vec{\mathbf{e}}_0^S = \tilde{F}(H_N)\vec{\mathbf{N}}, \tag{3.4.5}$$

where $\tilde{F}$ is the characteristic function of a material dependent on a plane distance from the origin of coordinates in the Ilyushin deviator space, H_N. As the initial distances are equal to $\sqrt{2}\tau_S$, $H_N - \sqrt{2}\tau_S$ gives the displacement of plane. An elementary plastic strain vector, $d\vec{\mathbf{e}}_N^S$, caused by the movements of planes whose normal vectors are inside the parallelepiped constituted by vectors $d\vec{\mathbf{N}}_\alpha$, $d\vec{\mathbf{N}}_\beta$, and $d\vec{\mathbf{N}}_\omega$ is assumed to be proportional to the volume dV of this parallelepiped:

$$d\vec{\mathbf{e}}_N^S = \tilde{F}(H_N)\vec{\mathbf{N}}dV\,. \tag{3.4.6}$$

The $d\vec{\mathbf{e}}_N^S$ vector components are

$$de_i^S = \tilde{F}(H_N)N_i dV\,. \quad (I = 1,2,3) \tag{3.4.7}$$

Total plastic strain components, e_i^S, are obtained as the sum of components (3.4.7), three-folded integral over those directions where the planes incur displacements on the endpoint of stress vector:

$$e_i^S = \kappa \iiint\limits_V \tilde{F}(H_N)N_i \cos\beta d\alpha d\beta d\omega. \tag{3.4.8}$$

In arriving at the result (3.4.8) we have taken into consideration the relation (3.4.4).

Each plane in (3.3.3) corresponds to certain slip system. Therefore, the integration over angles α, β, and λ in Eq. (3.4.8) can be replaced by integration over Ω and ω in terms of the slip concept, Ω determines the area on hemisphere of unit radius in *x*, *y*, *z* coordinate system where slips occur and ω gives the slip directions. In addition, the plane distance H_N (index *N* indicates on normal vector $\vec{\mathbf{N}}$) is equivalent to shear stress τ_{nl} in the appropriate *n-l* slip system. As a result of these remarks, Eq. (3.4.8) can be rewritten as

$$e_i^S = \kappa \iint\limits_\Omega \int\limits_\omega \tilde{F}(\tau_{nl})N_i \cos\beta d\alpha d\beta d\omega\,. \tag{3.4.9}$$

If to convert the strain vector components to strain tensor components by Eq. (1.7.13), then Eq. (3.4.9) results in Eq. (2.3.5) obtained in terms of the Budiansky's concept of slip. This result was obtained by V. Koiter by another method [64].

Therefore, two essentially different models of plastic deformation, the Budiansky theory of slip and the concept of the movements of planes, result in identical formulas. The important point here is that (i) the Sanders theory can be considered as a physical one as each tangent plane corresponds to appropriate slip system; (ii) the fact that the Budiansky theory possesses serious shortcomings induces us to modify the Tresca yield criterion.

3.5 Introduction of New Variables

Let us consider the three-dimensional subspace $S_1S_2S_3$, $\mathbf{R}^3$, of five-dimensional space S_1, S_2, …, S_5, $\mathbf{R}^5$. In $\mathbf{R}^3$, the trace of tangent plane (3.3.3) can be written as

$$l_x n_x \sqrt{3} S_1 + \left(l_y n_y - l_z n_z\right) S_2 + \left(l_x n_z + l_z n_x\right) S_3 = \sqrt{2}\tau_S . \qquad (3.5.1)$$

On the base of Eq. (2.3.8) it is possible to establish that the normalization factor of plane (3.5.1), d, is

$$d = \sqrt{1 - l_y^2 - n_y^2 + 4 l_y^2 n_y^2} \qquad (3.5.2)$$

Let us express Eq. (3.5.1) in terms of characteristics of $\mathbf{R}^3$. The normalized equation of plane has the form

$$m_1 S_1 + m_2 S_2 + m_3 S_3 = h_0 , \qquad (3.5.3)$$

where h_0 is the plane distance m_1, m_2, and m_3 are the direction cosines of unit vector $\vec{\mathbf{m}}$ normal to the plane. Let us define its components as follows

$$m_1 = \cos\hat{\alpha}\cos\hat{\beta}, \quad m_2 = \sin\hat{\alpha}\cos\hat{\beta}, \quad m_3 = \sin\hat{\beta} . \qquad (3.5.4)$$

Angles $\hat{\alpha}$ and $\hat{\beta}$ are shown in Fig. 3.3.

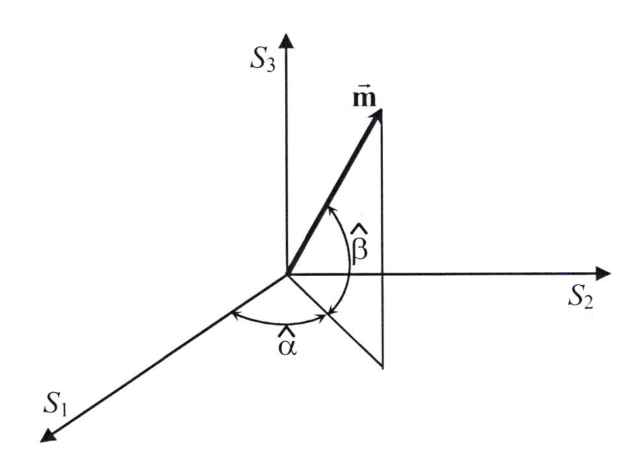

Fig. 3.3 Spherical coordinates of normal vector $\vec{\mathbf{m}}$

Let us take into consideration an angle λ, which is between vector $\vec{\mathbf{N}}$ normal to plane (3.3.3) and vector $\vec{\mathbf{m}}$ normal to trace-plane (3.5.1):

$$\cos\lambda = \vec{\mathbf{N}} \cdot \vec{\mathbf{m}} . \qquad (3.5.5)$$

From geometrical reasons it is possible to establish that the angle λ and the plane distance h_0 (3.5.3) are related to each other as

$$h_0 \cos\lambda = \sqrt{2}\tau_S\,. \tag{3.5.6}$$

Figure 3.4 shows the illustration of Eq. (3.5.6). For simplicity, the surface marked by **1** in Fig. 3.4 represents the part of the Tresca yield surface in $\mathbf{R}^5$ plane **2** with normal $\vec{\mathbf{N}}$ is tangential to this surface, line **3** is the trace of plane **2** in $\mathbf{R}^3$ that is marked by **4**, its orientation is given by vector $\vec{\mathbf{m}}$; OM_1 and OM_2 are the distances between the origin of coordinate and the planes **2** and **3**, respectively.

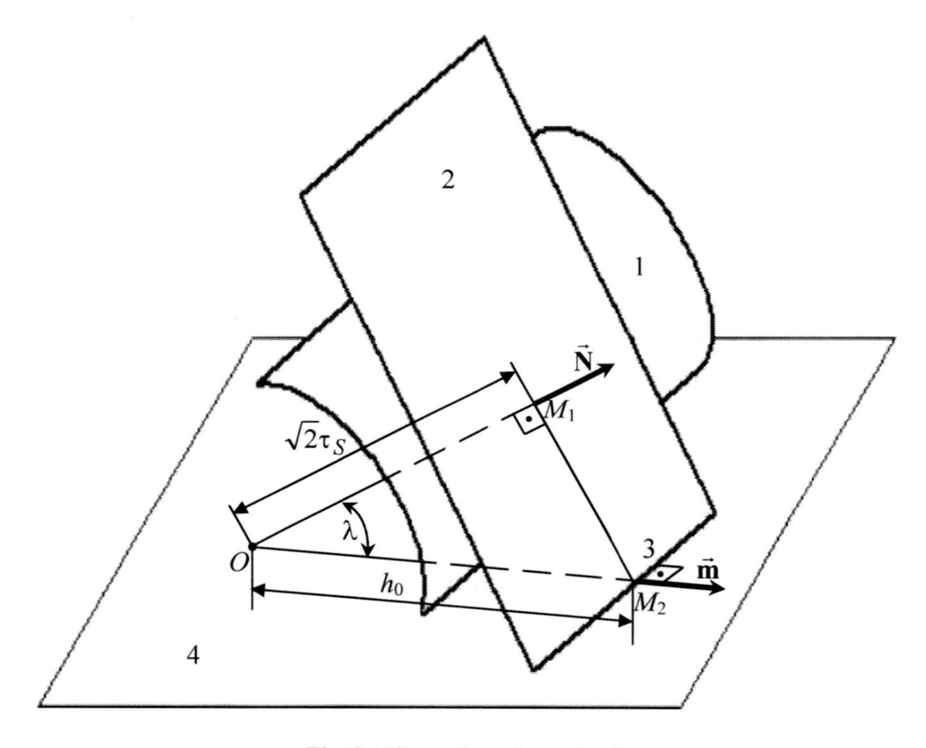

Fig. 3.4 Illustration of Eq. (3.5.6)

Since planes (3.5.1) and (3.5.3) are identical, we have

$$m_1 = \frac{l_x n_x \sqrt{3}}{d}, \quad m_2 = \frac{l_y n_y - l_z n_z}{d}, \quad m_3 = \frac{l_x n_z + l_z n_x}{d}. \tag{3.5.7}$$

In addition,

$$h_0 = \frac{\sqrt{2}\tau_S}{d}. \tag{3.5.8}$$

Therefore, Eqs. (3.5.6) and (3.5.8) give

$$d = \cos\lambda, \quad 0 \le d \le 1. \tag{3.5.9}$$

i.e.,

$$\begin{gathered} l_x n_x \sqrt{3} = m_1 \cos\lambda, \quad l_y n_y - l_z n_z = m_2 \cos\lambda, \\ l_x n_z + l_z n_x = m_3 \cos\lambda. \end{gathered} \tag{3.5.10}$$

Furthermore, owing to orthonormality of vectors $\vec{\mathbf{l}}$ and $\vec{\mathbf{n}}$ (Fig. 2.6), we have

$$l_x^2 + l_y^2 + l_z^2 = 1, \quad n_x^2 + n_y^2 + n_z^2 = 1, \; l_x n_x + l_y n_y + l_z n_z = 0. \tag{3.5.11}$$

We take Eqs. (3.5.10) and (3.5.11) as the system of six equations for six unknowns, $l_x, l_y, \ldots, n_z$. The following relations can be obtained from Eqs. (3.5.10) and (3.5.11)

$$l_x n_y + l_y n_x = \pm\sqrt{\frac{\sin^2\lambda}{2} + \chi}, \; l_z n_y + l_y n_z = \pm\sqrt{\frac{\sin^2\lambda}{2} - \chi}, \tag{3.5.12}$$

where

$$\begin{gathered} \chi = \frac{(\overline{m}_1 - m_2)(3\overline{m}_1 + m_2)\left\{1 - \left[3 - 2(\overline{m}_1 - m_2)^2\right]\cos^2\lambda\right\} - 4m_3\sqrt{1 - (\overline{m}_1 - m_2)^2}\sqrt{D}}{2\left[4 - 3(\overline{m}_1 - m_2)^2\right]}, \\ D = 1 - \left[2 - (\overline{m}_1 - m_2)^2\right]\cos^2\lambda + \left[1 - (\overline{m}_1 - m_2)^2\right]^2 \cos^4\lambda, \; \overline{m}_1 = m_1/\sqrt{3}. \end{gathered} \tag{3.5.13}$$

By inserting Eqs (3.5.10) and (3.5.12) into Eq. (3.3.3), we obtain

$$S_1 m_1 \cos\lambda + S_2 m_2 \cos\lambda + S_3 m_3 \cos\lambda + S_4\left(\pm\sqrt{\frac{\sin^2\lambda}{2} + \chi}\right) + S_5\left(\pm\sqrt{\frac{\sin^2\lambda}{2} - \chi}\right) = \sqrt{2}\tau_S. \tag{3.5.14}$$

Equations (3.3.3) and (3.5.14) represent one and the same plane, which is tangential to the Tresca surface. The difference in their records is that the factors at S_i in Eq. (3.3.3) are expressed through the Budiansky variables (α, β, and ω) and, in Eq. (3.5.14), those are defined by the new variables of the Ilyushin stress-deviator-space, $\hat{\alpha}$, $\hat{\beta}$, and λ.

3.6 Modification of Yield Criterion

Now we propose the following suggestion, the Tresca yield surface, which is the inner-envelop of planes given by Eq. (3.5.14), is replaced by a new surface, which is the inner-envelop of the following planes

$$S_1 m_1 \cos\lambda + S_2 m_2 \cos\lambda + S_3 m_3 \cos\lambda + S_4 \left(\pm\sqrt{\frac{\sin^2\lambda}{2} + \chi_0(\lambda)} \right) + S_5 \left(\pm\sqrt{\frac{\sin^2\lambda}{2} - \chi_0(\lambda)} \right) = \sqrt{2}\tau_S \quad (3.6.1)$$

Equation (3.6.1) differs from Eq. (3.5.14) in that the function $\chi(\hat{\alpha}, \hat{\beta}, \hat{\omega})$ of three variables is replaced by function $\chi_0(\lambda)$ of single variable. Let us designate through $\vec{\mathbf{N}}$ a normal vector to plane (3.6.1)

$$\vec{\mathbf{N}} = N_i \vec{\mathbf{g}}_i,$$

$$N_1 = m_1 \cos\lambda, \quad N_2 = m_2 \cos\lambda, \quad N_3 = m_3 \cos\lambda,$$

$$N_4 = \pm\sqrt{\frac{\sin^2\lambda}{2} + \chi_0(\lambda)}, \quad N_5 = \pm\sqrt{\frac{\sin^2\lambda}{2} - \chi_0(\lambda)} \quad (3.6.2)$$

and it is easy to see that $\left|\vec{\mathbf{N}}\right| = 1$.

Acting as in Sec. 3.4, we consider the increment in vector $\vec{\mathbf{N}}$ due to the increments in λ, $\hat{\alpha}$, and $\hat{\beta}$

$$d\vec{\mathbf{N}}_\lambda = \frac{\partial \vec{\mathbf{N}}}{\partial \lambda} d\lambda, \quad d\vec{\mathbf{N}}_{\hat{\alpha}} = \frac{\partial \vec{\mathbf{N}}}{\partial \hat{\alpha}} d\hat{\alpha}, \quad d\vec{\mathbf{N}}_{\hat{\beta}} = \frac{\partial \vec{\mathbf{N}}}{\partial \hat{\beta}} d\hat{\beta}. \quad (3.6.3)$$

Obviously that

$$d\vec{\mathbf{N}}_\lambda = \frac{\partial N_i}{\partial \lambda} \vec{\mathbf{g}}_i d\lambda, \quad d\vec{\mathbf{N}}_{\hat{\alpha}} = \frac{\partial N_i}{\partial \hat{\alpha}} \vec{\mathbf{g}}_i d\hat{\alpha}, \quad d\vec{\mathbf{N}}_{\hat{\beta}} = \frac{\partial N_i}{\partial \hat{\beta}} \vec{\mathbf{g}}_i d\hat{\beta}. \quad (3.6.4)$$

The volume of parallelepiped, dV^2, constructed on the vectors $d\vec{\mathbf{N}}_\lambda$, $d\vec{\mathbf{N}}_{\hat{\alpha}}$, and $d\vec{\mathbf{N}}_{\hat{\beta}}$ is calculated by the Gram determinant:

$$(dV)^2 = \begin{vmatrix} d\vec{\mathbf{N}}_\lambda \cdot d\vec{\mathbf{N}}_\lambda & d\vec{\mathbf{N}}_\lambda \cdot d\vec{\mathbf{N}}_{\hat{\alpha}} & d\vec{\mathbf{N}}_\lambda \cdot d\vec{\mathbf{N}}_{\hat{\beta}} \\ d\vec{\mathbf{N}}_{\hat{\alpha}} \cdot d\vec{\mathbf{N}}_\lambda & d\vec{\mathbf{N}}_{\hat{\alpha}} \cdot d\vec{\mathbf{N}}_{\hat{\alpha}} & d\vec{\mathbf{N}}_{\hat{\alpha}} \cdot d\vec{\mathbf{N}}_{\hat{\beta}} \\ d\vec{\mathbf{N}}_{\hat{\beta}} \cdot d\vec{\mathbf{N}}_\lambda & d\vec{\mathbf{N}}_{\hat{\beta}} \cdot d\vec{\mathbf{N}}_{\hat{\alpha}} & d\vec{\mathbf{N}}_{\hat{\beta}} \cdot d\vec{\mathbf{N}}_{\hat{\beta}} \end{vmatrix} \quad (3.6.5)$$

Equations (3.5.4), (3.6.2), and (3.6.4) give that the above determinant takes the following form

$$(dV)^2 = \begin{vmatrix} \frac{\partial \vec{\mathbf{N}}}{\partial \lambda} \cdot \frac{\partial \vec{\mathbf{N}}}{\partial \lambda} & 0 & 0 \\ 0 & \cos^2 \hat{\beta} \cos^2 \lambda & 0 \\ 0 & 0 & \cos^2 \lambda \end{vmatrix} \left(d\hat{\alpha} d\hat{\beta} d\lambda\right)^2 \tag{3.6.6}$$

i.e.,

$$(dV)^2 = \frac{\partial \vec{\mathbf{N}}}{\partial \lambda} \cdot \frac{\partial \vec{\mathbf{N}}}{\partial \lambda} \cos^2 \hat{\beta} \cos^4 \lambda \left(d\hat{\alpha} d\hat{\beta} d\lambda\right)^2. \tag{3.6.7}$$

Let us require that volume dV be

$$dV = \cos \hat{\beta} d\hat{\alpha} d\hat{\beta} d\lambda, \tag{3.6.8}$$

therefore,

$$\frac{\partial \vec{\mathbf{N}}}{\partial \lambda} \cdot \frac{\partial \vec{\mathbf{N}}}{\partial \lambda} = \frac{1}{\cos^4 \lambda}, \tag{3.6.9}$$

or

$$\frac{\partial N_i}{\partial \lambda} \frac{\partial N_i}{\partial \lambda} = \frac{1}{\cos^4 \lambda}. \tag{3.6.10}$$

The latter equality, on the base of Eq. (3.6.2), can be written in the following form:

$$[\chi_0'(\lambda)]^2 - 4\cot\lambda \chi_0(\lambda) \chi_0'(\lambda) + 4\left(\frac{1}{\cos^4 \lambda \sin^2 \lambda} - 1\right) \chi_0^2(\lambda) - \tan^2 \lambda \left(\tan^2 \lambda + \sin^2 \lambda\right) = 0 \cdot \tag{3.6.11}$$

This is the Ricatti differential equation for function $\chi_0(\lambda)$. Its solution [3,140] gives the function $\chi_0(\lambda)$, which appears in Eq. (3.6.1), as

$$\chi_0(\lambda) = \frac{1}{2} \sin^2 \lambda \sin y_0, \quad y_0 = 2\sqrt{2}\left[F_0\left(\lambda, \frac{1}{\sqrt{2}}\right) - E_0\left(\lambda, \frac{1}{\sqrt{2}}\right) + \sqrt{1 - \frac{1}{2}\sin^2 \lambda}\, \tan \lambda\right], \tag{3.6.12}$$

where F_0 and E_0 are incomplete elliptic integrals of the first and second kind:

$$F_0\left(\lambda, \frac{1}{\sqrt{2}}\right) = \int_0^\lambda \frac{d\lambda}{\sqrt{1 - 1/2 \sin^2 \lambda}}, \quad E_0\left(\lambda, \frac{1}{\sqrt{2}}\right) = \int_0^\lambda \sqrt{1 - 1/2 \sin^2 \lambda}\, d\lambda.$$

3.7 Modified Yield Surface

In subspace $\mathbf{R}^3$, we wish to construct the new (modified) yield surface, which is the inner-envelope of planes given by Eq. (3.6.1).

It is worthwhile to mention that the Tresca yield surface is not a smooth but piece-wise smooth surface, its parts are conjugated along some lines (see Fig. 3.1 and 3.2), for example, the cylinder is conjugated with two cones. Not all planes tangential to yield surface exists along the conjugate lines, this follows from Eq. (3.5.2) and (3.5.9) expressed in terms of the Budiansky variables. Indeed, since direction cosines $l_x, l_y, \ldots, n_z$ range from 0 to 1, Eq. (3.5.2) gives that the area of change of d is $0 \le d \le \sqrt{3}$, whereas the formula $d = \cos\lambda$ requires that the range of d be $0 \le d \le 1$. In terms of Ilyushin's variables, this situation is such that the radicand D in Eq. (3.5.13) can be negative at some values of m_1 and λ. As seen from Eqs. (3.6.2) and (3.6.12), the new yield surface, being the inner-envelope of planes (3.6.1), have tangent planes at all its points, i.e., it is a smooth surface that allows us to employ the standard procedure for the constructing of inner-envelope surface. The condition that the modified surface be the inner-envelope of planes from Eq. (3.6.1) is

$$\frac{d\tau}{d\hat{\alpha}} = \frac{d\tau}{d\hat{\beta}} = \frac{d\tau}{d\lambda} = 0, \tag{3.7.1}$$

where τ is the left-hand side in Eq. (3.6.1). First two equalities in Eq. (3.7.1) and Eqs. (3.5.4) and (3.6.1) give

$$\tan\hat{\alpha} = \frac{S_2}{S_1}, \quad \tan\hat{\beta} = \frac{S_3}{S_1\cos\hat{\alpha} + S_2\sin\hat{\alpha}}. \tag{3.7.2}$$

The third equality in (3.7.1) implies $\lambda = 0$ at $S_4 = S_5 = 0$. If the obtained values of the angles to substituted into Eq. (3.6.1), we obtain

$$S_1^2 + S_2^2 + S_3^2 = 2\tau_S^2 \tag{3.7.3}$$

at $S_4 = S_5 = 0$. For the case uniaxial tension along x-axis we have $S_1 = \sqrt{2}\,\sigma_x/\sqrt{3}$, $S_2 = S_3 = 0$, therefore the following relation between the yield limits under uniaxial tension, σ_S, and pure shear, τ_S, is obtained from Eq. (3.7.3):

$$\sigma_S = \sqrt{3}\tau_S. \tag{3.7.4}$$

Thus, in subspace $\mathbf{R}^3$, the surface, which is the inner envelope of planes (3.6.1), takes the form of the von-Mises-Guber sphere. Let us note that the form of

Eq. (3.7.3) does not depend on the form of function $\chi_0(\lambda)$. Consequently, the replacement of three-variable function $\chi(\hat{\alpha}, \hat{\beta}, \lambda)$ by single-variable function $\chi_0(\lambda)$ and the requirement that the Gram determinant degenerate to formula (3.6.8) give sufficient conditions to obtain the regular yield surface $\mathbf{R}^5$ whose trace in $\mathbf{R}^3$ is the von-Mises-Guber sphere (3.7.3). It is very important to emphasize once more that sphere (3.7.3) is the trace of the five-dimensional yield surface which is neither the Tresca nor von Mises-Guber yield surface.

The following question can arise, maybe, instead of the new surface of plasticity, it is easier to accept Guber's yield criterion in five-dimensional space. Then it would be possible to consider a general state of stress when all stress tensor components are distinct from zero. However, this variant is not suitable for two reasons. First, it is extremely difficult to watch the movements of tangential planes in five-dimensional space at curvilinear loading path, when the movement of some planes can be terminated and others start their displacements; the planes experiencing displacements are asymmetrically located above a stress vector etc. Secondly, planes tangential to five-dimensional Guber's hyper sphere are defined by four parameters requiring that a plastic strain be determined by four-fold integral, whereas the concept of slip prescribes that a plastic strain is calculated by three-fold integrals.

Another question to be decided is whether it is suitable to take the *three-dimensional* von Mises-Guber sphere (3.7.3) ($S_4 = S_5 = 0$) to be the yield surface and to determine an irreversible strain as the result of the movements of planes tangential only to (3.7.3). The answer to this question is also negative. In terms of all plastic flow theories, a plastic deformation is developed owing to the change of loading surface in *five-dimensional* space independently of the number of non-zero stress vector components.

3.8 Tangential Planes and Their Traces

The equation of any plane tangential to the new yield surface in $\mathbf{R}^5$ is defined by Eq. (3.6.1) and its trace in subspace $\mathbf{R}^3$ by Eq. (3.5.3). The planes are determined by three angles, spherical angles $\hat{\alpha}$ and $\hat{\beta}$ in $\mathbf{R}^3$ (Fig. 3.3) and angle λ between normal vectors $\vec{\mathbf{N}}$ and $\vec{\mathbf{m}}$. According to Eq. (3.5.6), the angle λ is expressed through the trace distance from the origin of coordinates (3.5.3) in $\mathbf{R}^3$, h_0. On the base of Eqs. (3.5.8) and (3.5.9) one can conclude that the distance h_0 can take values from the following range

$$\sqrt{2}\tau_S \le h_0 < \infty . \tag{3.8.1}$$

Therefore, the traces of the tangent planes, given by Eq. (3.5.3), fill up the whole space beyond the sphere (3.7.3) in $\mathbf{R}^3$. In other words, any plane located beyond the sphere (3.7.3) is the trace of some plane tangential to the yield surface in $\mathbf{R}^5$.

Let us consider planes which are located beyond yield surface in $\mathbf{R}^5$. Their equations can be written in the form of Eq. (1.19.3) and their traces in $\mathbf{R}^3$ are

$$N_{ij}S_i = H_j. \quad (i = 1,2,3) \tag{3.8.2}$$

It is obvious that planes (3.8.2) are also located beyond the sphere (3.7.3). Different N_{j4} and N_{j5} give different planes (1.19.3) with identical traces (3.8.2). Therefore, some fixed plane (3.8.2) locating beyond Guber's sphere is the trace of infinite number of planes in $\mathbf{R}^5$, one of these planes, for which $H_j = \sqrt{2}\tau_S$, is tangential to five-dimensional yield surface.

According to the initial rules of the Koiter-Sanders theory, a plastic deformation is developed due to the movements of tangent planes in five-dimensional stress-deviator space. Therefore, for investigating deformation in $\mathbf{R}^3$, it is necessary to take into account the displacements of planes tangential to sphere (3.7.3) as well as all planes filling the subspace $\mathbf{R}^3$ beyond this sphere; a stress vector can shift(push) all these planes.

3.9 Basic Equations

Further throughout, we will consider the case when the stress vector components S_4 and S_5 are zero, i.e., $\vec{\mathbf{S}} \in R^3$ and

$$\vec{\mathbf{S}} = S_1\vec{\mathbf{g}}_1 + S_2\vec{\mathbf{g}}_2 + S_3\vec{\mathbf{g}}_3. \tag{3.9.1}$$

Now, let us formulate the basic statements of the Koiter-Sanders theory in view of the discussion in Sec. 3.7 and 3.8. First, the Tresca yield criterion has been modified, its trace in $\mathbf{R}^3$ has the form of sphere (3.7.3). Second, at each point of the sphere (3.7.3), we draw a tangent plane whose normal vector is determined by angles $\hat{\alpha}$ and $\hat{\beta}$ (Fig. 3.3). Further, for each direction, parallel planes to the tangent plane are considered as well. These planes, being oriented identically, differ in their distances from the origin of coordinates h_0, which is related to angle λ as $h_0 \cos\lambda = \sqrt{2}\tau_S$ and $\lambda = 0$ implies $h_0 = \sqrt{2}\tau_S$ specifying plane tangential to sphere (3.7.3).

The following suggestion consists in that the stress vector (3.9.1) shifts planes tangential to the yield surface on its endpoint during loading. The movements of the planes on the endpoint of stress vector $\vec{\mathbf{S}}$ are translational, i.e without a change in their orientation. Those planes which are not reached by the endpoint of vector $\vec{\mathbf{S}}$ remain unmovable.

A movement of plane on the endpoint of stress vector symbolizes the development of plastic deformation. A plastic strain vector is assumed to be normal to the plane that incurs displacement.

The plastic strain increment, induced by the movement of an elementary system of planes due to an infinitesimal increment in stress vector , is assumed to be proportional to the parallelepiped volume dV (3.6.8) constructed on normal vector increments (3.6.3). Together with Eq. (3.6.2), this statement acquires the following form

$$de_k^S = \varphi_N m_k \cos\lambda dV \,, \qquad k = 1,2,3 \tag{3.9.2}$$

where φ_N is a new notion called the plastic deformation intensity.

The total plastic strain, produced by finite change in stress vector, is calculated as the sum of components (3.9.2), three-folded integration above all moving planes:

$$e_k^S = \int\limits_{\hat{\alpha}} \int\limits_{\hat{\beta}} \int\limits_{\lambda} \varphi_N m_k \cos\lambda dV \,, \qquad k = 1,2,3 \tag{3.9.3}$$

Making use of Eq. (3.6.8), we obtain[1]

$$e_k^S = \int\limits_{\alpha} \int\limits_{\beta} \int\limits_{\lambda} \varphi_N m_k \cos\beta \cos\lambda d\alpha d\beta d\lambda \,, \tag{3.9.4}$$

where m_k is the direction cosines given by Eq. (3.5.4). ***Formulae (3.9.4) are the basic relations of the synthetic theory of plasticity.*** They have been proposed in works [3,4,5,140]. Since the synthetic theory is based upon the concept of slip, this allows us to carry over the result $e_4^S = e_5^S = 0$.

To show the physical significance of irreversible strain intensity φ_N we rewrite Eq. (3.4.6) in absolute value. As $\left|\vec{\mathbf{N}}\right| = 1$,

$$\varphi_N = \frac{\left|d\vec{\mathbf{e}}^S\right|}{dV} \,, \tag{3.9.5}$$

---

[1] Further, for the simplicity of expressions, we will use abbreviations α and β for $\hat{\alpha}$ and $\hat{\beta}$, respectively. At the same time, one must remember that these angles are not to be confused with angles defined in terms of the concept of slip. Comparing Fig. 2.6 to Fig. 3.3, the spherical angles are seen to be determined in different coordinate systems.

where $\left|d\vec{\mathbf{e}}^S\right|$ is the absolute value of plastic-strain-vector-increment produced by the displacements of the family of planes within the parallelepiped of volume dV. The vector $d\vec{\mathbf{e}}^S$ is directed along vector $\vec{\mathbf{N}}$ normal to these planes. The ratio of $\left|d\vec{\mathbf{e}}^S\right|$ to dV is termed as the irreversible strain intensity of a plane oriented by vector $\vec{\mathbf{N}}$.

So far, the intensity φ_N is not defined. To do it, we notice that the initial plane distance , $\sqrt{2}\tau_S$, in five-dimensional space and its trace distance in subspace $\mathbf{R}^3$, h_0, are related to each other by Eq. (3.5.6). However, since the planes can experience only translational displacements, Eq. (3.5.6) holds for distances H_N and h_m during loading:

$$h_m \cos\lambda = H_N, \quad h_m \geq h_0, \quad H_N \geq \sqrt{2}\tau_S \tag{3.9.6}$$

The condition that a plane is situated on the endpoint of stress vector is expressed as

$$h_m = \vec{\mathbf{S}} \cdot \vec{\mathbf{m}}, \tag{3.9.7}$$

where scalar product $\vec{\mathbf{S}} \cdot \vec{\mathbf{m}}$ is the projection of the stress vector on the normal vector. On the other hand, the above relation, on the base of Eqs. (3.9.6) and (3.9.7), can be written as follows

$$H_N = \vec{\mathbf{S}} \cdot \vec{\mathbf{m}} \cos\lambda. \tag{3.9.8}$$

At a given moment of loading, a plane does not experience movement if

$$\vec{\mathbf{S}} \cdot \vec{\mathbf{m}} \cos\lambda < H_N. \tag{3.9.9}$$

In terms of the synthetic theory, a strain hardening is connected with (i) displacements of tangent planes and (ii) plastic deformation. Therefore, we need to establish a relation between φ_N and H_N. We propose to take relation $\varphi_N = \varphi_N\left(H_N\right)$ in the following forms, linear or quadratic,

$$\varphi_N = \frac{1}{r}\left(H_N - \sqrt{2}\tau_S\right), \tag{3.9.10}$$

$$\varphi_N = \frac{1}{r}\left(H_N^2 - 2\tau_S^2\right), \tag{3.9.11}$$

where r is the constant of material to be determined to best fit the experimentally obtained stress-strain curves (r have different values and units depending on the form of the above relations). It is obvious that $\varphi_N = 0$ as $H_N = \sqrt{2}\tau_S$.

Further, let us determine the range of angles α, β, and λ (the boundaries of integration in Eq. (3.9.4)) giving the positive value of φ_N. Regarding angle λ, the obvious fact is that the first plane to start its displacements on the endpoint of stress vector is tangential to Guber's surface, sphere (3.7.3). For this plane $\lambda = 0$ due to a plane tangential to the yield surface in $\mathbf{R}^3$ is, evidently, tangential to yield surface in $\mathbf{R}^5$, meaning that $\vec{\mathbf{N}} = \vec{\mathbf{m}}$ or $h_0 = \sqrt{2}\tau_S$ (see Eq (3.5.6)). During loading, together with the first plane, new planes take part in displacements that is caused by the increase in stress vector. For every fix direction, the stress vector $\vec{\mathbf{S}}$ shifts planes with growing value of λ. The boundary value of λ, λ_1, is calculated from the condition $\varphi_N = 0$

$$0 \le \lambda \le \lambda_1, \quad \cos\lambda_1 = \frac{\sqrt{2}\tau_S}{\vec{\mathbf{S}} \cdot \vec{\mathbf{m}}}. \tag{3.9.12}$$

Equation (3.9.12) can be obtained by equating the distance h_0 in Eq. (3.5.6) to $\vec{\mathbf{S}} \cdot \vec{\mathbf{m}}$ that expresses the condition that the vector $\vec{\mathbf{S}}$ reach boundary plane for a given direction.

To determine the range of angles α and β, we note, again, that $\lambda = 0$ for planes touching the sphere (3.7.3). Therefore, by letting $\lambda = 0$ in Eq. (3.9.12), we can derive the equation for the area of angles α and β on Guber's sphere in the following form

$$\vec{\mathbf{S}} \cdot \vec{\mathbf{m}} = S_1 m_1 + S_2 m_2 + S_3 m_3 = \sqrt{2}\tau_S. \tag{3.9.13}$$

Summarizing, we state that Eqs. (3.9.4), and (3.9.8) - (3.9.13) completely determine the plastic strain components e_1^S, e_2^S, and e_3^S produced by stress vector components S_1, S_2, and S_3.

Since the synthetic theory falls within the category of plastic flow theories, it is more preferable to rewrite Eq. (3.9.4) in an incremental form

$$\dot{e}_k^S = \int_\alpha \int_\beta \int_\lambda \dot{\varphi}_N m_k \cos\beta \cos\lambda \, d\alpha \, d\beta \, d\lambda. \tag{3.9.14}$$

It must be noted that the areas of integration in the integrals (3.9.4) and (3.9.14) can be different.

3.10 Pure Shear

As an example, this section is devoted to the determination of plastic strain under pure shear. Let $\tau_{xz} > 0$ and all other stress vector components are equal to zero. Equations (1.7.3) and (1.7.11) give $S_1 = S_2 = S_4 = S_5 = 0$ and $S_3 > 0$ meaning that stress vector is directed along S_3-axis. Therefore, Eq. (3.9.1) becomes

$$\vec{\mathbf{S}} = S_3 \vec{\mathbf{g}}_3, \quad S_3 = \sqrt{2}\tau_{xz}. \tag{3.10.1}$$

Consider a plane in $\mathbf{R}^3$ being at distance of h_0 from the origin of coordinate (dotted line 1 in Fig. 3.5a). The normal vector $\vec{\mathbf{m}}$ to this plane is fixed by angle β. During loading, the stress vector $\vec{\mathbf{S}}$ shifts the plane a distance of

$$M_1 M_2 = OM_2 - OM_1 = \sqrt{2}\tau_{xz} \sin\beta - h_0 \tag{3.10.2}$$

in the position 2 (solid line). The plane 1 in Fig. 3.5a is the trace of the plane 1', which is tangential to the yield surface in $\mathbf{R}^5$ (Fig. 3.5b). The displacement M_3M_4 of plane 1' (Fig. 3.5b) corresponds to the displacement of its trace, M_1M_2 in Fig. 3.5a. From Fig. 3.5 and Eq. (3.9.10) the following can be written

$$M_3 M_4 = M_1 M_2 \cos\lambda = \left(\sqrt{2}\tau_{xz} \sin\beta - h_0\right)\cos\lambda = \sqrt{2}(\tau_{xz} \sin\beta\cos\lambda - \tau_S),$$
$$\varphi_N = \frac{\sqrt{2}}{r}(\tau_{xz} \sin\beta\cos\lambda - \tau_S). \tag{3.10.3}$$

For the case of quadratic function $\varphi_N = \varphi_N(H_N)$ Eqs. (3.9.8) and (3.9.11) give

$$H_N = \sqrt{2}\tau_{xz} \sin\beta\cos\lambda, \; \varphi_N = \frac{2}{r}\left(\tau_{xz}^2 \sin^2\beta\cos^2\lambda - \tau_S^2\right). \tag{3.10.4}$$

According to Eq. (3.9.4), the plastic strain vector components are

$$e_1^S = \iint\limits_{\alpha\beta} \cos\alpha\cos^2\beta c(\beta) d\alpha d\beta,$$

$$e_2^S = \iint\limits_{\alpha\beta} \sin\alpha\cos^2\beta c(\beta) d\alpha d\beta, \tag{3.10.5}$$

$$e_3^S = \frac{1}{2}\iint\limits_{\alpha\beta} \sin 2\beta c(\beta) d\alpha d\beta,$$

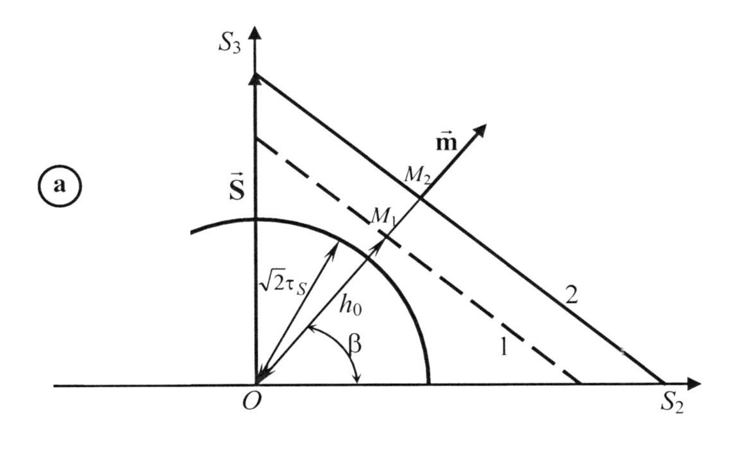

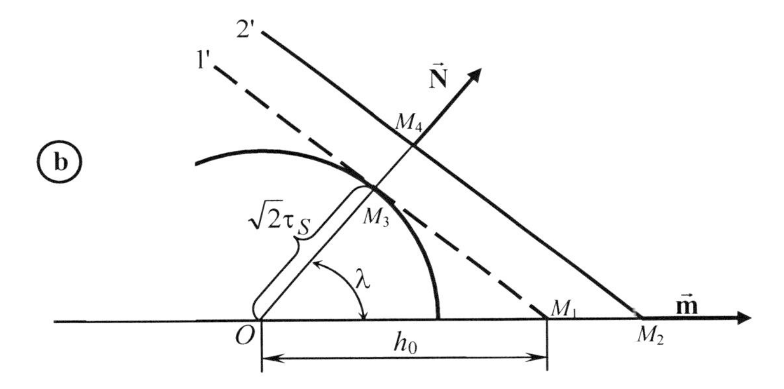

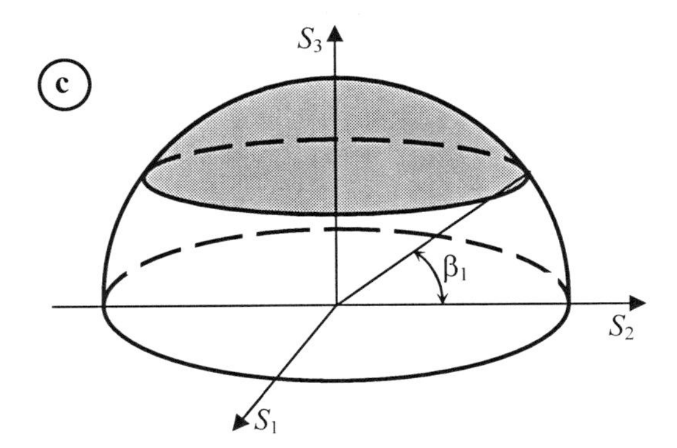

Fig. 3.5 Illustration of Eqs. (3.10.2) and (3.10.3) - (a) and (b). The domain of movable planes in pure shear - (c)

where

$$c(\beta) = \int_0^{\lambda_1} \varphi_N \cos\lambda d\lambda . \tag{3.10.6}$$

Equation (3.9.12) takes the form

$$\cos\lambda_1 = \frac{a}{\sin\beta}, \quad a = \frac{\tau_S}{\tau_{xz}} \tag{3.10.7}$$

and Eq. (3.9.13) becomes

$$\tau_{xz} \sin\beta = \tau_S . \tag{3.10.8}$$

Equation (3.10.8) does not contain α implying that

$$0 \le \alpha \le 2\pi, \quad \beta_1 \le \beta \le \pi/2, \quad \sin\beta_1 = a . \tag{3.10.9}$$

In a virgin state, the point of tangency of each plane to the sphere (3.7.3) is determined by spherical angles α and β. Since each tangent plane is displaced parallel to itself in a single direction, the values of α and β do not alter during the plane movement. This fact allows us to represent the planes, the range of angles α and β, which endure displacements on the endpoint of stress vector by the set of their points of tangency on the sphere (3.7.3). As such we will term this family ***the domain of movable planes***.

The domain of movable planes in pure shear, given by Eq. (3.10.9), is shaded in Fig. 3.5c; the planes that do not fall within the range (3.10.9) remain motionless.

Integrating in Eq. (3.10.5) above α gives:

$$e_1^S = e_2^S = 0, \quad e_3^S = \pi \int_{\beta_1}^{\pi/2} \sin 2\beta c(\beta) d\beta . \tag{3.10.10}$$

In the case of linear function $\varphi_N(H_N)$, the integral (3.10.6) is

$$c(\beta) = \frac{1}{2r}\left(S_3 \sin\beta \arccos\frac{\sqrt{2}\tau_S}{S_3 \sin\beta} - \sqrt{2}\tau_S \sqrt{1 - \frac{2\tau_S^2}{S_3^2 \sin^2\beta}} \right) \tag{3.10.11}$$

and if function $\varphi_N(H_N)$ is quadratic, Eq. (3.10.6) gives

$$c(\beta) = \frac{2}{3rS_3 \sin\beta}\left(S_3^2 \sin^2\beta - 2\tau_S^2\right)^{3/2} . \tag{3.10.12}$$

By integrating in Eq. (3.10.10), we obtain

$$e_3^S = a_0 F\left(\frac{\tau_S}{\tau_{xz}}\right). \tag{3.10.13}$$

If function $\varphi_N(H_N)$ is linear, then

$$a_0 = \frac{\pi\tau_S\sqrt{2}}{3r},$$

$$F(x) = \frac{\arccos x}{x} - 2\sqrt{1-x^2} + x^2 \ln\frac{1+\sqrt{1-x^2}}{x} \tag{3.10.14}$$

$$0 < x \le 1, \quad F(1) = 0, \quad F'(1) = 0, \quad F \to \infty \text{ as } x \to 0$$

If the function $\varphi_N(H_N)$ is square-law, then

$$a_0 = \frac{\pi\tau_S^2}{3r}$$

$$F(x) = \frac{1}{x^2}\left(2\sqrt{1-x^2} - 5x^2\sqrt{1-x^2} + 3x^4 \ln\frac{1+\sqrt{1-x^2}}{x}\right) \tag{3.10.15}$$

$$0 < x \le 1, \quad F(1) = 0, \quad F'(1) = 0, \quad F \to \infty \text{ as } x \to 0$$

Eqs. (3.10.13) - (3.10.15) determine the plastic strain vector component under pure shear, while the plastic strain tensor components are $\varepsilon_{xz}^S = e_3^S/\sqrt{2}$ (see Eq. 1.7.12) and $\gamma_{xz}^S = 2\varepsilon_{xz}^S = \sqrt{2}e_3^S$.

3.11 Proportional Loading

Consider the case when stress vector components, S_k, change proportionally to single parameter as in Eq. (3.3.4), where $0 \le \eta \le 1$ and S_{k0} is the final value of S_k. Under the proportional (simple) loading, a loading path is a direct line in $\mathbf{R}^3$ whose orientation is given by angles α_0 and β_0:

$$\cos\alpha_0 = \frac{S_{10}}{\sqrt{S_{10}^2 + S_{20}^2}} = \frac{S_1}{\sqrt{S_1^2 + S_2^2}}, \quad \sin\beta_0 = \frac{S_{30}}{\sqrt{S_{10}^2 + S_{20}^2 + S_{30}^2}} = \frac{S_3}{S}, \tag{3.11.1}$$

where $S = \left|\vec{\mathbf{S}}\right|$ is the module of the stress vector,

$$S = \sqrt{S_1^2 + S_2^2 + S_3^2}\ . \tag{3.11.2}$$

Let us find strain vector components under the proportional loading. Since the stress vector $\vec{\mathbf{S}}$ lengthens along a line arbitrary oriented in the subspace $\mathbf{R}^3$, we rotate the coordinate system (Fig. 3.6) to align S_3-axis with the vector $\vec{\mathbf{S}}$, i.e., to obtain an analog of pure shear (when stress vector possesses a single nonzero component, Sec. 3.10). Equations (3.10.10) and (3.10.13) determine plastic strain components, $e_3^{S\,'}$, in the indicated new coordinate system, S_1', S_2', S_3', if we replace the shear stress $\sqrt{2}\tau_{xz}$ by the modulus S:

$$e_{1'}^S = e_{2'}^S = 0, \quad e_{3'}^S = a_0 F\left(\frac{\sqrt{2}\tau_S}{S}\right), \tag{3.11.3}$$

where constant a_0 and function F are given by Eqs. (3.10.14) and (3.10.15).

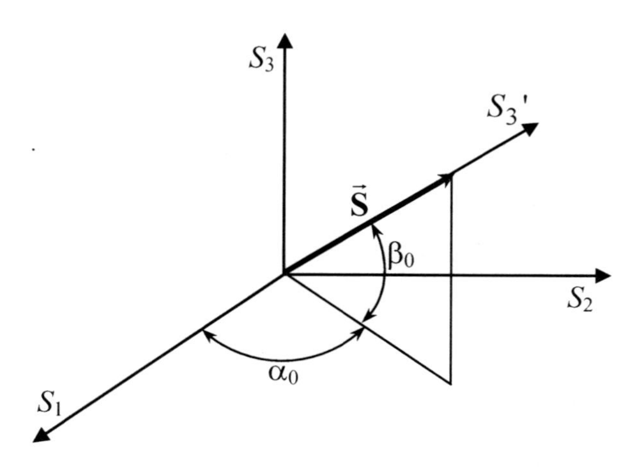

Fig. 3.6 The basic system of coordinate and auxiliary S_3' -axis

The strain vector components e_k^S in the original coordinate system are expressed in terms of $e_3^{S\,'}$ as follows. By projecting the plastic strain vector components $e_{3'}^S$ on S_1-, S_2- and S_3-axis, we obtain

$$e_1^S = e_{3'}^S \cos\alpha_0 \cos\beta_0, \quad e_2^S = e_{3'}^S \sin\alpha_0 \cos\beta_0,$$
$$e_3^S = e_{3'}^S \sin\beta_0. \tag{3.11.4}$$

Then, Eqs. (3.11.1) and (3.11.4) give

$$e_k^S = a_0 \frac{S_k}{S} F\left(\frac{\sqrt{2}\tau_S}{S}\right) \qquad (k = 1,2,3) \tag{3.11.5}$$

The above formula expresses the law of deviator proportionality.

The obtained result is of great importance, the concept of the movements of tangent planes leads to the deviator proportionality ($k = 1,2,3$) under a simple loading.

3.12 Cicala Formula

Consider a two-segment loading path, first, an element of a body is subjected to uniaxial tension, $\sigma_x > 0$, and then, holding σ_x constant, an infinitesimal additional shear stress $\Delta\tau_{xz}$ acts. The determination of plastic strains under specified loading regime constitutes the Cicala problem. Now, our goal is to compare strains in the vicinity of the break of the loading trajectory obtained in terms of synthetic model to those of the concept of slip.

We wish to determine the shear strain $\Delta\gamma_{xz}$ produced by the additional shear stress $\Delta\tau_{xz}$ and the ratio $\Delta\tau_{xz}/\Delta\gamma_{xz}$ called as additional shear module. First, we calculate a relative elongation, ε_x^S, under tension, when Eq. (1.7.11) gives the stress deviator components as $\overline{\sigma}_x = 2\sigma_x/3$, $\overline{\sigma}_y = \overline{\sigma}_z = -\sigma_x/3$ and Eqs. (1.7.3) and (1.7.12) lead to the following:

$$S_1 = \sqrt{\frac{2}{3}}\sigma_x, \quad S_2 = S_3 = 0, \quad S = \sqrt{\frac{2}{3}}\sigma_x \tag{3.12.1}$$

meaning that the stress vector is directed along S_1-axis. Hence Eqs. (3.9.8) and (3.9.11) take the form

$$H_N = S_1 \cos\alpha\cos\beta\cos\lambda, \quad r\varphi_N = S_1^2 \cos^2\alpha\cos^2\beta\cos^2\lambda - 2\tau_S^2. \tag{3.12.2}$$

Equating H_N to $\sqrt{2}\tau_S$, we obtain equations for the boundary value of angle λ:

$$0 \le \lambda \le \lambda_1, \quad \cos\lambda_1 = \frac{\sqrt{2}\tau_S}{S_1 \cos\alpha \cos\beta}. \tag{3.12.3}$$

As we have already discussed in Sec 3.9, $\lambda = \lambda_1 = 0$ on the Guber sphere (3.7.3). Therefore, on the sphere, the equation for the boundary of movable planes is

$$S_1 \cos\alpha \cos\beta = \sqrt{2}\tau_S. \tag{3.12.4}$$

As seen from this formula, the range of angle β is

$$-\beta_1 \le \beta \le \beta_1, \quad \cos\beta_1 = \frac{\sqrt{2}\tau_S}{S_1 \cos\alpha}. \tag{3.12.5}$$

By equating β_1 to 0 in Eq. (3.12.5), we get the range of angle α as

$$-\alpha_1 \le \alpha \le \alpha_1, \quad \cos\alpha_1 = \frac{\sqrt{2}\tau_S}{S_1}. \tag{3.12.6}$$

Equation (3.9.4) gives that the plastic strain component e_1^S takes the form

$$e_1^S = \int_{-\alpha_1}^{\alpha_1} \cos\alpha d\alpha \int_{-\beta_1}^{\beta_1} \cos^2\beta d\beta \int_0^{\lambda_1} \varphi_N \cos\lambda d\lambda \tag{3.12.7}$$

and, by integrating, we obtain

$$e_1^S = a_0 F\left(\frac{\sigma_S}{\sigma_x}\right). \tag{3.12.8}$$

It is worthwhile to note that the above formula can be obtained from the deviator proportionality, Eq. (3.11.5), as well.

Further, let an infinitesimal shear stress, $\Delta\tau_{xz}$, be applied at the constant tensile stress σ_x. Then

$$\Delta S_1 = \Delta S_2 = 0, \quad \Delta S_3 = \sqrt{2}\Delta\tau_{xz}. \tag{3.12.9}$$

At the additional loading, the shear strain component increment e_3^S (3.9.14) can be written as

$$\Delta e_3^S = \frac{1}{2} \int_\alpha \int_\beta \int_\lambda \sin 2\beta \cos \lambda \Delta\varphi_N \, d\alpha d\beta d\lambda . \tag{3.12.10}$$

A strain intensity increment, $\Delta\varphi_N$, given by Eq. (3.9.11), acquires the form

$$r\Delta\varphi_N = 2H_N \Delta H_N . \tag{3.12.11}$$

Plane distance (3.9.8) now is

$$H_N = (S_1\vec{\mathbf{g}}_1 + \Delta S_3\vec{\mathbf{g}}_3)(\cos\alpha\cos\beta\vec{\mathbf{g}}_1 + \sin\beta\vec{\mathbf{g}}_3)\cos\lambda = (S_1\cos\alpha\cos\beta + \Delta S_3\sin\beta)\cos\lambda \cdot \tag{3.12.12}$$

Therefore,

$$r\Delta\varphi_N = S_1 \Delta S_3 \cos\alpha \sin 2\beta \cos^2\lambda . \tag{3.12.13}$$

In Eq. (3.12.10), the borders of integration are to be determined. In additional loading(torsion), not all planes, which have incurred displacements in initial tension, move. The domain of movable planes, which are reached by the stress vector $S_1\vec{\mathbf{g}}_1 + \Delta S_3\vec{\mathbf{g}}_3$, is shaded in Fig. 3.7. Therefore, the range of angle α

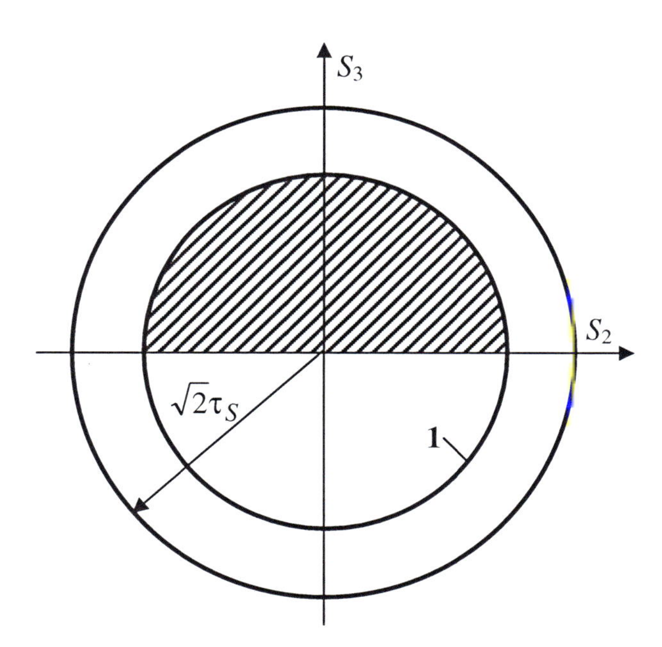

Fig. 3.7 The domain of movable planes in additional torsion

given by Eq. (3.12.6) has not altered. However, in contrast to Eq. (3.12.5), in this case, we have

$$0 \le \beta \le \beta_1, \quad \cos\beta_1 = \frac{\sqrt{2}\tau_S}{S_1 \cos\alpha}. \tag{3.12.14}$$

There is a complete analogy between the areas of integration in terms of the synthetic model and the concept of slip, both theories leads to that the area of integration under additional torsion is half the size than under the previous tension. The difference is that this result is seen immediately in terms of synthetic theory, while the concept of slip leads to this though bulky calculations.

Under the tension, the region of change of λ is determined by Eq. (3.12.3). They hold after the break of loading path as well. Indeed, the half of planes from range $0 \le \lambda \le \lambda_1$, which moved under tension, continue to move under the additional torsion as well resulting in that Eq. (3.12.10) becomes

$$\Delta e_3^S = \frac{1}{2r} S_1 \Delta S_3 \int_{-\alpha_1}^{\alpha_1} \cos\alpha d\alpha \int_0^{\beta_1} \sin^2 2\beta d\beta \int_0^{\lambda_1} \cos^3 \lambda d\lambda. \tag{3.12.15}$$

Integration in the above formula gives

$$\Delta e_3^S = \frac{\sqrt{3}}{2\sqrt{2}\sigma_x} a_0 F\left(\frac{\sigma_S}{\sigma_x}\right) \Delta S_3, \tag{3.12.16}$$

furthermore, the obtained relation is valid for the both, linear and square-law, forms of function. φ_N Equation (3.12.16), on the base of Eq. (3.12.8), takes the form

$$\Delta\gamma_{xz}^S = \frac{3\varepsilon_x^S}{2\sigma_x} \Delta\tau_{xz}. \tag{3.12.17}$$

The above equations, as a mater of fact coincides with Eq. (2.5.8) obtained in terms of the concept of slip, from which the Cicala formula is derived. Such concurrence of results in terms of the synthetic model and concept of slip justifies the modification of yield criterion proposed in Sec. 3.6.

3.13 Two-Segment Loading Trajectory

We propose the determination of plastic strains under two-segment loading path for the case of finite amount of additional loading. In contrast to the Budiansky slip theory, which is not feasible to describe this kind of loading at all, the synthetic theory will be turned out to be capable of describing the straining at two-segment loading paths [144].

Let an element of a body be loaded in tension up to some stress $\sigma_x = \sqrt{3}\, S_1 / \sqrt{2} > \sigma_S$ and then, at $\sigma_x = \mathrm{var}$, is subject to torsion of $\Delta\tau_{xz}$. On the second portion of trajectory, $\Delta\sigma_x > 0$ and $\Delta\tau_{xz} > 0$, is carried out so that the loading path is a straight line. Under combined loading, we have

$$\vec{\mathbf{S}} = \vec{\mathbf{S}}_1 + \Delta\vec{\mathbf{S}} = (S_1 + \Delta S_1)\vec{\mathbf{g}}_1 + \Delta S_3 \vec{\mathbf{g}}_3, \tag{3.13.1}$$

where $\vec{\mathbf{S}}_1$ is the stress vector at initial tension, while ΔS_1 and ΔS_3 are the stress vector increments due to additional tension and torsion, respectively. All these vectors lie in S_1S_3-plane. It is obvious that

$$\Delta S_1 = \sqrt{\frac{2}{3}}\Delta\sigma_x, \quad \Delta S_2 = 0, \quad \Delta S_3 = \sqrt{2}\Delta\tau_{xz}. \tag{3.13.2}$$

The mutual arrangement of the considered vectors is shown in Fig. 3.8, and Fig. 3.9 shows the projection of loading surface under initial tension on S_1S_3-plane. As seen from these figures, the angle at the cone apex is

$$\tan\delta_3 = \frac{\sqrt{2}\tau_S}{\sqrt{S_1^2 - 2\tau_S^2}}. \tag{3.13.3}$$

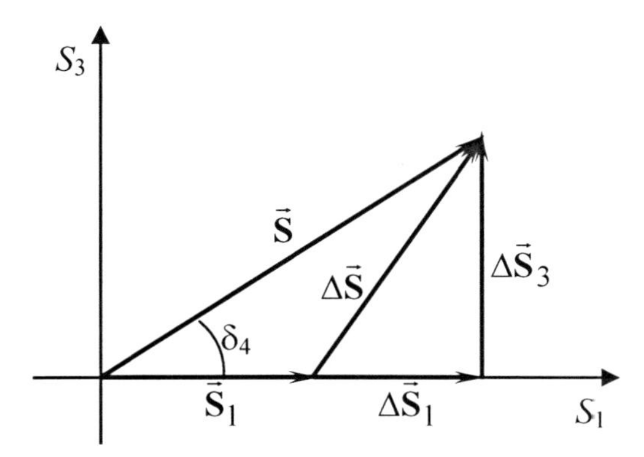

Fig. 3.8 Mutual arrangement of $\vec{\mathbf{S}}_1$, $\vec{\mathbf{S}}$, and $\Delta\vec{\mathbf{S}} = \Delta\vec{\mathbf{S}}_1 + \Delta\vec{\mathbf{S}}_3$ vectors

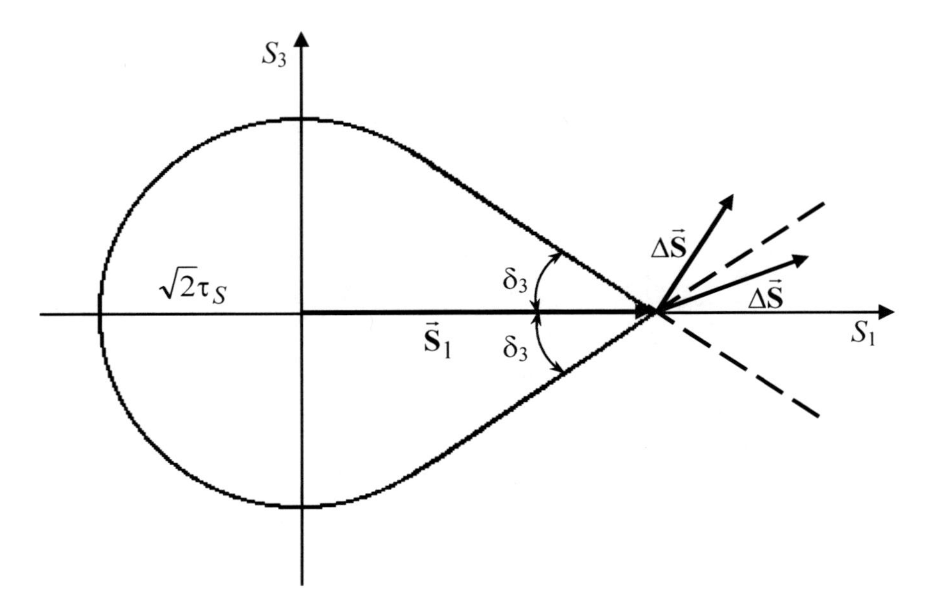

Fig. 3.9 Projection of loading surface in uniaxial tension onto S_1S_3-plane

Let us start with the case, when the stress vector increment $\Delta\vec{\mathbf{S}}$ is directed inward the cone (dotted line) constructed as the continuation of the cone generator, i.e.,

$$\frac{\Delta S_3}{\Delta S_1} \leq \tan \delta_3 . \tag{3.13.4}$$

In this case all planes, which moved under the action of $\vec{\mathbf{S}}_1$, continue to move under the combined stress state as well. This means the fulfillment of the deviator proportionality and, on the base of Eq. (3.11.5), we obtain

$$e_1^S = a_0 \frac{S_1 + \Delta S_1}{S} F\left(\frac{\sqrt{2}\tau_S}{S}\right), \; e_2^S = 0, \; e_3^S = a_0 \frac{\Delta S_3}{S} F\left(\frac{\sqrt{2}\tau_S}{S}\right),$$
$$S = \sqrt{(S_1 + \Delta S_1)^2 + (\Delta S_3)^2} . \tag{3.13.5}$$

Further, let the vector $\Delta\vec{\mathbf{S}}$ be directed outward the introduced cone marked by dotted line,

$$\frac{\Delta S_3}{\Delta S_1} \geq \tan \delta_3 . \qquad (3.13.6)$$

The planes incurring displacements on the endpoint of stress vector $\vec{\mathbf{S}}_1$ are at a distant of

$$h_m = S_1 \cos\alpha \cos\beta \quad \text{at} \quad S_1 \cos\alpha \cos\beta \geq \sqrt{2}\tau_S \qquad (3.13.7)$$

from the origin of coordinates. On the second portion of loading, the projection of stress vector $\vec{\mathbf{S}}$ on any normal vector $\vec{\mathbf{m}}$ is

$$\mathrm{pr}_{\mathbf{m}}\vec{\mathbf{S}} = h_m = \vec{\mathbf{S}} \cdot \vec{\mathbf{m}} = (S_1 + \Delta S_1)\cos\alpha \cos\beta + \Delta S_3 \sin\beta . \qquad (3.13.8)$$

Those planes, on which $\mathrm{pr}_{\mathbf{m}}\vec{\mathbf{S}} > h_m$, continue to move on the second portion of loading. If $\mathrm{pr}_{\mathbf{m}}\vec{\mathbf{S}} < h_m$, such planes are unmovable in combined loading, despite the fact they moved under uniaxial tension.

The boundary of the domain of movable planes in additional loading, where $\Delta\varphi \geq 0$, consists of two parts. Equating the right-hand sides in Eqs. (3.13.7) and (3.13.8) to each other, we obtain the equations for the first part of this boundary:

$$\Delta S_1 \cos\alpha \cos\beta + \Delta S_3 \sin\beta = 0 . \qquad (3.13.9)$$

It must be noted that Eq. (3.13.9) holds during the whole additional loading. Equating the projection (3.13.8) to $\sqrt{2}\tau_S$, we arrive at the equation of the second part of the boundary:

$$(S_1 + \Delta S_1)\cos\alpha \cos\beta + \Delta S_3 \sin\beta = \sqrt{2}\tau_S . \qquad (3.13.10)$$

Figure 3.10 illustrates the domains of movable planes (projection on S_2S_3-plane) in two-sectional loading. The shaded region corresponds to the additional loading. The double-shaded area gives those planes which terminate their movements during additional loading, although they moved in initial tension. The boundary marked by **1** is constructed by Eq. (3.13.9) and that marked by **2** is based on Eq. (3.13.10). The circle (3.12.4), uniaxial tension, is shown by dotted line and the circle (3.7.3) is marked by **4**.

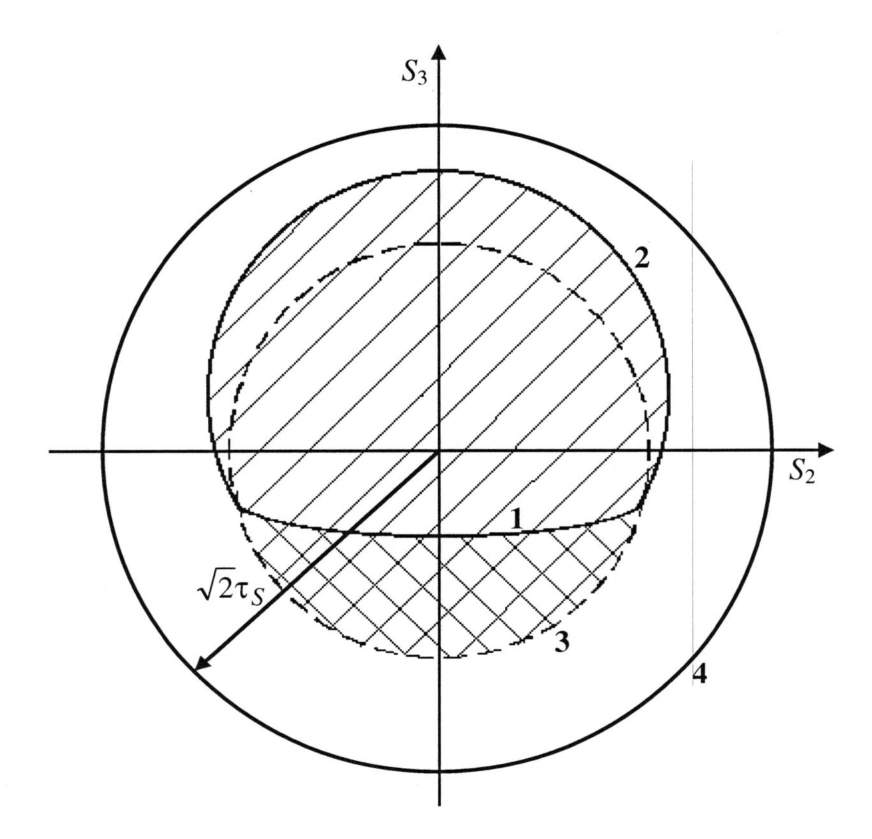

Fig. 3.10 The domain of movable planes in initial tension and additional loading

To calculate strain components, we introduce an auxiliary coordinate system, S_1', S_2', S_3'. S_3'-axis is aligned with vector $\vec{\mathbf{S}}$, S_2'-axis is aligned with S_2-axis, and S_1-axis is perpendicular to S_2'-S_3' coordinate plane (Fig. 3.11). S_1'- and S_1-axis and S_3'- and S_3-axis lie in the common coordinate plane. Similar to angles α and β (Fig. 3.3), we introduce angles α' and β' that are counted off from the new coordinate axes as shown in Fig. 3.12. A normal vector $\vec{\mathbf{m}}$ is expressed through angles α and β, and α' and β' as

$$\vec{\mathbf{m}} = \cos\alpha\cos\beta\vec{\mathbf{g}}_1 + \sin\alpha\cos\beta\vec{\mathbf{g}}_2 + \sin\beta\vec{\mathbf{g}}_3 = \cos\alpha'\cos\beta'\vec{\mathbf{g}}'_1 + \sin\alpha'\cos\beta'\vec{\mathbf{g}}'_2 + \sin\beta'\vec{\mathbf{g}}'_3 \quad (3.13.11)$$

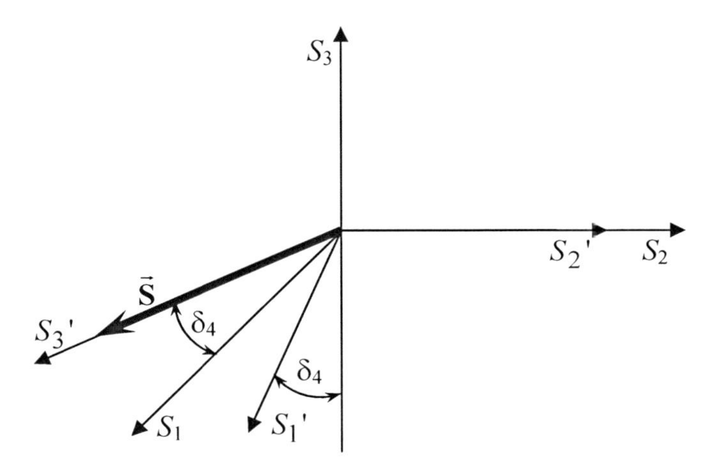

Fig. 3.11 Auxiliary system of coordinates

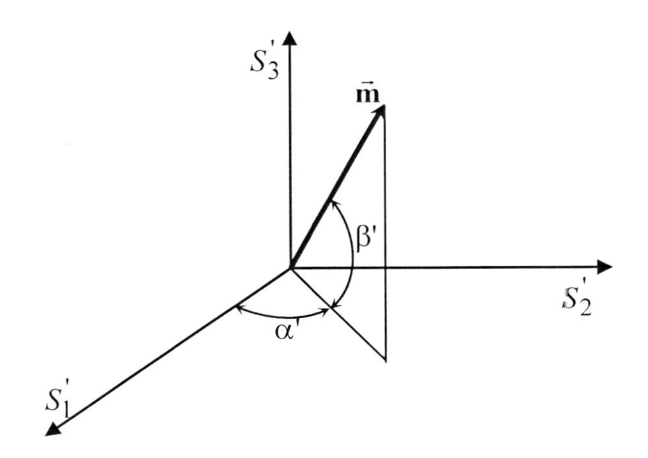

Fig. 3.12 Spherical coordinates of normal vector $\vec{\mathbf{m}}$

where $\vec{\mathbf{g}}'_1$, $\vec{\mathbf{g}}'_2$, and $\vec{\mathbf{g}}'_3$ are the unit base vectors in the new coordinate system. From Fig 3.8 it follows that

$$\tan \delta_4 = \frac{\Delta S_3}{S_1 + \Delta S_1}. \tag{3.13.12}$$

The scalar product of the left- and right-hand sides in Eq. (3.13.11) by vector $\vec{\mathbf{g}}_1$ and the relations

$$\vec{\mathbf{g}}_1 \cdot \vec{\mathbf{g}}'_1 = \sin \delta_4, \ \vec{\mathbf{g}}_1 \cdot \vec{\mathbf{g}}'_2 = 0, \text{ and } \vec{\mathbf{g}}_1 \cdot \vec{\mathbf{g}}'_3 = \cos \delta_4 \tag{3.13.13}$$

give

$$\cos\alpha\cos\beta = \cos\alpha'\cos\beta'\sin\delta_4 + \sin\beta'\cos\delta_4 . \qquad (3.13.14)$$

Further, since $\vec{g}_2 = \vec{g}'_2$, we have

$$\sin\alpha\cos\beta = \sin\alpha'\cos\beta' . \qquad (3.13.15)$$

Finally, multiplying Eq. (3.13.11) by $\vec{g}_3$, we obtain (Fig. 3.11)

$$\vec{g}_3\cdot\vec{g}'_1 = -\cos\delta_4, \quad \vec{g}_3\cdot\vec{g}'_2 = 0, \quad \vec{g}_3\cdot\vec{g}'_3 = \sin\delta_4 . \qquad (3.13.16)$$

Therefore,

$$\sin\beta = -\cos\alpha'\cos\beta'\cos\delta_4 + \sin\beta'\sin\delta_4 . \qquad (3.13.17)$$

The equations for the boundaries, Eqs. (3.13.9) (3.13.10), rewritten in terms of the new coordinate system[2] take the following form

$$\sin\beta = k\cos\alpha\cos\beta, \quad k = \frac{S_1\Delta S_3}{(S_1+\Delta S_1)\Delta S_1 + \Delta S_3^2}, \qquad (3.13.18)$$

$$\sin\beta = \sin\beta_1 = \frac{\sqrt{2}\tau_S}{S}, \qquad (3.13.19)$$

respectively.

3.14 Plastic Strain at Two-Segment Loading Path

We continue to investigate the loading regime described in the preceding section and wish to determine plastic strains in two-axial loading that follows the initial uniaxial tension. The stress vector components relative to the new coordinate system are $S'_1 = S'_2 = 0$, $S'_3 = S$ and the strain vector components, Eq. (3.9.4), are determined by Eq. (3.10.5), (3.10.6), and (3.10.12) implying that the function $\varphi_N(H_N)$ is quadratic, i.e., given by Eq. (3.9.11).

The domain of movable planes in the new coordinate system is shown in Fig. 3.13, it is limited by lines (3.13.18) and (3.13.19) marked by **1** and **2,** respectively. Further, we designate the value of β at point M_1, where $\alpha = 0$, through β_2. From Eq. (3.13.18) it follows that

$$\tan\beta_2 = k . \qquad (3.14.1)$$

[2] For the sake of simplicity, especially relative to the following section, the strokes above α and β are missed.

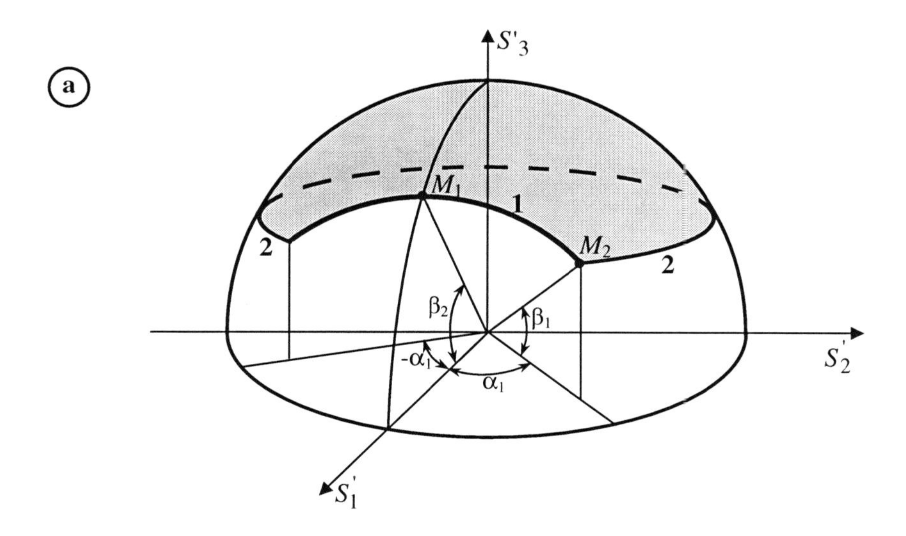

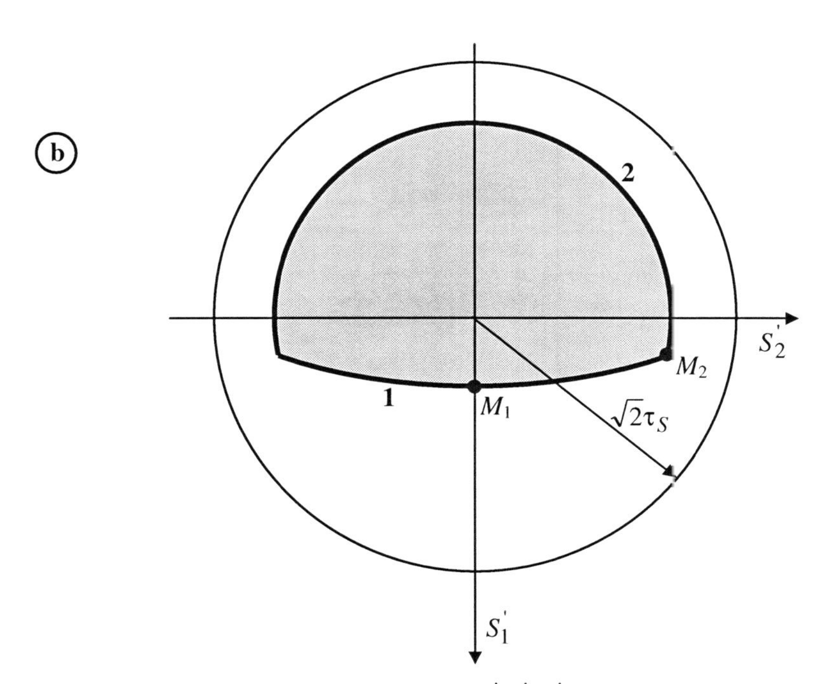

Fig. 3.13 The domain of movable planes in $S_1^{'}S_2^{'}S_3^{'}$ coordinate system (**a**) and on $S_1^{'}S_2^{'}$ coordinate planes (**b**)

Owing to the symmetry of the range of angles α and β over $S_1'S_3'$-plane, we have $e_2^{S}{}'=0$. By making use of Eqs. (3.10.5) and (3.10.12), we calculate the strain component increment $\Delta e_1^{S}{}'$ by the following iterated integral:

$$\Delta e_1^{S}{}'=\frac{4}{3rS}\int_{\beta_1}^{\beta_2}\frac{\cos^2\beta}{\sin\beta}\left(S^2\sin^2\beta-2\tau_S^2\right)^{3/2}d\beta\int_{\alpha(\beta)}^{\pi}\cos\alpha d\alpha+$$
$$+\frac{4}{3rS}\int_{\beta_2}^{\pi/2}\frac{\cos^2\beta}{\sin\beta}\left(S^2\sin^2\beta-2\tau_S^2\right)^{3/2}d\beta\int_{0}^{\pi}\cos\alpha d\alpha. \quad (3.14.2)$$

where angles β_1 and β_2 are given by Eqs. (3.13.19) and (3.14.1), respectively, and $\alpha(\beta)$ is the solution of Eq. (3.13.18) for α. By integrating the above relation over α,

$$\Delta e_1^{S}{}'=-\frac{4}{3rS}\int_{\beta_1}^{\beta_2}\frac{\cos^2\beta}{\sin\beta}\left(S^2\sin^2\beta-2\tau_S^2\right)^{3/2}\sqrt{1-\frac{\tan^2\beta}{k^2}}d\beta, \quad (3.14.3)$$

and then over β, we get

$$\Delta e_1^{S}{}'=-\frac{\pi S^2}{12r}\left(\frac{k}{\sqrt{k^2+1}}-\frac{\sqrt{2}\tau_S}{S}\right)^3\left(1+\frac{3\sqrt{2}\tau_S\sqrt{k^2+1}}{kS}\right). \quad (3.14.4)$$

Regarding strain vector increment $\Delta e_3^{S}{}'$, it is convenient first to carry out the integration over variable β:

$$\Delta e_3^{S}{}'=\frac{4S^2}{3r}\int_{0}^{\alpha_1}d\alpha\int_{\beta_1(\alpha)}^{\pi/2}\cos\beta\left(\sin^2\beta-\sin^2\beta_1\right)^{3/2}d\beta+$$
$$+\frac{4S^2}{3r}\int_{\alpha_1}^{\pi}d\alpha\int_{0}^{\pi/2}\cos\beta\left(\sin^2\beta-\sin^2\beta_1\right)^{3/2}d\beta \quad (3.14.5)$$

where $\beta_1=\beta_1(\alpha)$ is the equation of line **1** by Eq. (3.13.18); the angle β_1 is given by Eq. (3.13.19) and α_1 is the value of angle α at the point of the intersection of lines **1** and **2**, point M_2, (Fig. 3.13). From Eqs. (3.13.18) and (3.13.19) it follows that

$$\cos\alpha_1 = \frac{\sqrt{2}\tau_S}{k\sqrt{S^2 - 2\tau_S^2}}. \tag{3.14.6}$$

By integrating in (3.14.5), we obtain

$$\Delta e_3^{S}{}' = \frac{\pi\tau_S^2}{3r} F\left(\frac{\sqrt{2}\tau_S}{S}\right) - \frac{\pi S^2 \cos^3\beta_1}{12r\left(1+k^2\right)^{3/2}}\left[2\left(1+k^2\right)^{3/2} - \right.$$
$$\left. - \frac{1}{\cos\beta_1}\left(2+3k^2-\tan^2\beta_1\right)\right] + \frac{\pi S^2}{4r}\sin^2\beta_1\left(\cos\beta_1 - \frac{1}{\sqrt{1+k^2}}\right) - \frac{\pi S^2}{4r}\sin^4\beta_1 \ln\frac{k(1+\cos\beta_1)}{\left(1+\sqrt{1+k^2}\right)\sin\beta_1}. \tag{3.14.7}$$

Equations (3.14.4) and (3.14.7) determine the required strain vector increments relative to S_1'- and S_3'-axis. To obtain these increments relative to the initial axes S_1- and S_3-axis, it is necessary to project the components $\Delta e_1^{S}{}'$ and $\Delta e_3^{S}{}'$ on the initial axes (Fig. 3.11):

$$\Delta e_1^S = \Delta e_1^{S}{}'\sin\delta_4 + \Delta e_3^{S}{}'\cos\delta_4,\quad \Delta e_3^S = -\Delta e_1^{S}{}'\cos\delta_4 + \Delta e_3^{S}{}'\sin\delta_4, \tag{3.14.8}$$

where angle δ_4 is determined by Eq. (3.13.12). Equation (3.14.8) does not give total irreversible strains on the second segment of loading. To obtain them, we need to add the strain components, which are accumulated by the movements of planes due to stress vector $\vec{\mathbf{S}} = \vec{\mathbf{S}}_1$ from the double-shaded area in Fig. 3.10, to the strains from Eq. (3.14.8). To determine the strains from this area, we introduce, again, a new coordinate system aligning S_3'-axis with S_1-axis, see Fig. 3.13. The curve **1**, Eq. (3.13.9), in the introduced coordinate system has the form

$$\sin\beta = k_1\cos\alpha\cos\beta,\quad k_1 = \frac{\Delta S_3}{\Delta S_1}. \tag{3.14.9}$$

Now, the equation of curve **3** (Fig. 3.10) is

$$\sin\beta = \sin\beta_1 = \frac{\sqrt{2}\tau_S}{S_1}. \tag{3.14.10}$$

The strain component increment relative to S_1'- axis, $\Delta e_1^{S}{}''$, we obtain as

$$\Delta e_1^{S}{}'' = \frac{4}{3rS_1}\int_{\beta_1}^{\beta_2}\frac{\cos^2\beta}{\sin\beta}\left(S_1^2\sin^2\beta - 2\tau_S^2\right)^{3/2} d\beta \int_0^{\alpha(\beta)}\cos\alpha\, d\alpha, \tag{3.14.11}$$

where $\alpha(\beta)$ is the equations of the curve (3.14.9). Integrating over α leads to the radical similar to that of Eq. (3.14.3), therefore

$$\Delta e_1^S{}'' = \frac{\pi S_1^2}{12r}\left(\frac{k_1}{\sqrt{k_1^2+1}} - \frac{\sqrt{2}\tau_S}{S_1}\right)^3\left(1+\frac{3\sqrt{2}\tau_S\sqrt{k_1^2+1}}{k_1 S_1}\right). \tag{3.14.12}$$

The analogue of Eq. (3.14.5) has the form

$$\Delta e_3^S{}'' = \frac{4S_1^2}{3r}\int_0^{\alpha_1} d\alpha \int_{\beta_1}^{\beta_1(\alpha)} \cos\beta\left(\sin^2\beta - \sin^2\beta_1\right)^{3/2} d\beta. \tag{3.14.13}$$

The above formula and Eq. (3.14.7) give

$$\Delta e_3^S{}'' = \frac{\pi S_1^2\cos^3\beta_1}{12r\left(k_1^2+1\right)^{3/2}}\left[2\left(k_1^2+1\right)^{3/2} - \frac{1}{\cos\beta_1}\left(2+3k_1^2-\tan^2\beta_1\right)\right] -$$
$$-\frac{\pi S_1^2}{3r}\sin^2\beta_1\left(\cos\beta_1 - \frac{1}{\sqrt{k_1^2+1}}\right) + \frac{\pi S_1^2}{4r}\sin^4\beta_1\ln\frac{k_1(1+\cos\beta_1)}{\left(1+\sqrt{k_1^2+1}\right)\sin\beta_1}. \tag{3.14.14}$$

The total irreversible strain vector components are obtained if $\Delta e_1^S{}''$ and $\Delta e_3^S{}''$ are added to the right-hand sides of Eq. (3.14.8). In the view of the mutual arrangement of S_1- and S_3-axes and S_1'- and S_3'-axes, we obtain the relations giving the total plastic strain vector components under combined state of stress, i.e., on the second portion of the loading path.

$$e_1^S = \Delta e_1^S{}'\sin\delta_4 + \Delta e_3^S{}'\cos\delta_4 + \Delta e_3^S{}'',\quad e_3^S = -\Delta e_1^S{}'\cos\delta_4 + \Delta e_3^S{}'\sin\delta_4 - \Delta e_1^S{}''. \tag{3.14.15}$$

3.15 Partial Cases

Let us discuss Eq. (3.14.15) in more detail.

1. If under addition loading the vector $\Delta\vec{\mathbf{S}}$ is directed along the dashed line in Fig. 3.9,

$$\frac{\Delta S_3}{\Delta S_1} = \tan\delta_3 = \frac{\sqrt{2}\tau_S}{\sqrt{S_1^2-2\tau_S^2}}, \tag{3.15.1}$$

the coefficients in Eqs. (3.13.18) and (3.14.9) will take the form

$$k = \frac{\sqrt{2}\tau_S\sqrt{S_1^2-2\tau_S^2}}{S_1^2+S_1\Delta S_1-2\tau_S^2},\quad k_1 = \frac{\sqrt{2}\tau_S}{\sqrt{S_1^2-2\tau_S^2}}. \tag{3.15.2}$$

Now, the module of vector $\vec{\mathbf{S}}$, Eq. (3.13.5), is

$$S = \frac{\sqrt{S_1\left(S_1^3 + 2S_1^2\Delta S_1 + S_1\Delta S_1^2 - 2S_1\tau_S^2 - 4\Delta S_1\tau_S^2\right)}}{\sqrt{S_1^2 - 2\tau_S^2}}. \tag{3.15.3}$$

From Eqs. (3.15.2) and (3.15.3) it follows that

$$k = \frac{\sqrt{2}\tau_S}{\sqrt{S^2 - 2\tau_S^2}}. \tag{3.15.4}$$

In view of these relationships, Eqs. (3.14.4), (3.14.7), (3.14.12), and (3.14.14) give

$$\Delta e_1^S{}' = 0,\ \Delta e_3^S{}' = \frac{\pi\tau_S^2}{3r}F\left(\frac{\sqrt{2}\tau_S}{S}\right),\ \Delta e_1^S{}'' = 0,\ \Delta e_3^S{}'' = 0. \tag{3.15.5}$$

Therefore, Eq. (3.14.15) is resulted in the form of Eq. (3.13.5), i.e in the deviator proportionality.

The obtained result corresponds with the general ideas of the synthetic theory: under additional loading along the dashed line in Fig. 3.9, the movements of those planes that moved at initial uniaxial strain do not cease. Therefore, the deviator proportionality holds true.

2. For the case of an orthogonal break of loading trajectory, $\Delta S_1 = 0$, instead of Eqs. (3.15.2)–(3.15.4), we have

$$k = \frac{S_1}{\Delta S_3}, \quad k_1 \to \infty. \tag{3.15.6}$$

Equation (3.14.12) and (3.14.14) become simpler as $k_1 \to \infty$,

$$\begin{aligned} \Delta e_1^S{}'' &= \frac{\pi S_1^2}{12r}\left(1 - \frac{\sqrt{2}\tau_S}{S_1}\right)^3\left(1 + \frac{3\sqrt{2}\tau_S}{S_1}\right), \\ \Delta e_3^S{}'' &= \frac{\pi\tau_S^2}{6r}F\left(\frac{\sqrt{2}\tau_S}{S}\right) = \frac{a_0}{2}F\left(\frac{\sqrt{2}\tau_S}{S}\right). \end{aligned} \tag{3.15.7}$$

Comparing Eq. (3.15.7) to Eq. (3.12.18), one can see that $\Delta e_3^S{}'' = e_1^S/2$.

2a. Consider the initial stage of additional loading if $\Delta S_3 \to 0$. By the dropping of the quadratic and higher degree terms in ΔS_3, Eqs. (3.14.4) and (3.14.7) give

$$\Delta e_1^{S\,'} = -\frac{\pi S_1^2}{12r}\left(1-\frac{\sqrt{2}\tau_S}{S_1}\right)^3\left(1+\frac{3\sqrt{2}\tau_S}{S_1}\right),$$

$$\Delta e_3^{S\,'} = \frac{\pi\tau_S^2}{6r}F\left(\frac{\sqrt{2}\tau_S}{S_1}\right)+\frac{\pi S_1}{4r}\left(1-\frac{2\tau_S}{S_1^2}\right)^2\Delta S_3. \tag{3.15.8}$$

On the base of Eqs. (3.14.15), (3.15.7), and (3.15.8) the total plastic strain vector components are

$$e_1^S = \frac{\pi\tau_S^2}{3r}F\left(\frac{\sqrt{2}\tau_S}{S_1}\right)+\frac{\pi S_1}{6r}\left(1-\frac{\sqrt{2}\tau_S}{S_1}\right)^2\left(1+\frac{2\sqrt{2}\tau_S}{S_1}+\frac{6\tau_S^2}{S_1^2}\right)\Delta S_3,$$

$$e_2^S = 0, \quad e_3^S = \Delta e_3^S = \frac{\pi\tau_S^2}{6r}F\left(\frac{\sqrt{2}\tau_S}{S_1}\right)\frac{\Delta S_3}{S_1}, \tag{3.15.9}$$

i.e., regarding Δe_3^S, we arrive at Eq. (3.12.16) that results in Cicala formula.

2b. Let us consider a case opposite to that from point **2a**, namely, $S_1 << \Delta S_3$ meaning that, after initial tension, an element of a body is subjected to torsion of great magnitude. Letting that in initial tension $S_1 = \sqrt{2}\tau_S$, Eq. (3.14.15) gives:

$$e_3^S = \frac{\pi\tau_S^2}{3r}F\left(\frac{\sqrt{2}\tau_S}{\Delta S_3}\right), \tag{3.15.10}$$

i.e., at $S_1 << \Delta S_3$, the component e_3^S is seen to be equal to strain component calculated for the case of pure shear.

3. J. Andrusik's Ph.D. thesis concerns the comparisons of stress-strain diagrams constructed on the base of Eqs. (3.14.4), (3.14.7), (3.14.12), (3.14.14), and (3.14.15) to those recorded by the following experiments.

3a. The series of the Budiansky experiments [16] performed on thin-walled cylindrical specimens of aluminum alloy 14S-T4. After the plastic straining of specimens due to uniaxial pressure, twisting moment was applied thereby ensuring the break of loading trajectory. The compressive stress σ_x varied so that the relation $d\sigma_x/d\tau_{xz}$ remains constant for each experiment. The experiments were carried out at different values of the ratio $d\sigma_x/d\tau_{xz}$, both positive and negative.

3b. The P. Nahdi experiments performed on specimens of aluminum alloy 24S-T-4 [99]. The specimens were deformed plastically in uniaxial tension and then, for different amount of plastic prestrain, were subjected to torsion (naturally, holding the tensile stress unchangeable).

3c. The Vavakin experiments [179] on thin-walled tube samples of steel 30XГCA were carried out on system "Instron-1275". The influence of high-velocity effects on deformation was immaterial due to small strain rate, $\dot{\varepsilon}_x = 10^{-6}\ 1/\text{sec}$.

In all three cases, the comparison of the theoretical results to experimental data showed satisfactory agreement.

3.16 Loading Surface

In terms of the synthetic model, similarly to the concept of slip, the formula for plastic strain vector components, Eq. (3.9.4), is not connected to the notion of loading surface.

Following Sanders [165], a stress vector shifts planes tangential to the yield surface on its endpoint during loading. The movements of the planes located at the endpoint of vector $\vec{\mathbf{S}}$ are translational, i.e., without change in their orientations. Those planes which are not located on the endpoint of vector $\vec{\mathbf{S}}$ remain unmovable. The loading surface, which is constructed as an inner-envelope of tangent planes, takes the shape fully determined by the current positions of planes. It is easy to see (Fig. 1.26b) that a corner point arises on the loading surface at the endpoint of vector $\vec{\mathbf{S}}$. Therefore, in contrast to the plastic flow theories of, e.g., isotropic or kinematic hardening, the behavior of loading surface in terms of synthetic theory is not prescribed a priori, but is fully determined by a hodograph of stress vector.

For the case of pure shear the loading surface, see Fig. 3.14, consists of two parts:

- the sphere (3.7.3) constructed by means of tangent planes being at a distant of $H_N = \sqrt{2}\tau_S$ from the origin of coordinates, which do not move during loading;
- the cone whose generators are the boundary planes shifted by the stress vector $\vec{\mathbf{S}}$. On the top of this cone there are planes displaced by the vector $\vec{\mathbf{S}}$ during loading. Boundary angle β_1 is given by Eq. (3.10.9).

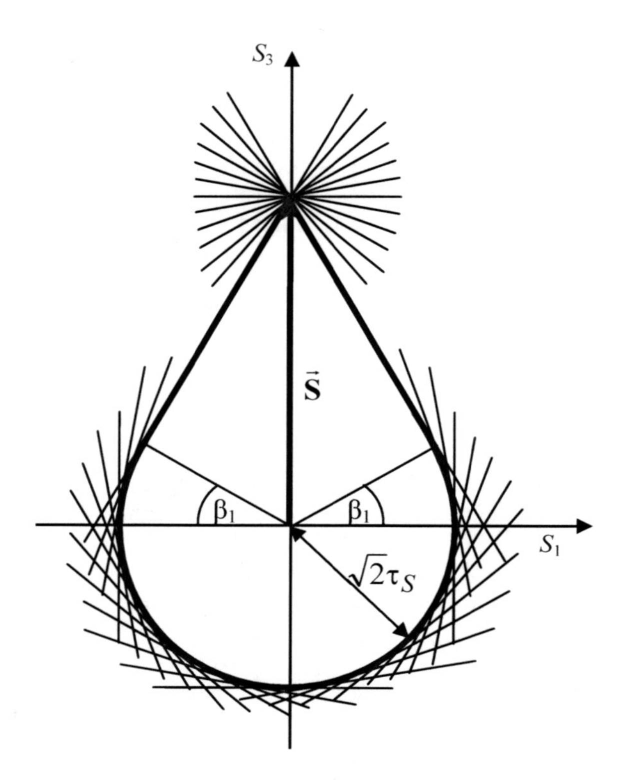

Fig. 3.14 Loading surface in pure shear

In what follows we extend the above regulations. Let us consider two parallel tangent planes, which, in $\mathbf{R}^5$, touch the yield surface on its opposite sides. Unit normal vectors to these planes, outward-pointing vectors with respect to the yield surface, we designate through $\vec{\mathbf{N}}$ and $-\vec{\mathbf{N}}$, respectively (Fig. 3.15). Suppose that a stress vector shifts (pushes) one of these planes away from the origin of coordinates. We introduce the following rule, a displacement of a plane with normal vector $\vec{\mathbf{N}}$ results in equal in magnitude displacement of the plane with oppositely directed normal vector $-\vec{\mathbf{N}}$:

$$\Delta H_{-N} = -\Delta H_N . \tag{3.16.1}$$

In other words: the distance between the specified planes does not change during their displacements. The increment in distance is assumed to be positive if a plane moves away from the origin of coordinates.

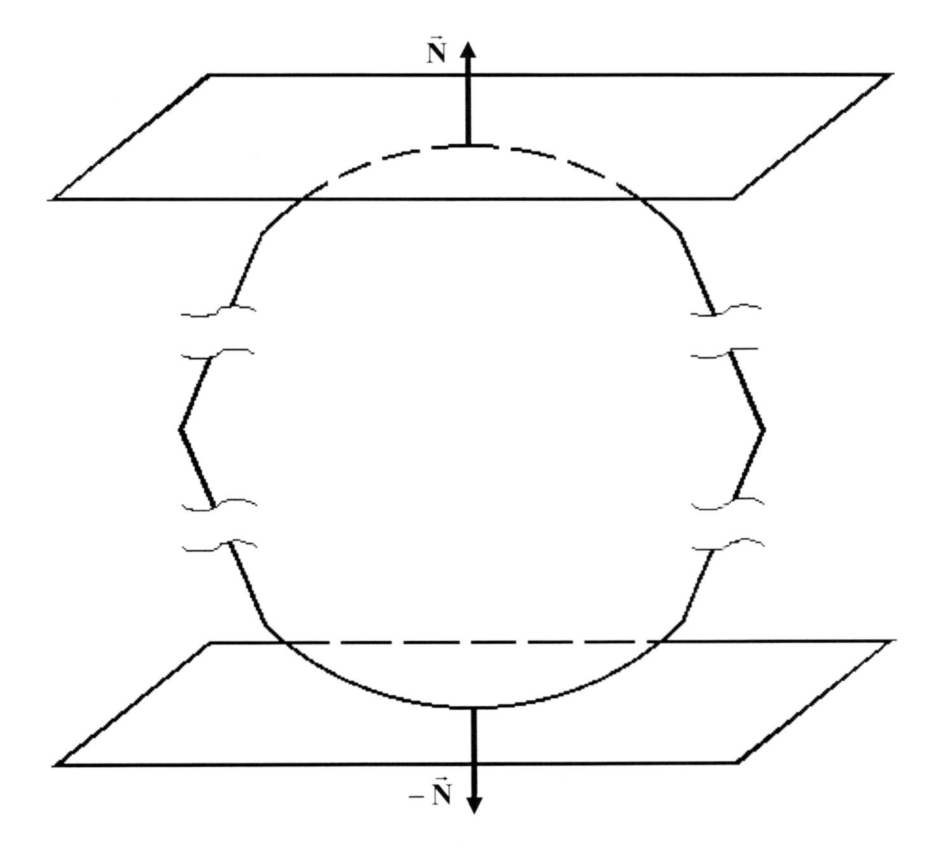

Fig. 3.15 Parallel tangent planes in R^5 with oppositely directed normal vectors

Employing the proposed rule, let us construct the loading surface for the case of pure shear, $S_1 = S_2 = 0$ and $S_3 > 0$. The component S_3 grows from zero to some maximal value S_0 and then an unloading follows. Let us consider the tangent plane that, prior to straining, touches the Guber sphere (3.7.3) at its apex point along S_3-axis, $\beta = \pi/2$ and $\lambda = 0$. In Fig. 3.16 this plane is shown by dotted line 1'. The stress vector $\vec{\mathbf{S}}$ shifts (pushes) the plane in the position 1, its displacement is

$$\Delta h = S_0 - \sqrt{2}\tau_S\,, \tag{3.16.2}$$

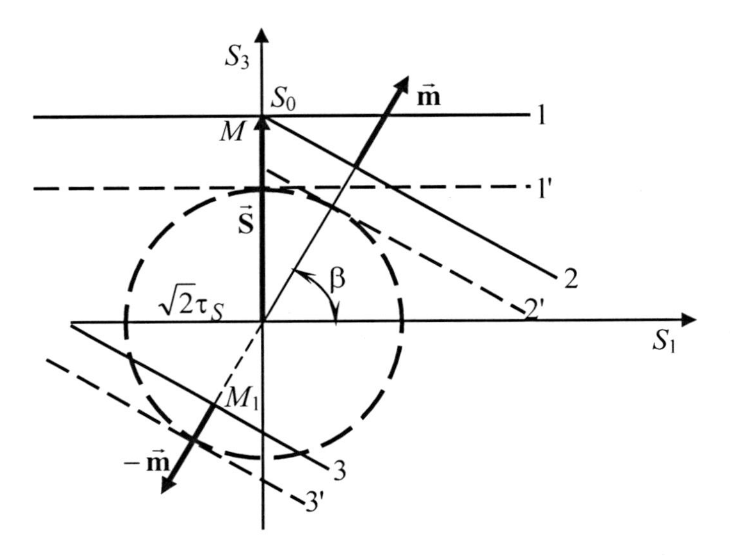

Fig. 3.16 Displacements of planes with opposite normals

where S_0 is the module of vector $\vec{S}$ and $\sqrt{2}\tau_S$ is the radius of the circle of plasticity. Together with plane 1', we consider tangential plane 2' with normal $\vec{m}$ given by angle β, which incurs displacement by amount of

$$\Delta h = S_0 \sin\beta - \sqrt{2}\tau_S . \tag{3.16.3}$$

As a result of this displacement, this plane takes the position marked by 2 (Fig. 3.16). Finally, our object is plane 3' parallel to 2, which goes over into position 3 during loading. According to Eqs. (3.16.1) and (3.16.3), plane 3 is at a distance of

$$h_3 = 2\sqrt{2}\tau_S - S_0 \sin\beta, \quad \beta > 0 \tag{3.16.4}$$

from the origin of coordinates. The equation of line 3 has the form

$$\left(S_1 - S_{M_1}\right)m_1 + \left(S_3 - S_{M_3}\right)m_3 = 0, \tag{3.16.5}$$

where S_{M1} and S_{M3} are the coordinates of point M_1, the point of intersection of line 3 and perpendicular to it radiating from the origin of coordinates. Therefore,

$$S_{M_1} = -h_3 m_1, \quad S_{M_3} = -h_3 m_3 . \tag{3.16.6}$$

The equation of line 3 takes the form

$$S_1 m_1 + S_3 m_3 + h_3 = 0, \tag{3.16.7}$$

where

$$m_1 = \cos\beta, \quad m_3 = \sin\beta . \tag{3.16.8}$$

Hence,

$$S_1 \cos\beta + S_3 \sin\beta + 2\sqrt{2}\tau_S - S_0 \sin\beta = 0. \tag{3.16.9}$$

Different values of angle β give different lines. To construct the inner-envelope of the family of lines (3.16.9), we need to differentiate Eq. (3.16.9) over parameter β,

$$-S_1 \sin\beta + S_3 \cos\beta - S_0 \cos\beta = 0, \tag{3.16.10}$$

and eliminate β from Eqs. (3.16.9) and (3.16.10),

$$S_1^2 + (S_3 - S_0)^2 = 8\tau_S^2. \tag{3.16.11}$$

This is a circle of radius twice more than circle (3.7.3) centered at point M (loading point) in Fig. 3.17. Therefore, in contrast to Sec. 3.16 and Fig. 3.14, now the loading surface consists of the following ***three*** parts, see Fig. 3.17:

- the initial sphere (circle) (3.7.3), portions AD and BC;
- the cone whose generators are the boundary planes on the endpoint of the vector $\vec{\mathbf{S}}_0$, segments AM and BM.
- the circle (3.16.11), portion DC.

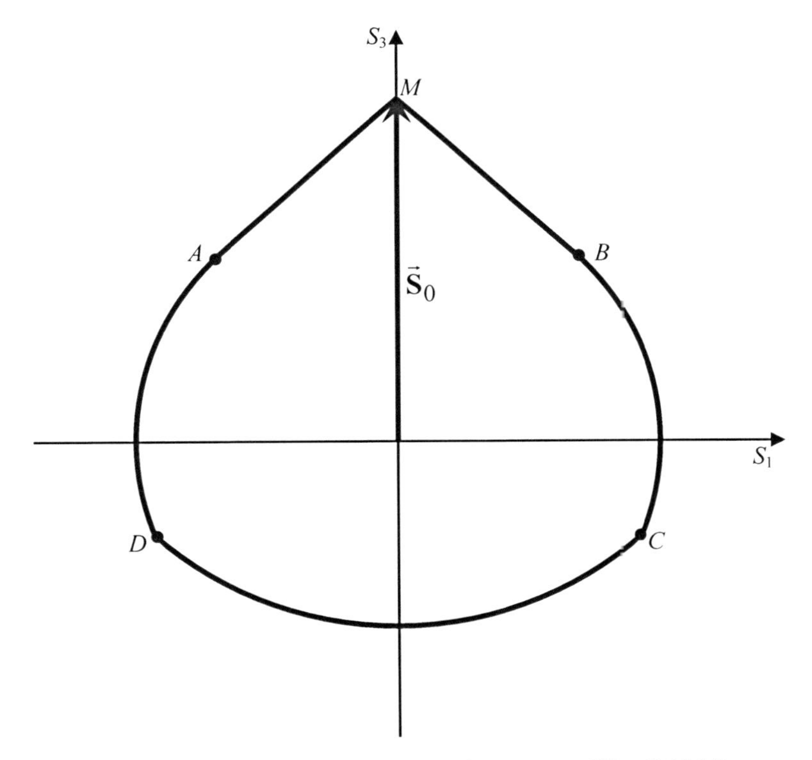

Fig. 3.17 Loading surface in pure shear with account of Eq. (3.16.11)

It is obvious that the circle (3.16.11) is the projection of the sphere

$$S_1^2 + S_2^2 + (S_3 - S_0)^2 = 8\tau_S^2 \tag{3.16.12}$$

onto S_1-S_3 coordinate plane. Under uniaxial tension, $S_1 > 0$ and $S_2 = S_3 = 0$, the loading surface is similar to (3.16.12),

$$(S_1 - S_0)^2 + S_2^2 + S_3^2 = 8\tau_S^2, \tag{3.16.13}$$

where S_0 is the value of S_1 at loading point. By comparing the surface (3.16.13) to the surface (2.23.5) constructed in terms of the concept of slip, one can see that they are similar. Such similarity of results obtained in terms of these theories justifies once more the transformation of yield criterion proposed in Sec. 3.6 and the validity of synthetic theory.

The fact that the potion of the loading surface *DC* differs from the initial sphere (3.7.3) and the center of (3.16.11) is not located at the origin of coordinates imply a deformation anisotropy of material with respect to the subsequent loading of opposite sign. Indeed, the condition for the onset of plastic deformation on the portion *DC* strongly depends on the orientation of stress vector.

3.17 Another Variant of the Deformation Anisotropy

Similarly to the previous Section, we consider two tangent planes in $\mathbf{R}^5$ with opposite normals, $\vec{\mathbf{N}}$ and $-\vec{\mathbf{N}}$ in Fig. 3.15. Now, we adopt the following relation

$$\varphi_{-N} = -\varphi_N. \tag{3.17.1}$$

For the case, when Eq. (3.9.10) governs function $\varphi_N(H_N)$, we get

$$H_N = \sqrt{2}\tau_S + r\varphi_N. \tag{3.17.2}$$

Then, on the base of Eq. (3.17.1), the plane of opposite normal is at a distance of

$$H_{-N} = \sqrt{2}\tau_S - r\varphi_N \tag{3.17.3}$$

from the origin of coordinates. For the quadratic form of function $\varphi_N(H_N)$, Eq. (3.9.11), the plane distance is

$$H_N = \sqrt{2\tau_S^2 + r\varphi_N} \tag{3.17.4}$$

or

$$H_{-N} = \sqrt{2\tau_S^2 - r\varphi_N}. \tag{3.17.5}$$

For the case of linear function $\varphi_N(H_N)$, Eqs. (3.17.2) and (3.17.3) lead to Eq. (3.16.1), meaning that two different formulations result in identical results. However, this is not the case under Eqs. (3.17.4) and (3.17.5), Eq. (3.16.1) does not hold implying that the distance between planes with normals $\vec{\mathbf{N}}$ and $-\vec{\mathbf{N}}$ changes during plastic straining, i.e., the generalizations of synthetic theory given here and in Sec. 3.16 result in different results.

Let us construct a yield surface under pure shear, we mark the stress vector component S_3 by S_0. We take the relation between the distances H_N and H_{-N} and φ_N to be given by Eqs. (3.17.4) and (3.17.5). Eqs. (3.10.4) and (3.17.4) give

$$H_N^2 - 2\tau_S^2 = r\varphi_N = S_0^2 \sin^2\beta\cos^2\lambda - 2\tau_S^2, \tag{3.17.6}$$

Therefore, Eq. (3.17.5) leads to the following expression for the distance H_{-N}

$$H_{-N} = \sqrt{4\tau_S^2 - S_0^2 \sin^2\beta\cos^2\lambda}\,. \tag{3.17.7}$$

Since Eq. (3.16.7) remains valid, the equation of line 3 (Fig. 3.16) at $\lambda = 0$ acquires the form

$$S_1\cos\beta + S_3\sin\beta + \sqrt{4\tau_S^2 - S_0^2\sin^2\beta} = 0\,. \tag{3.17.8}$$

Eq. (3.17.8) is an one-parametrical family of lines. The loading surface (curve) is constructed as the inner envelope of these lines. By differentiating Eq. (3.17.8) over parameter β,

$$-S_1\sin\beta + S_3\cos\beta - \frac{S_0^2\sin 2\beta}{2\sqrt{4\tau_S^2 - S_0^2\sin^2\beta}} = 0\,, \tag{3.17.9}$$

and eliminating parameter β from Eqs. (3.17.8) and (3.17.9), we obtain the equation of the inner-envelope of lines (3.17.8), i.e., the equation of loading surface, in the following form:

$$\frac{S_1^2}{4\tau_S^2} + \frac{S_3^2}{4\tau_S^2 - S_0^2} = 1\,. \tag{3.17.10}$$

In the three-dimensional Ilyushin space we have

$$\frac{S_1^2 + S_2^2}{4\tau_S^2} + \frac{S_3^2}{4\tau_S^2 - S_0^2} = 1\,. \tag{3.17.11}$$

Equations (3.16.13) and (3.17.11) can be obtained by other method, as shown in [5].

3.18 Alternating Torsion

Consider the case of pure shear at the following loading regime. First, a stress vector grows so that its component S_3 reaches some value $S_0 > 0$ producing plastic strain, e_0^S, determined by Eq. (3.10.13),

$$e_0^S = a_0 F\left(\frac{\tau_S \sqrt{2}}{S_0}\right). \tag{3.18.1}$$

Then, after unloading, the subsequent torsion of an opposite sign takes place, $S_3 < 0$, also inducing plastic straining. Since the plastic straining occurs in both loading directions, we can write that the strain intensities are positives,

$$\varphi_{N0} > 0, \quad \varphi'_{-N} > 0, \tag{3.18.2}$$

where φ_{N0} and φ'_{-N} correspond to the initial and subsequent loading, respectively; index '*N*' indicates on planes with normal vectors $\vec{\mathbf{N}}$ that are shifted by the stress vector of absolute value $S_0 > 0$ and index '-*N*' indicates on planes with normal vectors $-\vec{\mathbf{N}}$ that incur displacements on the endpoint of the stress vector of opposite sign.

For the case of the linear form of function $\varphi_N(H_N)$, Eq. (3.9.10) gives

$$r\varphi_{N0} = S_0 \sin\beta\cos\lambda - \sqrt{2}\tau_S. \tag{3.18.3}$$

If the dependence $\varphi_N(H_N)$ is a square-law one, the intensity φ_{N0} is determined by Eq. (3.17.6). According to Eqs. (3.17.3) and (3.17.5), the plane with normals $-\vec{\mathbf{N}}$ is at a distance of

$$H_{-N} = \begin{cases} \sqrt{2}\tau_S - r\varphi_{N0} + r\varphi'_{-N} \\ \sqrt{2\tau_S^2 - r\varphi_{N0} + r\varphi'_{-N}} \end{cases} \tag{3.18.4}$$

from the origin of coordinates and, together with Eq. (3.18.3) and Eq. (3.17.6), we obtain

$$H_{-N} = \begin{cases} 2\sqrt{2}\tau_S - S_0\cos\xi + r\varphi'_{-N} \\ \sqrt{4\tau_S^2 - S_0^2\cos^2\xi + r\varphi'_{-N}} \end{cases}, \quad \cos\xi = |\sin\beta|\cos\lambda \tag{3.18.5}$$

for the linear and quadratic function $\varphi_N(H_N)$, respectively.

According to the basic statement of synthetic theory, a plastic strain is produced by those planes which are reached by stress vector $\vec{\mathbf{S}}$, we have

$$S_3 \sin\beta\cos\lambda = H_{-N} \qquad (S_3 < 0, \beta < 0) \tag{3.18.6}$$

Planes do not move if

$$S_3 \sin\beta\cos\lambda < H_{-N} \,. \tag{3.18.7}$$

For small values of $|S_3|$, Eq. (3.18.7) holds for all planes with normals $-\vec{\mathbf{N}}$, the planes do not move in negative direction of S_3-axis implying that $\varphi'_{-N} = 0$. At increase in $|S_3|$, the first values of angles β and λ, to ensure Eq. (3.18.6) are $\beta = -\pi/2$ and $\lambda = 0$ indicating the first tangent plane to be reached by stress. Therefore, Eqs. (3.18.5) and Eq. (3.18.6) at $\beta = -\pi/2$ and $\lambda = 0$, and $\varphi'_{-N} = 0$ give the value of yield limit, S_S, in subsequent torsion:

$$S_S = -\sqrt{4\tau_S^2 - S_0^2} \,. \tag{3.18.8}$$

It is obvious that this formula can also be obtained from the equation for loading surface, Eq. (3.17.11), at $S_1 = S_2 = 0$.

During further loading of opposite sign, $|S_3| > S_S$, the stress vector shift the planes in negative direction of S_3-axis resulting in the increase of φ'_{-N} governed by Eqs. (3.18.5) and (3.18.6):

$$r\varphi'_{-N} = \left(S_3^2 + S_0^2\right)\cos^2\xi - 4\tau_S^2 \tag{3.18.9}$$

or

$$r\varphi'_{-N} = 4\tau_S^2\left(\frac{\cos^2\xi}{a_1^2} - 1\right), \tag{3.18.10}$$

where

$$a_1^2 = \frac{4\tau_S^2}{S_3^2 + S_0^2} \,. \tag{3.18.11}$$

Let us compare the obtained equality with intensity φ_{N0} determined by Eq. (3.17.6), which can be written as

$$r\varphi_{N0} = 2\tau_S^2\left(\frac{\cos^2\xi}{a_2^2} - 1\right), \tag{3.18.12}$$

where

$$a_2^2 = \frac{2\tau_S^2}{S_0^2}. \tag{3.18.13}$$

The formulae for φ'_{-N} and φ_{N0} are similar, they are dependent of angles β and λ in equal manner and independent of α. Therefore, the strain vector component e_3 caused by the movements of planes in negative directions of S_3-axis is calculated by equation similar to (3.10.13), namely

$$\left|e_3^{S\,\prime}\right| = 2a_0 F(a_1). \tag{3.18.14}$$

The strain vector component, e_3^S, under alternating torsion can be determined as

$$e_3^S = e_{3_0}^S - \left|e_3^{S\,\prime}\right|, \tag{3.18.15}$$

i.e.,

$$e_3^S = a_0 F\left(\frac{\sqrt{2}\tau_S}{S_0}\right) - 2a_0 F\left(\frac{\sqrt{2}\tau_S}{\sqrt{S_3^2 + S_0^2}}\right), \tag{3.18.16}$$

where the coefficient a_0 and the function F are determined by Eq. (3.10.15).

If the function $\varphi_N(H_N)$ is linear, the yield limit S_S in opposite loading is obtained from Eq. (3.16.11) as

$$S_S = S_0 - 2\sqrt{2}\tau_S. \tag{3.18.17}$$

There is an essential difference in the yield limits given by Eqs. (3.18.8) and (3.18.17). The yield limit given by (3.18.8) can take either negative values or $S_S = 0$ at $S_0 = 2\tau_S$, and Eq. (3.18.8) is not feasible for $S_0 > 2\tau_S$. This restriction is absent if the yield limit is given by Eq. (3.18.17). Now, the value of S_0 has no upper bound and S_S can take both negative and positive values (the latter requires that S_0 be greater than $2\sqrt{2}\tau_S$). The stress-strain diagram in alternating torsion and corresponding loading surface for the case $S_S > 0$ are shown in Fig. 3.18a and Fig. 3.18b, respectively. As one can see, the plastic deformation of

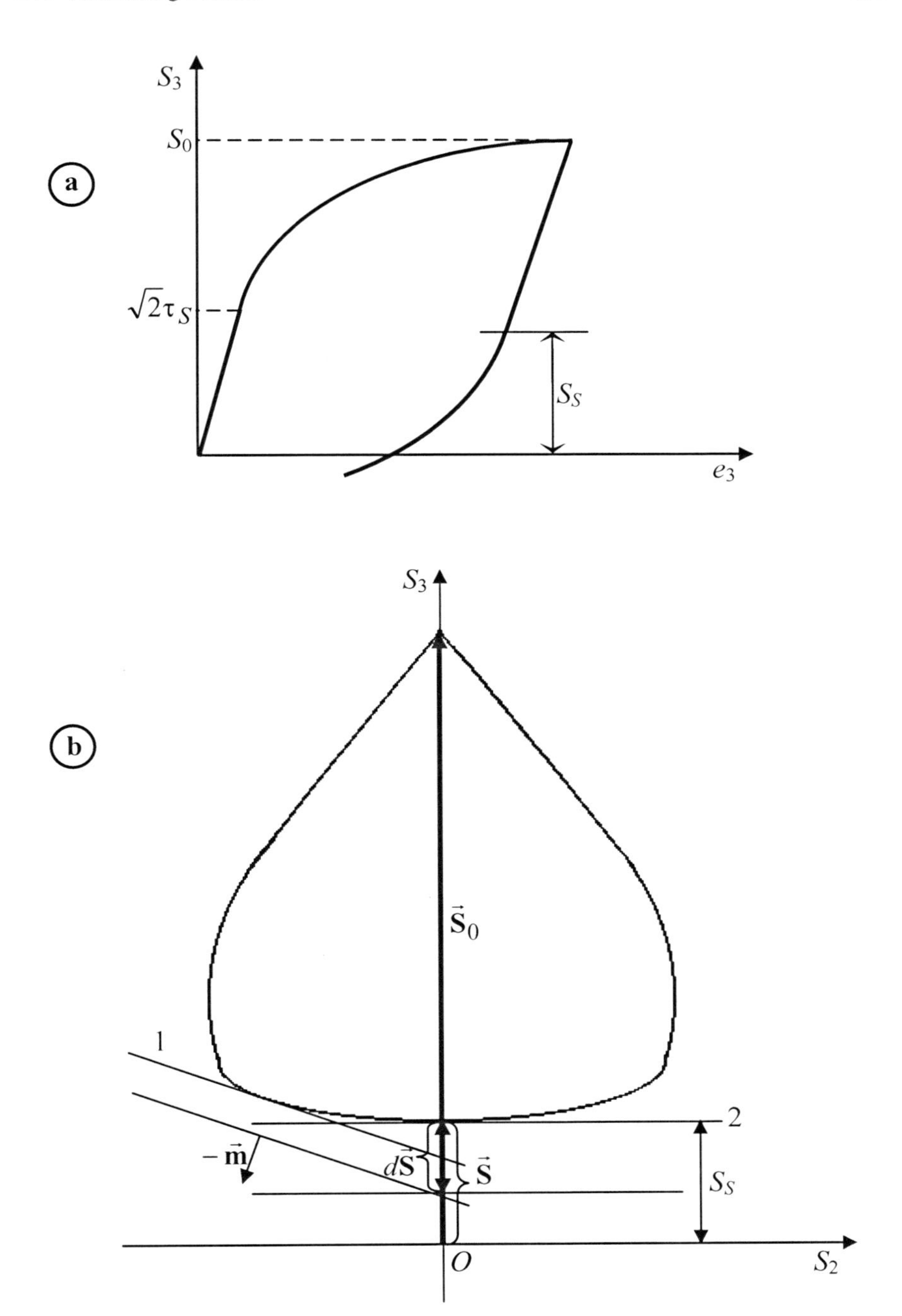

Fig. 3.18 Stress-strain diagram in alternating torsion (a) and the projection of loading surface onto S_2S_3-plane (b) for the case of the Bauschinger super-effect ($S_S > 0$)

negative sign starts occurring even at $S_3 > 0$. This phenomenon is referred to as the Bauschinger super-effect.

In order to provide the geometrical interpretation of plastic straining when the Bauschinger supper-effect takes place, we write down the condition for the development of plastic strain, Eq. (3.9.7), in an incremental form. If a stress vector, $\vec{\mathbf{S}}$, acquires an increment, $d\vec{\mathbf{S}}$, those plane experiences displacements for which

$$\left(\vec{\mathbf{S}} + d\vec{\mathbf{S}}\right) \cdot \vec{\mathbf{m}} > h_m, \tag{3.18.18}$$

where h_m is the plane distance that corresponds to the vector $\vec{\mathbf{S}}$. If

$$\left(\vec{\mathbf{S}} + d\vec{\mathbf{S}}\right) \cdot \vec{\mathbf{m}} < h_m, \tag{3.18.19}$$

the planes are not displaced by the increment $d\vec{\mathbf{S}}$. As $\vec{\mathbf{S}} \cdot \vec{\mathbf{m}} = h_m$ Eq. (3.1818), i.e., the condition for the progress of plastic strain, takes the form

$$d\vec{\mathbf{S}} \cdot \vec{\mathbf{m}} > 0 \tag{3.18.20}$$

and this inequality is valid for all planes. The inequality (3.18.20) says that an increment in plastic strain is caused by those planes which are displaced by stress vector increment. It must be noted that the condition (3.18.20) is of universal character and it is especial actual at the Bauschinger super-effect. Indeed, it is the vector $d\vec{\mathbf{S}}$ that shifts the planes with normal vectors $-\vec{\mathbf{m}}$, see Fig. 3.18b where the vector $d\vec{\mathbf{S}}$ makes to move planes **1** and **2** symbolizing the plastic deformation in the direction opposite to the acting stress.

For the case of the linear form of the function $\varphi_N(H_N)$, in place of Eqs. (3.18.10)–(3.18.13) we have

$$r\varphi'_{-N} = 2\sqrt{2}\tau_S\left(\frac{\cos\xi}{a_1} - 1\right), \quad a_1 = \frac{2\sqrt{2}\tau_S}{S_0 - S_3},$$
$$r\varphi_{N0} = \sqrt{2}\tau_S\left(\frac{\cos\xi}{a_2} - 1\right). \tag{3.18.21}$$

The plastic strain component e_3^S is determined by Eq. (3.18.15):

$$e_3^S = a_0 F\left(\frac{\sqrt{2}\tau_S}{S_0}\right) - 2a_0 F\left(\frac{2\sqrt{2}\tau_S}{S_0 - S_3}\right), \tag{3.18.22}$$

where the coefficient a_0 and the function F is given by Eq. (3.10.14).

Consider a case when the component S_3 at subsequent loading is equal in magnitude to the initial stress, $|S_3| = S_0$. Then, from Eqs. (3.18.16) and (3.18.22) it follows that

$$e_3^S = -a_0 F\left(\frac{\sqrt{2}\tau_S}{S_0}\right). \tag{3.18.23}$$

It is easy to see that Eq. (3.18.23) is identical, up to a sign, to Eq. (3.18.1) obtained for the initial torsion. Thus, the synthetic theory, similarly to the concept of slip, describes ideal-cyclic-materials.

3.19 Proportional Alternating Loading

Let us generalize the relation obtained in the preceding Section for the case of arbitrary proportional alternating loading, when a loading path is a straight line that is arbitrarily orientated in $\mathbf{R}^3$ and given by angles α_0 and β_0 (Fig. 3.6). The loading regime is the same as in Sec. 3.18, loading-unloading-reloading of opposite sign. Plastic strain vector components produced by stress vector components at initial loading, S_{10}, S_{20}, and S_{30}, can be determined by Eq. (3.11.5):

$$e_{k0}^S = a_0 \frac{S_{k0}}{S_0} F\left(\frac{\sqrt{2}\tau_S}{S_0}\right), \quad S_0 = \sqrt{S_{10}^2 + S_{20}^2 + S_{30}^2} \quad (k = 1,2,3) \tag{3.19.1}$$

For the case of linear function $\varphi_N(H_N)$, the loading surface Equation (3.16.12) can be generalized in the following way

$$(S_1 - S_{10})^2 + (S_2 - S_{20})^2 + (S_3 - S_{30})^2 = 8\tau_S^2. \tag{3.19.2}$$

Designating through S_{k_S} the stress vector components inducing plastic yield during subsequent loading of opposite sign, we obtain

$$S_{1S} = S_S \cos\alpha_0 \cos\beta_0, \quad S_{2S} = S_S \sin\alpha_0 \cos\beta_0, \quad S_{3S} = S_S \sin\beta_0. \tag{3.19.3}$$

In addition,

$$S_{10} = S_0 \cos\alpha_0 \cos\beta_0, \quad S_{20} = S_0 \sin\alpha_0 \cos\beta_0, \quad S_{30} = S_0 \sin\beta_0. \tag{3.19.4}$$

Inserting the above expressions into Eq. (3.19.2) gives

$$S_S = S_0 - 2\sqrt{2}\tau_S. \tag{3.19.5}$$

Let us draw an auxiliary axis, S_3', along the loading trajectory (Fig. 3.6) thereby obtaining an analog of pure shear and allowing to use the relations derived for pure shear. Therefore, according to Eq. (3.18.22), a plastic strain vector component relative S_3'-axis, $e^S_{3'}$, is

$$e^S_{3'} = a_0 F\left(\frac{\sqrt{2}\tau_S}{S_0}\right) - 2a_0 F\left(\frac{2\sqrt{2}\tau_S}{S_0 \pm S}\right), \tag{3.19.6}$$

furthermore, in the denominator, the plus sign appears as $S_S < 0$ and, if $S_S > 0$, we use the minus sign at unloading from S_S to $S = 0$.

Acting in the same way as in Sec. 3.11, i.e., by projecting the component $e^S_{3'}$ on S_k-axes, we obtain

$$e^S_k = e^S_{k'} \frac{S_k}{S}. \tag{3.19.7}$$

Therefore,

$$e^S_k = a_0 \left[F\left(\frac{\sqrt{2}\tau_S}{S_0}\right) - 2F\left(\frac{2\sqrt{2}\tau_S}{S_0 \pm S}\right)\right] \frac{S_k}{S}. \tag{3.19.8}$$

For the case of the quadratic form of function $\varphi_N(H_N)$, the yield limit which is opposite to initial loading direction is determined by a relation similar to (3.18.8), namely,

$$S_S = -\sqrt{4\tau_S^2 - \left(S_{10}^2 + S_{20}^2 + S_{30}^2\right)}, \tag{3.19.9}$$

which, together with Eq. (3.19.3), determines components S_{kS}. Equations (3.18.16) and (3.19.7) give the plastic strain vector components developed under reloading:

$$e^S_k = a_0 \left[F\left(\frac{\sqrt{2}\tau_S}{S_0}\right) - 2F\left(\frac{2\tau_S}{\sqrt{S_0^2 + S^2}}\right)\right] \frac{S_k}{S}. \tag{3.19.10}$$

3.20 Discussion

1. A new theory of plasticity, the synthetic theory, has been proposed. This theory falls within the category of plastic flow theories. A loading surface is constituted by an infinite number of loading surfaces, planes. At an initial undeformed state, these planes are tangential to the yield surface in the Ilyushin five-dimensional stress-deviator space, $\mathbf{R}^5$. During loading, a stress vector shifts the tangent planes moving away them from the origin of coordinates. The displacements of planes on the endpoint of stress vector symbolize the occurrence of irreversible deformation.

2. If to take the Tresca yield criterion for the condition for the onset of yielding, the proposed concept of the developing of plastic deformation through the displacements of tangent planes leads to the constitutive relations of the Budiansky slip concept. This results from that the movement of plane in terms of synthetic theory corresponds to plastic slip in an appropriative slip system at a point in a body in terms of the concept of slip. Consequently, the synthetic theory belongs to the class of physical theories, at least as the Budiansky theory.

3. The partial cases of stress states are considered, two stress tensor components, shear components, are equal to zero, $S_4 = S_5 = 0$. Therefore, a loading takes place in the three-dimensional subspace of $\mathbf{R}^5$, $\mathbf{R}^3$.

In terms of the synthetic theory, a new yield criterion in $\mathbf{R}^5$ has been introduced, which is intermediate between the Tresca and the Guber criteria. This criterion in $\mathbf{R}^3$ results in Guber's criterion. The movements of planes tangential to the new surface of plasticity are studied.

4. The proposed theory is a two-level one. The movement of single plane determines an elementary plastic deformation. The macro deformation is the sum of the elementary deformations produced by moving planes.

5. In terms of the plastic flow theories the following is accepted. A loading surface undergoes transformation in $\mathbf{R}^5$ independent of the number of non-zero stress tensor components. According to this, in terms of synthetic theory, the movements of planes occur in $\mathbf{R}^5$ as well. Such approach to the determination of plastic strain seems to be hopeless due to that it is very difficult to watch the displacements of planes on the endpoint of stress vector at complex loading path in $\mathbf{R}^5$. Nevertheless, it is possible to simplify the situation, we consider the movements of planes in $\mathbf{R}^5$ though those of their traces in $\mathbf{R}^3$. Therefore, one needs to take into account the following. Any plane in $\mathbf{R}^3$, which is beyond the Guber sphere, is tangential to yield surface in $\mathbf{R}^5$. Therefore, together with the movements of planes tangential to Guber's sphere in $\mathbf{R}^3$, those of planes located beyond the Guber sphere in $\mathbf{R}^3$ take part in the production of plastic deformation as well – at increase in loading, the stress vector shifts all the specified planes and formulae for total plastic strains allow for the displacements of all these planes.

As already specified, instead of the Tresca yield criterion, the new criterion is adopted. It would be possible to take the Guber-von-Mises criterion in $\mathbf{R}^5$ and to consider a general state of stress, when all stress tensor components are distinct from zero. However, this is not suitable due to a tangent plane to the Guber hypersphere in $\mathbf{R}^5$ is defined by four parameters resulting in that the plastic tensor components are expressed through four-folded integrals. At the same time, the concept of slip prescribes that the integrals determining plastic strain should be three-folded.

6. For the case of orthogonal break of loading trajectory both the concept of slip and synthetic theory lead to the identical formulas for the additional shear modulus, identical loading surface are obtained as well, provided the function

$\varphi_N(H_N)$ is linear. Such agreement justifies the result of the Tresca criterion and the synthetic theory as a whole.

7. The deviator proportionality and the universal shear-stress-intensity vs. shear-strain-intensity diagram are obtained for the case of a simple loading. This result is of great importance due to, first, the Hencky-Nadai deformation theory is a simplest, generally accepted theory of plasticity. Secondly, in terms of synthetic model, we have managed to remove the main accusations against the concept of slip; the Budiansky model is incapable of analytically describing experimental diagrams in uniaxial tension and in pure torsion with unique model constants.

Following the discussion of Sec. 1.12, the fact – the synthetic theory gives a loading surface with a conic singularity – means that the law of deviator proportionality can be employed not only in a simple loading, but also at some deviation from it.

8. Similar to the concept of slip, the synthetic theory describes the specific case of uniaxial stress system, when the tensile direction varies and the tensile stress decreases. As it was specified, the known theories of plasticity are incapable of describing this case. In contrast to this, the synthetic model predicts the occurrence of plastic deformation at such loading. Indeed, the change in stress vector orientation results in that new planes start to move on the endpoint of the stress vector (producing plastic deformation), even at slightly decreasing absolute value of stress vector.

Theoretical relations give satisfactory agreements with experimental data obtained by various authors at two-segment loading paths for different tensile/compressive-shear stress ratios (Sec. 3.15). Under alternating loading, such cases are embraced when, even during decrease in initial stress, a plastic deformation of opposite sign occurs.

9. At the same time, the prediction of synthetic theory not always corresponds with experiments. For example, the additional shear modulus at an orthogonal break of loading trajectory turns out to be less than its value obtained experimentally. Further, the sharp corner point on a loading surface predicted by the theory is not observed in experiments. Stress-strain diagrams in alternating loading have been described for only cyclically ideal materials.

Summarizing, the synthetic theory of plasticity agrees much better with experiments, comparing to the concept of slip, however it possesses some shortcomings as well specified, in particular, in Sec. 3.20 (point 9). The further researches are necessary. One of the ways to improve the synthetic theory is to take into account the influence of loading rate on irreversible straining.

Chapter 4
The Creep Theory

This chapter concerns the analytical description of plastic deformation (in particular, in cyclic loading) and unsteady/steady state creep deformation in terms of generalized synthetic theory. The following generalizations are proposed.

(i) The constitutive equation of the synthetic theory is written in such a way that it governs both plastic and creep deformation. The deformation is not divided into plastic and viscous parts; we will use throughout further the notion of irreversible (permanent) deformation that develops in time. For the case of great loading rate when a viscous, time-dependent, part of deformation (creep deformation) has no time to develop, the constitutive equation results in the relations of the theory of plastic deformation; if a load is constant in time, the equations of the creep theory are obtained.

(ii) Following the nature and mechanisms of irreversible deformation, a new parameter which depends on a loading rate-the integral of non-homogeneity-, is introduced thereby allowing us to embrace the whole spectrum of irreversible straining, from plastic deformation in various loading regimes to unsteady/steady state creep deformation and that is why some sections that concern with plastic deformation are presented within this chapter. The integral of non-homogeneity is of great importance at the modeling of both plastic and creep deformation.

4.1 General Remarks

Under certain force-temperature conditions, solids deform under constant stresses that is termed as creep. Creep is observed for a wide class of materials: metals, plastic, mountain rocks, concrete, and glasses.

The aim of creep theory is to describe strains responding to various stress states, magnitudes of stresses, and temperatures. Creep theory appeared not long ago as an independent division of engineering mechanics. It is the part of the *Solids Mechanics* and holds a worthy place among such sciences as the theory of elasticity and plasticity.

At present, the huge number of both experimental and theoretical works devoted to different aspects of creep has been published. With the exception of fundamental works, we will not dwell on the analysis of existing publications. Ju. Rabotnov's and J. Betten's monographs [122,11] give an excellent review of these works and show in full measure the state of the theory of creep.

A typical creep diagram in tension is shown in Fig. 4.1. It is possible to distinguish three portions marked as I, II, and III. Initially, the strain rate slows with increasing strain (I). This is known as primary (unsteady) creep. The strain rate, $\dot{\varepsilon}_x^p$, eventually reaches a minimum and becomes near-constant. This is known as secondary or steady-state creep (II). In tertiary creep (III), the strain-rate increases with strain.

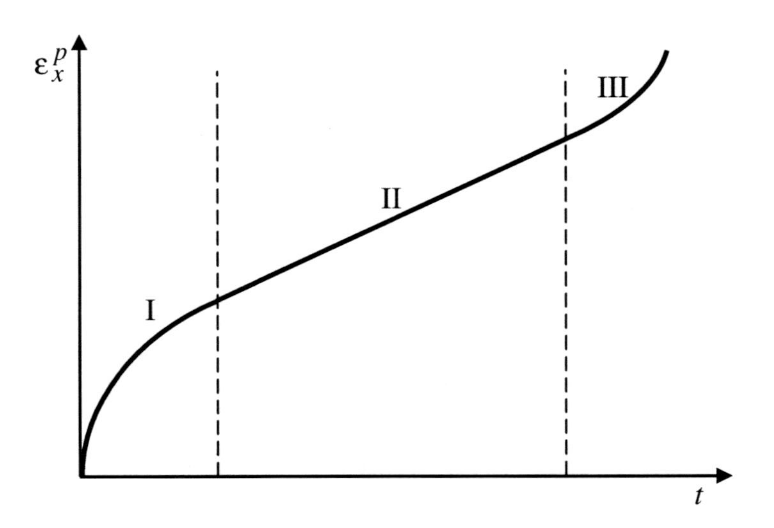

Fig. 4.1 Creep diagram in uniaxial tension

The creep deformation, to a considerable extent, depends on the regime of loading , by which we understand how the stress grows from zero to the value that remains constant in time. Under some conditions, the primary portion is absent and the strain-time diagram is linear from the very beginning. Under other conditions, there is no steady-state creep at all, i.e. the primary portion of creep diagram transits immediately into the portion with increasing strain rate (the tertiary creep). If the magnitude of the applied stress and its duration is insignificant, there is no tertiary creep on creep diagram. The tertiary creep is accompanied with fracture processes, which are not considered. Therefore, we limit ourselves in this book to investigate only the primary and steady-state creep of metals.

For the case of uniaxial tension experimental strain-rate versus stress relations can be expressed in the following forms

$$\dot{\varepsilon}_x^p = c\sigma_x^k, \tag{4.1.1}$$

where exponent k ranges from 3 to 7 for various metals, or

$$\dot{\varepsilon}_x^p = c_1 \exp(k_1 \sigma_x), (c_1, k_1 = \text{const}). \tag{4.1.2}$$

From the above formula it follows that we cannot employ linear visco-elastic models for the description of creep of metals, first, due to the nonlinearity of Eqs. (4.1.1) and (4.1.2). Another reason is that the specified theory prescribes that a material will return to its original shape when the applied stress is removed. In contrast to this, the creep deformation of metals is an irreversible one, i.e., it remains in a body after unloading.

Today, there is no unique creep theory suitable for all materials and we are sure that the working out of such theory is impossible. The creep mechanisms of, e.g., metals and polymers and their creep-macro-behaviors considerably differ. Therefore, when modeling creep, it is necessary to limit the class material to be modeled. For many metals and their alloys the basic mechanism of plastic or creep deformation is the dislocation movements that induce slips in crystal grains. Therefore, similar to plastic deformation, creep deformation can be modeled in terms of the concept of slip or the synthetic theory. Further we will use the term the synthetic theory of irreversible deformation meaning that plastic and creep deformation as well as their sum are considered.

Special emphasis must be placed on temperature and loading rate due to they strongly affect the behavior of metals. For example, carbon steel, whose deformation at room temperature is described by the theory of elasticity or plasticity depending on the magnitude of acting stress, at temperature close or above 450°C starts to behave in an absolutely different manner. Even under small stresses, the steel does not obey Hook's law and the stress-strain diagram in tension considerably depends on a loading rate. Therefore, in contrast to the theories of plasticity, we cannot take the stress-strain diagram as an initial characteristic of a material.

Therefore, we require that the theory of irreversible deformation describe not only creep strains but also how a yield limit and whole stress-strain diagram react to loading rate. We will then show that these problems relate to each other.

As it is well known, a homology temperature,

$$\Theta = \frac{T}{T_m}, \tag{4.1.3}$$

is used in the relation of creep theories, where T and T_m are considered absolute temperature and the melting point (both have the units of Kelvin), respectively. The homology temperature is convenient measure as it varies within identical limits for all bodies,

$$0 < \Theta < 1 \tag{4.1.4}$$

The development of creep theories of metals is influenced largely by the theory of plasticity. This is not by accident because the, fundamental principles of creep are worked out by scientists studying the theory of plasticity. The model presented in this monograph is not an exception either – it is developed on the base of the synthetic theory of plasticity. The parameters characterizing the processes responsible for creep deformation and loading-rate-effects are introduced into the constitutive equations of the synthetic theory.

4.2 Non-homogenous Distribution of Strains and Stresses of the Second Kind

Since the synthetic model is a two-level one, we will distinguish a macro-stress, the stress of the first kind, from that of the second and third kind. If an applied force acts on the area whose size is of the same order of magnitude as a crystal grain, we speak about the stress of the second kind. If the force is related to the area of lateral length equal to the several tens of interatomic distance, we arrive at the notion of the stress of the third kind.

Figure 4.2 shows experimental distribution diagrams of the strain and stress of the second kind obtained for specimens of pure copper (Fig. 4.2a), iron (b), and titan (c) loaded in uniaxial tension in elastic area (an average grain size 100 μm $=10^{-1}$mm) [66,67]. Equally spaced control points, at 10 μm distance from each other, were marked along the specimen, x-axis, and the changes in these distances were measured during loading. The values of the strains/stresses plotted in Fig. 4.2 are related to their average values. As seen from Fig. 4.2, the stress/strain of the second kind, curve 1 and 2, respectively, is distributed in a non-homogeneous manner. This non-homogeneity (non-uniformity) is observed not only from one grain to another but also within a grain. The magnitude of non-homogeneity is different for different metals, copper exhibits the highest level of heterogeneity, while titan exhibits the least. The non-homogeneity of the stress/strain distributions in elastic area is caused by the interaction of different-oriented grains possessing the anisotropy of elastic properties: the greater this anisotropy, the greater stress/strain non-homogeneity. Among investigated metals, copper (FCC structure) has the greatest anisotropy of elastic properties and titan (HCP structure) exhibits the least anisotropy [66,67]. Another effect observed in Fig. 4.2 for all the three metals is that if a strain larger than its average value, the corresponding stress is less than the stress-average and vice versa. If the deformation is limited by ,e.g., an unfavorable grain orientation relative to acting force, the stress concentration is observed.

The distributions of plastic strain related to its average value of 2% are plotted in Fig. 4.3. Comparing Fig. 4.2 and Fig. 4.3, it is clear that the non-homogeneity of plastic strain is higher than that of elastic strain. This can be explained by the difference in the mechanisms of elastic and plastic deformation of metals. The properties of crystals with respect to a plastic deformation have a strongly anisotropic character, plastic slips occur only along the close packed planes (slip systems), copper (FCC) has 12 slip systems and titan, a hexagonal close-packed metal, has only 4. The different quantity of slip systems results in the different conditions of the grain interaction and, consequently, in a different degree of strain-non-homogeneity.

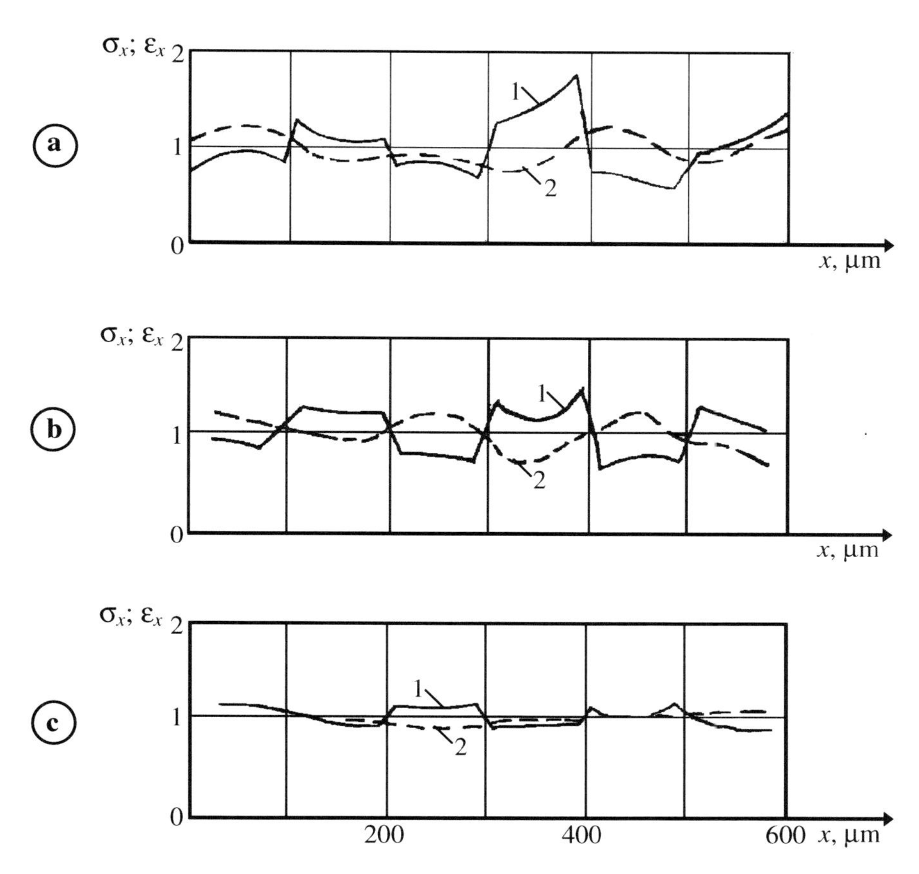

Fig. 4.2 Distribution of the strain (1) and stress (2) of the second kind through grains; (a) - copper, (b) - iron, (c) - titanium

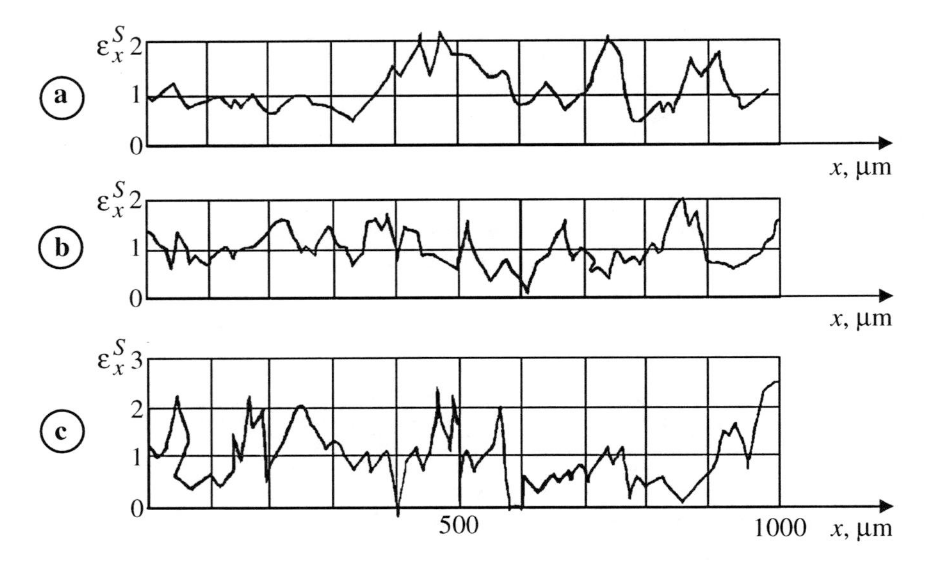

Fig. 4.3 Distribution of plastic strain throw grains;(a) - copper, (b) - iron, (c) - titanium

Non-homogenous deformations cause a non-homogenous stress distribution; one part of grain (I and III in Fig. 4.4a) is subjected to stresses that exceed the stress-average-value, while the other is under smaller stresses (II). The total over- and under-loading is equal, obviously, to zero. The smaller stress ($\sigma_x - \sigma_x''$) works on that part of grain (II), where the irreversible strain occurs easily, whereas greater stress ($\sigma_x + \sigma_x''$) acts where the straining is the most obstructed. Thus the driving force of plastic deformation is not stress σ_x, but only its part $\sigma_x - \sigma_x''$.

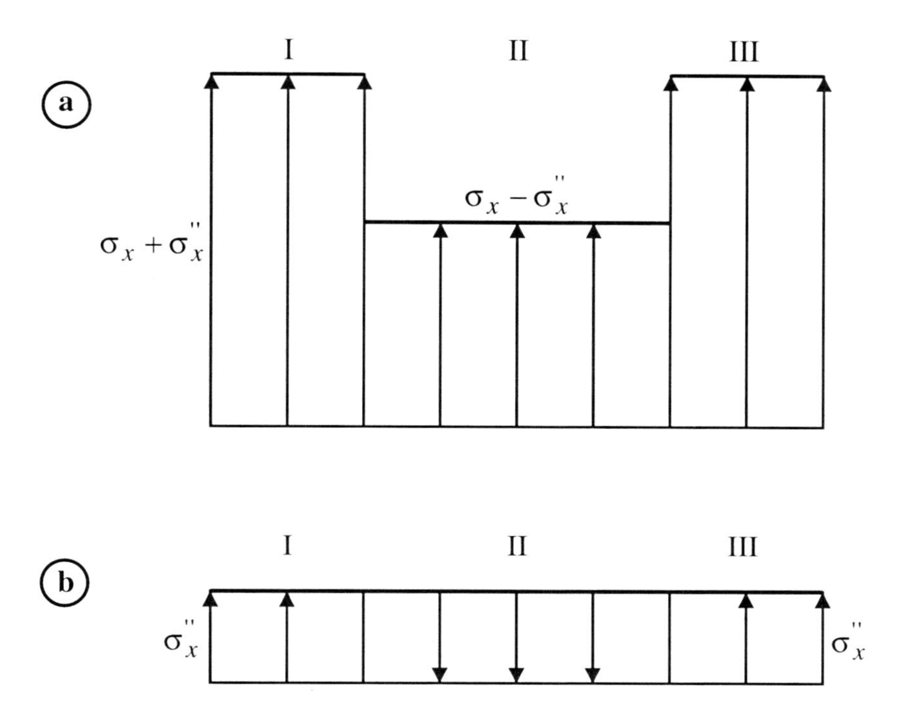

Fig. 4.4 Qualitative distribution of stresses (a) and residual stresses after unloading (b) through a grain

Let us note that, in terms of the Ishlinsky-Kadashevich-Novozsilov theory (Sec. 1.16), the macro-characteristic of micro-stress σ_x'' is modeled by the tension of left spring, which is responsible for the hardening of material (Fig. 1.17a).

As a plastic strain is irreversible, after unloading ($\sigma_x = 0$), the residual stress σ_x'' acts in a body (Fig. 4.4b). The energy of residual stress is called the latent energy. As it is generally known, during plastic straining, the most part of the strain energy is dissipated or "lost" as heat, but its some part is stored in a metal in the form of the latent energy [12,13].

As the driving force of plastic deformation is not stress σ_x, but only its part $\sigma_x - \sigma_x''$, the following question to be decided is whether theories ignoring the residual stress σ_x'', such as the concepts of slip, flow theories, and deformation theories, are acceptable for the description of the plastic properties of materials. The answer to this question is obtained as follows. In terms of the models allowing for stress σ_x'', a plastic strain in tension, ε_x^S, is determined as

$$\varepsilon_x^S = \tilde{f}_1(\sigma_x - \sigma_x''). \tag{4.2.1}$$

As components σ_x'' and ε_x^S are related to each other, then

$$\sigma_x'' = \tilde{f}_2\left(\varepsilon_x^S\right). \tag{4.2.2}$$

Let us note that in terms of the Ishlinsky theory the function $\tilde{f}_2$ is assumed to be linear, Eq. (1.16.3). Eliminating σ_x'' from Eqs. (4.2.1) and (4.2.2), we obtain

$$\varepsilon_x^S = \tilde{f}_1\left(\sigma_x - \tilde{f}_2\left(\varepsilon_x^S\right)\right). \tag{4.2.3}$$

By denoting the solution of this equation as $\tilde{f}_3$, we get

$$\varepsilon_x^S = \tilde{f}_3(\sigma_x). \tag{4.2.4}$$

Therefore, due to the existence of relation (4.2.2), Eq. (4.2.4) can be obtained from Eq. (4.2.1), i.e., they are equivalent. Consequently, both Eq. (4.2.4) and Eq. (4.2.1) can be used for the establishing of stress-strain relation, the kinematic flow models take into account stress σ_x'' explicitly, whereas the most other theories of plasticity do it in an implicit form. Thus, the kinematic flow models, in some specified sense, are more preferable as they are capable of describing the Feigin effect (Sec. 1.16) and allowing for the latent energy.

4.3 Local Peak Stresses

Conciderable non-homogeneity of the distribution of the stresses of the third kind as well as the stress concentration is observed.

As already specified, the plastic deformation develops due to the parts of crystal lattice glide(slip) along each other, which leads to the distortion of an initial geometry of crystal structure. These slips are caused, in general, by dislocation motions. The number of dislocations increases dramatically during plastic deformation. Dislocations spawn from existing dislocations, grain boundaries and surfaces. Dislocation densities can vary from 10^5–10^9 cm^{-2} in annealed metals to 10^9–10^{12} cm^{-2} in heavily deformed metals [44].

The average distance between dislocations decreases and dislocations start blocking the motion of each other as the consequent increase in overlap between the strain fields of adjacent dislocations gradually increases the resistance to further dislocation motion. This causes a hardening of the metal as deformation progresses. This effect is known as strain hardening (also "work hardening"). Tangles of dislocations are found at the early stage of deformation. The dislocation interactions are accompanied by the nucleation and movement of cellular structures, interstitial atoms, vacancies, jogs etc.; all these defects hinder the progress of plastic deformation as well. The described processes are well-known [69,93].

Let us distinguish the reversible (so called elastic) and irreversible distortions of lattice. The distortion of crystal lattice that can be liquidated only by anneal will be called the irreversible distortions, whereas reversible distortions can relax spontaneously (at temperature smaller than recrystallization temperature). Stresses inducing the local elastic distortions of lattice, similar to works [166,167], will be termed as local peak stresses. In view of the size of the area where the local peak stresses act, clear that they are the micro-stresses of the third kind. Despite the fact that the peak stresses are strongly localized, their magnitude can reach considerable values.

The local peak stresses possesses the following properties.

1. Local peak stresses correlate to the resistance of a material with respect to a subsequent plastic deformation. To show this, it is worthwhile to emphasize that the plastic deformation is governed not only by the stresses of the first and second kind, but also by the local peak stresses as well, i.e., by the states of those micro-substructures where the plastic deformation actually develops. Experimental works on many metals has amply justified that the distortion of crystal lattice increases the resistance to plastic deformation. Let us turn to work [69] stating that a large static distortions of lattice caused by the dissolution of impurity elements increases the yield limit of the investigated alloys. Furthermore, as shown in [31], the increase in yield limit is proportional to the number of atoms shifted from their places that results in the distortion of the crystal lattice.

We will not pay great attention to what is the cause and effect in the following: dislocation motions are impeded by the distortions of lattice produced by local stresses or, vice versa, local peak stresses that arise due to the dislocation movements, for some reason or other, are obstructed. The main fact is that the greater the local peak stress, the greater the resistance to plastic deformation.

2. Local peak stress grows with loading rate. Experimental work [130] concerned the relation between the distortion of lattice and loading rate. Investigated specimens were loaded in tension to the fixed value of plastic strain, $\varepsilon^S = 1{,}2 - 1{,}5\%$, at different rates of relative elongation, 0.005, 2.0, 8.0, and 32 mm/minute. The maximal values of the distortion of lattice were recorded at the maximum strain rate.

Since increase in loading rate (v) results in the growth of local distortions that, in turn, limits the development of plastic strain, the stress-strain curve with greater loading rate must be located higher than that obtained at smaller value of v. Experimental works on many metals has amply justified this statement for pure metals at the whole range of v, whereas for alloys within some diapason $v > v_0$ only.

Consider, for example, the results of experiment on titan alloy [180], which is sensitive to a change in strain rate. A tested specimen was loaded in tension at the following regime: a) loading with small strain rate (static testing) (curve 1 in Fig. 4.5, $\dot{\varepsilon}_x = 1{,}5 \cdot 10^{-5}\,\mathrm{s}^{-1}$), b) unloading, c) loading on impact tensile machine (curve 2) at $\dot{\varepsilon}_x = 0{,}5\,\mathrm{s}^{-1}$, d) unloading, e) loading on the machine for static testing (curve 3, $\dot{\varepsilon}_x = 1{,}5 \cdot 10^{-5}\mathrm{s}^{-1}$) and so on, curve 4 corresponds to the strain rate $\dot{\varepsilon}_x = 0{,}5\,\mathrm{s}^{-1}$, and curve 5 – $\dot{\varepsilon}_x = 1{,}5 \cdot 10^{-5}\mathrm{s}^{-1}$. Curve 6 corresponds to the loading with constant strain rate, $\dot{\varepsilon}_x = 1{,}5 \cdot 10^{-5}\mathrm{s}^{-1}$.

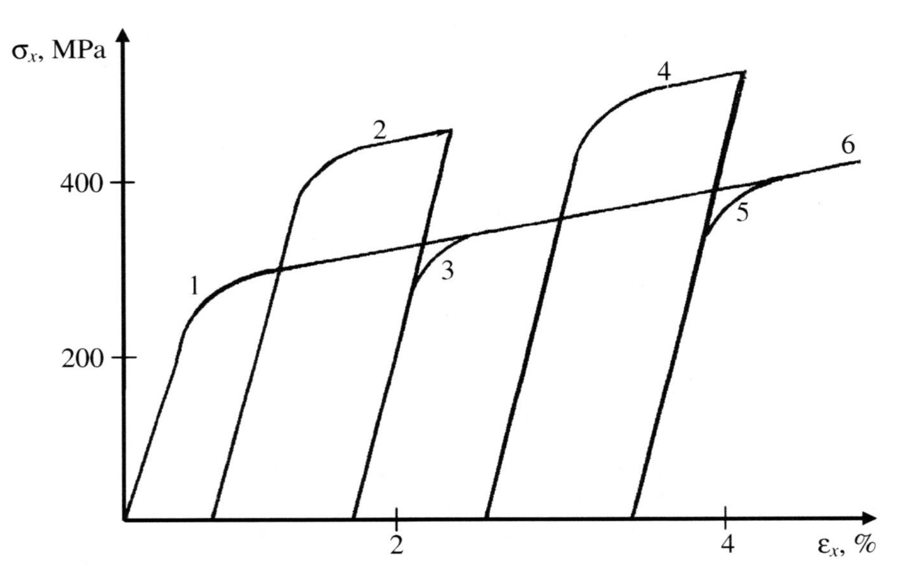

Fig. 4.5 Stress-strain diagrams of titanium with different strain rates

As seen from the experiment, the greater the stress/strain rate, the greater the strain hardening. In addition, it is the elastic distortions of lattice that causes the "strain-rate-hardening". The drop of v results in the decrease of these distortions and the stress-strain diagram has such form as if deformation occurs at $\dot{\varepsilon}_x = 1{,}5 \cdot 10^{-5}\mathrm{s}^{-1}$ from the very beginning.

3. Local peak stresses are unstable. Once a favorable condition arises, e.g., stresses cease to increase, the relaxation of the distortion of crystal lattice occurs.

In experimental works [128,129], (by X-ray diffraction method) the relaxation of the elastic deformation of lattice was studied. As shown, the relaxation is caused, in general, by spontaneous slips in grains, which are caused mainly, as follows from the huge amount of experimental material, by dislocation movements. Under thermal fluctuations, locked and tangled dislocations and the obstructions in their way themselves become progressively movable, meaning that an irreversible deformation develops due to relaxation of elastic distortions of lattice. At the same time, the specified relaxation can occur without being accompanied by residual deformation provided that diffusional processes promote the relaxation. The peak stress relaxation is observed under a slow loading as well.

The decrease in the elastic distortions of crystal lattice is essentially connected with the relaxation of macro-stresses (observed at an unchangeable relative elongation of specimen). The relaxation curves of specimens of tin, which endure identical strains but with different loading rate, join together [12] (see Fig. 4.6.). Since the local peak stresses, being responsible for the strain-rate-hardening of material, relax in time, the strain-rate-hardening also gradually vanishes in time.

Thus, summarizing, local peak stresses, being an increasing function of loading rate, are responsible for the so-called strain-rate hardening of material. On the other hand, local peak stresses can relax causing the softening of material. The macro-behavior of material is the result of rival processes interaction between the specified hardening and softening.

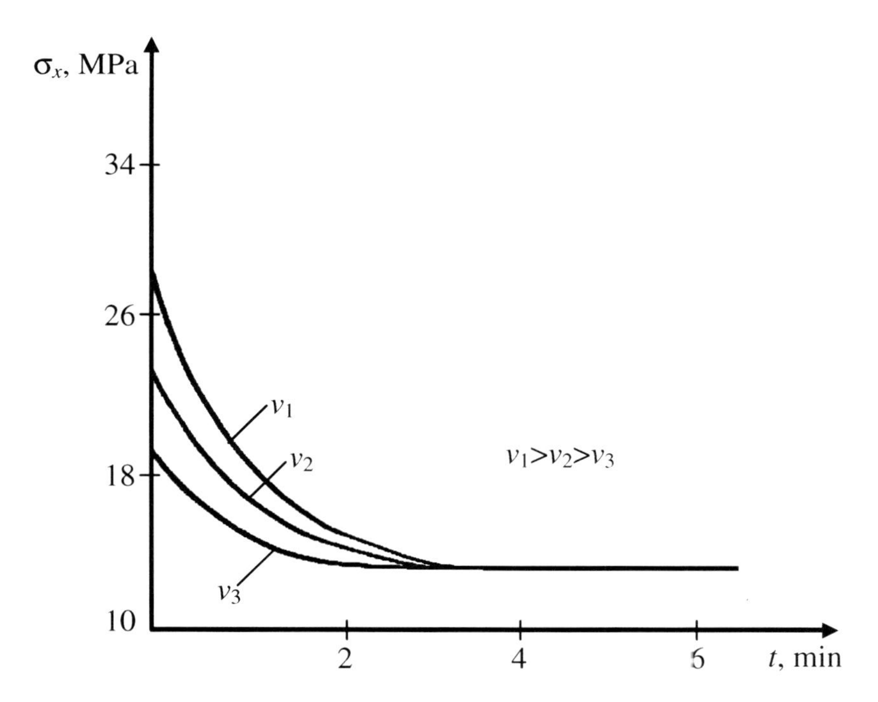

Fig. 4.6 Stress relaxation in specimens of tin pre-strained at various loading rates v_i, $i = 1,2,3$.

4.4 Elastic Area

Stresses well below elastic limit induce an intensive nucleation and movement of dislocations, i.e., the loading in elastic region is accompanied by micro-plastic deformation. It results in the arising of local peak stresses investigated in the works of V. Sarrak [166,167]. Similar to a macro-plastic region, the local peak stresses depend on loading rate and, under favorable conditions, can relax resulting in the decrease of crystal lattice distortion. In the case of small loading rate or high temperature the local-micro-stress relaxation occurs in the course of loading. Under great loading rates or low temperatures, the relaxation has no time to develop during loading, it manifest itself in the form of (i) stress relaxation at an unchangeable deformation of specimen or (ii) time-dependent irreversible deformation (creep) at a constant load.

It is worthwhile to consider the following experimentally recorded fact. The yield strength σ_S and stress relaxation behave similarly as a function of loading rate, in particular, if σ_S is not sensitive to loading rate, the macro-stress relaxation is absent as well [167]. This phenomenon is held in the whole temperature range (from - 40 to 200°C) where experiments were carried out. And what is more, the increase in the yield limit of, e.g., iron due to increase in $\dot{\varepsilon}$ is equal to the amount of relaxation. Figure 4.7 taken from [167] shows the schematic relaxation

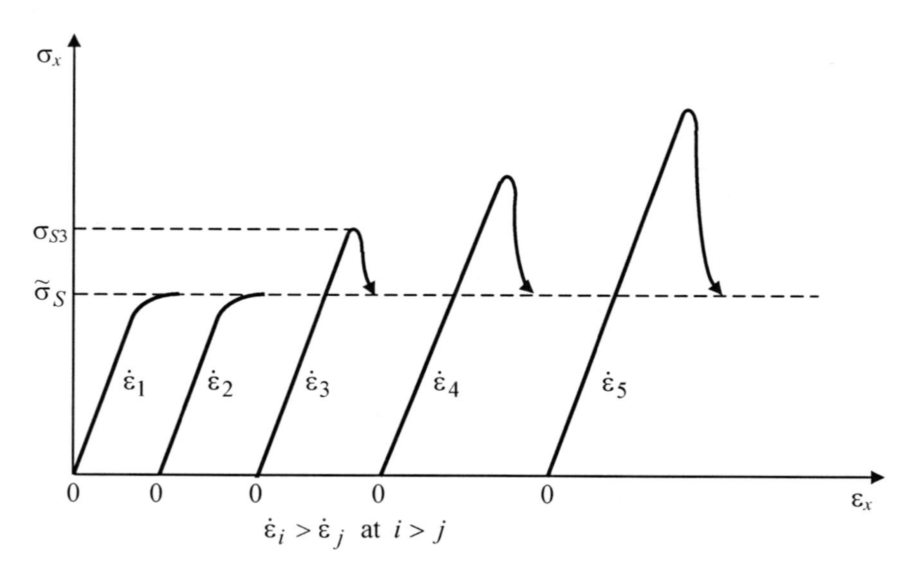

Fig. 4.7 The dependence of the material-hardening and stress relaxation on strain-rate

diagrams corresponding to different strain rates, $\dot{\varepsilon}_1 < \dot{\varepsilon}_2 < ... < \dot{\varepsilon}_6$; the peaks of the curves are the yield limits. Similar to Fig. 4.6, independent of strain rate, the acting stresses decrease during relaxation to the one and the same level. It is the local stresses that regulate the "high-rate hardening". After the load stops increasing, they reduce in time, therefore, the "high-rate hardening" diminishes and eventually all the relaxation curves arrive at the yield stress $\tilde{\sigma}_S$ corresponding to a smaller loading rate.

4.5 Mathematical Measure of Local Peak Stresses

Employing statistical technique [181], we wish to estimate the influence of the non-uniformity of microstress distribution upon elastic strain energy. For this purpose, consider an elementary volume of body (treated as a point) consisting of a large number of micro particles. The value of stress varies from particle to particle (i.e., it is of a random nature), although within each particle the stress is assumed to be distributed homogeneously. We assume every micro-particle to be an isotropy continuum for which the *Continuous Mechanics* is applicable.

If $\overline{\sigma}_{ij}^0$ denotes macro-stress deviator tensor components at a point in a body (the stress of the first kind or the average stress), then micro-stresses acting in the particles can be written as

$$\overline{\sigma}_{ij} = \overline{\sigma}_{ij}^0 + \overline{\sigma}_{ij}', \tag{4.5.1}$$

where $\overline{\sigma}_{ij}'$ are random quantities expressing an over-/under-loading in each particle. The reaction of $\overline{\sigma}_{ij}'$ on the change of the average stress we set as

$$d\overline{\sigma}_{ij}' = C_{ijkl} d\overline{\sigma}_{kl}^0, \tag{4.5.2}$$

where C_{ijkl} are random numbers that vary from particle to particle, they are assumed to be independent from $\overline{\sigma}_{ij}^0$. Let us suppose that all random numbers C_{ijkl} have an identical distribution function, F_*, and are independent of each other. Since $\overline{\sigma}_{ij}^{0}$ are macroscopic (average) deviator stress components, the mathematical expectation of parameters C_{ijkl} is

$$\int_{-\infty}^{\infty} C_{ijkl} F_{*}\left(C_{ijkl}\right) dC_{ijkl} = 0 . \quad \Sigma \tag{4.5.3}$$

meaning that the total over- and under-loading with respect to the average stress is equal to zero. Furthermore,

$$\int_{-\infty}^{\infty} F_{*}\left(C_{ijkl}\right) dC_{ijkl} = 1 . \quad \Sigma \tag{4.5.4}$$

As it was pointed out in the previous section, the local stresses are unstable and can relax. Let us present the equation to govern their time-dependent behavior in the form

$$d\overline{\sigma}_{ij}' = C_{ijkl} d\overline{\sigma}_{kl}^{0} - p\overline{\sigma}'_{ij}\, dt . \tag{4.5.5}$$

The first item in the right hand in the above formula characterizes the rise of $\overline{\sigma}_{ij}'$ given by Eq. (4.5.2), term $-p\overline{\sigma}_{ij}'dt$ governs the time-dependent decrease of the micro-stress, which is taken to be proportional to $\overline{\sigma}_{ij}'$. The solution of the obtained differential equation (4.5.5) for $\overline{\sigma}_{ij}'$ is

$$\overline{\sigma}'_{ij} = C_{ijkl} I_{kl}(t),$$

$$I_{kl}(t) = \int_{0}^{t} \frac{d\overline{\sigma}_{kl}^{0}}{ds} \exp(-p(t-s)) ds . \tag{4.5.6}$$

and Eq. (4.5.1) becomes

$$\overline{\sigma}_{ij} = \overline{\sigma}_{ij}^{0} + C_{ijkl} I_{kl}(t) . \tag{4.5.7}$$

The integrals $I_{kl}(t)$ are single-valued functions of time, whereas C_{ijkl} are random numbers.

Substituting $\overline{\sigma}_{ij}$ from Eq. (4.5.7) into the formula for elastic strain energy,

$$U=\frac{1}{12G}\left[\left(\overline{\sigma}_x-\overline{\sigma}_y\right)^2+\left(\overline{\sigma}_y-\overline{\sigma}_z\right)^2+\left(\overline{\sigma}_z-\overline{\sigma}_x\right)^2+6\left(\tau_{xy}^2+\tau_{yz}^2+\tau_{zx}^2\right)\right], \quad (4.5.8)$$

leads to the following expression

$$U=\frac{1}{12G}\left\{\left(\overline{\sigma}_x^0+C_{xxkl}I_{kl}-\overline{\sigma}_y^0-C_{yykl}I_{kl}\right)^2+\left(\overline{\sigma}_y^0+C_{yykl}I_{kl}-\overline{\sigma}_z^0-C_{zzkl}I_{kl}\right)^2+\right.$$
$$\left.+\left(\overline{\sigma}_z^0+C_{zzkl}I_{kl}-\overline{\sigma}_x^0-C_{xxkl}I_{kl}\right)^2+6\left[\left(\tau_{xy}^0+C_{xykl}I_{kl}\right)^2+\left(\tau_{yz}^0+C_{yzkl}I_{kl}\right)^2+\left(\tau_{xz}^0+C_{xzkl}I_{kl}\right)^2\right]\right\} \quad (4.5.9)$$

The mean value of U is determined by the following relation

$$\langle U\rangle=\int_{-\infty}^{\infty}\int_{-\infty}^{\infty}\ldots\int_{-\infty}^{\infty}UF_*\left(C_{xxxx}\right)F_*\left(C_{xyxx}\right)\ldots dC_{xxxx}dC_{xyxx}\ldots. \quad (4.5.10)$$

Now, we substitute U from (4.5.9) into the above formula. Taking into account that, due to Eq. (4.5.3), terms in the right side in Eq. (4.5.10) containing C_{ijkl} of first order equals to zero, we obtain

$$\langle U\rangle=U_0+\frac{2B_1}{G}\left(I_{xx}^2+I_{yy}^2+I_{zz}^2+2I_{xy}^2+2I_{yz}^2+2I_{zx}^2\right), \quad (4.5.11)$$

where U_0 is the elastic strain energy of ideal-homogeneous model,

$$U_0=\frac{1}{12G}\left[\left(\sigma_x^0-\sigma_y^0\right)^2+\left(\sigma_y^0-\sigma_z^0\right)^2+\left(\sigma_z^0-\sigma_x^0\right)^2+6\left({\tau_{xy}^0}^2+{\tau_{yz}^0}^2-{\tau_{zx}^0}^2\right)\right]. \quad (4.5.12)$$

B_1 is the variance of random numbers C_{ijkl}:

$$B_1=\int_{-\infty}^{\infty}C_{ijkl}^2F_*\left(C_{ijkl}\right)dC_{ijkl} \quad \Sigma \quad (4.5.13)$$

By subtracting from the right-hand side in Eq. (4.5.11) the expression $2B_1/3G\left(I_{xx}+I_{yy}+I_{zz}\right)^2$, which is equal to zero due to $\overline{\sigma}_x^0+\overline{\sigma}_y^0+\overline{\sigma}_z^0=0$, we obtain

$$\langle U\rangle=U_0+\frac{2B_1}{3G}\left[\left(I_{xx}-I_{yy}\right)^2+\left(I_{yy}-I_{zz}\right)^2+\left(I_{zz}-I_{xx}\right)^2+6\left(I_{xy}^2+I_{yz}^2+I_{zx}^2\right)\right] \quad (4.5.14)$$

Substituting I_{ij} from Eq. (4.5.6) into Eq. (4.5.14) and converting the stress tensor components σ_{ij} to the stress vector components S_n (see Eq. (1.7.10)),

$$\sigma_x - \sigma_y = \frac{1}{\sqrt{2}}\left(S_1\sqrt{3} - S_2\right), \quad \sigma_y - \sigma_z = S_2\sqrt{2},$$
$$\sigma_z - \sigma_x = -\frac{1}{\sqrt{2}}\left(S_1\sqrt{3} + S_2\right), \tag{4.5.15}$$

the expression for the mathematical expectation of the elastic strain energy is obtained as

$$\langle U \rangle = U_0 + \frac{2B_1}{3G}\sum_{i=1}^{5}\left[\int_0^t \frac{dS_i}{ds}\exp(-p(t-s))ds\right]^2. \tag{4.5.16}$$

The value of $\langle U \rangle$ is seen to consist of two parts, the term U_0 corresponds to homogeneous stress distribution and the second term characterizes the time-dependent deviation of stresses from their average value. If a body is ideally homogeneous, the distribution functions of random numbers C_{ijkl} degenerate in the Dirac delta-function and, according to Eq. (4.5.13), we obtain $B_1 = 0$. As seen from Eq. (4.5.16), $\langle U \rangle$ depends not only on the stress rate vector components $\dot{S}_i$ at a given instant, but also on its values for the whole history of loading.

For the case $\dot{S}_i = const$,

$$\langle U \rangle = U_0 + \frac{2B_1}{3G}\dot{S}_i\dot{S}_i\left[\int_0^t \exp(-p(t-s))ds\right]^2 \tag{4.5.17}$$

and, since

$$\dot{S}_i\dot{S}_i = \dot{S}^2 \tag{4.5.18}$$

(S denotes the length of stress vector), we get

$$\langle U \rangle - U_0 = \frac{2B_1}{3G}\left[\int_0^t \frac{dS}{ds}\exp(-p(t-s))ds\right]^2. \tag{4.5.19}$$

It must be noted that, for the case if a stress vector has a single non-zero component, Eqs. (4.5.16) and (4.5.19) are identical at variable loading rate as well.

We take the square root in the right-hand side in Eq. (4.5.19),

$$I = \int_0^t \frac{dS}{ds} Q(t-s)ds\,, \tag{4.5.20}$$

$$Q = B\exp(-p(t-s)), \quad B = \sqrt{\frac{2B_1}{3G}}\,. \tag{4.5.21}$$

We will term the I the ***integral (parameter) of non-homogeneity*** which provides the mathematical measure of local peak stresses. Together with the function Q given by Eq. (4.5.21), we introduce another form of this function,

$$Q = \frac{B}{(t-s)^p}\,. \tag{4.5.22}$$

The radical B_1 given by Eq. (4.5.13) and the factor p appearing in Eqs. (4.5.21) and (4.5.22) are supposed to be material constants.

4.6 Properties of the Integral of Non-homogeneity

To establish the properties of the integral I from (4.5.20), we will consider, first, a proportional loading of constant loading rate, v, for the following time-dependent regime (see Fig. 4.8a): $dS/dt > 0$ as $t \in [0, t_1]$ and $dS/dt = 0$ as $t > t_1$ where S is the absolute value of stress vector. During the time-diapason $t \in [0, t_1]$, Eqs. (4.5.20) and (4.5.21) give

$$I = Bv\int_0^t \exp(-p(t-s))ds = \frac{Bv}{p}[1-\exp(-pt)], \tag{4.6.1}$$
$$v = dS/dt\,.$$

For the second portion, $t > t_1$, we split the range of integration in (4.5.20) into two time-diapasons: from 0 to t_1 and from t_1 to t. Since $\dot{S} = 0$ for $t > t_1$, we have

$$I = Bv\int_0^{t_1} \exp(-p(t-s))ds = \frac{Bv}{p}[\exp(pt_1)-1]\exp(-pt). \tag{4.6.2}$$

If we take the function Q in the form of Eq. (4.5.22) (power function with polar singularity), Eqs. (4.5.20) and (4.5.22) give

$$I = Bv\int_0^t \frac{ds}{(t-s)^p} = \frac{Bv}{1-p}t^{1-p} \quad \text{for} \quad t < t_1 \tag{4.6.3}$$

and

$$I = Bv\int_0^{t_1} \frac{ds}{(t-s)^p} = \frac{Bv}{1-p}\left[t^{1-p} - (t-t_1)^{1-p}\right] \quad \text{for} \quad t > t_1. \tag{4.6.4}$$

Because $S = vt$ as $t < t_1$, Eq. (4.6.1) can be written as

$$I = \frac{Bv}{p}\left[1 - \exp(-pS/v)\right], \tag{4.6.5}$$

while Eq. (4.6.3) gives

$$I = \frac{Bv^p S^{1-p}}{1-p}. \tag{4.6.6}$$

Equations (4.6.2)–(4.6.6) indicate the following properties of the integral of non-homogeneity

1. Integral I grows during loading.
2. Integral I grows with the increase in loading rate.
3. Parameter I decreases in time at constant S.

Thus, the integral of non-homogeneity behaves similar to local peak stresses. A plot of I as a function of time for finite loading rates v is shown in Fig. 4.8.b, its qualitative form does not depend on the form of function Q, power or exponential. However, if $v \to \infty$, Eqs. (4.6.5) and (4.6.6) lead to different results. Indeed, expanding function $\exp(-pS/v)$ in Eq. (4.6.5) by its Tailor series expansion (dropping quadratic term in t), we obtain

$$I = BS \quad \text{as} \quad v \to \infty, \tag{4.6.7}$$

whereas Eq. (4.6.6) gives $I \to \infty$ as $v \to \infty$.

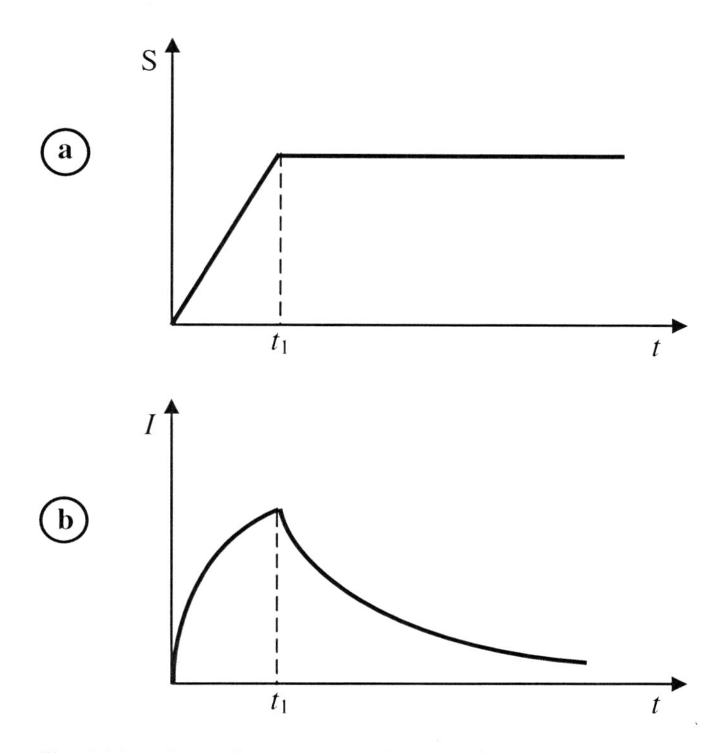

Fig. 4.8 Loading regime (a) and the integral of non-homogeneity (b)

4.7 Account of Loading Rate

As stated above, the synthetic theory is of two-level nature; a macro- and micro-level are considered. Similar to the concept of slip, an elementary act of irreversible deformation is that the parts of crystal grain glide(slip) along each other. Within any grain, there are one or several slip systems. One of the key point of the synthetic theory states that tangent planes correspond to appropriate slip systems. Further, in terms of both the concept of slip and the synthetic theory is supposed that the state of stress in micro-particles (grains) is identical to the macro-stress state at a point in a body, which is characterized by stress vector $\vec{\mathbf{S}}$. Following the concept of slip, the "driving force" of a non-reversible change of shape is not a whole stress tensor, but resolve shear stress τ_{nl} only. In a similar fashion, the synthetic model does not take a stress vector for the "driving force" of micro-plastic strains, but the projection of stress vector onto the direction of the possible movement of planes, scalar product $\vec{\mathbf{S}} \cdot \vec{\mathbf{N}}$. Therefore, instead of Eq. (4.5.20), we take the characteristic of local micro-stresses in the following form

$$I_N = \int_0^t \frac{d\vec{\mathbf{S}}}{ds} \cdot \vec{\mathbf{N}} Q(t-s) ds, \tag{4.7.1}$$

where Q is given by Eq. (4.5.21) or (4.5.22). In contrast to Eq.(4.5.20), now, the integral of non-homogeneity depends on angles α, β, and λ that determine the location of planes in Ilyushin's space.

The parameter of non-homogeneity for tangent planes with normal vectors $-\vec{\mathbf{N}}$ (Fig. 3.15), I_{-N}, is

$$I_{-N} = \int_0^t \frac{d\vec{\mathbf{S}}}{ds} \cdot \left(-\vec{\mathbf{N}}\right) Q(t-s) ds = -\int_0^t \frac{d\vec{\mathbf{S}}}{ds} \cdot \vec{\mathbf{N}} Q(t-s) ds = -I_N, \tag{4.7.2}$$

It is clear that if a stress vector translates planes with normals $\vec{\mathbf{N}}$, a strain-rate hardening does not take place for planes with normals $-\vec{\mathbf{N}}$ meaning that a "negative" strain-rate hardening is absent. Consequently, we require that the integral of non-homogeneity be only positive. Otherwise, we let it be equal zero. Therefore, we use Eq. (4.7.1) if it gives positive values of I_N and $I_{-N} = 0$. If I_N from Eq. (4.7.1) is negative, I_{-N} is calculated by Eq. (4.7.2) and I_N is assumed to be zero.

As known, in terms of synthetic theory, a strain hardening is determined by the change in a plane distance, H_N. Since local peak stresses govern the rate-hardening of material, we introduce their measure (I_N) into the formula for H_N, Eq. (3.17.2) or (3.17.4),

$$H_N = S_P + r\varphi_N + I_N \tag{4.7.3}$$

or

$$H_N = \sqrt{S_P^2 + r\varphi_N + I_N^2}, \tag{4.7.4}$$

where $S_P = \sqrt{2}\tau_P$ and τ_P is a material constant, so-called creep limit in shear. This notion will be discussed in more detail below.

As it may be seen from Eq. (4.7.1), the parameter I_N grows in certain directions of $\vec{\mathbf{N}}$ from the very beginning of loading implying that the distance H_N governed by Eq. (4.7.3) or (4.7.4) grows. This means that tangent planes move away from a creep surface and their displacements are not necessarily accompanied by plastic strain. Further, after an increase in load is terminated or after an unloading, the integral of heterogeneity I_N decreases leading to the decrease in distance H_N, i.e., planes start to come back to the creep surface.

Thus, the model, which is based on Eq. (4.7.3) or (4.7.4), differs considerably from the theory of plasticity formulated in chapter 3, where the increase in the planes distance is always related to the increment in plastic deformation. Alternatively, according to Eqs. (4.7.3) and (4.7.4), the plane movements can occur at $\Delta\varphi_N = 0$ as well. At the same time, Eq. (3.9.7) or (3.9.8) remains true, i.e., a given plane produces an irreversible strain, if a stress vector reaches it (the situation $\vec{\mathbf{S}} \cdot \vec{\mathbf{N}} > H_N$ is not permitted). If the inequality (3.9.9) holds, the plane does not develop an irreversible deformation, although, possibly, it moves. The basic relation of synthetic model (3.9.4), which determines the macro-strain components as the sum of plastic strains produced on micro level, holds true.

Equations (4.7.1), (4.7.3) or (4.7.4), (3.9.7) or (3.9.8), and (3.17.1) are the basic relations of the synthetic theory allowing for the influence of loading rate upon the diagrams of deformation. Let us specify the units of the introduced quantities. The units of the integral of non-homogeneity, Pa, are independent of relation used for H_N, Eq. (4.7.3) or (4.7.4). Therefore, if I_N is taken in the form of Eqs. (4.7.1) and (4.5.21), the constant B is dimensionless and the constant p has the units of sec^{-1}. If I_N is set by Eqs. (4.7.1) and (4.5.22), the constant p is dimensionless, whereas B has the units of sec^p.

Further, similarly to Sec. 3.17, the following is proposed. If a plane with normal vector $\vec{\mathbf{N}}$ moves, the plane with opposite normal vector $-\vec{\mathbf{N}}$ undergoes the displacement as well so that the plane distance, H_{-N}, is governed by the following expressions

$$H_{-N} = S_P + r\varphi_{-N} + I_{-N}, \quad S_P = \sqrt{2}\tau_P, \tag{4.7.5}$$

or

$$H_{-N} = \sqrt{S_P^2 + r\varphi_{-N} + I_{-N}^2}, \tag{4.7.6}$$

in addition, Eq. (3.17.1) is valid.

4.8 The Creep and Yield Limit

Consider the loading in pure shear, $\tau_{xz} > 0$, $S_3 = S = \sqrt{2}\tau_{xz}$, $S_1 = S_2 = 0$, with infinitesimal loading rate so that the integral of non-homogeneity (4.7.1) can be neglected, $I_N \approx 0$. If an irreversible strain is absent, $\varphi_N = 0$, Eqs. (4.7.3) and (4.7.4) give that the tangent planes are equidistant from the origin of coordinates,

$$H_N = \sqrt{2}\tau_P. \tag{4.8.1}$$

At pure shear, Eqs. (3.5.4) and (3.6.2) at $\beta = \pi/2$ and $\lambda = 0$ give that

$$\vec{\mathbf{S}} \cdot \vec{\mathbf{N}} = \sqrt{2}\tau_{xz}. \tag{4.8.2}$$

According to Eq. (3.9.8), a plastic strain is induced if the right-hand side in Eq. (4.8.2) becomes equal to the right-hand side in Eq. (4.8.1). Thus, the material constant τ_P is the yield limit for the infinitesimal loading rate. It is this limit that corresponds to the infinitesimal loading rate and will be termed as the creep limit. Therefore, τ_P is the creep limit of material in pure shear.

On the other hand, Eq. (4.8.1) implies that the condition for the onset of irreversible deformation, a stress vector reaches first tangent plane, for an arbitrary state of stress at $I_N = 0$ is

$$S = \sqrt{S_1^2 + S_2^2 + S_3^2} = \sqrt{2}\tau_P, \tag{4.8.3}$$

i.e., the creep surface has the form of the Guber sphere:

$$S_1^2 + S_2^2 + S_3^2 = S_P^2, \quad S_P = \sqrt{2}\tau_P. \tag{4.8.4}$$

Let us turn once more to the pure shear and determine the yield plasticity for a constant finite loading rate, $v = \dot{S}_3 = \dot{S} = const$, when Eqs. (4.5.21) and (4.7.1) become

$$I_N = B \sin\beta \cos\lambda \int_0^t \frac{dS}{ds} \exp(-p(t-s))ds. \tag{4.8.5}$$

As $v = dS/dt = const$,

$$I_N = I \sin\beta \cos\lambda, \tag{4.8.6}$$

where I is given by Eq. (4.6.1); I_N takes positive values for $0 < \beta < \pi/2$, $0 \le \lambda < \pi/2$.

The increase in stress causes the parameter I_N to grow as well, i.e., distance H_N increases and planes move away from the creep surface (4.8.4). It is assumed that the module of stress vector (3.10.1) grows faster than planes move away. The value of shear stress, τ_S, at which the endpoint of vector $\vec{\mathbf{S}}(0,0,\sqrt{2}\tau_S)$ reaches the first plane (perpendicular to S_3-axis, i.e., $\beta = \pi/2$ and $\lambda = 0$) gives the yield limit in pure shear. If t_S denotes the time needed for the stress vector to reach the first plane, then Eqs. (4.7.3) and (4.8.6) at $\varphi_N = 0$, $\beta = \pi/2$, and $\lambda = 0$ give the following equation

$$S_P + \frac{Bv}{p}(1 - \exp(-pt_S)) = S_S, \tag{4.8.7}$$

where $S_S = \sqrt{2}\tau_S$. In addition, S_S and t_S are related to each other by formula $S_S = vt_S$. S_S vs. v diagram plotted on the base of Eq. (4.8.7) is shown in Fig. 4.9 (curve 1), where the creep limit S_P corresponds to $v = 0$. As it follows from Eqs. (4.6.5), (4.6.7), and (4.8.7), the curve has horizontal asymptote that is at a distance of $S_P/(1-B)$, $0 \le B \le 1$, from the abscissa.

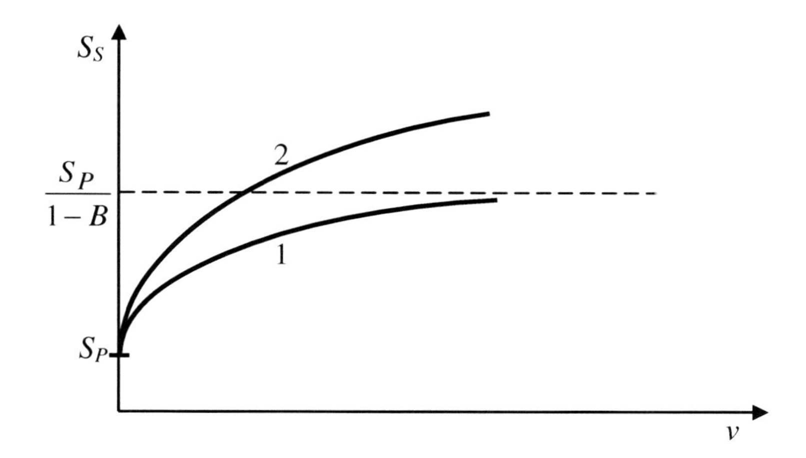

Fig. 4.9 Yield limit against loading rate dependence

If to take the core of the parameter of non-homogeneity (non-uniformity) in the form of Eq. (4.5.22), i.e., I is determined by Eq. (4.6.3), proceeding similarly as in the discussion before Eq. (4.8.7), we obtain

$$S_P + \frac{Bv}{1-p} t_S^{1-p} = S_S, \qquad S_S = vt_S. \tag{4.8.8}$$

The diagram of yield limit determined from this equation is shown in Fig. 4.9 by curve 2, as seen $S_S \to \infty$ as $v \to \infty$.

Now, our goal is to find yield limit for the case of an arbitrary, proportional loading of constant rate v. By inserting Eqs. (3.9.1), (3.5.4), and (3.6.2) into Eq. (4.7.1), we get

$$I_N = \cos\lambda \int_0^t \left(\frac{dS_1}{ds} \cos\alpha \cos\beta + \frac{dS_2}{ds} \sin\alpha \cos\beta + \frac{dS_3}{ds} \sin\beta \right) Q(t-s)ds\,\cdot \tag{4.8.9}$$

In the direction of the action of stress vector (the angles are denoted by α_0 and β_0 as in Fig. 3.6), Eq. (4.8.9), together with Eq. (3.11.1), takes the form

$$I_{N_0} = \cos\lambda \int_0^t \frac{S_i}{S} \frac{dS_i}{ds} Q(t-s)ds . \tag{4.8.10}$$

Since Eq. (3.11.2) gives that

$$v = \frac{dS}{dt} = \frac{S_i}{S} \frac{dS_i}{dt}, \tag{4.8.11}$$

Eq. (4.8.10) yields the form

$$I_{N_0} = \cos\lambda \int_0^t \frac{dS}{ds} Q(t-s)ds . \tag{4.8.12}$$

As seen, the integral I_{N_0} has the same form as that for pure shear at $\beta = \pi/2$, see Eq. (4.8.5). Therefore, the equation for yield limit has the form of Eq. (4.8.7) or (4.8.8), depending on the function Q, for the case of arbitrary, proportional loading as well.

Further, let us write down the equations for yield limit for the case arbitrary proportional loading if the planes distance, H_N, is defined by Eq. (4.7.4). In this case, we arrive at the following relationships

$$\begin{gathered} S_P^2 + \frac{B^2 v^2}{p^2}(1 - \exp(-pt_S))^2 = S_S^2 = v^2 t_S^2 , \\ S_P^2 + \frac{B^2 v^2}{(1-p)^2} t_S^{2(1-p)} = S_S^2 = v^2 t_S^2 \end{gathered} \tag{4.8.13}$$

for Eq. (4.5.21) and (4.5.22), respectively. The plot of S_S against v is shown in Fig.4.9, curves 1 and 2 represent the solution of the first and second equation in (4.8.13), respectively. Curve 1 has a horizontal asymptote whose ordinate equals to $S_P / \sqrt{1 - B^2}$, whereas curve 2 has no asymptote, $S_S \to \infty$ as $v \to \infty$.

Thus, in contrast to the classical theories of plasticity, where a yield limit (in tension or torsion) is taken for one of the basic constant of material, the yield limit in terms of the synthetic theory strongly depends on loading rate, see Eqs. (4.8.7), (4.8.8), and (4.8.13), and the creep limit becomes the basic material constant.

In technical computations, the creep limit in tension is the minimum tensile stress which produces permanent plastic deformation of, e.g., $\varepsilon^p = 0{,}002 = 0{,}2\%$, for, e.g., 1000 hours. Let us consider an experimental technique for determining creep limit proposed by G.S. Pisarenko. Figure 4.10a shows experimental creep diagrams constructed for different stresses,

$\sigma_1 < \sigma_2 < \sigma_3 \ldots$, at constant temperature. We draw a horizontal straight line being at a distance of $\varepsilon_x^p = 0{,}002$ from time-axis. The line cuts creep curves at points t_1, t_2, t_3,...whereby determining the time needed for the relative elongation to reach the value of $\varepsilon_x^p = 0{,}002$ under stresses σ_1, σ_2, σ_3,... The obtained pairs (t_1,σ_1), (t_2,σ_2), (t_3,σ_3), ... constitute a curve shown in Fig. 4.10b. Extrapolating this curve to point 1000 h, we can read the ordinate of point *P* that indicates the creep limit of the material: stress σ_P produces strain of $0{,}002$ for 1000 hours.

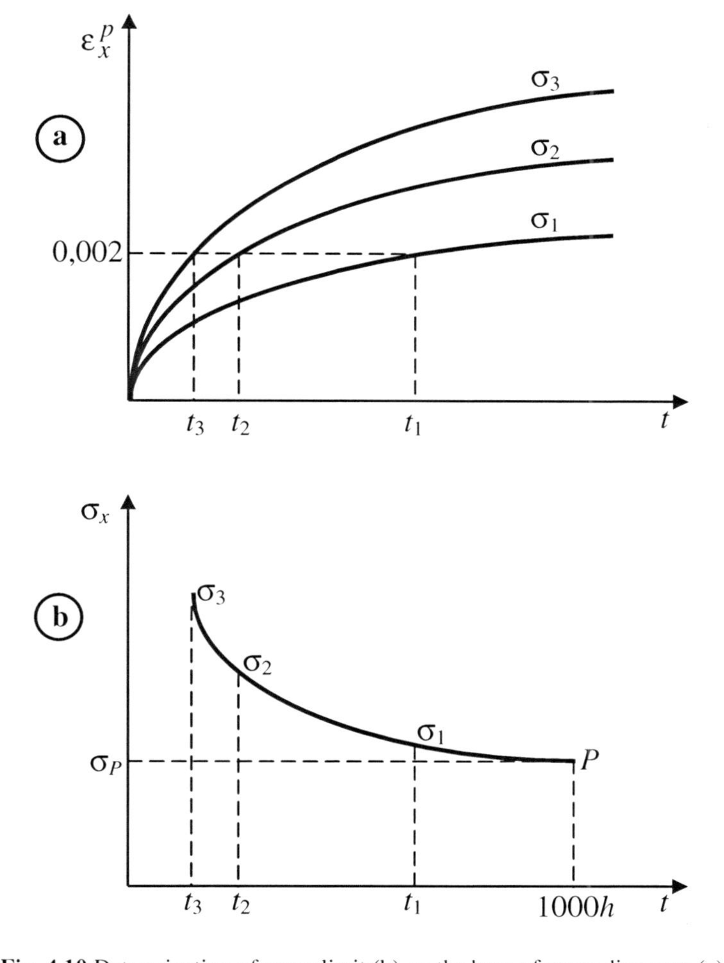

Fig. 4.10 Determination of creep limit (b) on the base of creep diagrams (a)

This method is a rough one: the extrapolation can introduce the error into the determination of σ_P. The larger the number of $\varepsilon_x^p \sim t$ diagrams close to σ_P, the more precise the result.

Reference books, for example [90], list the values of the creep limits, σ_P, and yield limits, σ_S, for standard loading rates of different materials. For the overwhelming majority of them $\sigma_P < \sigma_S$. However, there exist such alloys for which the creep limit is greater than yield limit. Such materials are not considered in this monograph.

4.9 Influence of Loading Rate on Stress-Strain Diagram in Pure Shear

Consider the case of pure shear, $S_3 = \sqrt{2}\tau_{xz} > 0$ and other stress vector components are equal to zero. The load is carried out with a constant rate, v, for elastic area and with the rate v' in plastic straining:

$$\dot{S}_3 = \begin{cases} v = const & \text{at } S_3 \le S_S \\ v' = const & \text{at } S_3 > S_S \end{cases} \tag{4.9.1}$$

where S_S is determined by Eq. (4.8.7) or (4.8.8).

Our goal is to calculate the plastic strain at $S_3 > S_S$. At $t > t_S = S_S / v$ the integral of non-homogeneity (4.8.5) is

$$I_N = B\left[v\int_0^{t_S} Q(t-s)ds + v'\int_{t_S}^{t} Q(t-s)ds\right]\cos\xi, \tag{4.9.2}$$

$$\cos\xi = \sin\beta\cos\lambda$$

or

$$I_N = (I_1 + I_2)\cos\xi, \tag{4.9.3}$$

where, for the case of exponential function Q, I_1 is determined by Eq. (4.6.2) at $t_1 = t_S$ and I_2 is

$$I_2 = \frac{Bv'}{p}\left[1 - \exp(-p(t - t_S))\right]. \tag{4.9.4}$$

If the function Q has the form of Eq. (4.5.22), I_1 is determined by Eq. (4.6.4) and I_2 takes the form

$$I_2 = \frac{Bv'}{1-p}(t - t_S)^{1-p}. \tag{4.9.5}$$

By making use of Eqs. (4.9.3)–(4.9.5) and Eq. (4.7.3), the relation for the irreversible strain intensity, Eq. (3.9.8), takes the form

$$r\varphi_N = \left(\frac{\cos\xi}{a} - 1\right) S_P, \quad a = \frac{S_P}{S_3 - (I_1 + I_2)} \tag{4.9.6}$$

The plastic strain vector component e_3^S is calculated by inserting φ_N from the above formula into Eq. (3.9.4),

$$e_3^S = \frac{S_P}{2r} \int\limits_{\alpha} \int\limits_{\beta} \int\limits_{\lambda} \sin 2\beta \left(\frac{\sin\beta\cos\lambda}{a} - 1\right) \cos\lambda d\alpha d\beta d\lambda, \tag{4.9.7}$$

whereas, as shown in Sec. 3.10, e_1^S and e_2^S are equal to zero. The integral and the limits in Eq. (4.9.7) are similar to that from Sec. 3.10, therefore,

$$e_3^S = a_0 F(a), \quad a_0 = \frac{\sqrt{2}\pi\tau_P}{3r}, \tag{4.9.8}$$

where function $F(a)$ is determined by Eq. (3.10.14).

Further, let us consider the case, when the planes distance is determined by Eq. (4.7.4). Now, together with Eq. (3.10.4), the irreversible strain intensity becomes

$$r\varphi_N = 2\tau_P^2 \left(\frac{\cos^2\xi}{a^2} - 1\right), \quad a^2 = \frac{S_P^2}{S_3^2 - (I_1 + I_2)^2}. \tag{4.9.9}$$

Therefore, similar to Eq. (3.10.13), we obtain

$$e_3^S = a_0 F(a), \qquad a_0 = \frac{\pi\tau_P^2}{3r}, \tag{4.9.10}$$

where the function F is determined by Eq. (3.10.15).

Eqs. (4.9.8) and (4.9.10) determine the strain vector component e_3^S in pure shear. Plots of shear stress as a function of total shear strain,

$$\gamma_{xz} = \frac{\tau_{xz}}{G} + \gamma_{xz}^S, \qquad \gamma_{xz}^S = \sqrt{2} e_3^S, \tag{4.9.11}$$

where e_3^S is given by Eq. (4.9.8) or (4.9.10), for different loading rates are shown in Fig. 4.11. The greater the loading rate, the greater the stress induces a given strain. If to take the integral of non-homogeneity in the form of Eqs. (4.9.3) and (4.9.4), the stress-strain curve (diagram) constructed for $v = v' \to \infty$ binds all other diagrams above. If Eqs. (4.9.3) and (4.9.5) are used, the diagram degenerates in a straight line as $v = v' \to \infty$ meaning that Hooke's law holds right up to failure. The comparison of the theoretical and experimental results can be found in writings [88,113,151,152,155,157,159].

4.10 Sudden Increase in Loading Rate

Consider the following loading mode in pure shear. First, stress vector component, S_3, grows to its final value, $S_0 = \sqrt{2}\tau_{xz}^0$, beyond the elastic limit with a constant rate, $v = dS_3/dt$, and then, at $S_3 = S_0$, the loading rate suddenly increases to the value of $v' > v$:

$$\dot{S}_3 = \begin{cases} v = const & \text{at } S_3 < S_0 \\ v' = const & \text{at } S_3 > S_0 \end{cases} \tag{4.10.1}$$

Let us investigate the strain response in such loading. If the loading rate $\dot{S}_3$ varies as in Eq. (4.10.1), the integral of non-homogeneity is determined by Eqs. (4.8.6), (4.6.1), and (4.6.3) at $S_3 < S_0$ and by Eqs. (4.9.3)–(4.9.5) at $S_3 > S_0$. In the specified formulae, in contrast to the previous Section, t_S is time needed for the component S_3 to reach the value of S_0, i.e., $S_0 = vt_S$.

To determine the irreversible strain intensity increment in time dt after the change in loading rate, we use Eq. (4.7.3) in an incremental form,

$$rd\varphi_N = dS_3 \cos\xi - dI_N . \tag{4.10.2}$$

If the core of the integral of non-homogeneity is exponential, Eqs. (4.9.3) and (4.9.4), at $t = t_S$, yield the following form

$$\frac{dI_N}{dt} = B(v' - v + v\exp(-pt_S))\cos\xi , \tag{4.10.3}$$

therefore,

$$rd\varphi_N = [v' - B(v' - v + v\exp(-pt_S))]\cos\xi dt . \tag{4.10.4}$$

As $0 \le B < 1$, the square bracket in the above formula is positive, i.e., the plastic straining is not terminated at $t = t_S$. At the same time, the increment (4.10.4) is less than that corresponding to the absence of the loading rate jump, i.e., for $v' = v$.

If the core of the integral I_N is taken as (4.5.22), then Eqs. (4.9.3) and (4.9.5) give

$$\frac{dI_N}{dt} = B\left[\frac{v' - v}{(t - t_S)^p} + \frac{v}{t_S^p}\right]\cos\xi. \tag{4.10.5}$$

Since $dI_N/dt \to \infty$ at $t = t_S$, then

$$dH_N > dS_3 \cos\xi. \tag{4.10.6}$$

This inequality shows that, due to the loading rate jump, planes move away faster than the module of stress vector increases, meaning that the plastic flow ceases temporarily beginning from $t = t_S$. Further, from Eqs. (4.9.3)–(4.9.5) follows that, as $t \to \infty$, the parameter I_N has the same form as if the loading develops with the constant rate v' right from the start. Therefore, the curve $\tau_{xz} \sim \gamma_{xz}$ constructed for the regime of Eq. (4.10.1) gets close to that corresponding to the constant loading rate $\dot{S}_3 = v'$.

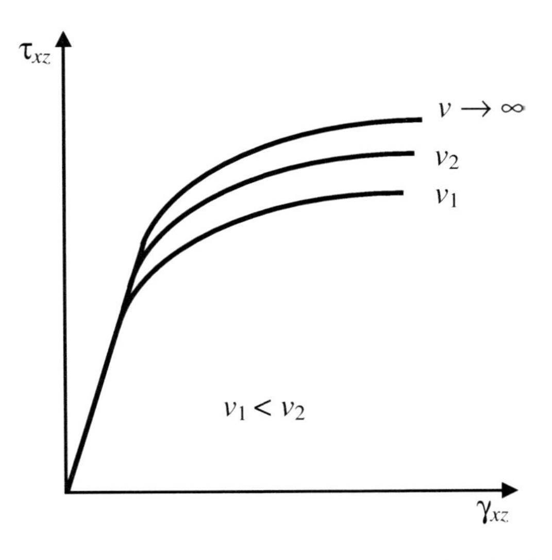

Fig. 4.11 Stress-strain diagrams in pure shear at various loading rates

Figure 4.12 shows stress-strain diagrams at different loading rates, curves marked by 1 and 2 constructed for constant loading rates, $v_2 > v_1$ and those marked by 3 and 4 correspond to the loading regime of Eq. (4.10.1). Curves 3 and 4 correspond to the integral of non-homogeneity with core (4.5.21) and (4.5.22), respectively. The use of Eq. (4.5.22) leads to a material that reacts elastically on the loading rate jump.

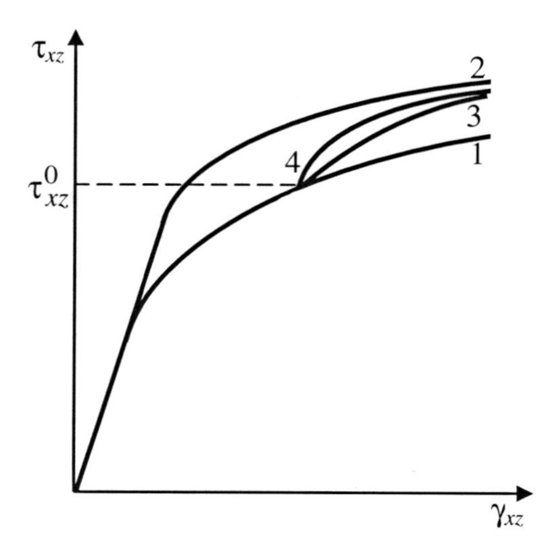

Fig. 4.12 Strain response to a sudden increase in loading rate

4.11 Proportional Loading

Consider a proportional loading, when the stress vector components S_k change proportionally to an unique common parameter as in Eq. (3.3.4). In this case, the loading trajectory is a straight line in the Ilyushin space whose orientation will be defined by angles α_0 and β_0 (Fig. 3.6) and Eqs. (3.11.1), (3.11.2), and (4.8.12) remain valid. Similar to Sec. 4.10, the loading rate is v and v' in elastic and plastic loading, respectively.

Let us determine plastic strain vector components in the proportional loading and compare obtained result with that for the case of pure shear (Sec. 4.9). In view of that, in both cases, the stress vector is co-directed with coordinate axis, S_3-axis at pure shear and an auxiliary axis S_3' at combined stress state (see Sec. 3.11), it is easy to conclude that they are identical. In addition, parameter I_N in torsion has the same form as in combined state of stress (4.8.12). Therefore, the strain vector component relative to auxiliary S_1'-, S_2'-, S_3'-axes is identical to that obtained at pure shear, i.e., $e_1' = e_2' = 0$, and the component e_3' is determined by Eqs. (4.8.12) and (4.9.8)–(4.9.10), in which S is the module of stress vector (3.11.2).

Projecting the plastic strain vector onto S_1-, S_2-, and S_3- axis, we arrive at the analog of Eq. (3.11.5),

$$e_i^S = a_0 \frac{S_i}{S} F(a), \tag{4.11.1}$$

where the argument a is given by Eqs. (4.9.6) or (4.9.9). Therefore, similar to Sec. 3.11, we have obtained the deviator proportionality.

Employing Eq. (4.11.1), let us determine the plastic strain vector components in uniaxial tension along x-axis, when $S_1 = S$, $S_2 = S_3 = 0$. Now, Eq. (4.11.1) gives

$$e_1^S = a_0 F(a), \quad e_2^S = e_3^S = 0. \tag{4.11.2}$$

The total strain tensor component ε_x, together with the relationship $\varepsilon_x^S = \sqrt{2} e_1^S / \sqrt{3}$, is

$$\varepsilon_x = \frac{\sigma_x}{E} + \sqrt{\frac{2}{3}} a_0 F(a). \tag{4.11.3}$$

Other components are

$$\varepsilon_y = \varepsilon_z = -\frac{\nu \sigma_x}{E} - \frac{a_0}{\sqrt{6}} F(a). \tag{4.11.4}$$

It is Eqs. (4.11.3) and (4.11.4) that determine strain tensor components at uniaxial tension.

4.12 Stress-Strain Diagram at Elevated Temperatures

An irreversible deformation of many alloys reacts poorly to the change in loading rate at room temperature. However, if a homologous temperature, Θ, is greater than $0.25 \div 0.3$, the influence of loading rate on plastic straining grows with the temperature increase and reaches its maximum at $\Theta = 0.5$; the role of temperature drops for $\Theta > 0.5$. The interval $0.3 \le \Theta \le 0.5$ is treated as an elevated temperature. Strain-stress diagrams of pure aluminum in uniaxial compression at temperatures $T = 20, 120, 240, 360$ °C are shown in Fig.4.13. The experiments were performed with the following strain rates: $|\dot{\varepsilon}_x| = 0{,}25 \sec^{-1}$ (Fig.4.13a), $|\dot{\varepsilon}_x| = 4 \sec^{-1}$ (b), $|\dot{\varepsilon}_x| = 63 \sec^{-1}$ (c). The change in strain rate does not affect the diagrams at $T = 20$ °C, whereas at larger temperatures the diagrams start reacting to the growth of strain rate $\dot{\varepsilon}_x$ (Fig. 4.13 a,b,c).

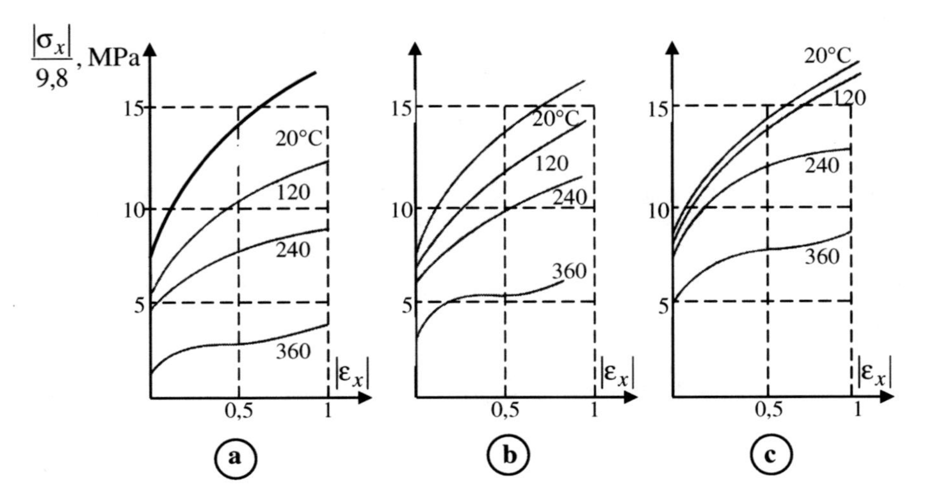

Fig. 4.13 The hardening of aluminum as a function of strain-rate and temperature

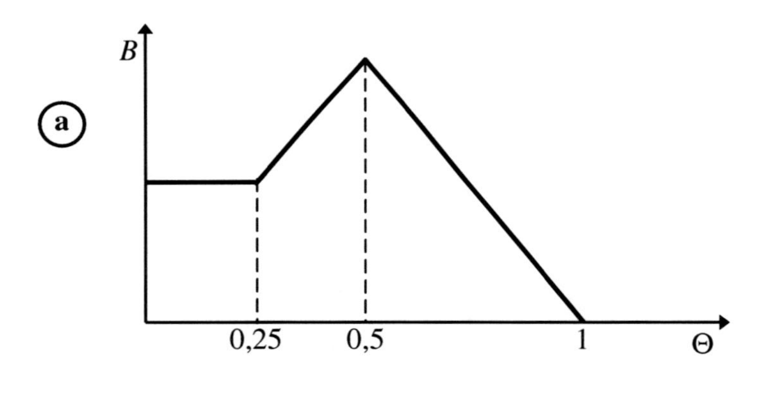

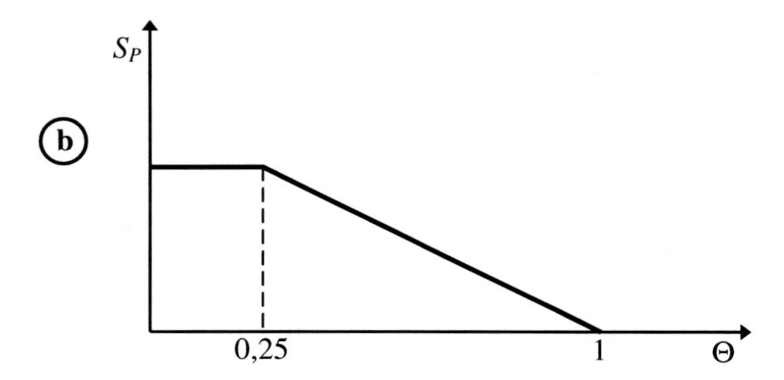

Fig. 4.14 Plots of B (Θ) (a) and S_P (Θ) (b)

Since the influence of loading rate is expressed through the integral of non-homogeneity I_N, it should take small values at $\Theta < 0.3$ and sharply grow at $0.3 \le \Theta \le 0.5$. Therefore, the coefficient B appearing in the integral of non-homogeneity should depend on temperature as shown in Fig. 4.14 a.

Another important point is that, as well known, creep limit, S_P, is a decreasing function of Θ. If the dependence $S_P(\Theta)$ was not decreasing, the behavior of $B(\Theta)$ from Fig. 4.14a would lead to that the yield limit increases with Θ that contradicts the experiments (Fig. 4.13 a, b, c). A schematic plot of the dependence between creep limit and temperature is shown in Fig. 4.14b. Although, there exist some alloys for which the dependence $S_P(\Theta)$ is not monotonous, such cases are not considered in the monograph.

It is logical to accept that the integral of non-homogeneity relaxes faster with increase of Θ. The relaxation rate is regulated by the parameter p appearing in Eqs. (4.5.21) and (4.5.22). Let us take a simplest assumption, the parameter p depends linearly on temperature:

$$p = p_0 \Theta , \tag{4.12.1}$$

where p_0 is a material constant.

Theoretical results obtained on the base of Eqs. (3.10.14) and (4.11.3) show satisfactory agreements with experimental diagrams [Slusarchuk's Ph.D. thesis] shown in Fig. 4.13 a, b, c.

It is worthwhile to note the following. In the above experiment, the strain rate, $\dot{\varepsilon}_x$, is given, whereas formulae in Sec. 4.9 contain the loading rate $\dot{S}_1$. When modeling the relationship between $\dot{S}_1$ and $\dot{\varepsilon}_x$, we consider the experimental diagrams in Fig. 4.13 as two-segment break-lines. Therefore,

$$\dot{S}_1 = \sqrt{\frac{2}{3}}\dot{\sigma}_x , \quad \dot{\sigma}_x = \begin{cases} E\dot{\varepsilon}_x & \text{at } \varepsilon_x \le \sigma_S / E \\ E_1\dot{\varepsilon}_x & \text{at } \varepsilon_x \ge \sigma_S / E \end{cases} , \tag{4.12.2}$$

where E is the Young modulus and E_1 is the slope of the second straight-line branch of the diagram. Therefore the integral of non-homogeneity is determined by formulae similar to those from Eqs. (4.9.1)–(4.9.5).

4.13 The Integral of Non-homogeneity in Alternating Torsion

Let a stress vector component, $S_3 = \sqrt{2}\tau_{xz}$, grow from 0 to its maximal value S_0. This stage of loading is marked as zero semi-cycle. Further, an unloading, from S_0 to 0, and successive loading (torsion) in opposite direction, from 0 to $S_3 = -S_0$, takes place which is treated as first semi-cycle. A second semi-cycle consists in unloading and loading in positive direction, etc. For simplicity, we suppose that a loading rate does not vary from one semi-cycle to another, $v = \dot{S}_3 = const$. The marks of the semi-cycles on the stress strain and loading diagrams are shown in Fig. 4.15a and 4.15d, respectively.

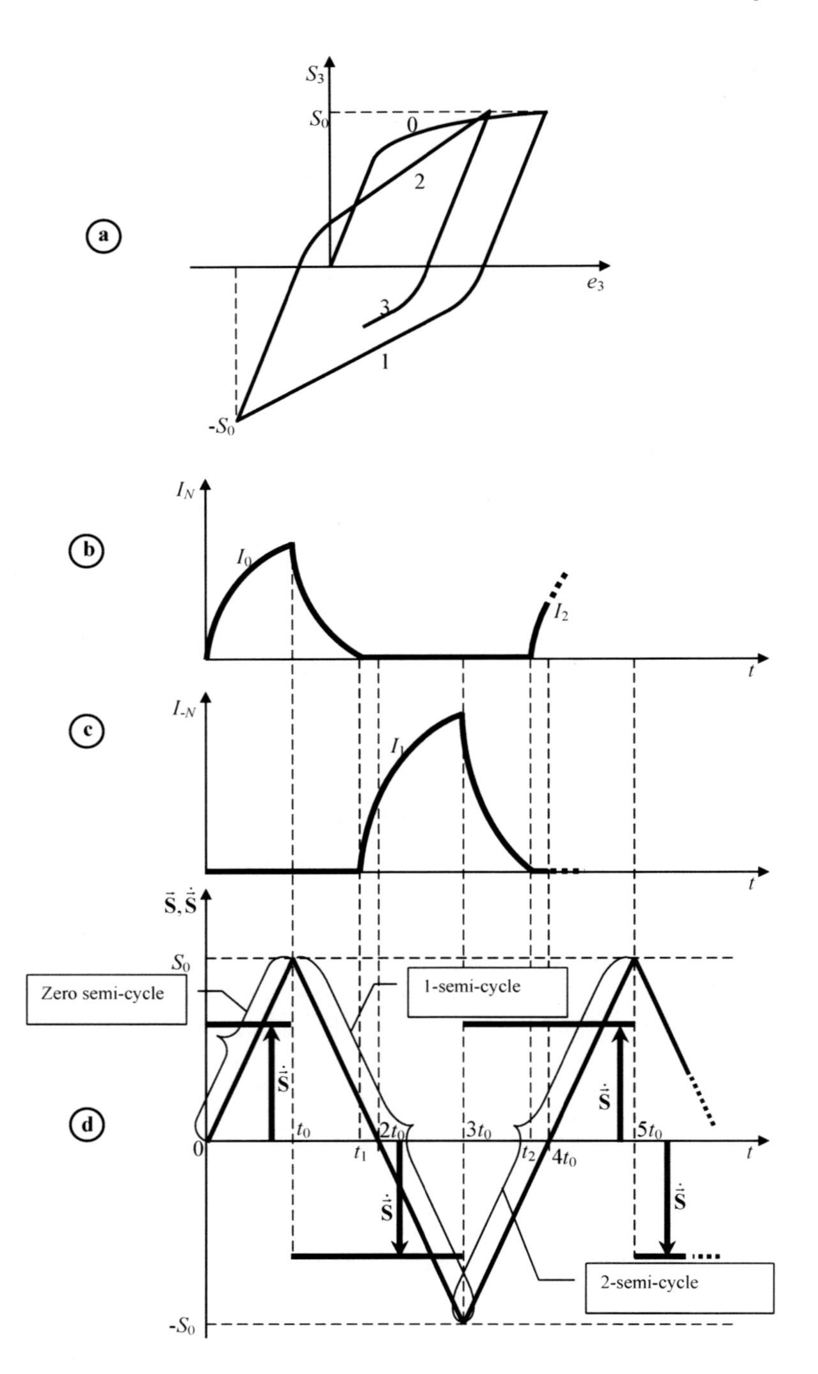

Fig. 4.15 Semi-cycles of loading, (a) and (d), and the behavior of the integral of non-homogeneity, (b) and (c)

Let us designate through $\vec{\mathbf{N}}$ the normal vector to the plane that produces irreversible strain at zero and pair semi-cycles of loading. The orientation of $\vec{\mathbf{N}}$ is defined by positive angle β. On the other hand, normal $-\vec{\mathbf{N}}$ indicates the plane producing plastic flow at unpaired semi-cycles; its orientation is given by negative angle β. In the case of alternate torsion, Eqs. (4.5.21) and (4.7.1) give

$$I_{\pm N} = I_k \cos\xi, \qquad \cos\xi = |\sin\beta|\cos\lambda, \tag{4.13.1}$$

where k is the number of semi-cycle; pair indexes k are related to planes with normal vector $\vec{\mathbf{N}}$, whereas odd indexes to $-\vec{\mathbf{N}}$.

At zero semi-cycles, the I_0 is determined by Eq. (4.6.1) and the zero semi-cycles time, t_0, is

$$0 \le t \le t_0, \qquad t_0 = \frac{S_0}{v}. \tag{4.13.2}$$

For the first semi-cycles we have

$$I_1 = Bv\left[-\int_0^{t_0}\exp(-p(t-s))ds + \int_{t_0}^{t}\exp(-p(t-s))ds\right] = \frac{Bv}{p}\left[1+\exp(-pt)-2\exp(-p(t-t_0))\right] \tag{4.13.3}$$

Equation (4.13.3) holds true for such time-interval, $t \ge t_1$, when $I_1 \ge 0$. Equating the expression in the square bracket to zero, we have

$$t_1 = \frac{1}{p}\ln\left(2\exp(pt_0)-1\right). \tag{4.13.4}$$

The requirement $I_1 \ge 0$ stems from the fact that the integral of non-homogeneity cannot take negative values (Sec. 4.7). The first and second integrals in Eq. (4.13.3) correspond to the zero and first semi-cycles of loading, respectively. The fact the stress-rate-vector $\dot{\vec{\mathbf{S}}}$ at $t \in [0, t_0]$ and $t \in [t_0, 3t_0]$ makes obtuse and acute angles with normal vectors $-\vec{\mathbf{N}}$, respectively, explains the minus and plus signs at the integrals in Eq. (4.13.3).

Second semi-cycle.

$$I_2 = Bv\left[\int_0^{t_0}\exp(-p(t-s))ds - \int_{t_0}^{2t_0}\exp(-p(t-s))ds + \int_{2t_0}^{t}\exp(-p(t-s))ds\right], \tag{4.13.5}$$

$$I_2 = \frac{Bv}{p}\left[1-\exp(-pt)+2\exp(-p(t-t_0))-2\exp(-p(t-2t_0))\right]. \tag{4.13.6}$$

$I_2 > 0$ at $t > t_2$, t_2 can be founded, in a similar way to Eq. (4.13.4), from the condition $I_2 = 0$. The plots of I_1 and I_2 as functions of time is shown in Fig. 4.15b,c. It is easy to see from Eqs. (4.13.3) and (4.13.5) that the plots of I_N and I_{-N} against time at $t \in [t_0, t_1]$ and $t \in [3t_0, t_2]$ are identical to $|I_1|$ and $|I_2|$, respectively.

The obtained formulas can be generalized to any k semi-cycle [111],

$$I_k = \frac{Bv}{p}\left[1 + \exp(-pt)(-1)^{k+1} + 2\sum_{j=1}^{k}\exp(-p(t - jt_0))(-1)^{j+k+1}\right]. \quad (4.13.7)$$

Equations (4.13.1) and (4.13.7) determine the integral of non-homogeneity for the arbitrary number of loading cycles in torsion.

4.14 Cyclic Properties of Material

In alternate loading, together with the Bauschinger effect (see Sections 3.16-3.19), other effects studied in Sec. 2.24 are observed. Stress-strain diagram varies from one cycle to other; a material can experience cyclic hardening, softening, or the stabilization of these processes. Hardening- and softening-processes can progress, reduce, or they appear alternately (Sec. 2.24).

Materials that cyclically harden in cyclic loading with constant stress amplitude (so-called soft loading) exhibit the gradual reduction of residual strains, i.e., a plastic hysteresis loop decreases. In the loading with controlled (fixed) strain amplitude (so-called rigid loading), stress amplitude usually varies; if it increases, the material is said to cyclically harden, if it decreases, the material is said to cyclically soften. However, this behavior tends to stabilize so that the variation in the stress amplitude is small after an initial period of transient hardening or softening. Once the behavior is stabilized, the stress-strain hysteresis loops fall close together in each strain cycle. Some materials experience that the area of hysteresis loop tends to zero in soft loading leading to linear stress-strain relation. For cyclically softening materials, the increase in residual strains and in the area of hysteresis loop (soft loading) or the reduction of stress amplitudes (rigid loading) are observed. The stabilization of these processes is possible as well [35].

Cyclically stable materials are characterized by unchangeable parameters of stress-strain diagram with the growth in cycles. As a rule, the cyclically stable materials are cyclically ideal ones as well, they are characterized by Eq. (3.18.23).

The goal of the Section 4.16 is to analytically describe the whole spectrum of the specified cyclical properties of material.

4.15 Peculiarities of Irreversible Straining in Alternate Loading

Consider an alternate loading in torsion with constant loading rate, $v = dS_3/dt$, in both elastic and plastic range as well as at unloading. On zero semi-cycle (Sec. 4.13) a yield limit can be determined by the one of Eqs. (4.8.7), (4.8.8), and (4.8.13) and the irreversible strain is determined by Eq. (4.9.8) or (4.9.10). The irreversible strain intensity due to the action of maximal stress $S_3 = S_0$ is obtained from Eq. (4.9.6) as

$$r\varphi_{N0} = \left[S_0 - \frac{Bv}{p}(1 - \exp(-pt_0)) \right] \cos\xi - S_P, \quad t_0 = \frac{S_0}{v} \qquad (4.15.1)$$

During unloading,

$$S_3 = S_0 - v(t - t_0), \quad t \in [t_0, 2t_0] \qquad (4.15.2)$$

the integral of non-homogeneity I_N for planes with $\beta > 0$ is determined by Eqs. (4.13.1) and (4.13.3) taken with minus sign. The planes with normal vectors $\vec{\mathbf{N}}$, i.e., $\beta > 0$, is at a distance of

$$H_N = S_P + r\varphi_{N0} + r\Delta\varphi_N + I_N, \qquad (4.15.3)$$

from the origin of coordinates, where $\Delta\varphi_N$ is the increment in φ_N at unloading. Inserting φ_{N0} and I_N into Eq. (4.15.3) and letting $\vec{\mathbf{S}} \cdot \vec{\mathbf{N}} = H_N$, we find the increment $\Delta\varphi_N$ as

$$r\Delta\varphi_N = -v(t - t_0)\cos\xi + \frac{Bv}{p}[2 + \exp(-pt) - 2\exp(-p(t - t_0))]\cos\xi. \qquad (4.15.4)$$

To analyze the obtained formula, we differentiate its left- and right-hand side with respect to time:

$$r\frac{d\Delta\varphi_N}{dt} = v[-1 - B\exp(-pt) + 2B\exp(-p(t - t_0))]\cos\xi. \qquad (4.15.5)$$

As $0 \le B < 1$, it is obvious that the derivative $d\Delta\varphi_N/dt$ is positive for some set of values B and pt_0. The increment of $\Delta\varphi_N$ grows within the range $t_0 < t < t_1$, where t_1 is determined by the condition of equality to zero of the right-hand side in (4.15.5),

$$t_1 = \frac{1}{p}\ln B(2\exp(pt_0) - 1). \qquad (4.15.6)$$

Since $d\Delta\varphi_N/dt > 0$ as $t_0 < t < t_1$, the plastic strain component e_3^S grows. Therefore, the synthetic model describes the increase in plastic deformation at unloading. This situation is represented by segment M_0M_1 in Fig. 4.16. The synthetic theory is capable of describing the observed behavior due to the integral of non-homogeneity, appearing in determining relation (4.7.3), decreases during unloading and planes come back to creep surface remaining on the endpoint of stress vector. Therefore, despite of the reduction of the module of stress vector $\vec{\mathbf{S}}$, the equality $\vec{\mathbf{S}} \cdot \vec{\mathbf{N}} = H_N$ remains valid meaning the irreversible straining is in progress.

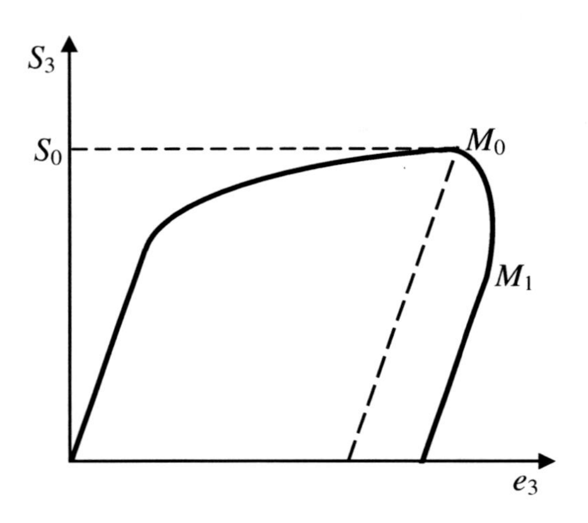

Fig. 4.16 Plastic strain increase, M_0M_1, during unloading

At the same time, on some set of parameters B and pt_0 the right-hand side in Eq. (4.15.5) is negative at $t = t_0$ (for example, if $B \leq 1/2$). Hence, $\Delta\varphi_N = 0$ and the unloading occurs elastically. Here and throughout further this more simpler variant is considered.

In first semi-cycle, by making use of Eq.(4.7.5), the tangent planes with vectors $-\vec{\mathbf{N}}$, $\beta < 0$, which produce plastic strain, is at a distance of

$$H_{-N} = S_P - r\left|\varphi_{N0}\right| + I_{-N}, \tag{4.15.7}$$

from the origin of coordinate, where φ_{N0} is the plastic strain intensity in directions $\vec{\mathbf{N}}$ accumulated in zero semi-cycle at $S_3 = S_0$. It is determined by Eq. (4.15.1). In addition, the integral of non-homogeneity in directions $-\vec{\mathbf{N}}$ is determined from Eqs. (4.13.1) and (4.13.3). Therefore,

$$H_{-N} = 2S_P + (I_0 + I_1 - S_0)\cos\xi, \tag{4.15.8}$$

where I_0 is the value of expression (4.6.1) at $t = t_0$ (the beginning of unloading). By equating H_{-N} at $\beta = -\pi/2$ and $\lambda = 0$ in (4.15.8) to the yield limit at first semi-cycle, S_{S1}, we obtain

$$2S_P + I_0 + I_1 - S_0 = S_{S1} > 0. \tag{4.15.9}$$

In this equation, unknown quantities are the yield limit S_{S1} to be reached in time t (the t appears in I_1 from (4.15.9)). They are related to each other as

$$t = 2t_0 + \frac{S_{S1}}{v}. \tag{4.15.10}$$

Equations (4.15.9) and (4.15.10) determine the yield limit at first semi-cycle.

If $|S_3| > S_{S1}$, the stress vector $\vec{\mathbf{S}}$ will shift tangent planes with normal vectors $-\vec{\mathbf{N}}$ producing an irreversible strain whose intensity we designate as φ_{-N1} and find from Eq. (4.7.5):

$$H_{-N} = 2S_P + (I_0 + I_1 - S_0)\cos\xi + r\varphi_{-N1}. \tag{4.15.11}$$

Equating H_{-N} from (4.15.11) to $|S_3|\cos\xi$, we obtain the equations for $r\varphi_{-N1}$:

$$r\varphi_{-N1} = (|S_3| + S_0 - I_0 - I_1)\cos\xi - 2S_P. \tag{4.15.12}$$

Now, consider the second semi-cycle (Fig. 4.15). Now, the planes with normal vectors $\vec{\mathbf{N}}$, $\beta > 0$, produce plastic strain. The distance of these planes is

$$H_N = S_P + r\varphi_{N0} - r|\varphi_{-N10}| + I_2\cos\xi, \tag{4.15.13}$$

where φ_{-N10} is the value of φ_{-N1} at the end of first semi-cycle. Therefore,

$$H_N = 2S_P + (-S_0 + I_2 + I_{10})\cos\xi. \tag{4.15.14}$$

Consequently, the yield limit on the second semi-cycle takes the form

$$2S_P - S_0 + I_2 + I_{10} = S_{S2}, \quad t = 4t_0 + \frac{S_{S2}}{v}. \tag{4.15.15}$$

Acting similarly to the first semi-cycle, $r\varphi_{N2}$ can be written as

$$r\varphi_{N2} = (S_3 + S_0 - I_2 - I_{10})\cos\xi - 2S_P. \tag{4.15.16}$$

The obtained equations can be generalized to any k semi-cycle, namely, the yield limit can be determined from the following equation

$$2S_P - S_0 + I_k + I_{(k-1)0} = S_{Sk}, \tag{4.15.17}$$

where I_k is given by Eq. (4.13.7). Unknown time, t, needed to reach the S_{Sk}, which appears in I_k, is

$$t = 2kt_0 + \frac{S_{Sk}}{v}. \tag{4.15.18}$$

The value of $I_{(k-1)0}$ is calculated by Eq. (4.13.7) at $t = (2k-1)t_0$,

$$I_{(k-1)0} = \frac{Bv}{p}\left[1 + \exp(-p(2k-1)t_0)(-1)^k + 2\sum_{j=1}^{k-1}\exp(-p(2k-j-1)t_0)(-1)^{k+j}\right]. \tag{4.15.19}$$

It is Eqs. (4.13.7) and (4.15.17)–(4.15.19) that determines the yield limit on arbitrary semi-cycle of alternate torsion.

Remark. As shown in Sec. 3.18, a yield limit can take both positive (the Bauschinger super-effect) and negative (the ordinary Bauschinger effect) values depending on the value of initial loading. It must be noted that within this Section only the ordinary Bauschinger effect is considered.

4.16 Deformation in Alternating Torsion

Eqs. (4.15.12) and (4.15.16) can be generalized to any k semi-cycle of alternate torsion, namely

$$r\varphi_{\pm Nk} = \left(|S_3| + S_0 - I_{(k-1)0} - I_k\right)\cos\xi - 2S_P, \tag{4.16.1}$$

where plus sign at index corresponds to pair index k and minus sign to unpaired; integrals I_k and $I_{(k-1)0}$ are given by Eqs. (4.13.7) and (4.15.19), respectively. Let us note that the above formula holds if the range of angles α, β, and λ at some semi-cycle is not greater than that at the preceding semi-cycle. Equation (4.16.1) can be rewritten as

$$r\varphi_{\pm Nk} = 2S_P\left(\frac{\cos\xi}{a_k} - 1\right), \tag{4.16.2}$$

where

$$a_k = \frac{2S_P}{|S_3| + S_0 - I_{(k-1)0} - I_k}. \tag{4.16.3}$$

It is easy to see that the irreversible strain intensity $r\varphi_{\pm Nk}$ in Eq. (4.16.2) depends on angles β and λ in the same manner as $r\varphi_N$ in Eq. (4.9.6). As a consequence, inserting $r\varphi_{\pm Nk}$ into Eq. (3.9.4), we arrive at integral (4.9.7) in which a is replaced by a_k. Therefore, performing integration, we obtain function F from (3.10.14) again, however now its argument is a_k. In view of this, the increment in the module of plastic strain vector component at k-semi-cycle, $\left|\Delta e_k^S\right|$ takes the form

$$\left|\Delta e_k^S\right| = 2a_0 F(a_k), \qquad k \geq 1. \tag{4.16.4}$$

Within this Section, since the strain vector has only one non-zero component in torsion, e_3^S, we omit the index "3" in the designation Δe_k^S. Total plastic strain is determined as an algebraic sum of strain increments developed in all semi-cycle:

$$e^S = \sum_{j=0}^{k-1} \left|\Delta e_j^S\right| (-1)^j + \left|\Delta e_k^S\right| (-1)^k, \tag{4.16.5}$$

where $\left|\Delta e_j^S\right|$ is determined by Eq. (4.16.4) at $|S_3| = S_0$, whereas Δe_k^S is determined by Eq. (4.16.4) at any $|S_3| \leq S_0$.

Therefore, Eqs. (3.10.14), (4.13.7), (4.15.18), and (4.16.3)–(4.16.5) determine irreversible strain in alternate torsion. If to neglect parameter I_Λ ($B = 0$) in these equations, they will describe stress-strain diagrams of cyclically ideal materials, namely, $e_3 = -e_0$ at $S_3 = -S_0$ and, independent of $k > 0$, the diagrams have identical form. If I_N appears in these equations, we obtain that $|e_3| < e_0$ at $S_3 = -S_0$, i.e., cyclically hardening materials are described (Figs. 4.15a, 4.17, and 4.18).

On the other hand, the integral of non-homogeneity is stabilized after comparatively small number of semi-cycles, which leads to the stabilization of stress-strain diagram as well. Therefore, it is necessary to introduce a parameter in the integral of non-homogeneity that changes with the increase of the number of semi-cycles. In terms of the classical theories of plasticity, the length of plastic straining path,

$$e^S = \int_0^t \left| d\vec{\mathbf{e}}^S \right|, \tag{4.16.6}$$

or the Odquist parameter [105] is taken for such parameter.

We propose, instead of (4.16.6), to take the integral

$$\varphi_m = \int_0^t \left| d\varphi_N \right|_{\max} \tag{4.16.7}$$

for the specified parameter, where the integrand is the absolute values of maximal strain intensities as a function of α, β, and λ at every loading cycle. Since the synthetic theory is a two-level one, the integral in Eq. (4.16.7) characterizes the amount of plastic deformation cumulated on micro-level. For the case of alternate torsion, Eq. (4.16.2) at $|S_3| = S_0$ and $\cos\xi = 1$ gives

$$\varphi_m = 2S_P \left(\sum_{j=0}^{k} \left(\frac{1}{a_j} \right) - k - 1 \right), \tag{4.16.8}$$

where a_j is given by Eq. (4.16.3). Let as introduce the φ_m from (4.16.8) into the formula for the integral of non-homogeneity in the following way

$$I_N = I_k \left(1 + \frac{k_1 \varphi_m}{1 + k_2 \varphi_m} \right) \cos\xi, \tag{4.16.9}$$

where k_1 and k_2 are material constants, and I_k is given by Eq. (4.13.7). We let $k_1 = 0$ at monotonous loading.

For the case of the very large number of semi-cycles, theoretically it tends to infinity, it is obvious from Eq. (4.16.7) it follows that $\varphi_m \to \infty$. Therefore, if $k_1 > 0$ and $k_2 = 0$, the square bracket in Eq. (4.16.9) also tends to infinity. This fact, in turn, implies that the integral of non-homogeneity in Eq. (4.16.9) grows with the number of loading-cycles causing the gradual increase in the yield limit. Consequently, the cyclically hardening materials, whose stress-strain diagram becomes eventually linear (Fig. 4.17), are described. For the case $k_1 > 0$ and $k_2 > 0$,

$$\lim_{\varphi_m \to \infty} \left(1 + \frac{k_1 \varphi_m}{1 + k_2 \varphi_m} \right) = 1 + \frac{k_1}{k_2}, \tag{4.16.10}$$

the model describes the diagrams of cyclically hardening materials which experience a stabilization into curvilinear diagram (Fig. 4.18). The values of constants $k_1 < 0$ and $k_2 = 0$ symbolizes the cyclically softening materials. If $k_1 < 0$, $k_2 > 0$, and $1 + k_1/k_2 > 0$, the model describes the stabilization of cyclically softening materials.

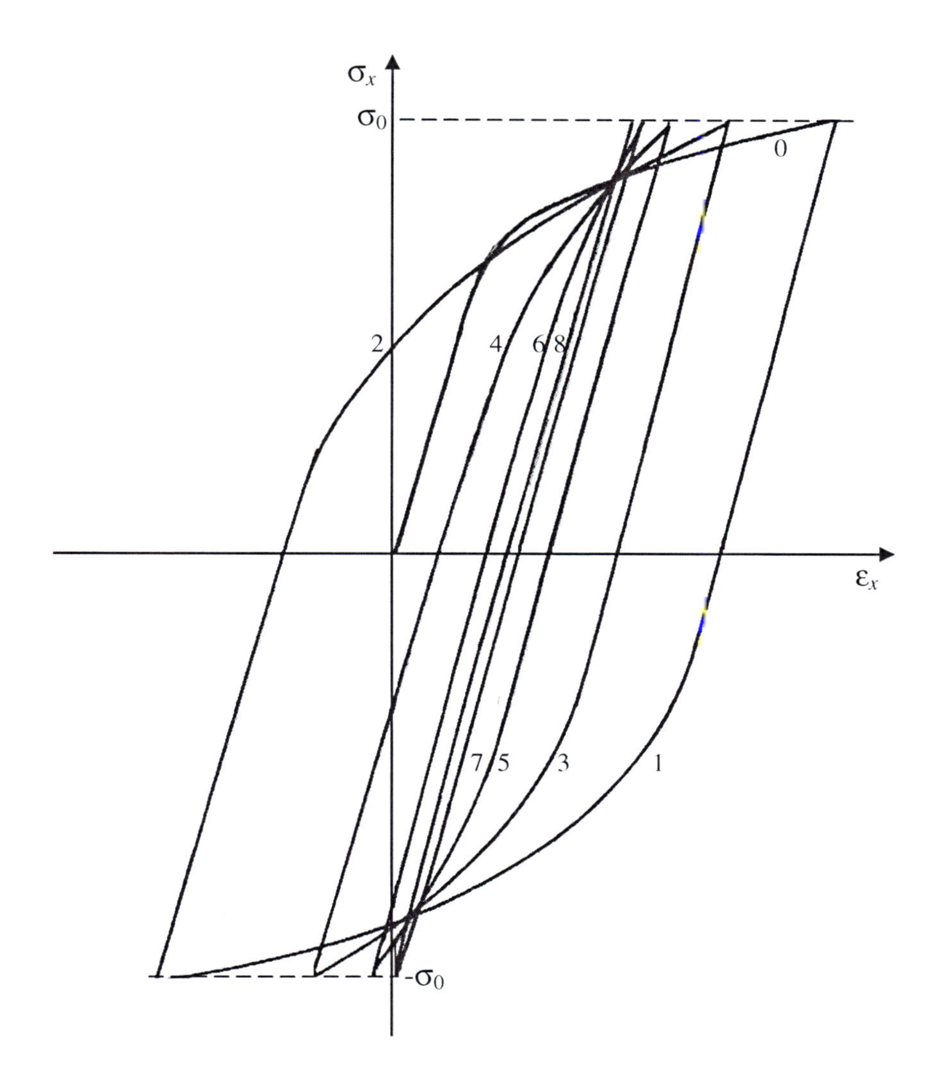

Fig. 4.17 Diagrams stabilization into strait line

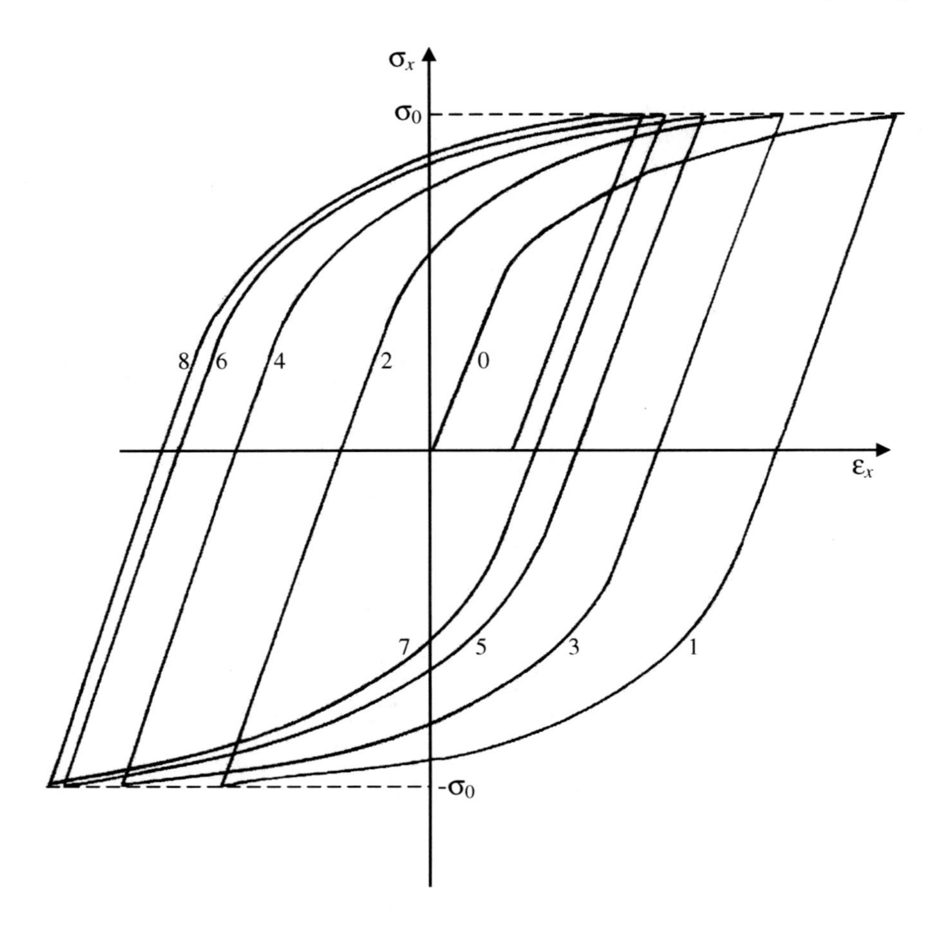

Fig. 4.18 Diagrams stabilization

4.17 Instantaneous Plastic and Creep Deformation

Let us proceed to the investigation and mathematical description of creep deformation. It is generally accepted that so-called "instant" plastic and creep deformation are of different nature and they must be modeled separately by different equations. On the one hand, creep deformation has a strengthening influence on plastic deformation. On the other hand, following [122,182,183], a small plastic deformation does not affect subsequent creep; the following experiment is cited as an example. Consider the creep diagram of two specimens, Specimen 1 and Specimen 2, in tension. Let Specimen 1 undergo creep deformation under the action of constant tensile stress σ_M (smaller or equal to yield limit), its creep diagram is shown by curve OM_1 in Fig. 4.19. Let us designate ε_M^P as the creep strain of Specimen 1 at some time t_M. Specimen 2 is loaded plastically with great loading rate until its plastic strain, ε^s, reaches the creep strain ε_M^P of Specimen 1. Then the partial unloading of Specimen 2 to the stress σ_M is carried out and Specimen 2 starts to produce a creep deformation under constant stress σ_M. The creep curve of Specimen 2 does not follow portion MM_1 of Specimen 1 and its form is such as if $\varepsilon^S = 0$, portion MM_2 of creep curve of Specimen 2 is similar to OM.

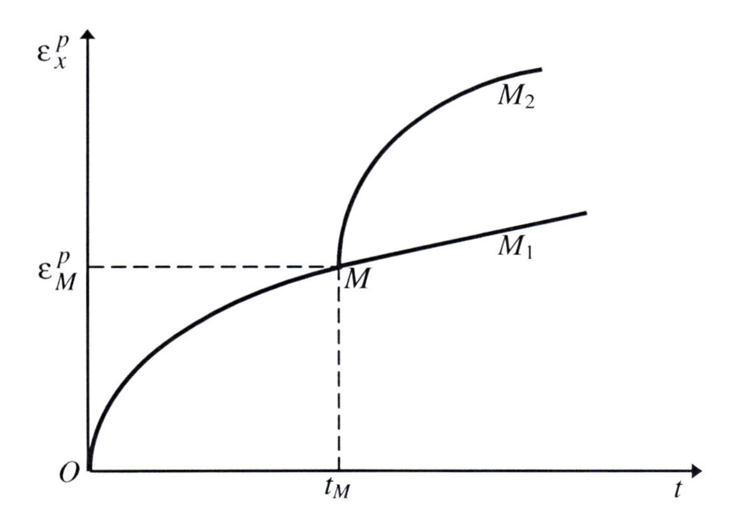

Fig. 4.19 Creep diagrams of two specimens

According to the result of the experiment above, creep deformation is assumed (it must be noted that incorrectly, see Sec. 4.38) to be not influenced by plastic deformation and the relationships of creep theories do not need to react to plastic deformations. This statement is often substantiated by the following. A plastic strain is of strongly localized nature, while the main volume of the grain remains undeformed. Creep deformation is resulted from irreversible slips as well, although their distribution through the grain is more uniform. If a metal undergoes small plastic strain, creep occurs basically in the unhardened volume, therefore plastic strain does not affect creep.

Remark. The form of the creep diagram of Specimen 2 (curve MM_2 in Fig. 4.19) can raise doubts as there is no creep delay after partial unloading. On the other hand, MacLean with the co-authors [86,87] state that the creep delay as a result of fast unloading is always observed. And what is more, during fast unloading, creep in opposite direction, i.e., decrease in ε_M^P, can be observed that will be considered in detail in Sections 4.40–4.42.

4.18 Classical Creep Theories in Uniaxial Tension

Ageing (time-hardening) theory. This theory gives that the equations of creep diagrams at a given tensile stresses, σ_x, and prevailing temperature, T, has the form

$$\varepsilon_x^P(t) = f(\sigma_x, T, t), \tag{4.18.1}$$

and the above equation is considered as a unique one due to it is assumed to be valid not only at constant σ_x, but at time dependent stress as well.

The employment of Eq. (4.18.1) can result in contradiction for structurally stable materials, i.e., materials whose structure and properties do not alter on prolonged exposure to the test temperature without load. Indeed, this equation is not invariant relative to how to measure time.

Further throughout, structurally stable materials will be only considered.

Let us consider an example demonstrating the incapability of Eq. (4.18.1) to describe creep at step change in stress shown in Fig. 4.20 (two creep curves, marked by 1 and 2, under constant stresses, σ_1 and σ_2, are shown in Fig. 4.20 by dotted lines). According to Eq. (4.18.1), stress jump at t_M causes an immediate increment in strain by amount MM_1 and point M_1 lies on the diagram 2. The creep curve under constant stress σ_2, $t > t_M$, follows diagram 2, portion M_1M_2. The distance from curve 1 to curve 2 grows with time implying that increment MM_1 can become considerable, which is not observed in experiments. A yet greater inconsistence arises for the case when the acting stress undergoes step-decrease, Eq. (4.18.1) leads to the strain drop by amount MM_1 that contradicts the irreversible nature of creep deformation.

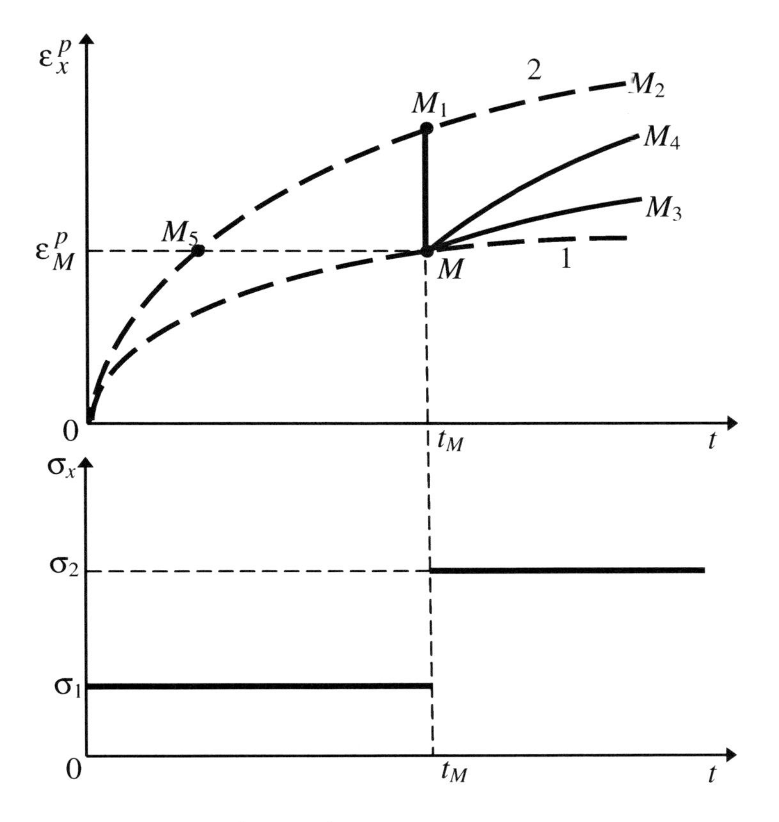

Fig. 4.20 Creep diagrams at step constant stress

2. ***Flow theory.*** According to this theory, a creep strain rate is a single-valued function of stress and time:

$$\dot{\varepsilon}_x^P = f(\sigma_x, T, t). \tag{4.18.2}$$

The serious shortcoming of this relation, similar to Eq. (4.18.1), is its invariance relative to count-up time. However, Eq. (4.18.2) is more acceptable, than Eq. (4.18.1) as creep strain, in terms of the flow theory, is of irreversible nature. Further, a sudden change in stress does not cause immediate change in creep strain, only the strain rate changes immediately.

The strain response under a step change in stress is the following (see Fig. 4.20). At point M, Eq. (4.18.2) gives that the creep strain rate takes the value corresponding to that for creep 2, therefore, starting from point M, the portion of creep curve MM_3 is parallel to curve 2. However, experiments show that the creep strain rate at point M is considerably greater than what Eq. (4.18.2) provides.

3. ***Strain-hardening theory.*** A creep rate at each instant is determined by the magnitude of stress, temperature, and cumulated creep deformation:

$$f\left(\dot{\varepsilon}_x^P, \sigma_x, \varepsilon_x^P, T\right) = 0 . \tag{4.18.3}$$

Since, according to Eq. (4.18.3), the creep rate at point M is determined by strain at this point, ε_M^P , and stress σ_2, the stress jump results in that the slope of creep curve at point M becomes the same as at point M_5. Therefore, in terms of the considered theory, the creep diagram for $t > t_M$, MM_4, is constructed by the horizontal translation of the portion of curve 2 from point M_5 to point M. Regarding the flow theory (point 2), to obtain the portion MM_3, one should to translate the portion of curve 2, for $t > t_M$, downward to point M. As seen, the difference in the results of these two theories is considerable; the strain-hardening theory is much better agreed with experiments.

If to represent Eq. (4.18.3) in the form

$$\dot{\varepsilon}_x^P \left(\varepsilon_x^P\right)^k = f(\sigma_x), \tag{4.18.4}$$

we will obtain the satisfactory description of unsteady-state creep curves. Indeed, integrating over time in (4.18.4) and holding σ_x constant, we obtain the power strain-time relation

$$\varepsilon_x^P = \sqrt[k+1]{(k+1) f(\sigma_x) t} . \tag{4.18.5}$$

At $k = 2$ we obtain degree 1/3 and the relationship $\varepsilon_x^P \sim t^{1/3}$ leads to satisfactory agreements with experimental data.

The fact that ε_x^P grows all over creep leads to that the strain rate in Eq. (4.18.4) progressively decrease, i.e. the theory is not sensitive to the transition from the unsteady- to steady-state of creep diagram. Therefore, instead of Eq. (4.18.4), the following two relations are proposed:

$$\begin{aligned} \dot{\varepsilon}_x^P \left(\varepsilon_x^P\right)^k &= f(\sigma_x) \quad \text{at} \quad \varepsilon_x^P \le \varepsilon_M^P \\ \dot{\varepsilon}_x^P \left(\varepsilon_M^P\right)^k &= f(\sigma_x) \quad \text{at} \quad \varepsilon_x^P \ge \varepsilon_M^P \end{aligned} \tag{4.18.6}$$

where ε_M^P is the strain on the boundary between steady and unsteady creep (it must be noted that this boundary can be defined conventionally only), ε_M^P is a function of acting stress and temperature.

Fundamental experimental verifications of the strain-hardening theory were carried out by V.S. Namestnikov. and A.A. Chvostunkov [101]. More than 130 specimens of aluminum alloy were tested under the condition of creep in uniaxial tension at temperatures 150°C and 200°C, test time about 100 hours. The great number and duration of tests ensured the high accuracy of obtained results. As a result, the satisfactory agreement between experiments and the strain-hardening theory was obtained. Besides, the cases of step constant stresses were considered as well. Strain-time relation derived from the strain-hardening theory gives satisfactory results for the case of constant load and step-decrease in stress. If the stress undergoes jump-increase, the creep develops faster, than it follows from the theory, especially for first minutes after the stress has been changed; this is usually observed when the lower value of stress induces no plastic strain prior to creep. For the case of large stresses, which produce an initial plastic deformation, the strain-hardening theory results in significant deviations from experiments, experimental strain rate is greater than calculated.

If we take Eq. (4.18.3) in the form

$$\dot{\varepsilon}_x^P h\left(\varepsilon_x^P\right) = f\left(\sigma_x\right), \tag{4.18.7}$$

then the commutative law is valid. Consider the case of step constant loading with stresses σ_i ($i = 1,\ldots,n$) acting during time-periods t_i ($\sigma_i \neq \sigma_j$ as $i \neq j$). Then, according to the commutative property, the total strain due to n-loadings does not depend on the order in which the stresses act. Indeed, integrating Eq. (4.18.7), we obtain

$$\int_0^{\varepsilon_x^P} h\left(\varepsilon_x^P\right) d\varepsilon_x^P = \int_0^t f\left(\sigma_x\right) dt = t_1 f\left(\sigma_1\right) + \ldots + t_n f\left(\sigma_n\right). \tag{4.18.8}$$

The left-hand side in the equation above represents single-valued function ε_x^P and its right-hand side does not depend on the order in which we number the loading cycles.

The experiments to verify of the commutative law were carried out by Odqvist [105] V.S. Namestnikov., and A.A. Chvostunkov. As it has turned out, the commutative law is not held, and, the total strain strongly depends on the loading-order. This fact is illustrated in Fig. 4.21, where the strain-response of a specimen of alloy *D16T*, OM_1M, and OM_2M, loaded by step constant tensile stresses in regimes $A_0A_1A_2A_3$ and $B_0B_1B_2B_3$ is shown, respectively; symbols ○ and ▫ indicate the experimental points. The experiment was carried out at temperature 200°C; $\sigma_1 = 120\,\text{MPa}$, $\sigma_2 = 160\,\text{MPa}$, and $t_1 = t_2$. As seen, calculated curve OM_1M satisfactory follows the experimental points for regime $A_0A_1A_2A_3$, whereas the experimental creep strains for the portion B_2B_3, symbols ▫, essentially distinguish from those provided by the strain-hardening theory.

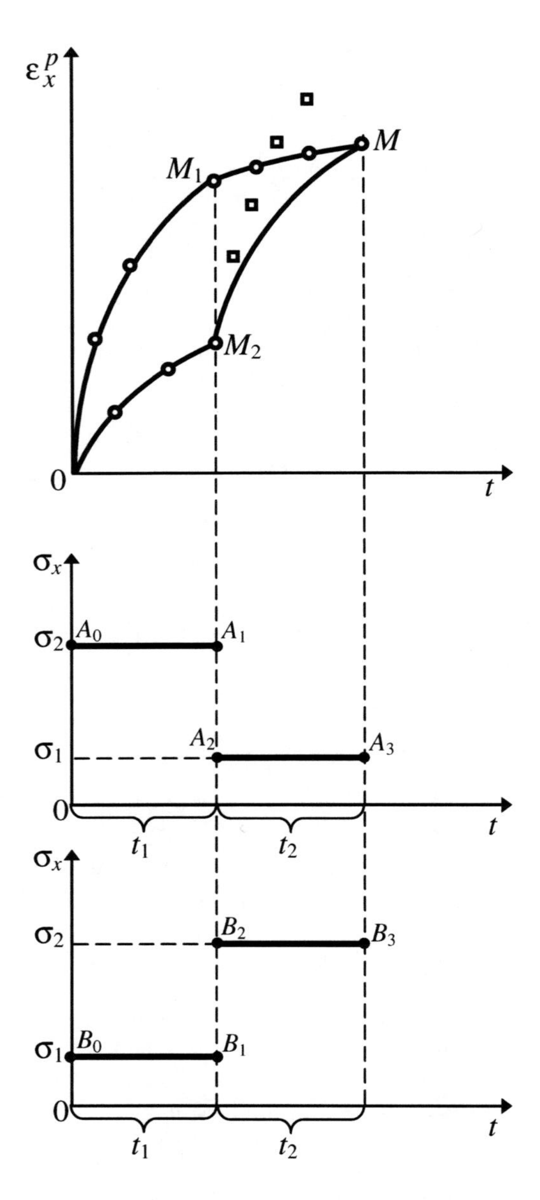

Fig. 4.21 Creep diagrams at step constant stresses with identical ranges, but different orders

4. *Kinetic creep equations.* In terms of this theory, creep rate is a function of stress, temperature, and state variables χ_i

$$\dot{\varepsilon}_x^P = f\left(\sigma_x, T, \chi_1, \chi_2, \ldots\right). \tag{4.18.9}$$

Parameters χ_i govern changes in the structure of metals that occur during creep, they are supposed to be defined by the following kinetic equation

$$d\chi_i = \chi_i' d\varepsilon_x^P + \chi_i'' d\sigma_x + \chi_i''' dT , \tag{4.18.10}$$

where χ_i', χ_i'', and χ_i''' are characteristic functions of a material.

Equations (4.18.9) and (4.18.10) provide ample opportunity to construct creep theories as accurate as necessary, however this perspective has been very poorly used. If to assume that there is the only structural parameter, $d\chi_1 = d\varepsilon_x^P$, i.e., $\chi_1 = \varepsilon_x^P$, we arrive at the strain-hardening theory. If we take the work performed by stress on creep strain for the single measure of hardening, Eq. (4.18.10) yields the form

$$d\chi_1 = \sigma_x d\varepsilon_x^P \quad \text{and} \quad \chi_1 = \int \sigma_x d\varepsilon_x^P . \tag{4.18.11}$$

Let us turn once more, to the experiment relative to the creep diagram under step constant stress, see Fig. 4.22 (curves marked by 1 and 2 indicate creep diagrams under time independent stresses σ_1 and σ_2, respectively). At time $t \to t_M - 0$, the parameter χ_1 is equal to $\sigma_1 \varepsilon_M^P$, where ε_M^P is the accumulated creep strain due to stress σ_1. The stress jump, determines the creep rate at point M, according to Eq. (4.18.9), as

$$\dot{\varepsilon}_x^P = f\left(\sigma_2, T, \sigma_1 \varepsilon_M^P\right). \tag{4.18.12}$$

One can find a point on curve 2 with ordinate ε_2^P, M_2, so that $\sigma_2 \varepsilon_2^P = \sigma_1 \varepsilon_M^P$ meaning that the strain rate at point M is equal to that at point M_2. Therefore, in order to construct the portion of creep curve for $t > t_M$, one needs to translate the portion of curve 2 from point M_2 to point M, along the line radiating from the point $\varepsilon_2^P = \varepsilon_1^P \sigma_1/\sigma_2$ on strain-axis. The strain rate determined in such way is greater than that in terms of the strain-hardening theory and better approximates experimental data.

As specified, the potential of the considered theory has not yet been realized to the full extent.

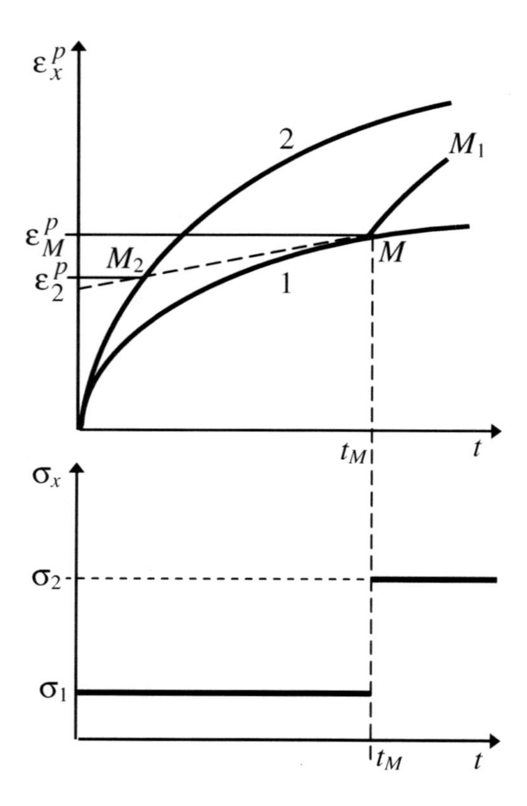

Fig. 4.22 Creep response to a step-wise increase in stress

4.19 Creep Potential

Creep rate at an arbitrary state of stress can by determined by means of creep potential. To introduce the creep potential, we define energy dissipation in creep [11,122]

$$\dot{U}^P = \sigma_{ij}\dot{\varepsilon}_{ij}^P. \tag{4.19.1}$$

Let us accept the hypothesis that steady creep rate is a single-valued function of the values of stresses independent of the way they have been reached. Then Eq. (4.19.1) becomes

$$\dot{U}^P = \dot{U}^P\left(\sigma_{ij}\right). \tag{4.19.2}$$

By equating $\dot{U}^P\left(\sigma_{ij}\right)$ to some constant, k, we obtain the surface of constant dissipation in the stress space (say, in the Ilyushin space). The surfaces related to different k do not intersect, for otherwise the energy dissipation would not be uniquely expressed in terms of stress; these surfaces enclose one another and are shown schematically in Fig. 4.23, here $k < k_1$. Now consider a certain loading path from the point M^* on the surface $\dot{U}^P = k_*$ to the point M_1 on the surface $\dot{U}^P = k_1$ so that this path intersects each of the constant dissipation surfaces $\dot{U}^P = k$, $k_* < k < k_1$, only once meaning that there is no diminution in the power dissipated during loading. If the points M and M^* lay on surface $\dot{U}^P = k$, then loading path M^*M completely follows this surface.

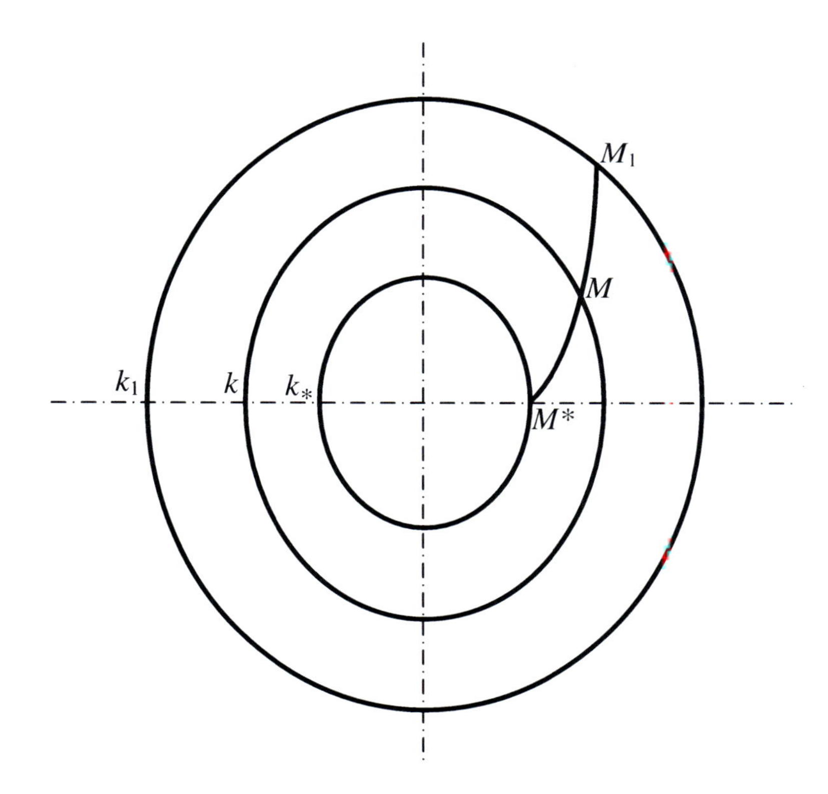

Fig. 4.23 Surfaces of constant energy dissipation

We require that along the loading path for which $dU^P \geq 0$ the following condition,

$$\left(\sigma_{ij} - \sigma_{ij}^*\right)\dot{\varepsilon}_{ij}^P \geq 0, \tag{4.19.3}$$

be fulfilled; or in vector symbols

$$\left(\vec{\sigma}-\vec{\sigma}^*\right)\dot{\vec{\varepsilon}}^P \geq 0\,. \tag{4.19.4}$$

In effect Eq. (4.19.4) is a rephrasing of Drucker's postulate (Sec. 1.10) to the case of steady-state creep. Just as in the theory of plasticity, condition (4.19.4) does not follow from the laws of thermodynamics and is not a thermodynamic postulate, this is a limitation which makes it possible to distinguish a class of materials (stable materials).

Following Drucker's postulate, because $\vec{\sigma}^*$ is an arbitrary vector, condition (4.19.4) implies that the surface $\dot{U}^P = k$ is convex and smooth and the deformation rate vector is in the direction of the normal to this surface,

$$\dot{\varepsilon}_{ij}^P = h\left(\sigma_{ij}\right)\frac{\partial \dot{U}^P}{\partial \sigma_{ij}}. \qquad \Sigma \tag{4.19.5}$$

If one can find a function, $W(\sigma_{ij})$, for which

$$h\left(\sigma_{ij}\right)\frac{\partial \dot{U}^P}{\partial \sigma_{ij}} = \frac{\partial W}{\partial \sigma_{ij}} \qquad \Sigma, \tag{4.19.6}$$

Eq. (4.19.5) becomes.

$$\dot{\varepsilon}_{ij}^P = \frac{\partial W}{\partial \sigma_{ij}}. \tag{4.19.7}$$

The function W is called the creep potential [11,71,122,174].

Now we wish to analyze a particular case when the condition (4.19.6) holds. Consider the case, when the argument of function h is the energy dissipation in creep,

$$h = h\left(\dot{U}^P\right). \tag{4.19.8}$$

To show that Eq. (4.19.6) holds true, we find the following complete differentials:

$$dW = \frac{\partial W}{\partial \sigma_{ij}} d\sigma_{ij}, \quad d\dot{U}^P = \frac{\partial \dot{U}^P}{\partial \sigma_{ij}} d\sigma_{ij}. \tag{4.19.9}$$

In view of Eqs. (4.19.6) and (4.19.9), we get

$$dW = h\left(\dot{U}^P\right)\frac{\partial \dot{U}^P}{\partial \sigma_{ij}} d\sigma_{ij} = h\left(\dot{U}^P\right) d\dot{U}^P, \tag{4.19.10}$$

$$W = \int h\left(\dot{U}^P\right) d\dot{U}^P . \tag{4.19.11}$$

Multiplying by σ_{ij} the both sides in Eq. (4.19.5), we obtain

$$\dot{U}^P = h\left(\sigma_{ij}\right) \frac{\partial \dot{U}^P}{\partial \sigma_{kl}} \sigma_{kl} . \tag{4.19.12}$$

Further, if $\dot{U}^P\left(\sigma_{ij}\right)$ is a homogeneous function of degree $n+1$, from Euler's theorem,

$$\frac{\partial \dot{U}^P}{\partial \sigma_{ij}} \sigma_{ij} = (n+1)\dot{U}^P , \tag{4.19.13}$$

together with Eq. (4.19.12), we get

$$h = \frac{1}{n+1} = const \tag{4.19.14}$$

and Eq. (4.19.11) gives

$$W = \frac{\dot{U}^P}{n+1} . \tag{4.19.15}$$

Therefore, accepting Eq. (4.19.8) and assuming that $\dot{U}^P$ is a homogeneous function of stresses, we have derived the formula for W through which the creep rate can be determined by Eq. (4.19.7).

If Eq. (4.19.7) holds, we can prove that there exists a function of creep strain rate tensor component, $V\left(\dot{\varepsilon}_{ij}^P\right)$, so that

$$\sigma_{ij} = \frac{\partial V}{\partial \dot{\varepsilon}_{ij}^P} . \tag{4.19.16}$$

Indeed, taking the function V in the form

$$V = \sigma_{ij}\dot{\varepsilon}_{ij}^P - W = \dot{U}^P - W \tag{4.19.17}$$

and supposing that the stress tensor components σ_{ij} are single-valued functions of $\dot{\varepsilon}_{ij}^P$, we obtain

$$\frac{\partial V}{\partial \dot{\varepsilon}_{ij}^{P}} = \sigma_{ij} + \dot{\varepsilon}_{kl}^{P} \frac{\partial \sigma_{kl}}{\partial \dot{\varepsilon}_{ij}^{P}} - \frac{\partial W}{\partial \sigma_{kl}} \frac{\partial \sigma_{kl}}{\partial \dot{\varepsilon}_{ij}^{P}}. \tag{4.19.18}$$

On the base of Eq. (4.19.7), the second and third terms in Eq. (4.19.18) are equal to each other and we arrive at Eq. (4.19.16).

In the framework of creep theory, the function $W(\sigma_{ij})$ plays the same role as a loading surface, $f(\sigma_{ij})$, in terms of plasticity theories. At the same time, there is an essential difference between $W(\sigma_{ij})$ and $f(\sigma_{ij})$. A loading surface separates plastic and elastic regions in stress space, an extra-loading into the elastic region does not result in the increment of plastic deformation, whereas that into a plastic one does. In contrast to this, a surface $\dot{U}^{P} = k$ does not split stress space into plastic and elastic regions, a creep straining does not terminates due to the decrease in acting stress and the creep rate is determined by Eq. (4.19.7) at the new value of stress.

In initially isotropy body, creep potential can depend on the invariants of tensor of stress only. However, for the most metals, no volume change occurs during creep meaning that W is a function of two invariants of stress tensor, τ_0^2 and $\overline{A}_3$, determined by Eqs. (1.2.29) and (1.6.8), respectively. The invariant $\overline{A}_3$ can be replaced by

$$\overline{A}_3{}' = \overline{\sigma}_{ij}\overline{\sigma}_{jk}\overline{\sigma}_{ki}. \tag{4.19.19}$$

Indeed, as the first invariant of stress deviator tensor $\overline{A}_1 = 0$, then [122]

$$\overline{A}_3 = -\frac{\overline{A}_3{}'}{3}. \tag{4.19.20}$$

Thus,

$$W = W\left(\tau_0^2, \overline{A}_3{}'\right). \tag{4.19.21}$$

Now, Eq. (4.19.7) will take the following form [122]:

$$\dot{\varepsilon}_{ij}^{P} = 3\frac{\partial W}{\partial \tau_0^2}\overline{\sigma}_{ij} + 3\frac{\partial W}{\partial \overline{A}_3{}'}\left(\overline{\sigma}_{ik}\overline{\sigma}_{kj} - \frac{2}{9}\tau_0^2\delta_{ij}\right). \tag{4.19.22}$$

By making use of designation

$$A_1 = 3\frac{\partial W}{\partial \tau_0^2}, \quad A_2 = 3\frac{\partial W}{\partial \overline{A}_3{}'}, \tag{4.19.23}$$

where A_1 and A_2 are functions of the second and third invariant of stress tensor, we obtain the formula for creep strain rate components in initially isotropy body:

$$\dot{\varepsilon}_{ij}^P = A_1\overline{\sigma}_{ij} + A_2\left(\overline{\sigma}_{ik}\overline{\sigma}_{kj} - \frac{2}{9}\tau_0^2\delta_{ij}\right). \tag{4.19.24}$$

The right-hand sides in Eq. (1.6.12) and Eq. (4.19.24) have identical structure. Equation (4.19.24) takes a more simple form if it is possible to neglect the influence of the third invariant on creep rate $\dot{\varepsilon}_{ij}^P$. In this case we obtain the relation

$$\dot{\varepsilon}_{ij}^P = A_1(\tau_0)\overline{\sigma}_{ij} \tag{4.19.25}$$

that expresses the law of the proportionality between strain rate and stress deviators.

The relationship (4.19.25) is a quasi-linear one to the effect that the strain rate components relate linearly to stress deviator components; nonlinearity is regulated by factor A_1 depending on the invariant τ_0. If $\overline{\sigma}_{ij}$ is time independent, the right-hand side in Eq. (1.19.25) does not depend on time and can be integrated:

$$\varepsilon_{ij}^P = A_1(\tau_0)\overline{\sigma}_{ij}t, \tag{4.19.26}$$

i.e., Eq. (4.19.25) describes steady state creep.

To model primer creep, the existence of creep potential is postulated, i.e., Eq. (4.19.7) is hold. In contrast to Eq. (4.19.21), the potential W depends on the second invariant τ_0 and some function of time, $\dot{\tau}(t)$. If we take $W = A_1(\tau_0)\dot{\tau}(t)$, then

$$\dot{\varepsilon}_{ij}^P = A_1(\tau_0)\overline{\sigma}_{ij}\dot{\tau}(t). \tag{4.19.27}$$

For time independent stress,

$$\varepsilon_{ij}^P = A_1(\tau_0)\overline{\sigma}_{ij}\tau(t). \tag{4.19.28}$$

The above equation is not invariant relative to the zero reference time-point.

From Eq. (4.19.25) an important conclusion can be drawn: contracting the strain and stress tensors on i and j indices,

$$\dot{\varepsilon}_{ij}^P\dot{\varepsilon}_{ij}^P = A_1^2(\tau_0)\overline{\sigma}_{ij}\overline{\sigma}_{ij}, \tag{4.19.29}$$

and, similar to Eq. (1.2.8), introducing the notion of strain rate intensity, $\dot{\gamma}_0^P$,

$$\dot{\gamma}_0^P = \frac{\sqrt{2}}{3}\left[\left(\dot{\varepsilon}_x^P - \dot{\varepsilon}_y^P\right)^2 + \left(\dot{\varepsilon}_y^P - \dot{\varepsilon}_z^P\right)^2 + \left(\dot{\varepsilon}_x^P - \dot{\varepsilon}_z^P\right)^2 + 6\left(\left(\dot{\varepsilon}_{xy}^P\right)^2 + \left(\dot{\varepsilon}_{xz}^P\right)^2 + \left(\dot{\varepsilon}_{zy}^P\right)^2\right)\right]^{1/2}. \tag{4.19.30}$$

we can show that

$$\left(\dot{\gamma}_0^P\right)^2 = \frac{2}{3}\dot{\varepsilon}_{ij}^P\dot{\varepsilon}_{ij}^P. \tag{4.19.31}$$

Therefore, Eq. (4.19.29) can be written as

$$\dot{\gamma}_0^P = \frac{2}{3}\tau_0 A_1(\tau_0). \tag{4.19.32}$$

Equation (4.19.32) assures the existence of unique(universal) function modeling steady-state creep for all states of stress. If to contract the tensors in Eq. (4.19.27), we obtain

$$\dot{\gamma}_0^P = \frac{2}{3}\tau_0 A_1(\tau_0)\dot{\tau}(t). \tag{4.19.33}$$

In Eqs. (4.19.32) and (4.19.33), factors $A_1(\tau_0)$ and $\tau(t)$ are the characteristic functions of material.

4.20 Experimental Verification of Creep Laws

At present, numerous experimental works have been published to verify the validity of the laws of creep for the case of combined states of stress. Most experiments are performed on thin-walled pipes subjected to the combined action of tension, torsion, and internal pressure. We put an x- and z-axis along the axis and radius of pipe, respectively, and y-axis lies in the plane perpendicular to the longitudinal axis and is perpendicular to the z-axis. Strain rate components, $\dot{\varepsilon}_x^P$, $\dot{\varepsilon}_y^P$, and $\dot{\varepsilon}_{xy}^P$, are measured in experiments and $\dot{\varepsilon}_z^P$ is determined from the condition of incompressibility:

$$\dot{\varepsilon}_x^P + \dot{\varepsilon}_y^P + \dot{\varepsilon}_z^P = 0. \tag{4.20.1}$$

Equation (4.19.27) gives the following strain rate components

$$\begin{aligned}
\dot{\varepsilon}_x^P &= \frac{1}{3}A_1(\tau_0)(2\sigma_x - \sigma_y)\dot{\tau}(t),\\
\dot{\varepsilon}_y^P &= \frac{1}{3}A_1(\tau_0)(2\sigma_y - \sigma_x)\dot{\tau}(t),\\
\dot{\gamma}_{xy}^P &= 2\dot{\varepsilon}_{xy}^P = 2A_1(\tau_0)\tau_{xy}\dot{\tau}(t).
\end{aligned} \tag{4.20.2}$$

From here

$$\frac{\dot{\varepsilon}_x^P}{\dot{\gamma}_{xy}^P} = \frac{2\sigma_x - \sigma_y}{6\tau_{xy}}, \quad \frac{\dot{\varepsilon}_y^P}{\dot{\gamma}_{xy}^P} = \frac{2\sigma_y - \sigma_x}{6\tau_{xy}}. \tag{4.20.3}$$

If the strain rate components are time independent, then the above relations are integrable and we get

$$\frac{\varepsilon_x^P}{\gamma_{xy}^P} = \frac{2\sigma_x - \sigma_y}{6\tau_{xy}}, \quad \frac{\varepsilon_y^P}{\gamma_{xy}^P} = \frac{2\sigma_y - \sigma_x}{6\tau_{xy}}. \tag{4.20.4}$$

The following questions to be experimentally verified are whether (i) there exists a unique relationship between creep strain rate and the second invariant (4.19.32) or (4.19.33) and (ii) does it hold true Eq. (4.20.4) and Eq. (4.19.28). We do not dwell at length on the techniques of experiments; their details are given in [122], but present their general results and conclusions.

1. Creep curves, in both uniaxial and combined state of stress are characterized by a considerable scatter in results. Therefore, the experiments giving reliable results, from which reliable averages can be obtained, were only considered.

2. The relationship between the shear stress intensity and steady creep rate turned out to be independent of the state of stress, the experimental points are close to analytical curve in $\log\dot{\gamma}_0 - \log\tau_0$ or $\log\dot{\gamma}_0 - \tau_0$ systems of coordinate. Some authors have reported that, at identical τ_0, $\dot{\gamma}_0$ in tension is greater than that in torsion.

3. Equation (4.20.4) turned out valid for different constant values of ratio $(2\sigma_x - \sigma_y)/6\tau_{xy}$ and $(2\sigma_y - \sigma_x)/6\tau_{xy}$.

4. Let us dwell in more detail on the experimental results given in work [100], where the existence of universal creep diagram has been studied,

$$\gamma_0^P = \gamma_0^P(t), \tag{4.20.5}$$

which would be independent of the state of stress. Here γ_0^P is the creep strain intensity:

$$\gamma_0^P = \frac{\sqrt{2}}{3}\left[\left(\varepsilon_x^P - \varepsilon_y^P\right)^2 + \left(\varepsilon_x^P - \varepsilon_z^P\right)^2 + \left(\varepsilon_z^P - \varepsilon_y^P\right)^2 + 6\left(\left(\varepsilon_{xy}^P\right)^2 + \left(\varepsilon_{yz}^P\right)^2 + \left(\varepsilon_{xz}^P\right)^2\right)\right]^{1/2} \tag{4.20.6}$$

By the uniqueness of creep diagram (4.20.5) we mean that it does not vary with the change in the state of stress, provided that the shear stress intensity τ_0 is constant.

First of all, we verify whether Eqs. (4.19.30) and (4.20.6) are not contradictory. For this purpose, we differentiate Eq. (4.20.6) with respect to time:

$$\dot{\gamma}_0^P = \frac{\sqrt{2}}{3} \frac{\left(\varepsilon_x^P - \varepsilon_y^P\right)\left(\dot{\varepsilon}_x^P - \dot{\varepsilon}_y^P\right) + \ldots + 6\varepsilon_{xy}^P \dot{\varepsilon}_{xy}^P + \ldots}{\left[\left(\varepsilon_x^P - \varepsilon_y^P\right)^2 + \left(\varepsilon_x^P - \varepsilon_z^P\right)^2 + \left(\varepsilon_z^P - \varepsilon_y^P\right)^2 + 6\left(\left(\varepsilon_{xy}^P\right)^2 + \left(\varepsilon_{yz}^P\right)^2 + \left(\varepsilon_{xz}^P\right)^2\right)\right]^{1/2}}. \quad (4.20.7)$$

As one can see, Eqs. (4.19.30) and (4.20.7) are identical owing to the presence of Eqs. (4.19.27) and (4.19.28). From this it follows that the time integral of $\dot{\gamma}_0^P$ from (4.19.30) is equal to γ_0^P from (4.20.6). Performing the time integration of the right- and left-hand side in Eq. (4.19.33), we obtain the function $\gamma_0^P = \gamma_0^P(t)$ in the form

$$\gamma_0^P = \frac{2}{3}\tau_0 A_1(\tau_0)\tau(t). \quad (4.20.8)$$

Therefore, the existence of universal diagram (4.20.8) follows from the concept of creep potential.

The existence of universal curve (4.20.8) with respect to both unsteady and steady stage of creep was subjected to experimental verifications. Let us consider creep diagrams obtained in the following creep tests ($T = 650°$, strain rate $|\dot{e}| = 5 \cdot 10^{-3}$ 1/min) that was carried out on the specimens of stainless steel 316 (yield strength $\tau_S = 100\text{MPa}$) in the following loadings: a) uniaxial tension, b) combined tension and torsion at $S_1 = S_3$, c) pure torsion, d) combined torsion and pressure, $S_1 = -S_3$, and e) uniaxial pressure. In all the cases the shear stress intensity was held constant, $\tau_0 = 140\text{ MPa}$. Strain vs. time curves constructed by Eq. (4.20.8) for the all states of stress were practically identical [100].

Thus, a theory founded on the hypothesis of creep potential has experimental justification for many materials at constant stress. In work [122] still more categorical statement is made, the behavior of initially isotropic structurally stable materials, governed by Eqs. (4.19.28) and (4.20.5), should be considered as a norm. The reasons of possible deviations from this norm can be caused by such factors as initial anisotropy, heterogeneity, and instability.

4.21 On the Advantages of the Physical Theories of Plasticity and Creep

Consider the results of creep tests in combined loading if the ratio σ_x/τ_{xy} varies during the test. When considering combined states of stress, the question to be decided is whether the strain hardening observed during primary creep is isotropic or anisotropic; in other words, whether it is possible to characterize the hardening by only one invariant parameter. To answer this question, we turn to Namestnikov's experiments [100]. Consider the straining of two specimens, marked by 1 and 2, under the condition of unsteady creep. Specimen 1 is subjected to uniaxial tension and Specimen 2 to torsion, in the both cases the intensities τ_0 are identical. At certain time, when due to the strain-hardening the strain rates of both specimens have decreased enough, Specimens 1 and 2 are unloaded. Further, Specimen 1 is loaded in torsion and Specimen 2 in uniaxial tension. One might hope to observe the decrease in the creep rate of previously deformed specimens. However, the experiment shows the contrary: both specimens deform in the same way as in initial loading, previous tension have no influence upon the subsequent torsion and vice versa. Thus, unsteady creep hardening is of strongly anisotropic nature.

The absence of the specified inter-influence follows immediately from the concept of slip. Indeed, at a given state of stress with ratio τ_{xy}/σ_x, creep occurs in those grains and slip planes, where great shear stresses act. Since the creep rate relates nonlinearly to stress, specified slip planes harden considerably, whereas low-stress slip planes do not produce the creep strain and, consequently, experience no hardening effects. Therefore, if, due to the change in stress state, the principal stress axes have been rotated enough the creep progresses in other slip systems, where the previous creep did not occurred. Thus, current creep is not affected by previous one produced in another type of loading.

On the other hand, under less contrast condition, if to require that ratios $\left(\tau_{xy}/\sigma_x\right)_1$ and $\left(\tau_{xy}/\sigma_x\right)_2$ be close in initial and subsequent loading, the hardening effect of the initial creep will be more marked [100]. In particular, for the case of proportional loading, $\left(\tau_{xy}/\sigma_x\right) = const$, the principal stress axes do not experience a rotation and the experiment [100] is satisfactorily coordinated with Eqs. (4.19.26) and (4.19.28), which are obtained from the theory of creep potential.

The fact that the creep in tension does not affect the subsequent creep in torsion can be easily explained in terms of the synthetic theory. In uniaxial tension, the domain of movable planes is concentrated around the S_1-axis (the smaller circle in Fig. 3.7), whereas, under the condition of torsion, this domain encloses the S_3-axis (Fig. 3.5c). Therefore, these domains do not intersect each other meaning that the planes, which have produced irreversible deformation in tension, are motionless in torsion and do not contribute to the current strain hardening. Since we work with the initial sphere (4.8.4), here and further throughout we will depict the domain of movable planes on the sphere of radius $\sqrt{2}\tau_P$.

The considered experiment testifies that in the case of a complex stress-state there is no universal curve (4.20.8), i.e., theories based on the hypothesis of creep potential are not applicable.

The similar situation is observed in terms of the theory of plasticity at the considered loading (loading in tension, unloading, and loading in torsion or in the opposite direction). The experiment data, as well as the concept of slip and the synthetic theory, give shear stress intensity against shear strain intensity plots shown in Fig. 4.24. The portion $0A$ corresponds to a tension and the portion BC – to a torsion that follows $0AB$ loading regime (or on the contrary); the curve $0AD$ indicates the monotonous loading in the uniaxial tension. As one can see, a given value of γ_0 corresponds to the different values of τ_0 meaning that the $\tau_0(\gamma_0)$ function is not a single-valued one. This fact, in turn, testifies to that the hypothesis of unique $\tau_0(\gamma_0)$ curve is not allowable in non-proportional loading.

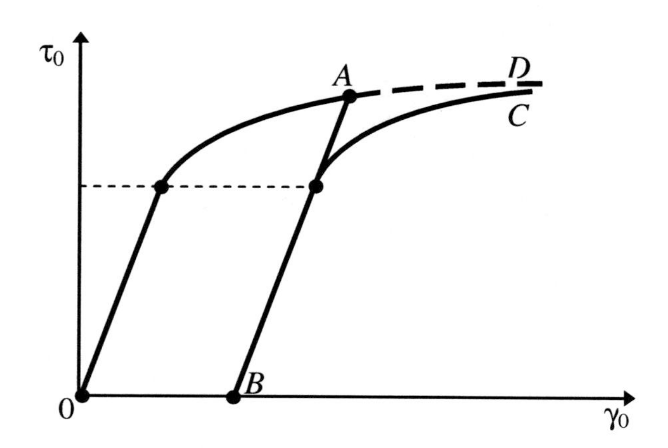

Fig. 4.24 Not single-valued relationship between shear stress intensity and shear strain intensity

The concept of slip accounts for another experimentally observed fact, creep strain intensity in tension is greater than that in torsion (see Fig. 2.11 in Sec. 2.12). Alas, the synthetic theory does not catch the difference between $\tau_0 \sim \gamma_0$ diagrams in tension and torsion, however it has not so many shortcomings as the Budiansky theory.

Summarizing, it is possible to state that Eq. (4.19.27) is not suitable to describe creep in complex stressed state, when the orientation of the principle stress axes varies considerably during loading. The unsteady creep hardening is of anisotropic nature, whereas Eq. (4.19.27) does not allow for this fact. The creep potential theory contradicts experiments not only quantitatively, but also qualitatively.

Another shortcoming of Eq. (4.19.27) to be emphasized is the following. As it is known, unsteady creep strongly depends on the loading rate, when the stress grows to reach a constant level. Unsteady creep is increasing function of the loading rate. At certain small loading rate the first stage of creep diagram is absent at all and the creep begins with stationary stage at once. Equation (4.19.27) is incapable of describing these well-known facts.

Let us note that Eq. (4.19.25), which determines steady state creep, holds true in complex loading as well, provided that the body is initially isotropy. On the other hand, if the creep is preceded by considerable plastic deformation that makes the body anisotropic, Eq. (4.19.25) leads to results contradicting experimental data.

The discrepancy between the theory and experiment for the cases of complex stressed states result from an excessive formalization of creep theories. Constitutive equations should be based on some average but real driving forces and mechanisms governing the creep. This idea is realized in the following sections paragraphs.

4.22 Physics of Unsteady-State Creep

Structural changes occurring in creep, in particular, at its unsteady stage have been investigated in numerous experimental works by the X-ray and metallographic methods. As it has turned out the mechanism of creep is much the same as in plastic deformation: creep deformation develops mainly due to the slips occurring crystal grains. These slips result from dislocation movements in preferred crystallographic directions.

Consider the following experimentally obtained facts showing the similarity between the mechanisms of instantaneous plastic and unsteady-state creep deformation [163].

1. In both cases material undergoes hardening. Let us note that the term "hardening" has a different sense for plastic and creep deformation. Namely, the strain-hardening in plastic deformation is manifested immediately during loading, whereas creep-strain hardening develops in time.
2. During straining of the both type, the number of crystal grains involved irreversible deformation grows.
3. An instantaneous plastic deformation is accompanied by the change in the grain orientations thereby increasing the number of grains with high values of resolved shear stress that promotes the further progress in irreversible straining. The unsteady creep is also accompanied by this phenomenon as well. Let us note that in terms of neither the concept of slip, nor the synthetic theory the specified effect is not taken into account.
4. The grain-size reduction and further hardening of substructure are observed.
5. Hydrostatic stress does not affect both creep and plastic deformation.
6. No volume change occurs.

Hence, there is a sufficient basis to model both instantaneous plastic deformation and unsteady-state creep in terms of the unified theory of irreversible deformation, the synthetic theory of irreversible deformation.

Further, as already specified (Sec. 4.3), the number of dislocations increases dramatically during plastic deformation. As the dislocations start to glide, dislocation sources develop and the dislocation density increases rapidly. But the increasing dislocation density results also in that they intersect each other frequently and the immobile jogs, the tangles of dislocations and other defects created this way hinder the motion of dislocations. As a consequence, the number of mobile dislocations decreases and therefore the material displays strain hardening. As a result, the crystal lattice becomes strongly distorted and possesses great stain energy associated with local peak stresses.

On the other hand, structural changes that have been occurred at fast loading are unstable. As soon as favorable conditions arise (e.g. acting stresses cease to increase, or high temperature), the relaxation of elastic distortions of lattice occurs. Numerous experimental works, for example [104] by making use of by X-ray method, have shown that elastic-distortion-relaxation causes time-dependent plastic slips within grains; the huge scope of experimental material allows to state that these slips are produced by dislocation motions. Under thermal fluctuations, locked and tangled dislocations and the obstructions in their way themselves become progressively movable, meaning that an irreversible (creep) deformation develops. The spontaneous nature of the pointed out processes is explained by the transition of the energy of crystal lattice on the more favorable level.

Another phenomenon observed in unsteady creep is a polygonization consisting in what follows. Dislocations take up positions that minimize their elastic energy; they slip or climb to the cell walls. As a result, a sub-grain structure with networks or groups of dislocations at the sub-grain boundaries is formed. With increasing time, polygonization becomes more nearly perfect and the sub-grain size gradually increases. In this stage, many of the subgrains appear to have boundaries that are free of dislocation tangles and concentrations. Dislocation structure formed in fast loading transforms into the polygonal substructure under subsequent constant stress for some length of time, whereas steady polygonization substructure is formed even during slow loading.

4.23 Unsteady Creep in Pure Shear

Although the background and the main derivations and calculations needed for a further analyze have been made in Sec. 3.9, 4.7, and 4.9, it is worthwhile, once more, to underline briefly the main points of the synthetic theory with respect to the determination of irreversible deformation, in particular, unsteady-state creep.

1. At a given instant those tangent planes produce irreversible deformation which move on the endpoint of stress vector $\vec{\mathbf{S}}$, i.e., the condition (3.9.8) is satisfied.

2. The irreversible strain, plastic or creep strain, is determined by Eq. (3.9.4) as the sum of elementary strains produced on movable planes.

3. Unsteady-state creep is closely connected to local peak stresses, precisely speaking, to their relaxation. Time-dependent behavior of the average value of these stresses is described by the integral of non-homogeneity, I_N ,introduced in Sec. 4.7; the I_N, together with irreversible strain intensity φ_N, defines the plane distance H_N through Eq. (4.7.3) or (4.7.4).

Consider unsteady creep strain in pure shear when all stress tensor components, except for $\tau_{xz} = S_3/\sqrt{2}$, are equal to zero. The S_3 against time diagram is the same as that in Fig. 4.8a; the loading time now is denoted by t_M, $t_M = S_3/v = S/v$ and $v = const$. Taking the function $\varphi_N(H_N)$ in the form of Eq. (3.9.11), the plastic strain vector component at $t = t_M$, e_{3M}^S, is determined by Eq (4.9.8):

$$e_{3M}^S = a_0 F(a_M), \quad a_0 = \frac{\pi S_P}{3r}, \quad a_M = \frac{S_P}{S_3 - I_M}. \tag{4.23.1}$$

Here I_M is determined by Eqs. (4.6.1) and (4.6.3) at $t = t_M$,

$$I_M = \frac{Bv}{p}\left(1 - \exp(-pt_M)\right), \tag{4.23.2}$$

or

$$I_M = \frac{Bv}{1-p} t_M^{1-p}, \tag{4.23.3}$$

respectively. If the function $\varphi_N(H_N)$ is defined by Eq. (3.9.10), the plastic strain vector component at $t = t_M$ is determined by Eq. (4.9.10),

$$e_{3M}^S = a_0 F(a_M), \quad a_0 = \frac{\pi S_P^2}{6r}, \quad a_M = \frac{S_P}{\sqrt{S_3^2 - I_M^2}}. \tag{4.23.4}$$

When the stress is constant, $t > t_M$, the integral of non-homogeneity begins to reduce according to Eq. (4.6.2) or (4.6.4). This leads to that the argument a in Eq. (4.9.6) begins to decrease as well implying that the total, plastic and creep, strain vector component, e_3^S, determined by Eqs. (4.9.8) or (4.9.10) grows. The difference between the strain developed at any instant from the interval $t > t_M$ and that at $t = t_M$ gives the creep strain vector component

$$e_3^P = e_3^S - e_{3M}^S. \tag{4.23.5}$$

The unsteady creep diagram constructed by Eqs. (4.23.1)–(4.23.5) is shown in Fig. 4.25. It has a horizontal asymptote (dotted line) to which the e_3^P tends as $t \to \infty$.

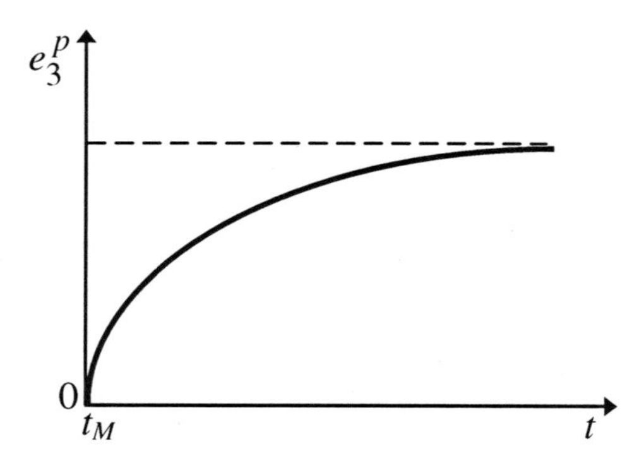

Fig. 4.25 Unsteady-state creep diagram constructed by Eqs. (4.23.1)-(4.23.5)

From the obtained formulas (4.23.1)–(4.23.5) it follows that the unsteady creep strain depends on the loading rate, the larger v, the larger creep strain. The unsteady creep is absent at small loading rates. Indeed, the integral of non-homogeneity is zero at $v = 0$ and a_M from (4.23.4) is equal to a from (4.9.9) implying that $e_3^S = e_{3M}^S$, therefore Eq. (4.23.5) gives $e_3^P = 0$. Hence, an "instantaneous" plastic deformation, as well as unsteady creep, depends on a loading rate. However, their sum does not depend on v if the exposure of specimen under constant stress is long-termed. Indeed, from Eq. (4.23.5) it follows that the specified sum is

$$e_{3M}^S + e_3^P = e_3^S. \tag{4.23.6}$$

The integral of non-homogeneity, which appears in the expression for e_3^S, relaxes in time and tends to zero as $t \longrightarrow \infty$, therefore

$$e_3^S = a_0 F\left(\frac{S_P}{S}\right). \tag{4.23.7}$$

Since this expression does not contain a loading rate v, the sum of the terms in the left-hand side in Eq. (4.23.6) does not depend on the rate v either. Thus, the material has a certain resource of irreversible straining that depends on the stress magnitude. This resource is shared by the plastic and creep deformation in the following proportion. Plastic deformation produced during loading, e_{3M}^S, decreases with the increase in v, whereas creep deformation under the subsequent constant stress, e_3^P, grows with the increase in v.

Summarizing, **in terms of the synthetic theory, the problems how stress-strain and strain-time diagrams are affected by loading rate are not different problems**. As seen from Eq. (4.23.5), the creep deformation is readily expressed from the total irreversible deformation. On the other hand, in the majority of cases there is no need to do this as it is not clear how to split γ_{xz} into plastic and viscous parts during the increase in stress τ_{xz}.

Let us dwell on another question. Under the condition of creep, $\tau_{xz}(t) = const$, an irreversible strain is produced by planes with normal vectors that create an area shaded in Fig. 3.5c, the area is symmetric above S_3-axis and its boundary is defined by angle β_1. According to Eqs. (3.10.9) and (4.9.6), the angle β_1, being dependent of I_N, decreases in time at $\tau_{xz}(t) = const$ meaning that the area increases in time. Therefore, the number of planes located on the endpoint of stress vector increases, meaning that new crystal grains are involved in the production of creep strain; this is justified by experiments (see Sec. 4.22).

4.24 Generalization to an Arbitrary State of Stress

The results obtained in the preceding Section are readily generalized to the case of an arbitrary state of stress. Let stress vector components, S_1, S_2, and S_3, increase proportionally with constant rate. In time t_M the module of the vector $\vec{\mathbf{S}}$, S, reaches some value, $S = vt_M$, where v is the loading rate. Further $S(t) = const$ for $t > t_M$.

On the basis of results obtained in Sections 4.11 and 4.23 the relationship between the acting constant stress and the creep strain vector components can be easily established as

$$e_i^P = a_0\left[F(a) - F(a_M)\right]\frac{S_i}{S}. \tag{4.24.1}$$

In the case of uniaxial tension along, say, OX-axis we have $S_1 = S$ and a and a_M are determined by Eqs. (4.9.6) or (4.9.9) and (4.23.1) or (4.23.4), respectively. The creep strain tensor component ε_x^P, due to $\varepsilon_x^P = \sqrt{2}e_1^P/\sqrt{3}$ by Eq. (1.7.12), is

$$\begin{aligned} \varepsilon_x^P &= a_0\sqrt{\frac{2}{3}}\left[F(a) - F(a_M)\right], \\ \varepsilon_y^P &= \varepsilon_z^P = -\frac{a_0}{\sqrt{6}}\left[F(a) - F(a_M)\right]. \end{aligned} \tag{4.24.2}$$

Consider another very important aspect to be stressed. As a rule, classical creep theories describe the creep in an elastic region. If constants of these theories appearing in their constitutive equations are determined to best fit the experimentally obtained creep curves in elastic region, the usage of these constants for creep beyond the yield strength leads to wrong results. This is because the classical theories do not take into account the anisotropy of plastic deformation.

Principally different situation is observed in terms of the synthetic theory. The only condition to be satisfied for the development of irreversible deformation, including creep deformation, is the module of a stress vector exceeds the yield limit S_S which is related to the base material constant, creep limit S_P, by Eq. (4.8.8) or (4.8.13). If $|\vec{\mathbf{S}}| < S_S$, there is no irreversible deformation, neither plastic, nor creep. In contrast to the classical creep theories which are very sensitive to whether the acting stress is greater or smaller than the yield limit in classical sense, we do not need to distinguish creep deformation with respect to the magnitude of stress. However, engineering practice needs models that describe a creep deformation in elastic region. The following Section is dedicated to this kind of creep which is accompanied with such phenomenon as creep delay.

4.25 Creep Delay

In contrast to generally accepted assumptions, in synthetic theory yield limit (strength) at some temperature is not constant for material, but it is depends on loading rate (Sec. 4.8, Fig. 4.9). Creep limit is taken for the material constant which depends on temperature only.

Consider the case of uniaxial tension, $\vec{\mathbf{S}}(S_1,0,0)$ and $S_1 = \sqrt{2}\sigma_x/\sqrt{3}$. In a virgin state, let us take the plane, *A*-plane, tangential to creep surface (4.8.4) which is perpendicular to S_1-axis. As the stress vector $\vec{\mathbf{S}}$ grows, the specified plane moves away from the surface (4.8.4) that is governed by Eq. (4.7.3) or (4.7.4) at $\varphi_N = 0$. The stress vector increase is not accompanied by irreversible straining until it reaches the first plane in its path, *A*-plane. The value of the module of the vector $\vec{\mathbf{S}}$ when it reaches the *A*-plane defines the yield limit, S_S, determined by Eqs. (4.8.7), (4.8.8), or (4.8.13). During further increase in σ_x, the stress vector shifts(moves) tangential planes and now their movements on the endpoint of the stress vector symbolize irreversible(plastic) strain increase. If the tensile stress σ_x beyond the yield limit is fixed and holds constant, the planes produce creep strain governed by Eq. (4.24.2).

Now, let the stress be fixed at such time t_M when the module of vector $\vec{\mathbf{S}}$ is larger than the radius of creep sphere, but does not reach the *A*-plane, i.e.,

$$S_P < S_1 < H_N(\alpha = \beta = \lambda = 0) \tag{4.25.1}$$

In this case, the irreversible strain does not occur at once after the termination of increase in stress, $\varphi_N(t_M) = 0$. However, as the integral of non-homogeneity decreases with time at constant stress, tangent planes, there is *A*-plane among them, come back to the creep surface so that their distances governed by, e.g., Eq. (4.7.3) at $\varphi_N = 0$ is

$$H_N = S_P + I_N, \quad \frac{dI_N}{dt} < 0. \tag{4.25.2}$$

At the instant, t_1, when the A-plane has found itself on the endpoint of the stress vector, $H_N(\alpha = \beta = \lambda = 0) = S_1 = S_S$, the developing of irreversible deformation begins. The time $t_1 - t_M$ is called the creep delay; its experimental observation is described in work [175]. The duration of creep delay grow with a loading rate, v, before the fixed stress σ_x has been achieved and the delay is absent at all at small, tending to zero, loading rate. The time t_1 is determined from Eq. (4.25.2) by letting in it $H_N = S_1 = S_S$:

$$S_P + v \int_0^{t_M} Q(t_1 - s) ds = S_S. \tag{4.25.3}$$

Due to the initial surface, creep surface has the form of sphere, Eq. (4.25.3) is readily generalized to an arbitrary state of stress.

Thus, the synthetic theory also models the creep strain for a stress-range $S_P < \left|\vec{\mathbf{S}}\right| < S_S$. As well known, the creep limit can be considerably smaller than the yield limit. Therefore, the synthetic theory is capable of describing the creep in elastic region, which starts after the creep delay.

4.26 Haazen-Kelly's Effect

Let a specimen be loaded in uniaxial tension to a given stress $\sigma_M > \sigma_S$ in time t_M, then the stress is held constant for some length of time, $t_M \le t \le t_1$, and for $t > t_1$ the loading process continue; the initial and subsequent portions of loading take place with equal loading rate, v. The stress-strain diagram corresponding to this loading regime is shown in Fig. 4.26, where the portion MM_1 represents creep deformation at $\sigma = \sigma_M = const$ and, beginning from strain value $\varepsilon(M_1)$, the stress grows again. The subsequent loading results in an interesting consequence: the strain response begins with an elastic portion, M_1M_4, on the stress-strain diagram. This phenomenon is referred to as Haazen-Kelly's effect [36]. The elastic portion length strongly depends on the magnitude of creep strain that has been accumulated at $\sigma = \sigma_M = const$.

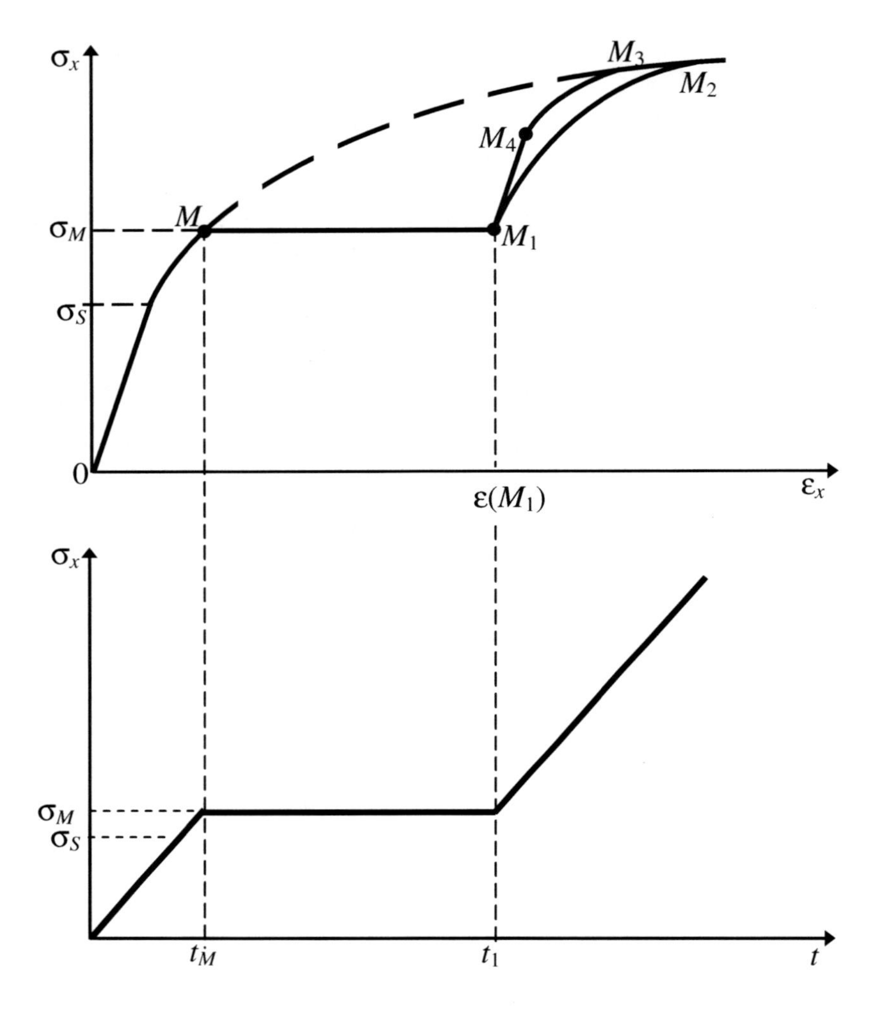

Fig. 4.26 Haazen-Kelly's effect

We wish to model this phenomenon in terms of the synthetic theory, adopting the linear form of function $\varphi_N\left(H_N\right)$. According to the problem formulation, the only stress vector component, $S_1=\sqrt{2}\sigma_x/\sqrt{3}$, is distinct from zero. Let us determine the irreversible strain intensity increment for infinitesimal small time-interval *dt* at the beginning of the second loading portion, at $t=t_1+dt$. On the base of Eqs. (3.9.8) and (4.7.3) we have

$$rd\varphi_N = (dS_1 - dI)\cos\xi, \quad \cos\xi = \cos\alpha\cos\beta\cos\lambda, \tag{4.26.1}$$

where I is given by Eq. (4.5.20) and at $t = t_1 + dt$ it takes the form

$$I + dI = v\int_0^{t_M} Q(t_1 + dt - s)ds + v\int_{t_1}^{t_1+dt} Q(t_1 + dt - s)ds. \tag{4.26.2}$$

If the core Q is exponential, we get

$$dI = B[1 - \exp(-p(t_1 - t_M)) + \exp(-pt_1)]dS_1. \tag{4.26.3}$$

In particular, if there is no constant stress-time-interval, $t_1 = t_M$, the above formula becomes

$$dI = B\exp(-pt_1)dS_1. \tag{4.26.4}$$

As one can see the right-hand side in Eq. (4.26.3) is greater than that in Eq. (4.26.4). Therefore, according to Eq. (4.26.1), the presence of time-interval with constant stress value leads to the $d\varphi_N$ decrease against the case when an acting stress monotonously grows along the whole loading. Thus, if the core Q is exponential, the Haazen-Kelly effect is described only partially, the plastic strain increment decreases in subsequent loading implying that the stress-strain diagram progresses in a way described by the OMM_1M_2 curve in Fig. 4.26.

If the core of the integral of non-homogeneity has a polar singularity, Eq. (4.26.2) becomes

$$I + dI = \frac{Bv}{1-p}\left[(t_1 + dt)^{1-p} - (t_1 - t_M + dt)^{1-p} + dt^{1-p}\right] \tag{4.26.5}$$

and at $t_1 = t_M$ we have

$$I + dI = \frac{Bv}{1-p}(t_1 + dt)^{1-p}. \tag{4.26.6}$$

By taking Taylor series expansion in the right-hand side of the above formula, we obtain

$$dI = B\frac{dS_1}{t_1^p}. \tag{4.26.7}$$

At the presence of constant-stress-time-interval, $t_1 > t_M$, from Eq. (4.26.5) it follows

$$dI = B\left[\frac{1}{t_1^p} - \frac{1}{(t_1 - t_M)^p} + \frac{1}{(1-p)dt^p}\right]dS_1. \tag{4.26.8}$$

For a period $t_M \le t \le t_1$ the creep strain develops on a set of planes locating on the endpoint of the stress vector, these planes are at the distances of

$$H_N = S_P + r\varphi_N + I\cos\xi = S_1\cos\xi \tag{4.26.9}$$

from the origin of coordinates. The presence or absence of irreversible strain increment at $t = t_1$ is determined by the relation between H_N and S_1, if $dH_N > dS_1$, the vector $\vec{\mathbf{S}}$ does not reach planes any more and the irreversible strain intensity φ_N acquires no increment meaning that an elastic strain will occur for some initial period of subsequent loading (the portion M_1M_4 in Fig. 4.26). This case is realized here as Eq. (4.26.8) gives that $dI >> dS_1$ as $dt \to 0$ and from Eq. (4.26.9) it follows that $\Delta H_N >> dS_1$. Thus, if the core of the integral of non-homogeneity has the form of Eq. (4.5.22), the synthetic theory is capable of describing Haazen-Kelly's effect shown by the $OMM_1M_4M_3$ curve in Fig. 4.26.

4.27 Creep with Variable Loading

As it was stated earlier, one of the basic conditions for the applicability of classical creep theories (Sec. 4.18) is the agreement between experimental and analytical results obtained in step-wise loading. Now we wish to model creep rate in terms of the synthetic theory for a more general case, variable loading with finite loading-rate. Let us consider the following problem. A specimen incurs creep strain due to the action of constant stress σ_1 for a time $t_2 - t_1$, the stress value is larger than creep limit, but less than yield limit. Then the stress increases to σ_2 and remains constant for $t > t_3$. The plot of the loading regime is shown in Fig. 4.27. Let us find the creep rate at $t = t_3 + 0$, i.e., when the stress σ_2 starts to act.

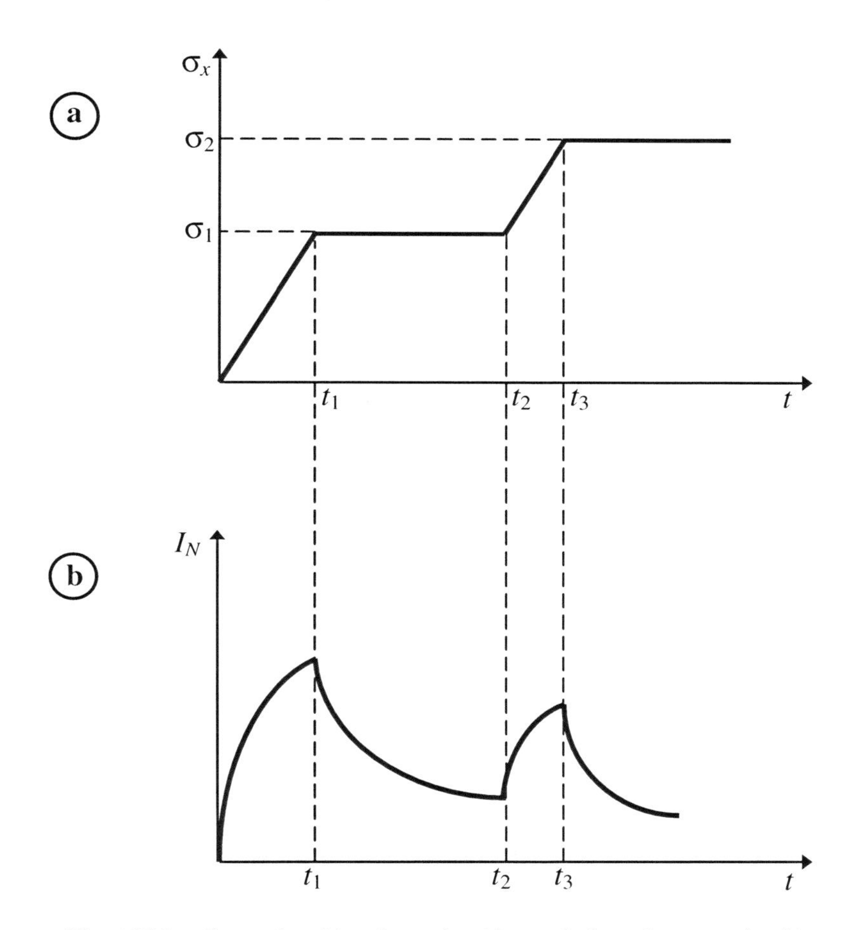

Fig. 4.27 Loading regime (a) and associated integral of non-homogeneity (b)

Since $\sigma_1 < \sigma_S$, the creep deformation starts developing not at $t = t_1$, but after some period of time (creep delay, see Sec. 4.25) and Eq. (4.7.3) transforms to the form

$$H_N = S_S + r\varphi_N + I_N, \quad S_S = \sqrt{2/3}\sigma_S \tag{4.27.1}$$

where the creep limit S_S is determined by Eq. (4.25.3). On the base of Eqs. (3.9.8), (4.6.4), and (4.27.1) the creep strain intensity at the action of σ_1, φ_{N1}, is

$$r\varphi_{N1} = \left\{ S_1 - \frac{Bv_1}{1-p}\left[t^{1-p} - (t-t_1)^{1-p}\right] \right\} \cos\xi - S_S ,$$
$$S_1 = \sqrt{2/3}\sigma_1 = v_1 t_1 \tag{4.27.2}$$

where t_1 is the time needed for the stress to grow from 0 to σ_1; v_1 is the loading rate. At $t = t_2$, we have

$$r\varphi_{N1}(t_2) = \left\{ S_1 - \frac{Bv_1}{1-p}\left[t_2^{1-p} - (t_2-t_1)^{1-p}\right] \right\} \cos\xi - S_S . \tag{4.27.3}$$

The stress growth from σ_1 to σ_2 induces plastic deformation and the action of constant stress σ_2 causes the development of creep deformation. The total (plastic + creep) strain intensity, φ_{N2}, can be found from the following formula

$$S_S + r\varphi_{N1}(t_2) + r\varphi_{N2} + I_N = S_2 \cos\xi, \quad S_2 = \sqrt{2/3}\sigma_2 . \tag{4.27.4}$$

The integral of non-homogeneity in the above formula is

$$I_N = B\left[v_1 \int_0^{t_1} \frac{ds}{(t-s)^p} + v_2 \int_{t_2}^{t_3} \frac{ds}{(t-s)^p} \right] =$$
$$= \frac{B}{1-p}\left\{ v_1\left[t^{1-p} - (t-t_1)^{1-p}\right] + v_2\left[(t-t_2)^{1-p} - (t-t_3)^{1-p}\right] \right\}\cos\xi \tag{4.27.5}$$

where v_2 is the loading rate for $t \in [t_2, t_3]$. Equations (4.27.3)–(4.27.5) give that for $t \geq t_3$

$$r\varphi_{N2} = \left\{ S_2 - S_1 + \frac{Bv_1}{1-p}\left[t_2^{1-p} - (t_2-t_1)^{1-p} - t^{1-p} + (t-t_1)^{1-p}\right] + \frac{Bv_2}{1-p}\left[(t-t_3)^{1-p} - (t-t_2)^{1-p}\right] \right\}\cos\xi \tag{4.27.6}$$

Equation (4.27.6) implies that

$$\dot{\varphi}_{N2} \to \infty \text{ as } t = t_3 + 0 \tag{4.27.7}$$

meaning that the creep strain rate tends to infinity. The obtained result is better agreed with experimental data (Fig. 4.21), than the prediction of the classical creep theories (Sec. 4.18).

Equation (4.27.7) follows from the property of the integral of non-homogeneity (4.27.5), namely $\dot{I}_N \to -\infty$ at $t = t_3 + 0$. The I_N against time diagram at $\cos\xi = 1$ is shown in Fig. 4.27b. At $t = t_2 + 0$ and $t = t_3 + 0$ the diagram is tangent to the vertical straight lines $t = t_2$ and $t = t_3$, respectively. It must be noted that the fact that $\dot{I}_N \to \pm\infty$ (Fig. 4.26) allows us to model both the Haazen-Kelly effect and the creep behavior at variable stresses considered in this section.

4.28 Generalization of the Cicala Formula

In terms of the concept of slip, the additional-loading modulus at orthogonal break of loading path, G_S, is determined by the Cicala formula, Eq. (2.5.11). The synthetic theory proposed in Chapter 3 leads to the same formula. This fact, on the one hand, shows the closeness of the predictions of these theories in complex stressed states that concerns to the positives of synthetic model. However, on the other hand, comparisons with experimental results show that the Cicala formula (2.5.11) gives low values for G_S.

Our goal is to determine the additional-loading shear modulus in terms of that version of the synthetic theory, see Sec. 4.7, where plastic properties of a body are influenced by loading-rate-hardening.

Consider the following problem. The element of a body is subjected to uniaxial tension, $\sigma_x = \sqrt{3}\,S_1/\sqrt{2}$. The component S_1 grows with constant velocity, $v_1 = \dot{S}_1$, from 0 and for some time t_M reaches a given value, $S_1 = v_1 t_M$ and $\sigma_M = \sqrt{3} v_1 t_M / \sqrt{2}$. Further, the constant stress σ_M acts on the element of body for the time-range $t_M \le t \le t_1$ and then, holding σ_M constant, an infinitesimal tangential stress $\Delta\tau_{xz} = \Delta S_3/\sqrt{2} = v_3 \Delta t/\sqrt{2}$ is applied in a time Δt, where v_3 is the additional loading rate. Let us find the shear strain component, $\Delta\gamma_{xz}$, due to the action of $\Delta\tau_{xz}$ and the ration $\Delta\tau_{xz}/\Delta\gamma_{xz}$ which is called the additional shear modulus.

To expand the arsenal of synthetic theory, we show another way to solve the given problem differing from that proposed in Sec. 3.12. Let us designate through $\Delta\varphi_N$ the irreversible strain intensity due to the additional loading ΔS_3. According to Eqs. (3.9.8) and (4.7.4), we have

$$r\Delta\varphi_N = \Delta\left(\vec{\mathbf{S}} \cdot \vec{\mathbf{m}}\right)^2 \cos^2\lambda - \Delta I_N^2 . \tag{4.28.1}$$

Similarly to Sections 3.11 and 3.13, we introduce an auxiliary system of coordinate, S_1', S_2', and S_3'. We put the S_3'-axis along the initial S_1-axis, the axes S_2 and S_2' are co-directed, and the S_1'-axis is introduced in the direction opposite to S_3-axis. Therefore, the stress vector in additional loading can be expressed as

$$\vec{\mathbf{S}} = S_1 \vec{\mathbf{g}}_1 + \Delta S_3 \vec{\mathbf{g}}_3 = -\Delta S_3 \vec{\mathbf{g}}_1' + S_1 \vec{\mathbf{g}}_3', \tag{4.28.2}$$

where $\vec{\mathbf{g}}_1$ and $\vec{\mathbf{g}}_3$, and $\vec{\mathbf{g}}_1{}'$ and $\vec{\mathbf{g}}_3{}'$ are the base vectors in S_1-S_3 and S_1'-S_3' coordinate plane, respectively. Equation (4.28.2), together with Eq. (3.13.11), gives

$$\vec{\mathbf{S}} \cdot \vec{\mathbf{m}} = S_1 \sin\beta - \Delta S_3 \cos\alpha \cos\beta \,, \tag{4.28.3}$$

where the angles α and β are measured similarly to Figs. 3.3 and 3.12. Neglecting the square of ΔS_3, we obtain

$$\Delta\left(\vec{\mathbf{S}} \cdot \vec{\mathbf{m}}\right)^2 = -S_1 \Delta S_3 \cos\alpha \sin 2\beta \,. \tag{4.28.4}$$

The integral of non-homogeneity (4.7.1) will be written as

$$I_N = B\left[v_1 \sin\beta \int_0^{t_M} Q(t_1 - s)ds - v_3 \cos\alpha \cos\beta \int_{t_1}^{t_1+\Delta t} Q(t_1 - s)ds\right]\cos\lambda \,. \tag{4.28.5}$$

If the core Q is exponential,

$$I_N = B\left[\frac{v_1}{p}\sin\beta \exp(-pt_1)(\exp(pt_M)-1) - \cos\alpha \cos\beta \Delta S_3\right]\cos\lambda \,. \tag{4.28.6}$$

Therefore,

$$\Delta I_N^2 = -\frac{B^2 v_1}{p}\exp(-pt_1)(\exp(pt_M)-1)\cos\alpha \sin 2\beta \cos^2\lambda \Delta S_3 \,. \tag{4.28.7}$$

Hence, the irreversible strain intensity increment (4.28.1) is

$$\begin{aligned} r\Delta\varphi_N &= -a_1 S_1 \Delta S_3 \cos\alpha \sin 2\beta \cos^2\lambda \,, \\ a_1 &= 1 - \frac{B^2 v_1}{S_1 p}\exp(-pt_1)(\exp(pt_M)-1) \,. \end{aligned} \tag{4.28.8}$$

In particular, if the initial loading occurs at enough high rate so that product pt_M is small compared to unit, approximating the term $\exp(pt_M)$ in Eq. (4.28.8) by a Tailor series approximation, we obtain

$$a_1 = 1 - B^2 \exp(-pt_1) \,. \tag{4.28.9}$$

Equations (4.28.8) and (4.28.9) determine the irreversible strain intensity increment due to the tangential stress increment $\Delta\tau_{xz}$.

Now we proceed to the determination of the strain increment Δe_3^S caused by the action of ΔS_3. According to the introduction of the auxiliary system of coordinate, S_1'- and S_3-axis have opposite orientations, therefore

$$\Delta e_3^S = -\Delta e_1'^S, \tag{4.28.10}$$

where $-\Delta e_1'^S$ is determined by Eq. (3.9.14) at $k = 1$. Hence, on the base of Eq. (3.5.4) we have

$$r\Delta e_3^S = -\int\limits_{\alpha}\int\limits_{\beta}\int\limits_{\lambda} \cos\alpha\cos^2\beta\cos\lambda\Delta\varphi_N d\alpha d\beta d\lambda \tag{4.28.11}$$

and, taking (4.28.8) into account, we obtain

$$r\Delta e_3^S = a_1 S_1 \Delta S_3 \int\limits_{\alpha}\int\limits_{\beta}\int\limits_{\lambda} \cos^2\alpha\sin 2\beta\cos^2\beta\cos^3\lambda d\alpha d\beta d\lambda. \tag{4.28.12}$$

Bearing in mind that the intensity $\Delta\varphi_N$ cannot take negative values, the range of the angle α is $\pi/2 \le \alpha \le 3\pi/2$. The boundary values of this diapason can be obtained from Eq. (3.13.18); letting $\Delta S_1 = 0$ and $\Delta S_3 \to 0$ in Eq. (3.13.18), which implies that $k \to \infty$, we obtain $\cos\alpha = 0$. The shaded semicircle in Fig. 3.7 (where $\sqrt{2}\tau_S$ is replaced by $\sqrt{2}\tau_P$) is the domain of movable planes in additional loading that produce the plastic strain increment Δe_3^S. Beyond this semicircle we have $(\vec{\mathbf{S}} \cdot \vec{\mathbf{m}}) < H_N$ at the action of ΔS_3 and an irreversible strain does not occur. Angles β and λ range in the same limits as in torsion. Indeed, half of the planes, which have incurred displacements in torsion and for which $\beta_1 \le \beta \le \pi/2$ and $0 \le \lambda \le \lambda_1$, continue to be translated by the stress vector at the additional loading ΔS_3.

It is convenient to rewrite the three-folded integral in Eq. (4.28.12) as an iterated integral,

$$r\Delta e_3^S = a_1 S_1 \Delta S_3 \int\limits_{\pi/2}^{3\pi/2} \cos^2\alpha d\alpha \int\limits_{0}^{\lambda_1} \cos^3\lambda d\lambda \int\limits_{\beta_1}^{\pi/2} \sin 2\beta\cos^2\beta d\beta. \tag{4.28.13}$$

The limit β_1 relates to λ as

$$\sin\beta_1 = \frac{a}{\cos\lambda} \tag{4.28.14}$$

and the parameter a is determined by Eqs. (4.9.9) and (4.6.2) which in the framework of the given problem takes the form

$$a^2 = \frac{S_P^2}{S_1^2 - I^2}, \quad I = \frac{Bv_1}{p}(\exp(pt_M) - 1)\exp(-pt_1). \tag{4.28.15}$$

Integrating in (4.28.13) gives

$$r\Delta e_3^S = \frac{\pi}{12} a_1 S_1 \Delta S_3 a^2 F(a). \tag{4.28.16}$$

Let us connect the increment Δe_3^S from the above formula with the component e_1^S in tension,

$$r e_1^S = \frac{\pi \tau_P^2}{3} F(a). \tag{4.28.17}$$

By eliminating the function $F(a)$ from Eqs. (4.28.16) and (4.28.17), we get

$$\Delta e_3^S = \frac{a_2}{2} e_1^S \frac{\Delta S_3}{S_1}, \tag{4.28.18}$$

where

$$a_2 = \frac{S_1 p \left\lfloor S_1 p - B^2 v_1 \exp(-pt_1)(\exp(pt_M) - 1) \right\rfloor}{S_1^2 p^2 - B^2 v_1^2 \exp(-2pt_1)(\exp(pt_M) - 1)^2}. \tag{4.28.19}$$

Converting the strain/stress vector components into strain/stress tensor components, Eq. (4.28.18) takes the form

$$\Delta\gamma_{xz}^S = \frac{3 a_2 \varepsilon_x^S}{2\sigma_x} \Delta\tau_{xz}. \tag{4.28.20}$$

If to write the residual strain components as the difference between total and elastic ones, see Eq. (2.5.9), we arrive at the following formula,

$$G_S = \frac{\Delta\tau_{xz}}{\Delta\gamma_{xz}} = \frac{G}{1 + \frac{3}{2} a_2 G \left(\frac{1}{E_S} - \frac{1}{E} \right)}, \tag{4.28.21}$$

which we will term the generalized Cicala formula. It differs from the Cicala formula, first, by the presence of the multiplier a_2 in the denominator of Eq. (4.28.21). Second, the relative elongation, appearing in the secant modulus $E_S = \sigma_M / \varepsilon_M$, consists of elastic, instantaneous plastic as well as creep components. The portion MM_1 in Fig. 4.28 represents creep strain at $\sigma_x = const$.

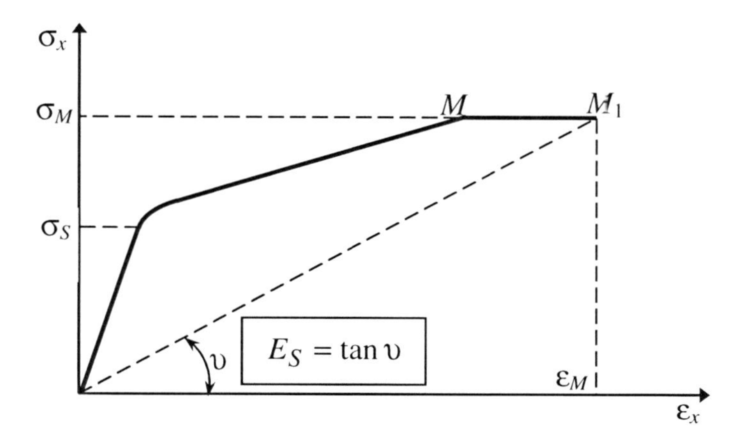

Fig. 4.28 Stress-strain diagram in uniaxial tension (segment MM_1 corresponds to creep deformation)

4.29 Analysis of the Generalized Cicala Formula

Let us analyze Eq. (4.28.21) in more detail. If the action of constant tensile stress lasts for the great enough length of time ($t_1 \to \infty$ and $\exp(-pt_1) \to 0$), we have $a_2 = 1$ and Eq. (4.28.21) becomes identical with the Cicala formula. At finite values of t_1 we have $a_2 < 1$ meaning that G_S calculated by Eq. (4.28.21) is larger than that obtained by the Cicala formula. If the velocity of the initial loading is large, $v_1 \to \infty$ ($t_M \to 0$),

$$a_2 = \frac{1 - B^2 \exp(-pt_1)}{1 - B^2 \exp(-2pt_1)}, \tag{4.29.1}$$

i.e., $a_2 < 1$ again.

Now, consider the case if the core of the integral of non-homogeneity has a polar singularity, see Eq. (4.5.22). Here, we suppose that the integral of non-homogeneity is of isotropic nature, i.e., it does not depend on angles α, β, and λ, namely

$$I = B\sum_{i=1}^{3}\int_{0}^{t}\frac{dS_i}{ds}\frac{ds}{(t-s)^p}. \tag{4.29.2}$$

In our case, under the orthogonal additional loading,

$$I = B\int_{0}^{t}\left(\frac{dS_1}{ds}+\frac{dS_3}{ds}\right)\frac{ds}{(t-s)^p}. \tag{4.29.3}$$

The components S_1 and S_3 grow in time-intervals $[0, t_M]$ and $[t_1, t_1+\Delta t]$, respectively, so that

$$I = B\left[v_1\int_{0}^{t_M}\frac{ds}{(t-s)^p}+v_3\int_{t_1}^{t_1+\Delta t}\frac{ds}{(t_1+\Delta t-s)^p}\right]. \tag{4.29.4}$$

Hence,

$$I = \frac{B}{1-p}\left\{v_1\left[t^{1-p}-(t-t_M)^{1-p}\right]+v_3\Delta t^{1-p}\right\} \tag{4.29.5}$$

and, because $v_3 = \Delta S_3/\Delta t$,

$$\Delta I^2 = \frac{2B^2}{(1-p)^2}v_1\left[t^{1-p}-(t-t_M)^{1-p}\right]\frac{\Delta S_3}{\Delta t^p}. \tag{4.29.6}$$

Letting $\Delta t \to 0$, we obtain

$$\Delta I^2 >> \Delta S_3. \tag{4.29.7}$$

The scalar product $\vec{\mathbf{S}}\cdot\vec{\mathbf{N}}$ due to the action of ΔS_3 changes by the amount of (see Eq. (4.28.3))

$$\Delta\left(\vec{\mathbf{S}}\cdot\vec{\mathbf{N}}\right) = -\Delta S_3\cos\alpha\cos\beta. \tag{4.29.8}$$

Equation (4.29.7) gives that $H_N >> \vec{\mathbf{S}}\cdot\vec{\mathbf{N}}$ at the action of ΔS_3. This inequality holds true for all values of angles α, β, and λ. Therefore, in the additional torsion, none of planes produce irreversible deformation, $\Delta\varphi_N \equiv 0$ and $\Delta e_3^S = 0$, and the additional modulus G_S is equal to the elastic modulus G.

There is no consensus among experimental researchers regarding the value of G_S, Sveshnikova [176] insists that $G_S < G$, whereas most researchers say that $G_S = G$. The generalized Cicala formula can lead to the both results.

4.30 Intermediate Discussion

1. The working out of creep theory shows that satisfactory agreements with experiments – at different temperatures, states of stress, complex loading, etc – is an extremely difficult problem. The problem of the mathematical description of creep in uniaxial tension stress can be considered, with certain reservations, as solved. In particular, the kinetic creep equations lead to results that good fit experimental curves in step constant loading.

2. For the case of a combined state of stress the situation is as follows. In time-independent loading, the comparison of the theory based on the hypothesis of existence of a creep potential to experimental results have showed satisfactory agreement. At the same time, if the directions of principal axes vary considerably with time, the creep potential theory gives results that contradict experimental data, this inconsistence is not only of quantitative but also of qualitative nature as well.

Therefore, the urgent problem in the creation of a more powerful new theory remains to be solved.

3. This chapter concerns with the generalization of synthetic theory for the modeling of unsteady-state creep and instantaneous plastic deformation influenced by loading rate. On the base of the physics of unsteady creep, we have introduced the integral of non-homogeneity, I_N, which depends on a loading-rate-history, into the formula defining a degree of strain-hardening, i.e., into the equation for the tangent planes distance. The presence of I_N in the formula for tangent planes distance, H_N, governs this distance as follows:

a) H_N grows with an increase in the module of stress vector;

b) H_N grows with an increase in loading rate;

c) H_N decreases at constant and decreasing stress.

4. According to the initial formulation of the synthetic theory, Chapter 3, planes tangential to yield surface can move parallel to themselves if they are shifted("pushed") by stress vector (i.e., if they are located at the point of loading). The introduction of the integral of non-homogeneity considerably changes this statement:

(i) Now, during an increase in acting stress, planes can move without being at point of loading. Indeed, according to Eq. (4.6.1) or (4.6.3), the integral of non-homogeneity grows due to the increase in stress thereby leading to the increase in the plane distance (the planes move away) that is governed by Eq. (4.7.3) or (4.7.4). The plain movements are not accompanied by irreversible strain increments until the stress vector reaches a tangential plane. This is due to the fact that the basic thesis of the synthetic theory – ***only those planes which are reached by stress vector $\vec{\mathbf{S}}$ at a given instant produce irreversible deformation, independently of whether the stress vector varies or remain unchangeable in,***

e.g., time; and vice versa, if planes are not located at the endpoint of the stress vector they do not take part in the developing of irreversible strain – remains valid. The stress value when the endpoint of stress vector $\vec{\mathbf{S}}$ reaches a first plane is defined as the yield limit; the latter strongly depends on loading rate. Similar to Chapter 3, the increments in irreversible deformation are calculated for each plane separately and added.

(ii) In a virgin state, the planes are tangential not to a yield surface, but to a creep surface. The creep limit is defined as the yield limit of a material at small (infinitesimally small) loading rate.

5. We have managed to describe analytically stress-strain diagrams in uniaxial tension diagrams at high temperatures. They considerably differ from those obtained at room temperature when the influence of loading rate, v, is immaterial. Yield limit at elevated temperatures is not a material constant but strongly influenced by v. In addition, the whole stress-strain diagram depends on v: the smaller v, the closer the diagram lies to strain-axis.

The integral of non-homogeneity is a decisive factor regulating high-temperature stress-strain diagrams.

6. The synthetic theory is capable of describing the effect (see Fig. 4.16) when the increment in irreversible strain occurs during unloading. This phenomenon fully depends on the unload-velocity.

7. The experimental verification of the theories of plasticity in alternating-cyclic loading is considered as the fullest one as the straining-path direction repeatedly changes during such type of loading. That is why we pay a special attention to the cyclic properties of material.

Owing to the integral of non-homogeneity, we have managed to model the behavior of cyclically hardening, softening, and stabilizing materials. The cyclic stabilization, when the stress-strain curves become unchangeable after the certain number of cycles, has been described for both hardening and softening materials. The case when the stabilization occurs if the hysteresis loop tends to zero (stress-strain relationships becomes linear) has also taken into account. If to neglect the parameter of non-homogeneity, the synthetic theory gives the diagrams of cyclically ideal materials.

8. One of the main parts that the integral of non-homogeneity plays is the modeling of unsteady-state creep. After the stress becomes constant, the integral of non-homogeneity starts decreasing that leads to an increase in irreversible strain intensity, i.e., an unsteady creep develops. Unsteady creep strain increments are "produced" by the family of immovable planes that are located on the endpoint of stress vector. The synthetic theory takes into account that unsteady creep is strongly influenced by the velocity of the load preceding steady stress, v. The increase in v leads to the reduction in instantaneous plastic deformation, whereas the creep strain grows. At the same time, the sum of the specified deformations does not depend on v, provided that the constant-stress-duration is large enough. It means that a material possesses certain resource of irreversible straining which depends on stress-value.

The phenomenon of creep delay, when the development of unsteady creep starts not immediately as a constant stress acts, but after a certain period of time, has been allowed for.

9. In terms of the synthetic theory, the formulae for strain components are derived from the unique set of equations irrespective of whether a load progresses or remains constant in time. The key point is that we use a single notion, ***irreversible deformation***, – ***we do not split a deformation into its plastic and viscous parts***.

10. Unsteady state creep in step-wise loading has been described. Taking the polar core of the integral of non-homogeneity the synthetic theory gives that the creep rate tends to infinity due to a jump-increase in stress. This result is in better accord with experiments, than in terms of the classical theories.

11. In contrast to the theories founded on the hypothesis of a potential of creep, the synthetic model does not accept an universal relationship $\dot{\gamma}_0 = \dot{\gamma}_0(\tau_0(t), t)$ for different states of stress; a loading history strongly affects the behavior of creep diagram.

12. The generalized Cicala formula has been derived, which gives the larger values for G_S, compared to G_S calculated by the classical Cicala formula, meaning the better agreements with experiments. The usage of the core of integral of non-homogeneity with a polar singularity leads to that the additional shear modulus G_S is equal to elastic shear modulus, which is observed in experiments on many materials.

The great number of non-classical problems solved successfully in terms of the synthetic and the possibility to model plastic and creep straining in a complex way makes the synthetic theory a perspective one.

4.31 Physics of Steady-State Creep

On the base of numerous crystallographic experiments we can conclude the following.

1. Stationary creep rate depends only on temperature and stress value. The metallographic structure and creep-strain rate of specimens in steady-state creep is practically identical at equal stress and temperature.

2. Steady creep kinetics is controlled by the nucleation, movement (thermally activated glide, e.g., via cross-slip or more often observed climb assisted glide), and interaction of dislocations [33, 34,59,104,138]. The huge amount of experiments confirm to the dislocation nature of steady creep meaning that the main mechanism of irreversible strain is slips within crystal grains. Therefore, the synthetic theory, as well as the concept of slip, is applicable to the analytical description of steady-state creep.

3. Diffusion-controlled creep is observed only at the temperatures near melting point; further throughout, creep at such temperatures will be not considered. On the other hand, it is worthwhile to note that the diffusion of atoms and vacancies can assist the movements of dislocations; it will be discussed in more detail in Sec. 4.36.

Bailey has offered to consider steady-state creep as a result of the balance between work-hardening and thermally activated recovery (softening) processes. At present this thesis is accepted by many researchers [1,21,59,164,178].

4. The nature of work-hardening in creep is the same as in plastic deformation. The increase in the number of dislocations results in the formation of the three-dimensional network of tangled dislocations and other defects that act as obstacles to dislocation motion. However, at high temperatures vacancies in the crystal can diffuse to the location of a dislocation and cause the dislocation to move to an adjacent slip plane. By climbing to adjacent slip planes dislocations can get around obstacles to their motion, allowing further deformation to occur. In other words, thermally activated recovery (softening) processes are observed. Because it takes time for vacancies to diffuse to the location of a dislocation this results in time-dependent strain, or creep.

Another example of recovery is the breakdown of polygonal sub-grain boundary. As shown in Sec. 4.22, during unsteady creep, dislocations slip or climb to the cell (sub-grain or block) walls, i.e., the polygonization develops. Dislocations cannot easily cross sub-grain boundaries and, as a consequence, stop their propagations. Once the boundaries break down the dislocations can continue their motions through the grain and make a contribution to creep strain progress. The breakdown of subgrains has been often observed in high-temperature creep experiments on many materials [93,94,173]. The steady-state creep is characterized by the equilibrium between the formation (hardening) and breakdown (softening) of polygonal subgrains.

4.32 Generalized Synthetic Theory of Irreversible Deformation

Since, as stated above, the main mechanism of both instantaneous plastic and unsteady/steady-state creep deformation in a polycrystalline body is shears (slips) within crystal grains, we perform the modeling of irreversible deformation in terms of the synthetic theory which is based (at least regarding physical aspect of irreversible straining) on the concept of slip.

The basic equation of synthetic theory, Eq. (3.9.4), determining irreversible strain vector components remains valid. Another unchangeable point of synthetic theory is the fact that a tangent plane distance, H_N, symbolizes a strain hardening of material.

In what follows, we propose other way to determine an irreversible strain intensity, φ_N, which appears in the integrand in Eq. (3.9.4), so that Eq. (3.9.4) gives both plastic and unsteady/steady creep strain components. The key point is that we

(i) introduce a new quantity, ***defects intensity***, ψ_N, and

(ii) establish the relationship between the plane distance and defects intensity, instead of the $\varphi_N = \varphi_N(H_N)$ from (4.7.3) or (4.7.4).

Any crystal lattice imperfections caused by non-elastic deformations that obstruct the increment in irreversible straining we be called the defects of crystal structure. The defects intensity is an average scalar continuous measure of defects inherent in real bodies (dislocations, vacancies, etc.). Therefore, the object whose irreversible deformation we model is a continuous medium experiencing the nucleation and evolution of defects during irreversible straining. In a virgin state the defects intensity is assumed to be equal to zero. We deliberately disregard an actually type of defects, important the fact that irreversible straining is inevitably accompanied (or caused) with defects of crystal lattice which impede the progress of irreversible deformation.

Similarly to φ_N, we define ψ_N as a function angles α, β, and λ determining the orientation of vector $\vec{\mathbf{N}}$ normal to a tangent plane. Since in terms of synthetic theory every tangent plane corresponds to an appropriate slip system, ψ_N symbolizes the defects intensity generated in a given slip system. Further throughout, we will use the expression as "defects intensity on a given plane" or "defects intensity generated on some plane". Furthermore, we suppose that ψ_N is a time-dependent quantity. So, finally, $\psi_N = \psi_N(\alpha, \beta, \lambda, t)$. Let us designate through $d\psi_N$ and $d\varphi_N$ an increment of ψ_N and φ_N in time dt for a fixed plane, i.e., at $d\alpha = d\beta = d\lambda = 0$. According to the well-known fact that the defects of crystal structure of a metal are the *carriers* of irreversible deformation, we relate the strain intensity to the defects intensity by the following equation [137,143,149]

$$\boxed{d\psi_N = r d\varphi_N - K\psi_N dt}, \tag{4.32.1}$$

where *K* is a function of shear stress intensity and homological temperature, Θ, that regulates steady-state creep rate. As long as in work [1], it was noted that the relations of creep theories should include the homology temperature, i.e., the current and melting temperature ratio. The form and physical significance of the function *K* will be discussed in more details in Sec. 4.36 and 4.37.

Equation (4.32.1) gives that as an irreversible strain occurs, $d\varphi_N > 0$, the terms $d\psi_N > 0$ symbolize the increase in the number of defects and the negative term $-K\psi_N dt$ indicates on their simultaneous relaxation. Consider the particular forms of Eq. (4.32.1).

(A) In the case of fast plastic deforming, when the duration of loading is usually small, $dt \to 0$, Eq. (4.32.1) takes the form

$$d\psi_N = r d\varphi_N . \tag{4.32.2}$$

This equation shows that the defects develop on those planes that produce an irreversible deformation. In particular, if some plane does not produce the irreversible deformation increment, $d\varphi_N = 0$, then $d\psi_N = 0$. The latter is fully coordinated with experiments: if slip systems of a crystal grain (presented by tangential planes in terms of synthetic theory) are oriented relative acting stresses so that there are no plastic shears in them, no structural changes are observed in these slip systems.

(B) After complete or partial unloading, when the increment in irreversible deformation is terminated, $d\varphi_N = 0$, Eq. (4.32.1) is

$$d\psi_N = -K\psi_N dt, \quad K = -\frac{\dot{\psi}_N}{\psi_N}. \tag{4.32.3}$$

The first formula in (4.32.3) describes defects relaxation (see Sec. 4.31) – the breakdown of subgrains, the reduction of the efficiency of obstacles to impede dislocation motions, the annihilation of dislocations of opposite signs, etc. Equation (4.32.3) is frequently applied to the description of relaxation processes in creep. In work [122], the process which is described by Eq. (4.32.3) is called the thermally activated softening. The second formula in Eq. (4.32.3) gives the physical significance of the function *K* – defects relaxation-rate related to their number.

As it was stated earlier, we relate a tangent plane distance, H_N, to defects intensity and not to strain intensity. We propose the following formulae

$$H_N = S_P + \psi_N + I_N \tag{4.32.4}$$

or

$$H_N = \sqrt{S_P^2 + \psi_N + I_N^2} \tag{4.32.5}$$

which are obtained from Eq. (4.7.3) or (4.7.4) by the substitution of ψ_N for φ_N.

It is worthwhile to repeat once more the basic point of synthetic theory: at a given instant a plane produces an irreversible strain if the stress vector, $\vec{\mathbf{S}}$, reaches this plane, i.e., Eq. (3.9.8) holds,

$$H_N = \vec{\mathbf{S}} \cdot \vec{\mathbf{N}}. \tag{4.32.6}$$

If the stress vector does not reach a plane ($H_N > \vec{\mathbf{S}} \cdot \vec{\mathbf{N}}$), this plane does not produce an irreversible strain, $d\varphi_N = 0$.

Similar to Eqs. (4.7.5) and (4.7.6), we introduce the equation for the distance between the origin of coordinates and tangential planes with normal vectors $-\vec{\mathbf{N}}$,

$$H_{-N} = S_P + \psi_{-N} + I_{-N} \tag{4.32.7}$$

or

$$H_{-N} = \sqrt{S_P^2 + \psi_{-N} + I_{-N}^2}\,. \tag{4.32.8}$$

We suppose that the defect intensities on oppositely oriented planes are related to each other as

$$\psi_{-N} = -\psi_N\,. \tag{4.32.9}$$

It must be noted that one should supplement Eq. (4.32.9) by Eq. (3.17.1), otherwise Eq. (4.32.1) would not be of general nature, i.e., valid for an arbitrary direction.

The physical significance of Eq. (4.32.9) is as follows. During unloading, repulsive forces, which act within dislocations conglomeration generated in initial loading, induce the dislocation to move in the direction opposite to acting stress, thereby changing the strain direction and hastening the onset of the plastic deformation of opposite sign. In terms of the Ishlinsky model, the part of the repulsive forces is played by the force which the left spring in Fig. 1.17 exerts. As we have already discussed in Sec. 4.2, when working out of a model it is not necessary to take these forces into account in explicit form. In terms of the synthetic theory, Eqs. (4.7.5) and (4.7.6) or (4.32.7)–(4.32.9) allow for the specified forces that are sufficient for the description of deformation in alternate loading. Equations (4.32.7)–(4.32.9) or (4.7.5) and (4.7.6) allow us to model the Bauschinger effect. In the past, this effect was supposed to be caused by non-even stress distribution over crystal grains in polycrystalline aggregate. However, experiments on *mono-crystals* have denied this point of view showing that the Bauschinger effect on mono-crystals in alternate loading is observed regularly.

(C) Operating on two functions, φ_N and ψ_N, allows us, together with plastic and primary creep deformation, to describe steady-state creep as well. Indeed, under the action of constant stress the integral of non-homogeneity decreases and at some time its value can be neglected. The condition $I_N = 0$ symbolizes the end of transformations occurring in the crystal lattice under primary creep and the transition to the steady-state stage of creep. Therefore, Eqs. (4.32.4) and (4.32.6) at $\vec{\mathbf{S}} = const$ and $I_N = 0$ give that that $\psi_N(t) = const$, i.e., $d\psi_N = 0$. Consequently, Eq. (4.32.1) is

$$r d\varphi_N = K\psi_N dt \quad \text{or} \quad r\dot{\varphi}_N = K\psi_N \tag{4.32.10}$$

and

$$r\varphi_N = r\overset{0}{\varphi}_N + K\psi_N t, \tag{4.32.11}$$

where $r\overset{0}{\varphi}_N$ is the irreversible strain intensity accumulated prior to the steady-state creep; the time t in Eq. (4.32.11) is counted from the instant when the integral of non-homogeneity is taken as zero. The obtained intensity φ_N relates linearly to time t, hence, it describes the stationary stage of creep. The condition $\psi_N(t) = const$ symbolizes a balance between nucleated and relaxed defects in steady creep.

If to let $K = 0$, we arrive at Eq. (4.32.2), $\psi_N = r\varphi_N$, and the equations for distance, Eq. (4.32.4) or (4.32.5), degenerate into Eq. (4.7.3) or (4.7.4), respectively. Therefore, we return to the version of the synthetic theory formulated only for the description of instantaneous plastic and unsteady creep deformation.

According to Eqs. (4.32.10) and (3.9.4), the function K regulates steady-state creep rate which, as well known from experiment, takes small value. Therefore, we can conclude that the function K takes small values as well and the term $K\psi_N t$ in Eq. (4.32.11) exerts essential influence for a long period of time only.

Summarizing, formulae (3.9.4), (3.17.1), (4.5.21) or (4.5.22), (4.7.1), (4.32.1), and (4.32.4)–(4.32.9) constitute the family of the basic equations of the generalized synthetic theory of irreversible deformation:

Macro-level			
$e_k^S = \int\limits_{\alpha}\int\limits_{\beta}\int\limits_{\lambda} \varphi_N m_k \cos\beta \cos\lambda \, d\alpha \, d\beta \, d\lambda$	**(3.9.4)**		
Micro-level			
$d\psi_N = r d\varphi_N - K\psi_N dt$	**(4.32.1)**		
$H_N = S_P + \psi_N + I_N$	**(4.32.4)**	$H_N = \sqrt{S_P^2 + \psi_N + I_N^2}$	**(4.32.5)**
$H_N = \vec{\mathbf{S}} \cdot \vec{\mathbf{N}}$	**(4.32.6)**		
$I_N = \int\limits_0^t \frac{d\vec{\mathbf{S}}}{ds} \cdot \vec{\mathbf{N}} Q(t-s) ds$	**(4.7.1)**		
$Q = B\exp(-p(t-s))$ or	**(4.5.21)**		
$Q = \frac{B}{(t-s)^p}$	**(4.5.22)**		
$H_{-N} = S_P + \psi_{-N} + I_{-N}$	**(4.32.7)**	$H_{-N} = \sqrt{S_P^2 + \psi_{-N} + I_{-N}^2}$	**(4.32.8)**
$\psi_{-N} = -\psi_N$	**(4.32.9)**		
$\varphi_{-N} = -\varphi_N$	**(3.17.1)**		

Now, for the case of arbitrary (curvilinear) loading paths, we wish to formulate the criterion for the developing of irreversible straining. Let the current stress vector $\vec{\mathbf{S}}$ have produced irreversible strain, i.e., some set of tangent planes are on the endpoint of this vector. For these planes Eqs. (4.32.4) and (4.32.6) give

$$S_P + \psi_N + I_N = \vec{\mathbf{S}} \cdot \vec{\mathbf{N}}, \tag{4.32.12}$$

If the vector $\vec{\mathbf{S}}$ acquires increment $d\vec{\mathbf{S}}$, for planes that are on the endpoint of vector $\vec{\mathbf{S}} + d\vec{\mathbf{S}}$ we have

$$S_P + \psi_N + d\psi_N + I_N + dI_N = \vec{\mathbf{S}} \cdot \vec{\mathbf{N}} + d\vec{\mathbf{S}} \cdot \vec{\mathbf{N}}. \tag{4.32.13}$$

Therefore, together with Eq. (4.32.12), we obtain that

$$d\psi_N = d\vec{\mathbf{S}} \cdot \vec{\mathbf{N}} - dI_N. \tag{4.32.14}$$

We propose the following criterion – the planes that produce irreversible strains under the vector $\vec{\mathbf{S}}$ continue to do this under the vector $\vec{\mathbf{S}} + d\vec{\mathbf{S}}$ if $d\psi_N \geq 0$ for these planes:

$$d\vec{\mathbf{S}} \cdot \vec{\mathbf{N}} - dI_N \geq 0. \tag{4.32.15}$$

If a plane distance is given by Eq. (4.32.5), we have

$$\left(\vec{\mathbf{S}} \cdot \vec{\mathbf{N}}\right) d\vec{\mathbf{S}} \cdot \vec{\mathbf{N}} - I_N dI_N \geq 0. \tag{4.32.16}$$

This is inequality (4.32.15) or (4.32.16) that must be applied to the determination of integration limits in Eqs. (3.9.4) and (3.9.14)

As seen from Eq. (4.32.4) or (4.32.5), a plane distance depends on two quantities: the defects intensity and the integral of non-homogeneity. The sum $\psi_N + I_N$ or $\psi_N + I_N^2$ has a simple physical significance, it characterizes the straining state of material and determines the stress value to be applied for the onset/progress of irreversible deformation. The parameters ψ_N and I_N have a common trait, they can relax in time under some conditions. On the other hand, there is an essential difference between these quantities. The defects intensity ψ_N expresses the number of defects producing irreversible deformation, while the integral I_N characterizes the obstacles to develop the deformation. The integral I_N behaves in a various way depending on a loading regime: a) under fast loading – the greater the loading-rate, the greater number of tangled and locked dislocations – the integral I_N symbolizes the load-rate strengthening of material; b) under constant stresses – favorable conditions arise to unlock the dislocations from their obstruction and rearrange them in the low-energy dislocation configurations (polygonization), – I_N drops expressing the relaxation of lattice distortion and, consequently, the time-dependent progressive development of irreversible deformation. The behavior of ψ_N and I_N is governed by different equations, I_N depends on a loading-rate history, Eq. (4.7.1), whereas ψ_N, in general case, is related to irreversible deformation by Eq. (4.32.1).

The difference between ψ_N and I_N is especially vivid in unsteady-state creep. The decrease in I_N symbolizes the reduction of strain-rate-hardening due to the decrease in a crystal lattice distortion caused by the mechanism described above (unlocking of dislocation tangles, polygonization, etc.). At the same time, dislocations themselves remain in body and are capable of producing creep strain. Therefore, in contrast to I_N, the average characteristic of defects, ψ_N, grows in unsteady-state creep and becomes steady in steady creep.

Finally, let us specify the units of K and ψ_N. If a plane distance is governed by Eq. (4.32.4), $[\psi] = \text{Pa}$. If we use Eq. (4.32.5), $[\psi] = \text{Pa}^2$. For the both cases $[K] = \sec^{-1}$.

4.33 Irreversible Deformation in Pure Shear

Let us apply the equations presented in the preceding Section for the determination of irreversible strain in pure shear, $\tau_{yz} = S_3/\sqrt{2}$ and $S_1 = S_2 = 0$. The component $S_3 = |\vec{S}|$ is an arbitrary function of time, $dS_3/dt \geq 0$.

Before the yield limit of material is reached, there are no strain intensity and its increment, $\varphi_N = 0$ and $d\varphi_N = 0$. Therefore, Eq. (4.32.1) degenerates in Eq. (4.32.3) whose solution is $\psi_N = 0$ due to there are no defects in a virgin-state body. Consequently, Eq. (4.32.4) gives the plane distance as

$$H_N = S_P + I_N \,. \tag{4.33.1}$$

This relationship can be obtained from Eq. (4.7.3) at $\varphi_N = 0$ as well, therefore the yield limit, S_S, is determined from Eq. (4.8.7) or (4.8.8). When the acting stress exceeds the yield limit, $S_3 > S_S$, irreversible deformation starts developing and Eqs. (4.32.4) and (4.32.6) give that

$$\begin{gathered} \psi_N = (S_3 - I)\cos\xi - S_P, \quad d\psi_N = (dS_3 - dI)\cos\xi, \\ \cos\xi = \sin\beta\cos\lambda, \end{gathered} \tag{4.33.2}$$

$$I = B\int_0^t \frac{dS_3}{ds} Q(t-s)ds\,.$$

Defect intensity from Eq. (4.33.2) is positive if

$$\beta_1 \le \beta \le \pi/2\,,\quad 0 \le \lambda \le \lambda_1\,,$$
$$\cos\lambda_1 = \frac{\sin\beta_1}{\sin\beta}\,,\quad \sin\beta_1 = a = \frac{S_P}{S_3 - I}\,. \tag{4.33.3}$$

Because Eq. (4.32.1), the irreversible strain intensity increment is written as:

$$r d\varphi_N = (dS_3 - dI)\cos\xi + S_P K\left(\frac{\cos\xi}{a} - 1\right)dt\,. \tag{4.33.4}$$

Beyond the angles-diapason given by (4.33.3) we have

$$\psi_N = d\psi_N = 0\,,\quad \varphi_N = d\varphi_N = 0\,. \tag{4.33.5}$$

Finally, Eq. (3.9.14) gives the total irreversible strain-vector component increment, Δe_3^S, as

$$\Delta e_3^S = \frac{1}{2r}\int_0^{2\pi} d\alpha \int_{\beta_1}^{\pi/2} \sin 2\beta d\beta \int_0^{\lambda_1} \Delta\varphi_N \cos\lambda d\lambda\,, \tag{4.33.6}$$

when $\Delta\varphi_N$ is given by Eq. (4.33.4). Here and further within this section we introduce the symbol Δ to distinguish the strain/strain-intensity increment due to the stress and time increment from the variables over which the integration is carried out. By the integration over α, β, and λ in (4.33.6), we obtain

$$\Delta e_3^S = -\frac{\pi}{3r} a^2 F'(a)(\Delta S_3 - \Delta I) + \frac{\pi S_P}{3r} K F(a)\Delta t\,, \tag{4.33.7}$$

where $F'(a) = dF/da$ and the function F is determined by Eq. (3.10.14). On the basis of Eq. (4.33.3), the formula (4.33.7) takes the form

$$\Delta e_3^S = a_0\left[F'(a)\Delta a + K F(a)\Delta t\right], \tag{4.33.8}$$

where the constant a_0 is given by Eq. (4.9.8). By integrating over time in Eq. (4.33.8), we obtain the formula for the irreversible strain component in pure shear:

$$\cdot e_3^S = a_0\left[F(a) + \int_{t_S}^{t} KF(a)dt\right], \tag{4.33.9}$$

where t_S is the time when the acting stress has achieved the yield limit. To evaluate the integral in Eq. (4.33.9), one needs to know the form of the function $K(S_3(t))$. However, as the function K takes small values (see the discussion after Eq. (4.32.11)) we can neglect the time integral in Eq. (4.33.9) proviso typical loading rates, when time-interval $[t_S, t]$ is relatively small. In this case we arrive at Eq. (4.9.8). On the other hand, it is important to emphasize that the role of the time integral in (4.33.9) grows with increase in acting-stress-time.

Equation (4.33.9) is of generalized character; it is applicable for the modeling of any deformation – both plastic and unsteady/steady state creep.

Summarizing, the procedure of the calculation of irreversible deformation is the following: (i) by integrating of Eq. (3.9.14) with respect to angles α, β, and λ, we find the strain rate vector components $\dot{e}_i$; (ii) to obtain e_i components, we integrate the obtained $\dot{e}_i$ with respect to time.

4.34 Creep Deformation in Pure Shear

The procedure to calculate creep strain component (both unsteady and steady) in pure shear in terms of the generalized synthetic theory is outlined in this section. In contrast to the preceding Section, when determining strain vector components we change the integration order, first, we integrate $d\varphi_N$ with respect to time (as a result we know φ_N) and then e_i components are calculated by integrating with respect to α, β, and λ in Eq. (3.9.4).

Let a loading regime be such as shown in Fig. 4.29a. Consider the case if the relation for a plane distance, H_N, is given by Eq. (4.32.5). Equating the H_N to scalar product $\vec{\mathbf{S}} \cdot \vec{\mathbf{N}}$, we obtain the following equation for defect intensity:

$$\psi_N = \left(S^2 - I^2\exp(-2pt)\right)\cos^2\xi - S_P^2,$$
$$S = S_3 = const, \quad I = \frac{Bv}{p}\left(\exp(pt_M) - 1\right), \tag{4.34.1}$$

which is valid for the angles β and λ giving the positive values of ψ_N. The boundary values of β and λ are determined by Eq. (4.33.3) in which the time-dependent factor a has the form

$$a = a(t) = \frac{S_P}{\sqrt{S^2 - I^2 \exp(-2pt)}}. \tag{4.34.2}$$

From Eq. (4.34.1) we obtain

$$d\psi_N = 2pI^2 \exp(-2pt)\cos^2 \xi dt. \tag{4.34.3}$$

Therefore, Eq. (4.32.1) gives the irreversible strain intensity increment as

$$rd\varphi_N = 2pI^2 \exp(-2pt)\cos^2 \xi dt + K\left[\left(S^2 - I^2 \exp(-2pt)\right)\cos^2 \xi - S_P^2\right]dt \tag{4.34.4}$$

To integrate the above relationship, we split the range of angles β and λ into two domains. The first of them corresponds to the termination of the increase in stress, $t = t_M$ (Fig. 4.29a),

$$a_M \le \cos \xi \le 1 \tag{4.34.5}$$

and is marked by **I** in Fig. 4.29b. The second domain, $S_3 = S = const$ as $t > t_M$ (Fig. 4.29b), is

$$a(t) \le \cos \xi \le a_M, \tag{4.34.6}$$

region **II** in Fig. 4.29b. If to designate the boundary values of β and λ for the region **I** and **II** as β_1 and β_2, and λ_1 and λ_2, respectively, we can write down that

$$\sin \beta_1 = \cos \lambda_1 = a_M \quad \text{and} \quad \sin \beta_2 = \cos \lambda_2 = a(t), \tag{4.34.7}$$

where

$$a_M = \frac{S_P}{\sqrt{S^2 - I^2 \exp(-2pt_M)}}. \tag{4.34.8}$$

As seen from Eqs. (4.34.2), (4.34.5), (4.34.6), and (4.34.8), the domain **II** is the expansion of the domain **I** as $t > t_M$. Taking the time integral of Eq. (4.34.4) for $t > t_M$, the creep strain intensity developed in the region **I** is obtained as

$$r\varphi_{N1} = I^2\left(\exp(-2pt_M) - \exp(-2pt)\right)\left(1 - \frac{K}{2p}\right)\cos^2 \xi + K\left(S^2 \cos^2 \xi - S_P^2\right)(t - t_M) \tag{4.34.9}$$

Now, let us determine the instant, t_0, when a plane from the region **II** starts producing creep strain. From the inequalities (4.34.6) it follows that

$$a(t_0) = \cos \xi \tag{4.34.10}$$

and, in view of Eq. (4.34.2), we obtain

$$t_0 = \frac{1}{2p} \ln \frac{I^2 \cos^2 \xi}{S^2 \cos^2 \xi - S_P^2}. \qquad (4.34.11)$$

The range of time t_0 is

$$t_M \le t_0 \le t \qquad (4.34.12)$$

and the lower value t_M corresponds to the boundary curve **2** and the upper value is related to the current boundary curve **1** (Fig. 4.29b).

Since the creep deformation is produced by planes from the region **II** for the period of time from t_0 to t, the integration in Eq. (4.34.4) for the range $t \in (t_0, t)$ gives the creep strain intensity:

$$r\varphi_{N2} = I^2(\exp(-2pt_0) - \exp(-2pt))\left(1 - \frac{K}{2p}\right)\cos^2 \xi + K\left(S^2 \cos^2 \xi - S_P^2\right)(t - t_0). \qquad (4.34.13)$$

In view of Eq. (4.34.11), the above formula can be written as

$$r\varphi_{N2} = -I^2 \exp(-2pt)\left(1 - \frac{K}{2p}\right)\cos^2 \xi + \left[1 + K\left(t - t_0 - \frac{1}{2p}\right)\right]\left(S^2 \cos^2 \xi - S_P^2\right) \qquad (4.34.14)$$

From (4.34.9), (4.34.11), (4.34.12), and (4.34.14) it is clear that the intensity φ_N changes continuously at the transition from the region **I** to **II**. A total creep strain, unsteady and steady creep strain, is determined by the integration of φ_{N1} and φ_{N2} over the regions **I** and **II**,

$$\begin{aligned} e_3^P &= \pi \int_{\beta_2}^{\pi/2} \sin 2\beta d\beta \int_0^{\lambda_2(\beta)} \varphi_{N1} \cos \lambda d\lambda + \\ &+ \pi \int_{\beta_1}^{\beta_2} \sin 2\beta d\beta \int_0^{\lambda_1(\beta)} \varphi_{N2} \cos \lambda d\lambda + \\ &+ \pi \int_{\beta_2}^{\pi/2} \sin 2\beta d\beta \int_{\lambda_2(\beta)}^{\lambda_1(\beta)} \varphi_{N2} \cos \lambda d\lambda \end{aligned} \qquad (4.34.15)$$

where the angles β_1 and β_2 are determined by Eq. (4.34.7). Equations (4.33.2) and (4.34.6) give expressions for the $\lambda_2(\beta)$ and $\lambda_1(\beta)$ as

$$\cos\lambda_2(\beta)=\frac{a_M}{\sin\beta} \quad \text{and} \quad \cos\lambda_1(\beta)=\frac{a}{\sin\beta}. \tag{4.34.16}$$

The integrals in (4.34.15) cannot be expressed as elementary functions but can be reduced to the dilogarithm. Therefore, one comes to the point where numerical computation is the only practicable way of the integration in (4.34.15).

4.35 Creep in a State of Complex Stress

The formulae obtained for the case of pure shear are easily generalized to a state of complex stress. To this effect, consider a simple loading, i.e., stress vector components, S_1, S_2, and S_3 vary in such a way that they have a constant ratio. The loading regime is the same as in the preceding Section (Fig. 4.29a). Acting in a similar manner as in Sec. 3.11, from Eq. (4.33.9) we obtain the deviator proportionality as

$$e_i^S = a_0\left[F(a)+\int_{t_S}^{t} KF(a)dt\right]\frac{S_i}{S}, \tag{4.35.1}$$

where t_S is the time needed to reach the yield limit and $S=\left|\vec{\mathbf{S}}\right|$, and a is given by Eq. (4.34.2)

In spite of the fact that the module S and the stress vector components S_k do not change with time, the time-dependent strain vector components (4.35.1), i.e., creep deformation, continue to progress that is caused by that (i) the argument a of the function F decreases due to the relaxation of the integral of non-homogeneity and (ii) the integral in Eq. (4.35.1), as $K>0$, grows. To find creep strain components alone, one must subtract the plastic strain developed for $t\in[0,t_M]$ from the right-hand side of Eq. (4.35.1):

$$e_i^P = a_0\left[F(a)-F(a_M)+K\int_{t_M}^{t} F(a)dt\right]\frac{S_i}{S}. \tag{4.35.2}$$

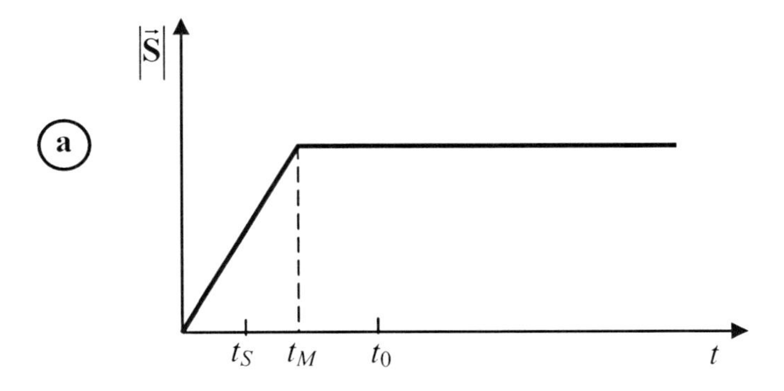

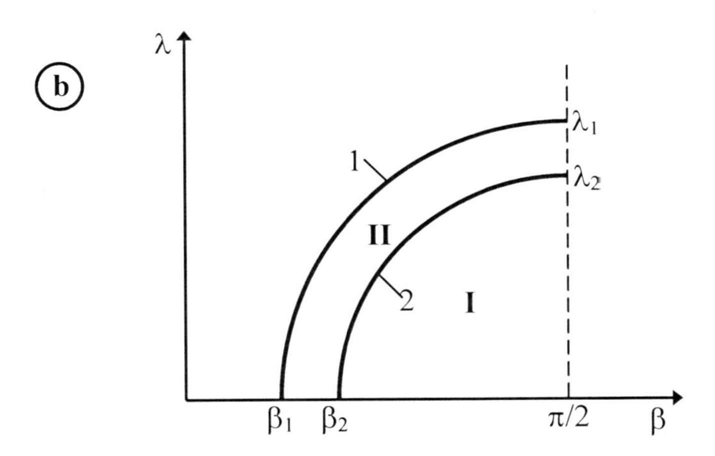

Fig. 4.29 Loading regime (a). Ranges of angles β and λ in pure sear (b)

The above formula can be written as

$$e_i^P = a_0 \left\{ F(a) - F(a_M) + K \int_{t_M}^{t} \left[F(a) - F\left(\frac{S_P}{S} \right) \right] dt + KF\left(\frac{S_P}{S} \right)(t - t_M) \right\} \frac{S_i}{S}. \quad (4.35.3)$$

The total creep strain e_i^P can be presented as $e_i^P = e_i^{P(1)} + e_i^{P(2)}$, where:

$$e_i^{P(1)} = a_0 \left\{ F(a) - F(a_M) + K \int_{t_M}^{t} \left[F(a) - F\left(\frac{S_P}{S} \right) \right] dt \right\} \frac{S_i}{S}, \quad (4.35.4)$$

$$e_i^{P(2)} = a_0 K \cdot F\left(\frac{S_P}{S} \right)(t - t_M) \frac{S_i}{S}. \quad (4.35.5)$$

The component $e_i^{P(1)}$, which changes with time in a nonlinear way, is the unsteady-state creep strain components and $e_i^{P(2)}$, which is a linear function of time, is the steady-state creep strain components.

If to denote the instant of the transition from unsteady to steady creep as $\tilde{t}$, when $I_N(\tilde{t}) \approx 0$, Eq. (4.35.4) at $t \geq \tilde{t}$,

$$e_i^{P(1)} = a_0 \left\{ F(a) - F(a_M) + K \int_{t_M}^{\tilde{t}} \left[F(a) - F\left(\frac{S_P}{S} \right) \right] dt \right\} \frac{S_i}{S}, \quad (4.35.6)$$

gives the final (maximal) value of unsteady-state creep strain components. As it was stated earlier, since the function K regulating steady creep rate takes small values, we can ignore it comparing to the terms $F(a)$ and $F(a_M)$ in Eq.(4.35.4):

$$e_i^{P(1)} = a_0 [F(a) - F(a_M)] \frac{S_i}{S}, \quad (4.35.7)$$

In this case we can express the total creep strain as

$$e_i^P = a_0\left[F(a) - F(a_M) + KF\left(\frac{S_P}{S}\right)(t - t_M)\right]\frac{S_i}{S}, \tag{4.35.8}$$

The differentiation of $e_i^{P(2)}$ from Eq. (4.35.5) with respect to time, together with Eq. (1.7.12), gives the *steady-state*-creep-strain-rate tensor components, for simplicity we will denote them with "P" index. For the case of, e.g., uniaxial tension we have

$$\dot{\varepsilon}_x^P = \sqrt{\frac{2}{3}} a_0 KF\left(\frac{\sigma_P}{\sigma_x}\right), \quad \dot{\varepsilon}_y^P = \dot{\varepsilon}_z^P = -\frac{\dot{\varepsilon}_x^P}{2} \tag{4.35.9}$$

where σ_P is the creep limit of material in uniaxial tension.

4.36 Definition of Function *K*

According to the model offered in Sec. 4.32, the function K regulates the steady-state creep rate. In generally, the function K depends on homology temperature Θ and stress. Since only initially isotropic materials whose creep is not accompanied by volume change are considered, we suggest that $K = K(\Theta, \tau_0)$, where τ_0 is the second invariant of stress tensor or, in other words, the shear stress intensity. We propose to define the function K as the product of two functions

$$K = K_1(\Theta)K_2(\tau_0). \tag{4.36.1}$$

To establish $K_1(\Theta)$, we turn to the physics of steady creep. Since the steady creep mechanism strongly depends on temperature, we split the whole homology temperature range into the segments that correspond to different driving forces in steady-state creep. The temperature range

$$0 < \Theta \le \Theta_1 \tag{4.36.2}$$

is known as the low-temperature; $\Theta_1 \approx 0{,}25$ for pure metals and $\Theta_1 \approx 0{,}3$ for alloys. The diapason

$$\Theta_1 \le \Theta \le \Theta_2, \tag{4.36.3}$$

where $\Theta_2 \approx 0{,}5$ for pure metals and $\Theta_2 \approx 0{,}55$ for alloys, is termed the elevated temperature. Finally,

$$\Theta_2 \le \Theta \le \Theta_3, \tag{4.36.4}$$

where $\Theta_3 \approx 0{,}7$ for pure metals and $\Theta_3 \approx 0{,}75$ for alloys, is referred to as the high-temperature. The temperature range

$$\Theta_3 \le \Theta < 1 \tag{4.36.5}$$

we will not consider for two reasons. First, at such temperature the creep is not controlled by dislocations movement but mainly by diffusional processes, secondly, the model of ideal-plastic body is more suitable for the description of deformational properties at such temperature.

Low temperatures (4.36.2) are insufficient to provide thermally activated dislocation glide via cross-slip or climb so that blocked dislocations are incapable of producing further deformation. So, in this case, steady-state creep is absent:

$$K_1(\Theta) = 0 \quad \text{at} \quad 0 < \Theta < \Theta_1. \tag{4.36.6}$$

Now consider the temperature diapason of Eq. (4.36.4). At high temperatures, there is an alternative mechanism of dislocation motion, first proposed by Mott, fundamentally different from slip that allows an edge dislocation to move out of its slip plane, known as dislocation climb. Dislocation climb allows an edge dislocation to move perpendicular to its slip plane.

The driving force for dislocation climb is the movement of vacancies through a crystal lattice. If a vacancy moves next to the boundary of the extra half plane of atoms that forms an edge dislocation, the atom in the half plane closest to the vacancy can "jump" and fill the vacancy. This atom shift "moves" the vacancy in line with the half plane of atoms, causing a shift, or positive climb, of the dislocation. The process of a vacancy being absorbed at the boundary of a half plane of atoms, rather than created, is known as negative climb. Since dislocation climb results from individual atoms "jumping" into vacancies, climb occurs in single atom diameter increments.

During positive climb, the crystal shrinks in the direction perpendicular to the extra half plane of atoms because atoms are being removed from the half plane. Since negative climb involves an addition of atoms to the half plane, the crystal grows in the direction perpendicular to the half plane. Therefore, compressive stress in the direction perpendicular to the half plane promotes positive climb, while tensile stress promotes negative climb. This is one main difference between slip and climb, since slip is caused only by shear stress.

One additional difference between dislocation slip and climb is the temperature dependence. Climb occurs much more rapidly at high temperatures than low temperatures due to an increase in vacancy motion. Slip, on the other hand, has only a small dependence on temperature. By climbing to adjacent slip planes dislocations can get around obstacles to their motion, allowing further deformation to occur.

Since the climb assisted glide is controlled by diffusional processes, it is natural to relate the function $K_1(\Theta)$, which regulates the dependence of steady creep rate from temperature, to the number of migrating point defects [122]. The probability of the atom energy being in U and $U + dU$ is pdU, where, utilizing the Maxwell-Boltzmann energy distribution,

$$p = \frac{1}{RT} \exp\left(-\frac{U}{RT}\right), \tag{4.36.7}$$

where R is gas constant, and

$$\int_0^\infty p dU = 1. \tag{4.36.8}$$

If the atom energy is above certain value U_0, the atom-migration-activation energy, this atom starts diffusing. The atom activation probability or, in other words, the relative quantity of activated atoms, η_0, is determined as follows

$$\eta^0 = \int_{U_0}^\infty p dU = \exp\left(-\frac{U_0}{RT}\right). \tag{4.36.9}$$

Therefore,

$$K_1(\Theta) = \exp\left(-\frac{U_0}{RT}\right) \quad \text{at} \quad \Theta_2 \le \Theta \le \Theta_3. \tag{4.36.10}$$

Other physical theories, see [122], result in more complicate formulae, e.g.,

$$K_1(\Theta) = \exp\left(-\frac{U_0 - \gamma_* \tau_0}{RT}\right), \tag{4.36.11}$$

where γ_* is constant. If to let that the activation energy in (4.36.10) depends on stress, we arrive at Eq. (4.36.10). From Eq. (4.36.11) it follows that the hypotheses about the representation of the function K in the form of Eq. (4.36.1) is impermissible. On the other hand, however, experimental investigation shows better agreements with results from Eq. (4.36.10) for the range (4.36.4).

At elevated temperature, Eq. (4.36.3), different mechanisms work in creep, none of them dominate above others. Hence, the establishment of $K_1(\Theta)$ based on physical significances cannot result in reliable results. Therefore, we define the function $K_1(\Theta)$ in an empirical way, it is a direct line connecting the points $K_1(\Theta_1) = 0$ and $K_1(\Theta_2)$, the $K_1(\Theta_2)$ is given by Eq. (4.36.10) at $\Theta = \Theta_2$:

$$K_1(\Theta) = \frac{\Theta - \Theta_1}{\Theta_2 - \Theta_1} \exp\left(-\frac{U_0}{R\Theta_2 T_m}\right) \quad \Theta_1 < \Theta < \Theta_2, \tag{4.36.12}$$

where T_m is the melting temperature.

The K_1 against (Θ) plot constructed on the base of Eqs. (4.36.6), (4.36.10), and (4.36.12) is shown by solid line in Fig. 4.30; this plot has a horizontal asymptote $K_1 = 1$. The dotted line describes the curve $K_1(\Theta)$ obtained from Eq. (4.36.10) for $0 \le \Theta \le \Theta_2$.

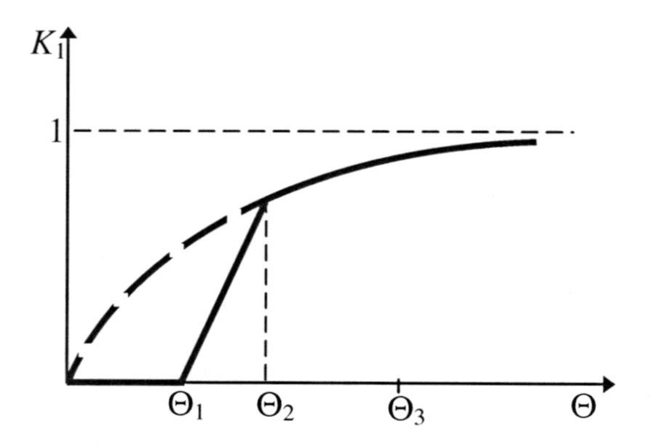

Fig. 4.30 K_1 as a function of homology temperature

4.37 Definition of Function *K*: Continuation

Steady-state creep rate in uniaxial tension is determined by empirical formulae, Eq. (4.1.1) or (4.1.2), and by Eq. (4.35.9) in terms of the synthetic theory. Equation (4.35.9) gives zero creep-rate when the acting stress is equal to the creep limit and, it is obvious, that $\dot{\varepsilon}_x^P$ grows with an increase in stress. Let us require that Eq. (4.35.9) leads to Eq. (4.1.1) at $\sigma_P/\sigma_x \to 0$ that, together with the $\dot{\varepsilon}_x^P = \dot{\varepsilon}_x^P(\Theta)$ dependency, can be expressed as

$$\dot{\varepsilon}_x^P = cK_1(\Theta)\sigma_x^k. \tag{4.37.1}$$

If $\sigma_P/\sigma_x = x \to 0$, Eq. (3.10.14) and (3.10.15) imply the following formulae

$$F = \frac{\pi\sigma_x}{2\sigma_P} \quad \text{and} \quad F = \frac{2\sigma_x^2}{\sigma_P^2}, \tag{4.37.2}$$

respectively. The first of them corresponds to Eq. (4.32.4) and the second to Eq. (4.32.5). The obtained formulae for *F* lead to the creep rate (4.35.9)

$$\dot{\varepsilon}_x^P = \frac{\pi^2}{9r} K_1(\Theta) K_2(\sigma_x)\sigma_x \quad \text{and}$$
$$\dot{\varepsilon}_x^P = \frac{2\sqrt{2}\pi}{9\sqrt{3}r} K_1(\Theta) K_2(\sigma_x)\sigma_x^2, \tag{4.37.3}$$

respectively. Comparing Eqs. (4.37.1) to (4.37.3) we conclude that

$$K_2(\sigma_x) = \frac{9cr}{\pi^2}\sigma_x^{k-1} \quad \text{and} \quad K_2(\sigma_x) = \frac{9\sqrt{3}cr}{2\sqrt{2}\pi}\sigma_x^{k-2}, \tag{4.37.4}$$

respectively. For the case of arbitrary state of stress, we define the function $K_2(\tau_0)$

$$K_2 = \frac{9cr}{\pi^2}\tau_0^{k-1} \quad \text{and} \quad K_2 = \frac{9\sqrt{3}cr}{2\sqrt{2}\pi}\tau_0^{k-2}, \tag{4.37.5}$$

respectively.

Equations (4.36.6), (4.36.10), (4.36.12), and (4.37.5) fully define the function K, which supplements the model of irreversible deformation formulated in Sec. 4.32. More complicated forms of the functions K_1 and K_2 are presented in the works [132-136] enable to allow for the influence of preliminary ultrasound and mechanical-thermal treatments on the stationary creep of metals.

4.38 Creep with Different Plastic Pre-strains

Creep at the decrease in acting stress has been considered partially in Sec. 4.15; as was shown, under certain conditions the stress reduction can be accompanied by plastic straining of the sign of initial loading (Fig. 4.16). Let us consider this effect from the point of view of the interaction between plastic and creep deformation. In addition, steady creep will be taken into account.

Consider the experiments performed in [106] on specimens of stainless steel 316. The scheme of loading is shown in Fig. 4.31. A specimen No 1 is loaded in tension for time t_M to some stress σ_M and, for $t > t_M$, the specimen creeps at the constant stress σ_M. A specimen No 2 is loaded for time t_2 to a greater stress, $\sigma_2 > \sigma_M$, and then the partial unloading is carried out so that the stress value returns to the value of σ_M; point N_2. Further, the specimen No 2 creeps under constant stress σ_M. Specimen No 3 and 4 are first loaded to σ_3 and σ_4 stress values, respectively, and then creep at stress σ_M beginning from points N_3 and N_4. Therefore, all the four specimens, after different previous loading, creep under the same stress value, σ_M.

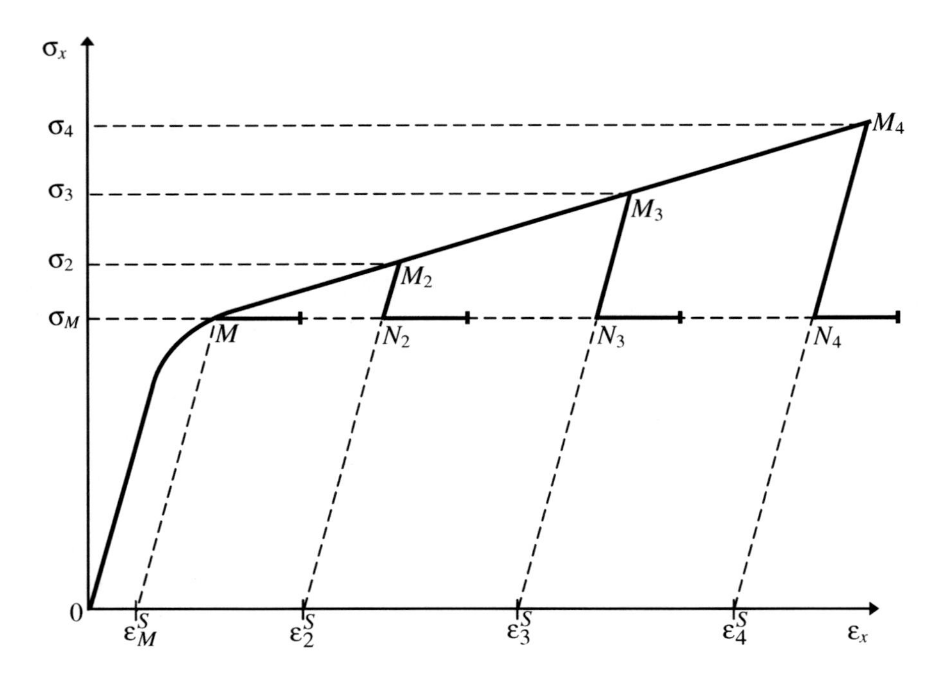

Fig. 4.31 Loading regimes with partial unloading

The results of the experiments – creep diagrams – are shown in Fig. 4.32, where the numbers above the curves indicate the specimen-numbers. The maximal stresses achieved in the initial loading and induced plastic strains are: $\sigma_M = 140\,\text{MPa}$ and $\varepsilon_M^S = 0.28\%$, $\sigma_2 = 174\,\text{MPa}$ and $\varepsilon_2^S = 1.0\%$, $\sigma_3 = 196\,\text{MPa}$ and $\varepsilon_3^S = 2.0\%$, $\sigma_4 = 218\,\text{MPa}$ and $\varepsilon_4^S = 3.0\%$.

As seen from Fig. 4.32 the creep deformation of Specimen No 2, which begins from plastic deformation $\varepsilon^S = 1\%$, is considerably smaller, by 40 %, than that of Specimen No 1 whose creep follows the initial plastic loading, i.e., without intermediate unloading. The decrease in the creep deformation becomes more and more considerable with the increase in the previous plastic deformation. Hence, ***previous plastic deformation considerably increases the resistance of material with respect to subsequent creep***, and this effect is of a progressive character.

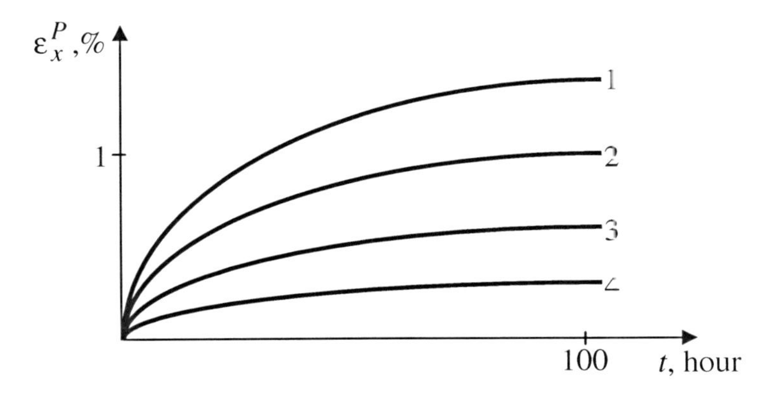

Fig. 4.32 Creep diagrams with various pre-strains

In terms of the generalized synthetic theory, let us find the creep strains in the above specified loading. Within this section, we neglect the function K and Eq. (4.32.1) takes the form $d\psi = r d\varphi$ that allows us to calculate the boundary values of angles α, β, and λ by relation $\varphi_N = 0$ or $\Delta\varphi_N = 0$.

The deformation of Specimen No 1 is a conventional creep that is determined by Eq. (4.35.8) employed for the case of uniaxial tension, $S_1 = S = S_M = \sqrt{2/3}\sigma_M$. In these formulae

$$a = \frac{S_P}{\sqrt{S_M^2 - I^2 \exp(-2pt)}},$$
$$a_M = \frac{S_P}{\sqrt{S_M^2 - I^2 \exp(-2pt_M)}}, \quad I = \frac{Bv}{p}(\exp(pt_M) - 1). \tag{4.38.1}$$

The plastic strain intensity of Specimen No 2 at point M_2 is

$$r\varphi_{N(M_2)} = S_P^2\left(\frac{\cos^2\xi}{a_1^2} - 1\right), \quad \cos\xi = \cos\alpha\cos\beta\cos\lambda, \tag{4.38.2}$$

where

$$a_1 = \frac{S_P}{\sqrt{S_{M2}^2 - B^2 v^2 (1 - \exp(-pt_2))^2 / p^2}}. \tag{4.38.3}$$

v is the loading rate, and the intensity $\varphi_N(M_2)$ is positive in the range of angles α, β, and λ bounded by the surface whose equation is $\varphi_N(M_2) = 0$, i.e.,

$$\cos\xi = a_1. \tag{4.38.4}$$

If to assume that the unloading occurs with the velocity v, then, similar to Eq. (4.15.2), we have

$$S_1 = S_{M2} - v(t - t_2) \tag{4.38.5}$$

and integral of non-homogeneity has the form

$$I_N = Bv\left[\int_0^{t_2} \exp(-p(t-s))ds - \int_{t_2}^{t} \exp(-p(t-s))ds\right]\cos\xi = I_1\cos\xi,$$
$$I_1 = \frac{Bv}{p}[-1 - \exp(-pt) + 2\exp(-p(t-t_2))]. \tag{4.38.6}$$

In particular, at time $t = 2t_2 - t_M$ when the stress drops to the value of σ_M, point N_2, we have

$$I_1 = \frac{Bv}{p}[-1 - \exp(-p(2t_2 - t_M)) + 2\exp(-p(t_2 - t_M))]. \tag{4.38.7}$$

During the stress reduction there exist two variants: the partial unloading is accompanied by unsteady creep (Fig. 4.16) or obeys to the elastic law. In this Section, we consider the first variant and suppose that the irreversible straining does not terminate during unloading to the level of σ_M:

$$\varphi_N = \varphi_N(M_2) + \Delta\varphi_N, \tag{4.38.8}$$

where $\Delta\varphi_N$ is the φ_N increment in the unloading. Therefore, the tangent plane distance due to the stress reduction can be written as (see Eqs. (4.32.2) and (4.32.5))

$$H_N = \sqrt{S_P^2 + r\varphi_N(M_2) + r\Delta\varphi_N + I_N^2}. \tag{4.38.9}$$

Equating this distance to the scalar product $\vec{\mathbf{S}} \cdot \vec{\mathbf{N}}$, we obtain the equations for the increment of $\Delta\varphi_N$ at the end of the partial unloading:

$$r\Delta\varphi_N = \left(S_M^2 - \frac{S_P^2}{a_1^2} - I_1^2 \right) \cos^2 \xi \,. \tag{4.38.10}$$

If the bracket in (4.38.10) is positive, i.e., $\Delta\varphi_N > 0$, the irreversible deformation does not terminate during the unloading. This is possible as the integral I_N decreases at the stress reduction and, as a result, the plane distance H_N decreases meaning that tangential planes start coming back to creep surface. If, at certain values of v and constants B and p, the decrease in the distance (4.38.9) at $\Delta\varphi_N = 0$ is faster than that in the stress vector, hence, the tensile creep deformation will progress even during the stress reduction. From Eq. (4.38.10) it follows that in this case all planes, which produce plastic deformation at the point M_2, will continue to take part in the progress of strain during the decrease in stress.

Equation (4.38.10) determines irreversible strain intensity increment within the domain bounded by the surface (4.38.4). However, during unloading, the irreversible strain occurs beyond the specified domain, where $\varphi_N(M_2) = 0$, as well. From the conditions $H_N = \vec{\mathbf{S}} \cdot \vec{\mathbf{N}}$ and $\varphi_N(M_2) = 0$ Eq. (4.38.9) gives the intensity $\Delta\varphi_N$ as follows

$$r\Delta\varphi_N = \left(S_M^2 - I_1^2 \right) \cos^2 \xi - S_P^2 \tag{4.38.11}$$

and $\Delta\varphi_N = 0$ at

$$\cos \xi = \frac{S_P}{\sqrt{S_M^2 - I_1^2}} \,. \tag{4.38.12}$$

So, we have two regions (Fig. 4.29) where Eq. (4.38.10) and Eq (4.38.11) hold. It is easy to see that the function $\Delta\varphi_N$ changes continuously at the transition between these regions.

As it follows from Eqs. (4.38.8), (4.38.10), and (4.38.11), the sum of plastic strain intensity at the point M_2 and unsteady creep strain intensity accumulated during the unloading is identical in the both specified regions and determined by the following formula

$$r\varphi_N = \left(S_M^2 - I_1^2 \right) \cos^2 \xi - S_P^2 \,. \tag{4.38.13}$$

The above formula can also be obtained from the following expression:

$$H_N = \sqrt{S_P^2 + r\varphi_N + I_N^2}\ . \tag{4.38.14}$$

Let us proceed to the determination of unsteady creep strain that occurs at the stress value σ_M to which the unloading is carried out. After the specified unloading, the integral of non-homogeneity yields the form

$$I_N = Bv\left[\int_0^{t_2}\exp(-p(t-s))ds - \int_{t_2}^{2t_2-t_M}\exp(-p(t-s))ds\right]\cos\xi = I_2\cos\xi,$$

$$I_2 = \frac{Bv}{p}\left[2\exp(pt_2) - \exp(p(2t_2 - t_M)) - 1\right]\exp(-pt) \tag{4.38.15}$$

From Eq. (4.38.14) we can find the total strain intensity φ_N consisting of three parts: induced by plastic loading, developed in unloading, and produced by constant stress σ_M . Hence

$$r\varphi_N = \left(\frac{\cos^2\xi}{a_2^2(t)} - 1\right)S_P^2, \quad a_2(t) = \frac{S_P}{\sqrt{S_M^2 - I_2^2}} \tag{4.38.16}$$

and $\varphi_N = 0$ at

$$\cos\xi = a_2(t). \tag{4.38.17}$$

To obtain the creep strain intensity of the specimen No 2 at $\sigma_M = const$ (to the right of point N_2), one needs to subtract the strain intensity given by (4.38.13) from that given by (4.38.16):

$$r\varphi_N = \left(\frac{\cos^2\xi}{a_2^2(t)} - 1\right)S_P^2 - \left(\frac{\cos^2\xi}{a_2^2(2t_2 - t_M)} - 1\right)S_P^2. \tag{4.38.18}$$

In arriving at the result (4.38.18) we have taken into account that $I_1 = I_2$ at $t = 2t_2 - t_M$ (I_1 and I_2 appear in the expressions for a_1 and a_2). The integration of the intensity (4.38.18) within the ranges (4.38.12) and (4.38.17) gives the following unsteady creep strain

$$e_1^P = a_0\{F[a_2(t)] - F[a_2(2t_2 - t_M)]\}, \tag{4.38.19}$$

Comparing Eq. (4.35.8) at $S_1 = S$ to Eq. (4.38.19), it follows that the unsteady creep of Specimen 1 and Specimen 2 is different. This difference results from the different values of the integral of non-homogeneity caused by different loading regimes – the larger the previous plastic deformation, the smaller the integral of non-homogeneity after the partial unloading, which implies the smaller unsteady creep.

The creep strain of specimens No 3 and 4 is calculated by Eq. (4.38.19) if to replace in it t_2 by t_3 and t_4, which are the times needed to reach the stress σ_3 and σ_4, respectively. At the same time, the synthetic model, independently on the value of pre-strain, gives identical steady-state creep rates for all specimens.

4.39 Creep with Different Plastic Pre-strains: Continuation

Let us continue the investigations of the previous Section. However, now, we analyze the influence of previous plastic deformation on the following creep provided that the partial unloading occurs elastically. In contrast to the previous Section, taking into account the function K is necessary.

In the considered experiments, the creep deformation develops during a few hundreds of hours (see Fig. 4.32), whereas the loading together with partial unloading occurs for several minutes. This allows us to assume that the stress changes occur with infinitely large velocity. At such suggestion the integral of non-homogeneity at point M in Fig. 4.31 takes the form

$$I_N = BS_M \exp(-pt)\cos\xi \tag{4.39.1}$$

and at point M_2 we have

$$I_N = BS_{M2} \exp(-pt)\cos\xi . \tag{4.39.2}$$

Therefore, the plastic strain intensity of Specimen No 2 at point M_2 at $t = 0$ can be written as

$$r\varphi_{M2} = \psi_{M2} = S_P^2 \left(\frac{\cos^2\xi}{a_{M2}^2} - 1 \right), \quad a_{M2} = \frac{S_P}{S_{M2}\sqrt{1-B^2}} . \tag{4.39.3}$$

The integral of non-homogeneity at point N_2 is determined by Eq. (4.38.6) which is reduced to Eq. (4.39.1) as $t_M \to 0$, $t_{M2} \to 0$, and $v \to \infty$. This means that the integral of non-homogeneity gives equal values at point N_2, N_3, and N_4 independently from the value of pre-strain. Therefore, Eqs. (4.39.1) and (4.39.3) gives the plane distances, Eq. (4.32.5), at $t = 0$, i.e., at point N_2, as follows

$$H_N = \sqrt{S_{M2}^2 - B^2\left(S_{M2}^2 - S_M^2\right)}\cos\xi . \tag{4.39.4}$$

Since $0 \le B < 1$,

$$H_N > \vec{\mathbf{S}} \cdot \vec{\mathbf{N}} = S_M \cos\xi \tag{4.39.5}$$

meaning that the basic condition for the onset/development of irreversible straining is not satisfied. Therefore, the irreversible deformation at point N_2 is terminated temporally and the effect of creep delay is observed. Since the module of vector $\vec{\mathbf{S}}$ drops from S_{M2} to S_M on the segment M_2N_2, the integral of non-homogeneity experiences the stepwise change from (4.39.2) to (4.39.1). Therefore, tangent planes make jump-displacement in the direction of creep surface but the stress vector $\vec{\mathbf{S}}$ does not reached them.

If an irreversible strain does not occur, Eq. (4.32.1) leads to Eq. (4.32.3), which describes defects relaxation. The solution of Eq. (4.32.3) for the initial condition (4.39.3) is

$$\psi_N = S_P^2 \left(\frac{\cos^2\xi}{a_{M2}^2} - 1 \right) \exp(-Kt). \tag{4.39.6}$$

Inserting this ψ_N and I_M from Eq. (4.39.1) into Eq. (4.32.5), we obtain the following plane distances

$$H_N = \sqrt{S_P^2 + S_P^2 \left(\frac{\cos^2\xi}{a_{M2}^2} - 1 \right) \exp(-Kt) + B^2 S_M^2 \exp(-2pt) \cos^2\xi} \cdot \tag{4.39.7}$$

Hence, the planes continue to move back toward the creep surface, while the scalar product $\vec{\mathbf{S}} \cdot \vec{\mathbf{N}} = S_M \cos\xi$ remains constant. Therefore, at some time, t_d, and at $\cos\xi = 1$ the distance H_N becomes equal to $\vec{\mathbf{S}} \cdot \vec{\mathbf{N}}$, i.e.,

$$S_P^2 + \left[S_{M2}^2 \left(1 - B^2\right) - S_P^2 \right] \exp(-Kt_d) + B^2 S_M^2 \exp(-2pt_d) = S_M^2 \,. \tag{4.39.8}$$

The time t_d, which is determined from the above transcendental equation, is called the creep delay.

The creep delay due to unloading is often observed in experiments (Sec. 4.17). Furthermore, McLean and co-authors [86,87] state that any fast enough decrease in acting stress always results in creep delay.

Since the vector $\vec{\mathbf{S}}$ reaches the plane with $\cos\xi = 1$ at $t = t_d$, this plane starts producing irreversible deformation and the defects relaxation terminates. For other planes, $\cos\xi < 1$, the defects intensity continues to decreases according to Eq. (4.39.6). Let us designate through t_z the instant of time when the plane with coordinates β and λ resumes the production of irreversible straining. The time t_0 is determined by the following equation

$$S_P^2 + \left[S_{M2}^2\left(1-B^2\right)\cos^2\xi - S_P^2\right]\exp(-Kt_z) + B^2 S_M^2 \exp(-2pt_z)\cos^2\xi = S_M^2\cos^2\xi. \quad (4.39.9)$$

The defects intensity on the planes that develop irreversible deformation can be found from the following equation

$$H_N = \sqrt{S_P^2 + \psi_N + B^2 S_M^2 \exp(-2pt)\cos^2\xi} = S_M \cos\xi. \quad (4.39.10)$$

From here

$$\psi_N = S_M^2\left(1 - B^2\exp(-2pt)\right)\cos^2\xi - S_P^2,$$
$$d\psi_N = 2pS_M^2 B^2 \exp(-2pt)\cos^2\xi dt. \quad (4.39.11)$$

Therefore, the creep strain intensity increment is

$$rd\varphi_N = 2pS_M^2 B^2\exp(-2pt)\cos^2\xi dt + K\left[S_M^2\left(1 - B^2\exp(-2pt)\right)\cos^2\xi - S_P^2\right]dt \quad (4.39.12)$$

The intensity φ_N is obtained by the time integration of $d\varphi_N$ in the limits from t_z, to t:

$$\varphi_N = \int_{t_z}^{t} d\varphi_N. \quad (4.39.13)$$

So

$$r\varphi_N = S_M^2 B^2\left(1 - \frac{K}{2p}\right)\left(\exp(-2pt_z) - \exp(-2pt)\right)\cos^2\xi + K\left(S_M^2\cos^2\xi - S_P^2\right)(t - t_z) \quad (4.39.14)$$

Since the time t_z is determined by the transcendental equation (4.39.9), the creep deformation can be calculated only by means of numerical methods. At the same time, the steady-state creep strain rate is readily determined from Eq. (4.39.12) or (4.39.14); these equation, as $t \to \infty$, give

$$r\dot{\varphi}_N = K\left(S_M^2 \cos^2 \xi - S_P^2\right), \tag{4.39.15}$$

i.e., we have arrived at the same formula as in the previous Section.

Let us compare the creep strain intensity of Specimen No 1 given by Eqs. (4.34.9) and (4.34.13) at $t_M = 0$ ($I = BS_M$) to that of Specimen No 2 given by Eq. (4.39.14). The difference between these formulae consists only in the value of times t_0 and t_z – for Specimen No 1, $t_0 = 0$ within the region **I** and t_0 is given by (4.34.11) within the region **II** (Fig. 4.29); for Specimen No 2, t_z can be determined from Eq. (4.39.9). From the analysis of Eqs. (4.34.11) and Eq. (4.39.9) it may be inferred that $t_z > t_0$. This fact implies that the creep strain intensity of Specimen No 2 is smaller than that of Specimen No 1. Therefore, independently of whether partial unloading (M_2N_2, M_3N_3, ...) is accompanied by creep deformation (Sec. 4.38) or not (Sec. 4.39), the synthetic theory leads to the decrease in creep deformation after plastic pre-strain.

4.40 Reverse Creep

The phenomenon of reverse creep is observed under the following loading regime (Fig. 4.33b):

1) (portion 0-1) a specimen is subjected to uniaxial tension, a stress, σ_x, grows from zero to the values of $\sigma_1 > \sigma_S$ (the loading time is assumed to be zero);

2) (portion 1-2) the action of the constant stress σ_1 for $t \in (0, t_n)$;

3) (portion 2-3) partial unloading by the amount of $\Delta\sigma$ at $t = t_n$ (the unloading rate is infinitely large);

4) the stress $\sigma_1 - \Delta\sigma$ is hold constant for $t > t_n$.

This loading regime differs from that considered in Secs. 4.38 and 4.39 in that there is the time-period t_n with constant stress between the initial loading and partial unloading.

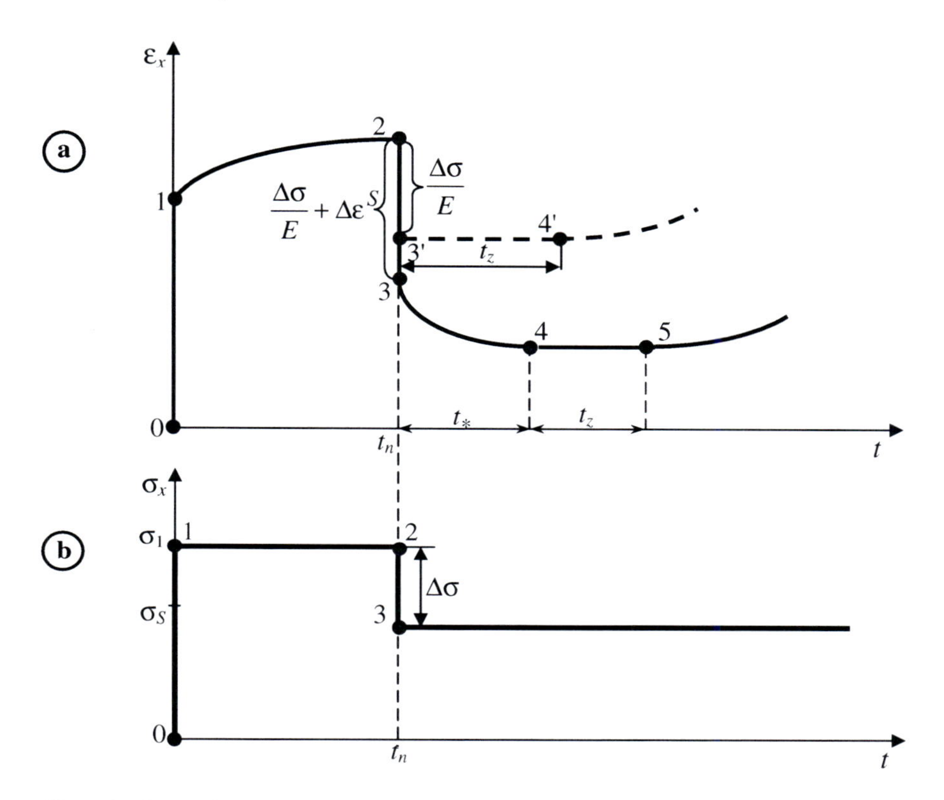

Fig. 4.33 Creep diagram (a) of a specimen loaded in regime (b); portion 3-4 reverse creep

The strain-time diagram of the specimen due to the loading regime from Fig. 4.33b is presented in Fig. 4.33a. The portion 1–2 represents the creep deformation of the specimen due to the action of the stress σ_1. The partial unloading at $t = t_n$ results in the decrease in the elastic relative elongation and, if the Bauschinger super-effect takes place, the specimen can undergo instantaneous plastic compressive deformation at $t = t_n$, $\Delta\varepsilon^S$, (portion 2-3). However, the most outstanding effect is that the specimen undergoes ***the creep deformation of sign opposite to the acting stress***, i.e., ***compressive creep deformation***, for some period of time, t_*, under the action of the constant stress $\sigma_1 - \Delta\sigma$ (portion 3-4). This phenomenon will be called as the ***reverse creep***. Further, one can see that after the development of the reverse creep is terminated ($t = t_n + t_*$), the relative elongation of the specimen does not progress for time-period t_z (creep delay – portion 4-5). Finally, the creep deformation of positive sign is resumed at $t = t_n + t_* + t_z$.

The duration of reverse creep t_* depends on both the amount of unloading $\Delta\sigma$ and the duration of the initial creep t_n. Figures 4.34 and 4.35 show the experimental plots t_* vs. $\Delta\sigma$ at different values of t_n and t_* vs. t_n at different values of $\Delta\sigma$. It is easy to see Fig. 4.35 can be constructed on the basis of Fig. 4.34 and vice versa. Figures 4.34 and 4.35 have been obtained in the experiments [107] carried out on specimens of aluminum alloy RA4 in uniaxial tension at room temperature; the maximal tensile stress $\sigma_1 = 227$ MPa .

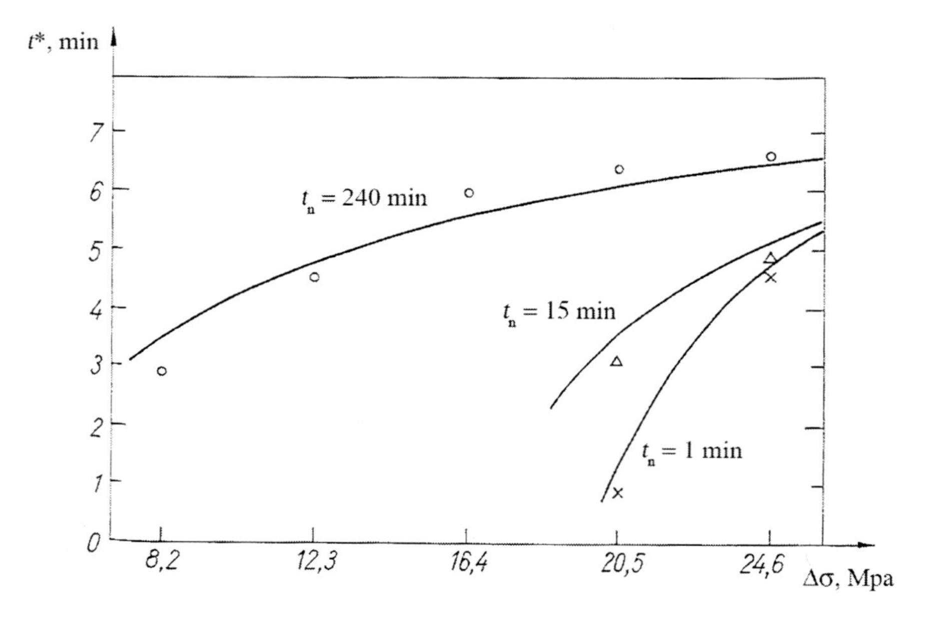

Fig. 4.34 *Reverse creep time* t^* vs. *partial unloading* $\Delta\sigma$ plot at different values of t_n

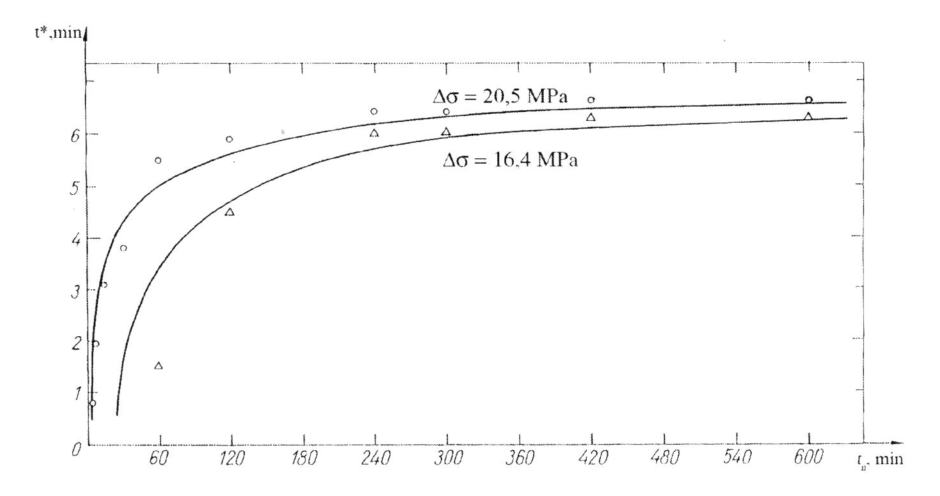

Fig. 4.35 Reverse creep time t^* vs. initial creep duration t_n plot at different values of $\Delta\sigma$

4.41 Duration of Creep Delay and Reverse Creep

In virgin state, let us consider two planes, which are tangent to the creep surface – circle of radius S_P in S_1–S_2 stress-deviator plane – and perpendicular to S_1-axis, with opposite normals $\vec{\mathbf{N}}$ and $-\vec{\mathbf{N}}$. These planes are shown by dotted lines in Fig. 4.36 and marked by **I** and **II**, respectively. The initial distance between the planes is $2S_P$. The normal vectors $\vec{\mathbf{N}}$ and $-\vec{\mathbf{N}}$ for planes **I** and **II** are set by angles $\alpha = 0$, $\beta = 0$, and $\lambda = 0$ and $\alpha = \pi$, $\beta = 0$, and $\lambda = 0$, respectively. Our goal is to describe the movements of the planes **I** and **II** on each portion in Fig. 4.33a and determine the periods of time t_z and t_*. The indexes in further formulae, 1,2,..., follow the marks in Fig. 4.33a.

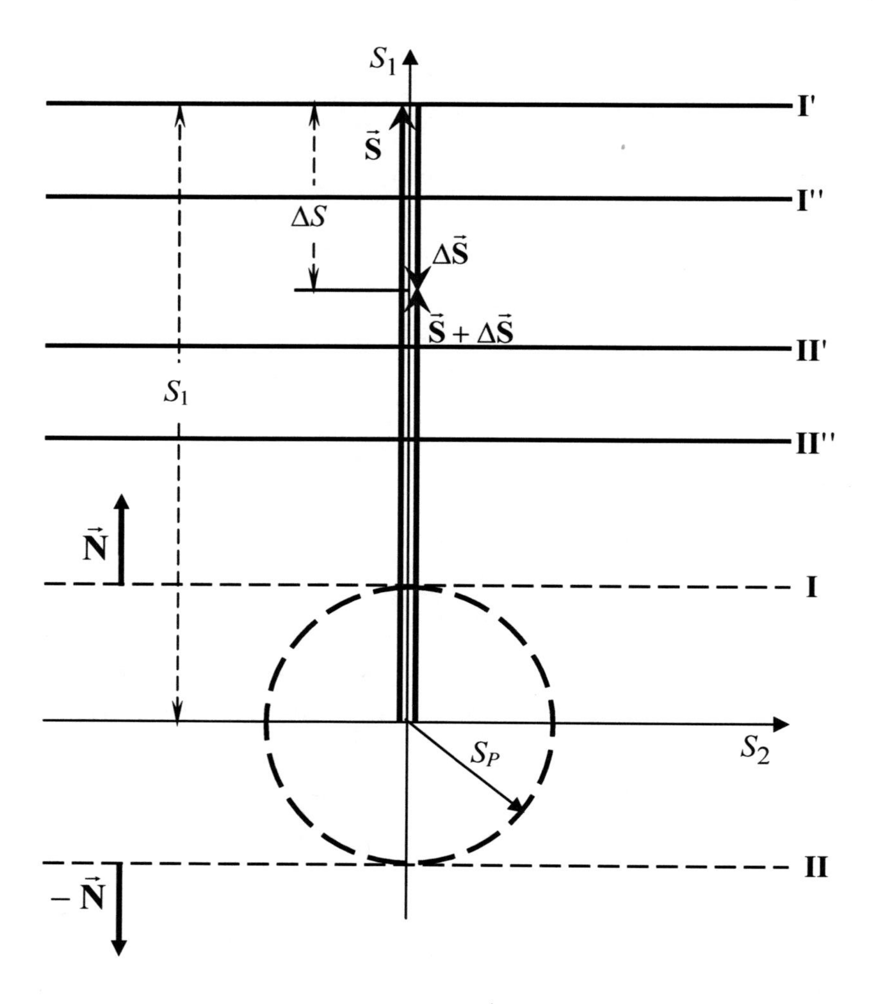

Fig. 4.36 Positions of planes with normal vectors $\vec{\mathbf{N}}$ and $-\vec{\mathbf{N}}$ due to the loading regime illustrated in Fig. 4.33b (reverse creep is not observed)

As stated in Sec. 4.40, reverse creep is observed only if the Bauschinger super-effect takes place. Therefore, to allow for the Bauschinger super-effect, we adopt the linear relation between plane distance and defect intensity, Eq. (4.32.4).

Portion 0-1. The plane **I** is displaced by stress vector $\vec{\mathbf{S}}(S_1,0,0)$ and takes the position marked by **I'** in Fig. 4.36. Equation (4.32.6) for $\alpha=0$, $\beta=0$, and $\lambda=0$ gives that $H_{N1}=S_1$ ($S_1=\sqrt{2}\sigma_1/\sqrt{3}$).

Portion 1-2. Since the loading rate is assumed to be infinitely large, Eqs. (4.32.4), (4.32.6), and Eq. (4.39.2) give that

$$I_{N(1-2)} = BS_1 \exp(-pt)\cos\xi,$$
$$\psi_{N(1-2)} = S_1(1 - B\exp(-pt))\cos\xi - S_P,$$
$$\cos\xi = \cos\alpha\cos\beta\cos\lambda, \quad 0 \le t \le t_n.$$

In particular, for the plane **I'**, $\alpha = 0$, $\beta = 0$, and $\lambda = 0$, we have

$$\psi_{N(1-2)} = S_1(1 - B\exp(-pt)) - S_P, \quad 0 \le t \le t_n. \tag{4.41.1}$$

The distance to plane **I** remains unchangeable for $0 \le t \le t_n$, whereas the plane **II** moves that is governed by Eqs. (4.32.7) and (4.32.9). The integral of non-homogeneity I_{-N} from Eqs. (4.32.7) is zero due to $I_{N(1-2)} > 0$ and the distance to planes with negative normals at $t = t_n$ are

$$H_{-N2} = S_P - \psi_{N2} = 2S_P - S_1(1 - B\exp(-pt_n))\cos\xi,$$
$$\cos\xi = |\cos\alpha|\cos\beta\cos\lambda$$

In particular, the above formula for $\alpha = \pi$, $\beta = 0$, and $\lambda = 0$,

$$H_{-N2} = S_P - \psi_{N2} = 2S_P - S_1(1 - B\exp(-pt_n)) \text{ [1]}, \tag{4.41.2}$$

gives the position of the plane with normal $-\vec{\mathbf{N}}$ at $t = t_n$ marked by **II'** in Fig. 4.36. If $S_1 \gg S_P$, the H_{-N2} becomes negative meaning that plane **II** gets over the origin of coordinates.

Further, we consider the decrease in stress by the amount of $\Delta\sigma$ that results in that the planes **I'** and **II'** move in the directions to their initial positions, marked by **I''** and **II''**, respectively.

Depending on the magnitude of the $\Delta\sigma$, we obtain different strain-time curves, **A)** 2-3'-4' (reverse creep is not observed) and **B)** 2-3-4-5 (portion 3-4 represents reverse creep) in Fig. 4.33a.

A) Let us consider the case when the endpoint of the vector $\vec{\mathbf{S}} + \Delta\vec{\mathbf{S}}$ is located between planes **I''** and **II''** as shown in Fig. 4.36. This means that the partial unloading induces neither the further development of positive deformation, nor compressive plastic deformation (segment ***2-3'*** in Fig. 4.33a represents only elastic

---

[1] Further throughout within this Section, we will write all formulae for $\alpha = 0$, $\beta = 0$, and $\lambda = 0$ and $\alpha = \pi$, $\beta = 0$, and $\lambda = 0$ meaning that $\cos\xi = 1$.

strain decrease.). Therefore, $d\varphi_N = 0$ implies that $d\psi_N = 0$ because of Eq. (4.32.1). The fact that defect intensity does not change due to the partial unloading can be expressed as

$$\psi_{N2} = \psi_{N3'} = S_1(1 - B\exp(-pt_n)) - S_P. \tag{4.41.3}$$

Since the condition $d\varphi_N = 0$ holds not only at $t = t_n$ but also for some period of time $t > t_n$ (portion ***3'-4'*** in Fig. 4.33a), Eq. (4.32.1) degenerates into the equation of defect relaxation, Eq. (4.32.3), whose solution at initial condition (4.41.3) takes the form

$$\psi_{N(3'-4')} = [S_1(1 - B\exp(-pt_n)) - S_P]\exp(-K(t - t_n)),\\ t > t_n. \tag{4.41.4}$$

The integral of non-homogeneity for plane **I**" is determined by Eq. (4.38.6) which at $\nu \to \infty$ takes the form

$$I_{N(3'-4')} = B[S_1 - \Delta S\exp(pt_n)]\exp(-pt),\\ \Delta S = \sqrt{2}\Delta\sigma/\sqrt{3}, \quad t > t_n. \tag{4.41.5}$$

In the case of small value of ΔS, when the square handle in the above formula is positive, the distance between the plane **I**" and origin of coordinate is

$$H_{N(3'-4')} = S_P + \psi_{N(3'-4')} + I_{N(3'-4')} =\\ = S_P + \{S_1[1 - B\exp(-pt_n)] - S_P\}\exp[-K(t - t_n)] + B[S_1 - \Delta S\exp(pt_n)]\exp(-pt) \tag{4.41.6}$$

Equation (4.41.6) describes the displacement of plane **I**" in the direction to the origin of coordinate. At some instant, $t = t_n + t_z$, (point 4' in Fig. 4.33a) the distance (4.41.6) becomes equal to $S_1 - \Delta S$ meaning that plane **I**" is on the endpoint of the vector $\vec{\mathbf{S}} + \Delta\vec{\mathbf{S}}$ and starts to produce time-dependent irreversible deformation. The period of time t_z is called the creep delay. The condition $H_{N4'} = S_1 - \Delta S$ gives the following equation for t_z:

$$S_P + [S_1(1 - B\exp(-pt_n)) - S_P]\exp(-Kt_z) +\\ + B[S_1 - \Delta S\exp(pt_n)]\exp[-p(t_n + t_z)] = S_1 - \Delta S \tag{4.41.7}$$

For $t > t_n + t_z$, strain intensity progresses in time symbolizing the development of creep deformation.

It is worthwhile to mention that the irreversible strain is produced by a set of tangent planes. On the other hand, if we restrict ourselves to the determination of the duration of creep delay alone, Eq. (4.41.7), it is suffice to calculate plane distances for $\alpha = 0$. Indeed, the plane **I** is first to be shifted by stress vector during loading along portion ***0-1*** and the plane **I**'' is first to return on the endpoint of the vector $\vec{\mathbf{S}} + \Delta\vec{\mathbf{S}}$ (point 4'). Another important conclusion is that the irreversible strain is produced only by tangent planes with positive normals $\vec{\mathbf{N}}$, i.e., planes with opposite normals do not take part in the irreversible straining at all.

B) Let us consider a case when the stress decreases such that the stress increment vector $\Delta\vec{\mathbf{S}}$ shifts a set of planes with normals $-\vec{\mathbf{N}}$, there is plane **II**'' among them, thereby producing a compressive plastic deformation, $\Delta\varepsilon^S$, (segment 2-3 in Fig. 4.33a). Since the compressive strain $\Delta\varepsilon^S$ is produced under the action of the tensile stress $\vec{\mathbf{S}} + \Delta\vec{\mathbf{S}}$, it is clear that the Bauschinger supper effect takes place.

Let us find a yield limit in the partial unloading, S_S. To do this, we need to determine the distance between the plane **II**'' and origin of coordinates at the instant of time when the vector $\Delta\vec{\mathbf{S}}$ reaches this plane. At $\Delta S = S_1 - S_S$, the compressive plastic strain starts developing (Fig. 4.37).

Since the reverse creep is of compressive and unsteady-state nature, it must be produced by the movements of planes with normals $-\vec{\mathbf{N}}$ whose distances are regulated by the integral of non-homogeneity I_{-N}. To meet this condition, I_{-N} for the plane **II**'', I_{-N3}, must be positive. The value of I_{-N3} is calculated by Eq. (4.41.5) at $t = t_n$:

$$I_{-N3} = B[\Delta S - S_1 \exp(-pt_n)] \tag{4.41.8}$$

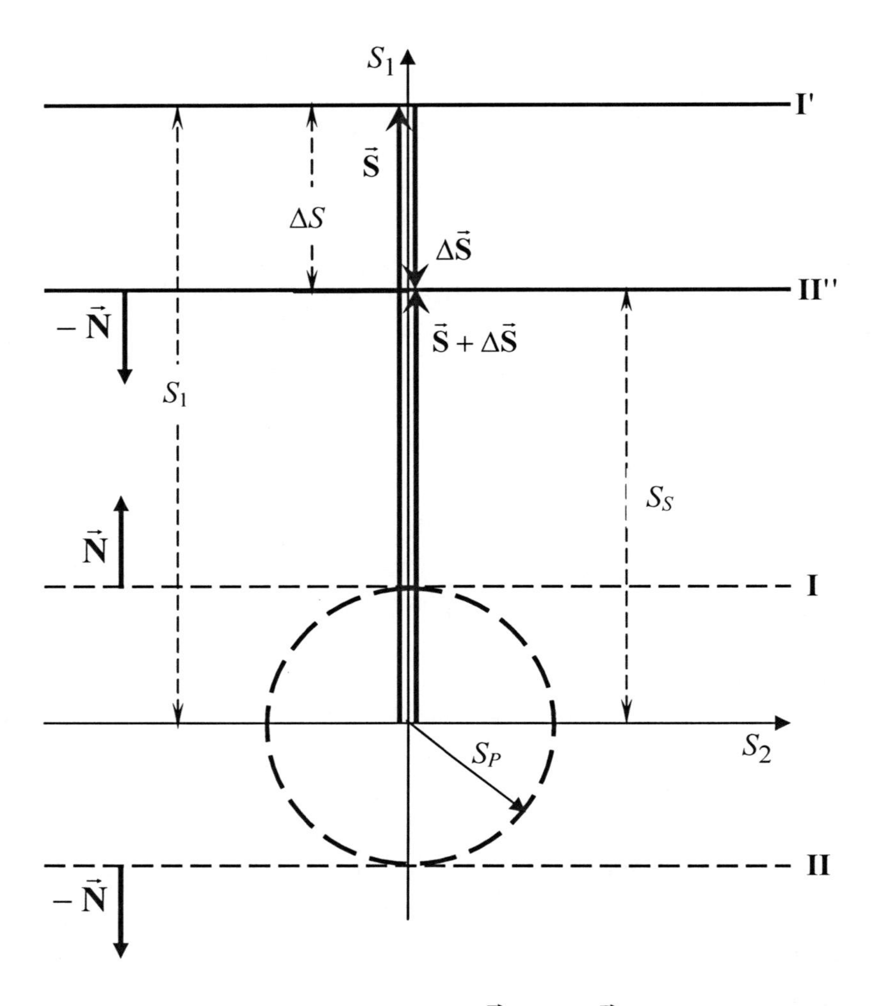

Fig. 4.37 Positions of planes with normal vectors $\vec{\mathbf{N}}$ and $-\vec{\mathbf{N}}$ due to the loading regime illustrated in Fig. 4.33b (reverse creep is observed)

and this formula holds true for

$$\Delta S > S_1 \exp(-pt_n). \tag{4.41.9}$$

The yield limit S_S is equal to the module of stress vector when it reaches the plane **II**" which is at a distance of

$$H_{-N} = \vec{\mathbf{S}} \cdot \left(-\vec{\mathbf{N}}\right) = -S_S \tag{4.41.10}$$

from the origin of coordinate. On the other hand, Eqs. (4.41.3) and (4.41.8) give at $\Delta S = S_1 - S_S$ that

$$H_{-N} = S_P + \psi_{-N} + I_{-N} = 2S_P - S_1(1-B) - BS_S. \tag{4.41.11}$$

Equating H_{-N} to $-S_S$, we obtain the equation for the yield limit in the partial unloading:

$$S_S = S_1 - \frac{2S_P}{1-B}. \tag{4.41.12}$$

It must be noted once more that the occurrence of reverse creep requires that magnitude of the S_S to be positive and the stress $S_1 - \Delta S$ to be less than S_S. These conditions, in the view of Eq. (4.41.12), take the form

$$S_1 > \frac{2S_P}{1-B} \text{ and } \Delta S > \frac{2S_P}{1-B}. \tag{4.41.13}$$

As $\Delta S \leq S_1$, the fulfillment of the second inequality in Eq. (4.41.13) provides the fulfillment of the first inequality.

Portion 3-4 (reverse creep). We obtain the integral of non-homogeneity for $t > t_n$ if to multiply I_{-N3} from Eq. (4.41.8) by $\exp[-p(t-t_n)]$:

$$I_{-N(3-4)} = B[\Delta S - S_1 \exp(-pt_n)]\exp[-p(t-t_n)]. \tag{4.41.14}$$

Since plane **II**" is on the endpoint of the vector $\vec{\mathbf{S}} + \Delta\vec{\mathbf{S}}$ (Fig. 4.37), we can write that

$$H_{-N(3-4)} = S_P + \psi_{-N(3-4)} + I_{-N(3-4)} = -(S_1 - \Delta S) \tag{4.41.15}$$

from which

$$\psi_{-N(3-4)} = -(S_1 - \Delta S) - I_{-N(3-4)} - S_P, \tag{4.41.16}$$

$$\dot{\psi}_{-N(3-4)} = -\dot{I}_{-N(3-4)} = pI_{-N(3-4)}. \tag{4.41.17}$$

Defect intensity (4.41.16) for plane **II**" is negative as $\psi_{N(1-2)} > 0$ for plane **I**" and $\psi_{-N} = -\psi_N$.

Reverse creep strain rate intensity, $\dot{\varphi}_{-N(3-4)}$, can be found from the basic relationship of synthetic theory (4.32.1):

$$r\dot{\varphi}_{-N(3-4)} = \dot{\psi}_{-N(3-4)} + K\psi_{-N(3-4)} = pI_{-N(3-4)} + K\psi_{-N(3-4)}. \quad (4.41.18)$$

Now, Eqs. (4.41.14) and (4.41.16)–(4.41.18) give that

$$r\dot{\varphi}_{-N(3-4)} = B(p-K)[\Delta S \exp(pt_n) - S_1]\exp(-pt) - K[(S_1 - \Delta S) + S_P]. \quad (4.41.19)$$

If the right-hand side in Eq. (4.41.18) or Eq. (4.41.19) is positive, tangent planes with negative normals produce irreversible(creep) strain at constant $S_1 - \Delta S$. In addition, if to neglect *K* in (4.41.19), $\dot{\varphi}_{-N(3-4)} > 0$ for any $t > t_n$ meaning that the reverse creep is never terminated. On the other hand, Eq. (4.41.19) for $p > K > 0$ gives that the reverse creep lasts for a finite period of time, t_*. By letting $\dot{\varphi}_{-N(3-4)} = 0$, we can calculate the duration of reverse creep t_* as

$$t_* = \frac{1}{p}\ln\frac{B(p-K)(\Delta S - S_1\exp(-pt_n))}{K(S_1 - \Delta S + S_P)} \quad (4.41.20)$$

and $t_* > 0$ if the numerator is greater than the denominator in the above formula:

$$[B(p-K) + K]\Delta S > [B(p-K)\exp(-pt_n) + K]S_1 + KS_P. \quad (4.41.21)$$

Therefore, the formulae we have derived are valid if the value of partial unloading ΔS satisfies the following three conditions: Eqs. (4.41.9), (4.41.13), and (4.41.21).

Let us analyze Eq. (4.41.20). As seen from this equation, holding S_1 and t_n fixed, the reverse creep time t_* grows with ΔS. This is true for the whole range of ΔS, from $S_1 - S_S$ to S_1. Another result is the reverse creep time t_* grows with the initial creep duration t_n at fixed values of S_1 and ΔS. Furthermore, the function $t_*(t_n)$ is bounded above by horizontal asymptote

$$\max t_* = \frac{1}{p}\ln\frac{B(p-K)\Delta S}{K(S_1 - \Delta S + S_P)} \quad \text{as } t_n \to \infty. \quad (4.41.22)$$

These results agree with experiments presented in Figs. 4.34 and 4.35.

Portion 4-5 (creep delay). Defect intensity for the plane **I**" at $t = t_n + t_*$, ψ_{N4}, is determined by formulae $\psi_N = -\psi_{-N}$, (4.41.14), and (4.14.16) at $t = t_n + t_*$:

$$\psi_{N4} = -\psi_{-N4} = \frac{p}{p-K}\left(S_1 + S_P - \Delta S\right). \tag{4.41.23}$$

Since irreversible straining does not occur for $t > t_n + t_*$, $d\varphi_N = 0$, we arrive at defects relaxation equation (4.32.3) that, together with initial condition (4.41.23), takes the form

$$\psi_{N(4-5)} = \frac{p}{p-K}\left(S_1 + S_P - \Delta S\right)\exp\left(-K\left(t - t_n - t_*\right)\right). \tag{4.41.24}$$

As $I_{N(4-5)} = 0$, the distance to plane **I**" is

$$H_{N(4-5)} = S_P + \frac{p}{p-K}\left(S_1 + S_P - \Delta S\right)\exp\left(-K\left(t - t_n - t_*\right)\right) \tag{4.41.25}$$

meaning that plane **I**" moves in the direction of the origin of coordinates. The instant of time when the plane is on the endpoint of vector $\vec{\mathbf{S}} + \Delta\vec{\mathbf{S}}$, t_z, can be obtained from Eq. (4.41.25) by letting

$$H_{N(4-5)} = S_1 - \Delta S\,. \tag{4.41.26}$$

As a result, we obtain that

$$t_z = \frac{1}{K}\ln\frac{p\left(S_1 + S_P - \Delta S\right)}{\left(p-K\right)\left(S_1 - S_P - \Delta S\right)} \tag{4.41.27}$$

and t_z is the creep delay time. For $t > t_n + t_* + t_z$, it is clear that the creep of positive sign is resumed. It is worthwhile to note that t_z holds true if

$$S_1 - \Delta S > S_P\,. \tag{4.41.28}$$

This inequality expresses an obvious condition for the occurring of the positive creep deformation, the acting stress $S_1 - \Delta S$ must be greater than the creep limit S_P.

4.42 Discussion

As seen from the previous Section, the generalized synthetic theory successfully describes the phenomenon of reverse creep; in particular, the equations for the duration of reverse creep and creep delay, Eqs. (4.41.20) and (4.41.27), are in satisfactory agreement with experiments. We have managed to do this as the synthetic theory reflects real physical processes occurring in a material during loading/unloading.

1. The duration of reverse creep t_* is determined by Eq. (4.41.20) which provides the relations $t_*(t_n)$ and $t_*(\Delta S)$ at fixed ΔS and t_n, respectively. The function $t_*(t_n)$ at $\Delta S = const$ has the horizontal asymptote (4.41.22) as it is observed in experiments (Fig. 4.35).

2. The decisive part in the modeling of reverse creep plays the specific definition of the relation between I_N and I_{-N} that appears in Eqs. (4.32.4) and (4.32.7) – if $I_N > 0$, $I_{-N} = 0$ and vice versa the condition $I_{-N} > 0$ implies that $I_N = 0$.

3. Different creep diagrams ($\varepsilon_x^P(t)$ for $t > t_n$) depending on the magnitude of partial unloading can be obtained (see Fig. 4.38). Line 1 in Fig. 4.38 represents a standard creep diagram at constant stress ($\Delta S = 0$); line 2 is a creep diagram at insignificant unloading, a creep in the direction of acting stress follows creep delay; line 3 represents the case of reverse creep, after which there is a delay with following positive creep; line 4 corresponds to the case of complete unloading ($\Delta S = S_1$) when the relative elongation of specimen does not occur after reverse creep.

4. In works [107], the phenomenon of reverse creep is modeled in terms of the concept of slip. It is of interest to note that the concept of slip leads to the formulae (4.41.20) and (4.41.27). This fact means that the synthetic and slip theories give similar results not only for plastic, but also for creep deformations.

5. The phenomenon of reverse creep is of great importance relatively for the theories based on the hypothesis of creep potential. According to the concept of potential, a creep rate is a single-valued function of the state of stress and stress values meaning that a loading prehistory does not affect the creep rate. On the other hand, the reverse creep observed experimentally shows that the creep rate strongly depends on a loading regime. Furthermore, the sign of reverse creep is opposite to that of acting stress.

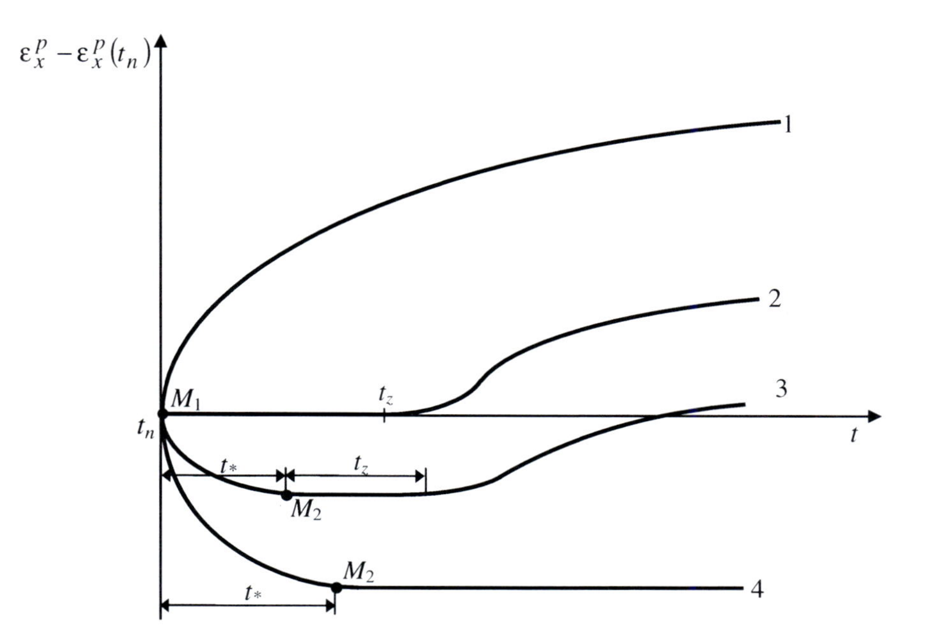

Fig. 4.38 Creep diagrams: 1 – standard; at different values of $\Delta\sigma$ from Fig. 4.33b: 2 – $\Delta\sigma < \sigma_1 - \sqrt{3/2}S_S$ (S_S is given by Eq. (4.41.12)), reverse creep is not observed; 3 – $\Delta\sigma > \sigma_1 - \sqrt{3/2}S_S$, reverse creep is observed; 4 – complete unloading, $\Delta\sigma = \sigma_1$, reverse creep is observed. M_1M_2 is the reverse creep portion

4.43 Light Alloys Creep

Numerous experiments testify to that light alloys, which offer equal resist to tension and compression in fast enough elastic-plastic straining, creep in a different way in tension and compression. Let us describe the creep deformation of light alloys by means of introducing a hydrostatic stress, σ from Eq. (1.2.2), into the basic relations of the synthetic theory.

Since stress-strain diagrams are identical in a fast loading, the yield limit does not depend on the σ. On the other hand, since the yield limit is determined via a creep limit, σ_P, and the integral of non-homogeneity, I_N, we can conclude that neither I_N nor σ_P depends on σ and are determined by conventional formulae, Eqs. (4.5.21) or (4.5.22) and (4.7.1).

The deformation of light alloys can be described in terms of the generalized synthetic theory if to replace Eq. (4.32.1) by the following [112]:

$$d\psi_N = r d\varphi_N - K(1 + 3\sigma C)\psi_N dt\,, \qquad (4.43.1)$$

where C is a material constant.

Utilizing Eq. (4.43.1), let us study a steady-state creep, when defects intensity is constant in time value. Therefore, Eq. (4.43.1) gives that

$$r\dot{\varphi}_N = K(1+3C\sigma)\psi_N. \tag{4.43.2}$$

Since $\sigma > 0$ in tension and $\sigma < 0$ in compression, Eq. (4.43.2) gives the creep rate in tension is larger than that in pressure. On the other hand, in the case of fast loading, if creep deformation does not develop, Eq. (4.43.1) can be replaced by Eq. (4.32.2), which does not include the hydrostatic pressure σ; this provides identical stress-strain diagrams in tension and compression.

Summarizing, except Eq. (4.32.1), all equations of the generalized synthetic theory, together with I_N and σ_P discussed above, remain valid: the formulae for plane distance (4.32.4) and (4.32.5), the condition for irreversible straining (3.9.7) or (4.32.6), and the formula for the summation (integration) of elementary shape changes (3.9.4).

Consider secondary creep when Eq. (4.43.2) gives

$$r\dot{\varphi}_N = K(1+C\sigma_x)\psi_N \tag{4.43.3}$$

and

$$r\dot{\varphi}_N = K(1-C|\sigma_x|)\psi_N \tag{4.43.4}$$

for the case of uniaxial tension and pressure, respectively. Because the integral of non-homogeneity is zero in steady-state creep, the defects intensity, which is determined by Eq. (4.32.4) (tangent plane distance) and Eq. (4.32.6) (the condition for the development of irreversible strain), is

$$\psi_N = S_P\left(\frac{\cos\xi}{a} - 1\right), \quad a = \frac{S_P}{S_1}. \tag{4.43.5}$$

Equations (4.43.3)–(4.43.5), together with Eq. (3.9.14), determine the steady-state creep rate of light alloys:

$$\dot{\varepsilon}_x = \sqrt{\frac{2}{3}}a_0 K(1 \pm C|\sigma_x|)F\left(\frac{\sigma_P}{\sigma_x}\right), \tag{4.43.6}$$

the sign plus indicates on tension, whereas the sign minus does on pressure. In arriving at the result (4.43.10), we have used Eq. (1.7.13).

4.44 Conclusions to the Chapter 4

1. As a rule, different theories founding on different hypotheses describe plastic and creep deformations separately. If the specified types of deformation do not affect each other, such approach proves its value, otherwise not. Therefore, the experiments considered in Sec. 4.38 are of great, principal importance. As it follows from them, a previous plastic deformation considerably raises the resistance of material with respect to a subsequent creep and this effect has a progressive character.

Thus, such theory is needed that (i) describes both plastic and creep deformation via the unique family of equations, (ii) allows for their possible inter-influence, and (iii) takes into account loading-rate-effects. It is the generalized synthetic theory that meets these requirements.

The thesis about the necessity of the working out of a universal theory proves to be true for the following reasons. If we deal with independent theories of plasticity and creep for the case of small enough loading rate, then it is absolutely unclear which theory should be used: on the one hand, it must be the theory of plasticity as the load grows, however, on the other hand, as the load practically does not change, the theory of creep must be utilized. Another question is what loading rate should be taken as a boundary one delimiting the sphere of the theory of plasticity or creep?

2. The difference between the generalized synthetic theory that has been formulated in Sec. 4.32 and the model which describes the plastic deformation and unsteady-state creep (see Sec. 4.30) consist in that how to define the plane distance H_N: via the intensity of defects ψ_N or irreversible strain intensity φ_N (in both variants H_N depends on the integral of non-homogeneity I_N). It is safe to say that of two relationships,

$$H_N = H_N(\psi_N, I_N), \quad \frac{\partial H_N}{\partial \psi_N} > 0, \tag{4.44.1}$$

$$H_N = H_N(\varphi_N, I_N), \quad \frac{\partial H_N}{\partial \varphi_N} > 0, \tag{4.44.2}$$

Eq. (4.44.1) is more physical. Indeed, if to expose a cold-worked body to high temperature (e.g., at recrystallization temperature), the drastic reduction in the number of defects generated during irreversible straining is observed ($\psi_N \approx 0, I_N \approx 0$), whereas, as known, the irreversible deformation is irrecoverable at any homology temperature $\Theta < 1$. Therefore, it remains valid that $\varphi_N > 0$ at heating. So, in contrast to Eq. (4.44.2), Eq. (4.44.1) gives a correct result, the plane distance takes its initial value $H_N = H_N(0,0)$.

The intensities φ_N and ψ_N are related to each other by Eq. (4.32.1) which imply that the defects intensity can relax by the law of Eq. (4.32.3). Due to the defects relaxation, the theory based on Eq. (4.44.1) is capable of describing a steady-state creep; Equation (4.44.2) does not provide this. The usage of Eq. (4.44.2) does not enable to model a steady-state creep as the strain intensity grows monotonously, whereas the plane distance is constant.

3. From the generalized synthetic theory, just as from the classical theories based on the concepts of creep potential, it follows that there exist universal, strain deformation vs. time, diagram at constant stress, Eq. (4.20.5) or (4.20.8). Experimental works on many metals and at various states of stress have amply justified the existence of the universal diagram [100]. However, in combined loading, if stress components ratio varies during loading, the generalized synthetic theory gives that there is no universal creep diagram; this result is confirmed experimentally.

4. The generalized synthetic theory models the reverse creep, the creep strain of the sign opposite to that of acting stress. The theories founded on the concept of creep potential are not capable of describing this phenomenon.

5. High-temperature stress-strain diagrams, when a stress-strain relation is strongly influenced by loading rate, have been successfully described – a slow loading produces greater deformation than fast one. This is caused by the considerable relaxation of the integral of non-homogeneity at slow loading rate; in fast loading, the integral has no time to relax, which results in smaller deformation.

Summarizing, the generalized synthetic theory of irreversible deformation describes analytically quite a number of phenomena whose modeling is impossible in terms of existing theories. Analytical results obtained within the synthetic theory show good agreement with experimental data, providing a promising outlook for engineering application.

Similar to the Budiansky concept of slip, where the interaction between different slip systems is ignored, the synthetic does not allow for the interaction between the displacements of different planes either (except for planes with opposite normal vectors). A possible way to improve the synthetic model is the account of the specified interaction.

Chapter 5
New Problems of Plasticity and Creep

In previous chapters, the laws of plastic and creep deformation of metals due to applied forces have been studied. At the same time, irreversible strains can be induced by other, non-force impacts. This chapter concerns with the mathematical modeling of irreversible deformation due to jump-wise temperature changes, irradiation, and phase transformations.

5.1 Temperature After-Effect Strain

Let temperature T cyclically change with time as shown in Fig. 5.1, i.e., it experiences jump-wise (step-wise) changes between its minimum (T_1) and maximum (T_2) values; the temperature-constant periods, $t_2 - t_1$, $t_3 - t_2, \ldots$, can be either equal or different. Many experiments show that specimens deform in such temperature field. For example [23], the length of the specimens of uranium in 1300 temperature cycles ($T_1 = 50°$ and $T_2 = 550°$) grows in 2.5 times, while in 3000 cycles – almost in 6 times.

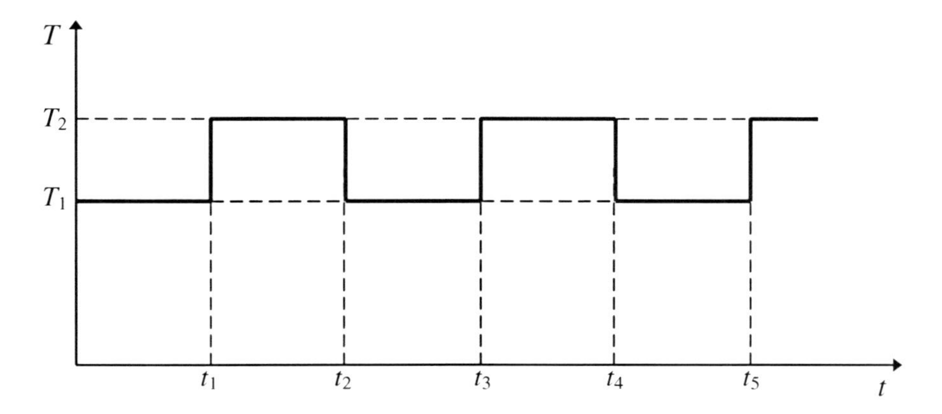

Fig. 5.1 Step-wise temperature changes

Figure 5.2 illustrates *deformation* vs. *time* plot in step-wise changing temperature field ($T_1 = 14°C$, $T_2 = 70°C$) for a specimen of zinc in load-free state [23].

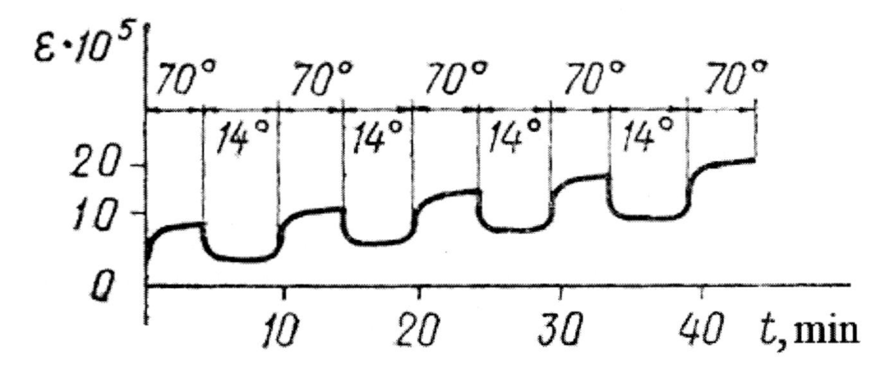

Fig. 5.2 Temperature after-effect diagram of load-free specimen

Regarding creep (when a specimen is loaded by time independent stresses) in the temperature field of Fig. 5.1, experiments record creep curves similar to those from Fig. 5.3 [23]. Curve 1 in Fig. 5.3 is the creep diagram for cadmium in torsion ($\tau = 0.35$ MPa) in changing temperature field ($T_1 = 13°$, and $T_2 = 86°$); curves 2 and 3 are the isothermal ($T(t) = const$) creep diagrams at $T = 86°C$ and $T = 13°C$, respectively.

The strain induced by the temperature regime as in Fig. 5.1 is called the temperature after-effect strain[1]. The notion "*after-effect*" arises due to the fact that the strain change is observed not only immediately at temperature jumps but for some period of time following the temperature changes. The strain increment due to the temperature jump is shown by segment M_1M_2 in Fig. 5.3. As seen from Fig. 5.3, the sign of the strain increment does not change upon heating and cooling, the temperature after-effect strain increment is co-directed with creep strain caused by the action of external loading. In contrast to this, the response of load-free specimen due to abrupt temperature changes has different signs at heating and cooling (Fig. 5.2), in addition, the strain arises only in the materials having texture. Non-textured materials do not experience deformation at the temperature changes as $\sigma_{ij} = 0$. On the other hand, if a specimen is subjected to a considerable load, the influence of texture upon the after-effect magnitude is immaterial.

[1] Further throughout, we also will use the notion temperature after-effect.

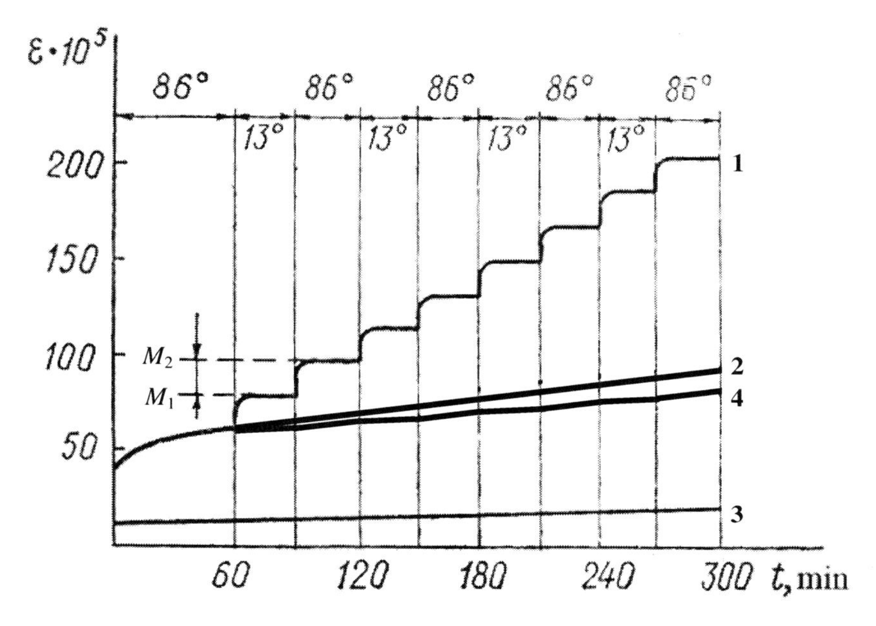

Fig. 5.3 Temperature after-effect diagram (1), isothermal creep diagrams (2,3), and non-isothermal creep diagram for the absence of thermal effects (4)

The after-effect, together with uranium and zinc, has been reliably recorded in experiments on specimens of α-iron, steel, titanium, copper, aluminum, and nickel. The increment in after-effect strain for one cycle depends on the number of previous temperature changes, it decreases with the number of temperature changes and, eventualy, the after-effect strain increments do not change with temperature changes. The *total after-effect strain* vs. *the number of temperature cycles* diagram has the same form as an isothermal creep diagram, i.e., it contains unsteady- and steady-state portions.

The increment in after-effect strain is sensitive to many factors, it (a) considerably grows with the increase of temperature jump, $T_2 - T_1$, and stress, (b) depends, although to a lesser extent, on the values of T_1 and T_2, and (c) depends on the temperature change rate, $\dot{T}$, ($\dot{T}$ is not necessarily infinite large as in Fig. 5.1). Further, temperature after-effect strains occur at elevated homology temperatures, $\Theta > 0.35 \div 0.5$.

The detailed analysis of temperature after-effect and the review of investigations regarding this phenomenon can be found in the publications of Lichachev [23,82].

5.2 Physical Nature of Temperature After-Effect

A rapid temperature change of any sign generates stresses in a body. There are many different reasons for their arising [23] such as (i) temperature-dependent dispersion of elastic moduli above grains, (ii) the anisotropy and dispersion of thermal-expansion coefficient above grains, and (iii) different orientations of the crystallographic axes of grains, etc. Therefore, the rapid temperature change should have produced strains of different orientations within adjacent micro-volumes (grains) of a body. However, as the adjacent grains are forced to ***compatible*** deforming in polycrystalline aggregate, the temperature change results in the arising of thermal stresses. These stresses can reach considerable values, both in polycrystalline and mono-crystals. They belong to the stress of the second kind and strongly depend on the amount of temperature change $\Delta T = T_2 - T_1$. The thermal stresses caused by abrupt temperature change decreases with time and, eventualy, vanishes. It is the thermal stresses that produce the temperature after-effect strain; it acquires a finite increment at temperature jump and then its rate decreases tending to zero.

In the case of the absence of external forces, first of all, it must be noted that the condition for the occrence of after-effect strain is the texture[2] of specimen that directs a deformation. Further, experiments show that the temperature change (either heating or cooling) smaller than critical magnitude, ΔT_l, produces no after-effect strain. Cyclical-temperature jumps cause alternate thermal stresses which, in turn, produce alternate strains in an element of a body. The reason for the occurrence of the considerable temperature after-effect strain at the absence of external forces is the following [23]. A polycrystalline aggregate consists of "strong" and "weak" crystal grains. By "weak" grains we mean ones whose slip systems are favorably oriented with respect to thermal shear stresses.

Let temperature acquire a positive increment, $\Delta T > 0$ and $T = T_2$. As stated above, the temperature jump results in the arising of thermal stresses. As the macro stress of a load-free specimen is zero, the thermal stresses over strong and weak grains are distributed in such a way that they balance each other (i.e., the "strong" and "weak" grains are subjected to stresses of opposite signs). On the other hand, since stress-strain relations within weak and strong grains are different, the total (macro) strain of specimen is non-zero. Further, it is obvious that the fraction of weak and strong grains remains unaltered at the subsequent cooling of specimen, $\Delta T < 0$ and $T = T_1$. The temperature jump of opposite (negative) sign changes the stress signs within both strong and weak grains against those at $\Delta T > 0$. As the increment and development of deformation at $\Delta T > 0$ and $T = T_2$ differ from those at $\Delta T < 0$ and $T = T_1$, the total strain for cycle (heating and cooling) is non-zero and can be reproduced any number of times.

[2] If the crystallographic orientations of a specimen are not random, but have some preferred orientation, then the sample has a weak, strong, or moderate texture. The degree is dependent on the percentage of crystals that have the preferred orientation.

In spite of a great variety of models describing the development of strain in non-loaded body due to cyclical temperature jumps, they all can be reduced to that presented above and differ only in the assumptions concerning the laws of weak- and strong-grain straining. If to suggest that strong grains incur no deformation at all, the mechanism of strain-development is called the mechanism of thermal ratchet-wheel.

The mechanism of the temperature after-effect strain of specimen loaded by time-independent stress (see Fig. 5.3) is the following. As well-known, an isotherm steady-state creep strain results from slips occurring in the activated slip systems of crystal grains (as well as at any type of irreversible deformation). Depending on the sum of thermal stresses produced by a temperature change and time-independent stresses due to external load, the one part of grain becames overloaded and, consequently, they produce larger deformation against that at constant temperature. In contrast, the other part of grain is subjected to smaller stresses with comparison to the isothermal creep and they cease to produce irreversible strains. Since the stress-strain relation of grain is strongly non-linear character and the thermal stresses can reach considerable values, the additional strains produced in the overloaded grains exceed the termination of straining in the underloaded grains. As a result, the total strain (macro strain of the first kind) is greater than that at $T = const$. The following temperature change of opposite sign results in that crystal grains “reverse roles”, those of which were overloaded became underloaded and vice versa. Therefore, again, the macro strain of specimen grows.

It is important to stress that the temperature after-effect phenomenon can be explained without any specific mechanisms and assumption to be applied, its manifestation obeys well-known facts about the micro-structure of real (non-uniform) solids.

The important point here is that the condition for the occurrence of temperature after-effect strains is, together with a considerable high-temperature increment, high-temperature change rate (in experiments [23], the specimen was heated by $\Delta T = 50°C$ faster than in 1 second). Slow heating or cooling does not cause temperature after-effect, experiments [50] record no after-effect strains if the temperature varies as sinusoidal function (even at large amplitudes). The main reason for this is that crystal grains have time to adapt for slow temperature changes and no thermal stresses (thermal effect) arises.

5.3 Modeling of Temperature After-Effect

Our goal is to model the temperature after-effect strain in terms of the generalized synthetic theories developed in Chapter 4. Let us introduce an average measure of thermal stresses (the stresses of the second kind), I_T, arising due to rapid temperature changes. This measure must satisfy the following requirements:

A) I_T must be of isotropic character, i.e., independent of angles α, β, and λ, due to the temperature change is assumed to be identical in all particles(grains) of body; (if I_T was dependent on angles α, β, and λ, the theory would contradict to this fact),

B) I_T must be dependent on temperature change rate, current temperature, and the temperature history,

C) I_T must decrease with time at $T = const$.

The fact that the permanent strains do not occur in the load-free specimen having no texture means that the introducing of I_T is not sufficient for the modeling of after-effect in load-free state. In the monograph [143], this problem is solved by the introducing of initial defects representing texture. In the framework of the given monograph, this problem is not considered.

To satisfy the above requirements, we establish the I_T in the following form [139,143,145,150]:

$$I_T = B_T \int_0^t \left|\dot{T}(s)\right| \exp(-p_T(t-s))ds, \tag{5.3.1}$$

where B_T and p_T are the model constants. The B_T characterizes the magnitude of thermal stresses ($[B] = Pa/°C$) and the p_T expresses the rate of their relaxation, $[p] = 1/\sec$. We will call the integral I_T the ***thermal stress parameter***.

Since, in terms of the synthetic theory (Chapters 3 and 4), the strain hardening of material is determined by the distance to tangent planes, we introduce the thermal stress parameter into Eq. (4.32.4) and Eq. (4.32.5) in the following way:

$$H_N = S_P + \psi_N + I_N - I_T, \tag{5.3.2}$$

or

$$H_N = \sqrt{S_P^2 + \psi_N + I_N^2 - I_T}; \tag{5.3.3}$$

in Eq. (5.3.3), $[B] = Pa^2/°C$.

The proposition, the displacement of plane with normal vector $\vec{\mathbf{N}}$ on the endpoint of stress vector, induces that plane with normal vector $-\vec{\mathbf{N}}$, now takes the form

$$H_{-N} = S_P - \psi_N + I_{-N} - I_T \tag{5.3.4}$$

or

$$H_{-N} = \sqrt{S_P^2 - \psi_N + I_{-N}^2 - I_T}. \tag{5.3.5}$$

Other basic relationships of the generalized synthetic theory, Eqs (3.9.4) (3.9.8), and (4.32.1), remain unchangeable.

Therefore, Eqs. (5.3.1)–(5.3.5), (3.9.8), (3.9.4), and (4.32.1) constitute the model of temperature after-effect strains in terms of the synthetic theory.

As seen from Eqs. (5.3.2) and (5.3.3), the integral of non-homogeneity I_N (Eqs. (4.7.1) and (4.5.21)) and the thermal stresses parameter I_T have different signs. This mean that the parameter I_N increases the plane distance H_N, i.e., expresses the strain-hardening of material, whereas I_T reduces H_N thereby assisting in an irreversible straining. The different influence of I_N and I_T upon the deformation properties of material can be explained as follows. The integral of non-homogeneity I_N is the measure of the local peak stresses of the third kind which cause the distortions of crystal lattice thereby hindering the movement of dislocations. In contrast, the thermal stresses parameter I_T belongs to the stresses of the second kind distributed in more or less regular manner within crystal grain. Thermal stresses together with the stresses caused by external load stimulate both plastic and creep deformation.

5.4 Temperature After-Effect in Pure Shear

Consider the case when a specimen incurs a steady-state creep strain at constant temperature T_1 in pure shear: $\vec{\mathbf{S}}(0,0,S_3)$, $S_3=\sqrt{2}\tau_{xz}$, and $\tau_{xz}(t)=const$. Further, let the temperature change in step-wise manner to the value of T_2 at $t=t_1$ be as shown in Fig. 5.4b; $T_2(t)=T_2=const$ as $t>t_1$.

Let us determine the temperature after-effect strain. The steady-state creep strain at lower constant temperature T_1, according to Eqs. (4.9.10) and (4.35.3), is

$$e_3^P=\frac{\pi\tau_P^2(T_1)}{3r}K(T_1)F\left[\frac{\tau_P(T_1)}{\tau_{xz}}\right]t\,, \tag{5.4.1}$$

where $K(T_1)$ and $\tau_P(T_1)$ are the values the function K and the creep limit at pure shear at $T=T_1$, respectively. The function K is given by Eqs. (4.36.1), (4.36.10), and (4.36.12) and the plot of $\tau_P(T)=S_P(T)/\sqrt{2}$ is shown in Fig. 4.14. The steady-state creep diagram at $T(t)=T_1=const$ is shown by the straight **1** in Fig. 5.4a.

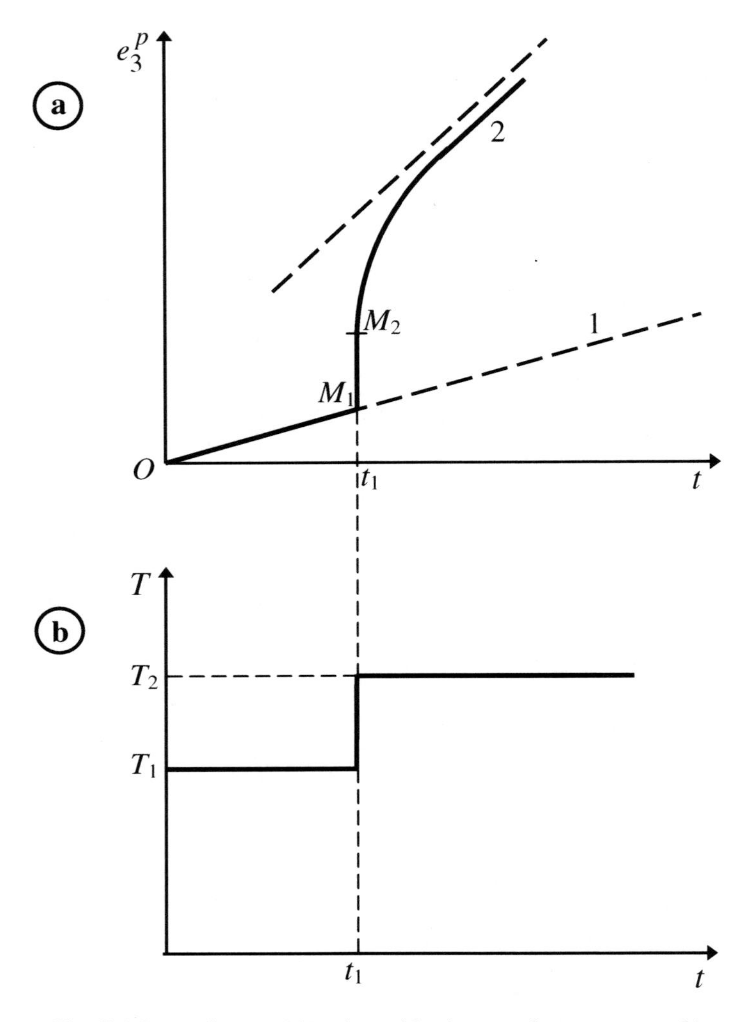

Fig. 5.4 Creep diagram (a) at the sudden increase in temperature (b)

The thermal stress parameter $I_T = 0$ prior to the temperature change, $t < t_1$. In addition, since a steady-state creep is considered, $I_N = 0$ as well. Therefore, Eq. (5.3.3), which determines the defects intensity $\psi_N(t) = const$, at $I_T = 0$, and $I_N = 0$ for $t < t_1$ becomes as

$$\psi_N = S_3^2 \cos^2 \xi - S_P^2(T_1), \quad \cos\xi = \sin\beta \cos\lambda \tag{5.4.2}$$

and ψ_N takes positive values for

$$0 \le \lambda \le \lambda_1,\ \beta_1 \le \beta \le \pi/2,\ \cos\lambda_1 = \frac{S_P(T_1)}{S_3 \sin\beta},\ \sin\beta_1 = \frac{S_P(T_1)}{S_3}; \tag{5.4.3}$$

$\psi_N = 0$ beyond the range of Eq. (5.4.3). The range of the angles β and λ is shown in Fig. 5.5 by the region I.

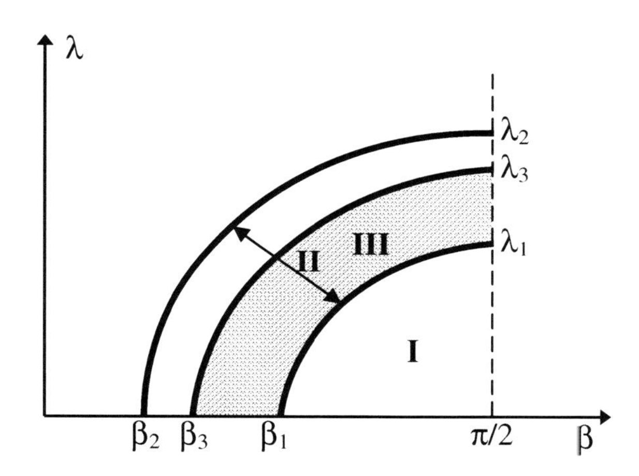

Fig. 5.5 The regions of angles β and λ from Eqs. (5.4.3) and (5.4.8)

Now, let temperature suddenly increase by an amount $\Delta T = T_2 - T_1$ at $t = t_1$ and remain constant for $t > t_1$ (Fig. 5.4b). As a result, the thermal stress parameter I_T given by Eq. (5.3.1) takes the form:

$$I_T = B_T \Delta T \exp(-p_T(t - t_1)),\quad t \ge t_1. \tag{5.4.4}$$

In particular, we have $I_T = B_T \Delta T$ at $t = t_1$. As the shear stress τ_{xz} remains unaltered at heating, $I_N = 0$ for $t > t_1$ as well. Therefore, Eq. (5.3.3) at $I_N = 0$ take the following form

$$H_N = \sqrt{S_P^2(T_1) + \psi_N} \quad t < t_1, \tag{5.4.5}$$

$$H_N = \sqrt{S_P^2(T_2) + \psi_N - B_T \Delta T} \quad t = t_1 + 0. \tag{5.4.6}$$

Now we wish to compare the distances H_N from the above two formulae. If to assume that the defects intensity takes the same value at both $t < t_1$ and $t = t_1 + 0$, the right-hand side of Eq. (5.4.6) is less than that of Eq. (5.4.5). This fact is due to the following reasons: (i) because $T_2 > T_1$, $S_P(T_2) < S_P(T_1)$ and (ii) in contrast to Eq. (5.4.5), there is the negative term $B_T \Delta T$ in Eq. (5.4.6). Therefore, the assumption that the defects intensity remains unaltered due to the temperature increase gives that tangent planes are near the origin of coordinates. However, this situation is impossible because the tangent planes that produce steady-state creep strain prior to the temperature jump are located on the endpoint of time-independent stress vector (region **I** in Fig. 5.5) and their displacements toward the origin of coordinates are blocked by the vector itself (the situation when the stress vector "pierce" through a plane is unallowable). Therefore, Eqs. (5.4.5) and (5.4.6) give that the defects intensity ψ_N acquires a finite increment at $t = t_1$.

Equating the right-hand side in (5.4.6) to $S_3 \cos\xi$, we obtain the equations for ψ_N at $t = t_1 + 0$ as

$$\psi_N = S_3^2 \cos^2 \xi - S_P^2(T_2) + B_T \Delta T \tag{5.4.7}$$

and, now, $\psi_N > 0$ in regions **I** and **II** shown in Fig. 5.5 whose ranges are

$$0 \le \lambda \le \lambda_2, \quad \beta_2 \le \beta \le \pi/2, \quad \cos\lambda_2 = \frac{\sin\beta_2}{\sin\beta},$$
$$\sin\beta_2 = \frac{\sqrt{S_P^2(T_2) - B_T \Delta T}}{S_3}. \tag{5.4.8}$$

Defects intensity increment induced by the temperature change, $\Delta\psi_N$, is calculated as the difference of the right-hand sides of Eqs. (5.4.7) and (5.4.2) for the region **I** and by Eq. (5.4.7) alone for the region **II** due to ψ_N within region **II** at $t < t_1$ is zero. Therefore, $\Delta\psi_N$ within region **I** is

$$\Delta\psi_N = S_P^2(T_1) - S_P^2(T_2) + B_T \Delta T\,. \tag{5.4.9}$$

In terms of the generalized synthetic model, crystal lattice defects (ψ_N) are related to irreversible strains. Let us designate through $\Delta\varphi_N$ the irreversible strain intensity increment which causes the defects intensity increment $\Delta\psi_N$. These two quantities are related to each other by one of the basic relationships of the generalized synthetic model, Eq. (4.32.1). As the heating occurs suddenly, $dt = 0$, Eq. (4.32.1) becomes

$$r\Delta\varphi_N = \Delta\psi_N . \tag{5.4.10}$$

Equations (5.4.3) and (5.4.7)–(5.4.10) determine the increment in the strain intensity at the temperature jump.

As stated above, the tangent planes producing creep strain at $t < t_1$ remain motionless at $t = t_1$. On the other hand, Eq. (5.4.8) gives that the set of planes on the endpoint of the stress vector $\vec{\mathbf{S}}$ increases at $t = t_1$ meaning that planes from region **II** move (in a jump-wise manner) toward the origin of coordinates at heating.

The finite increment in strain vector component at $t = t_1 + 0$, Δe_3^P is determined by Eq. (4.34.15) where one need to replace φ_{N1} and φ_{N2} by $\Delta\varphi_N$ from Eq. (5.4.10) and integrate between the limits given by Eqs. (5.4.3) and (5.4.8). As a result, we obtain

$$\Delta e_3^P = \frac{\pi S_3^2}{3r}\left[\sin^2\beta_2 F(\sin\beta_2) - \sin^2\beta_1 F(\sin\beta_1)\right], \tag{5.4.11}$$

where the function F is given by Eq. (3.10.15). The increment Δe_3^P is shown in Fig. 5.4a by the portion M_1M_2.

As shown in Sec. 4.34, the integrals in (4.34.15) cannot be expressed as elementary functions. At the same time, when calculating Δe_3^P we have managed to calculate the integrals (4.34.15). This situation can be explained by the strain intensity φ_N in Sec. 4.34 is a time integral with respect to $\dot{\varphi}_N$ that considerably complicates the integration. In this Section, the strain intensity increment $\Delta\varphi_N$ is obtained not by integrating, but immediately by Eq. (5.4.10).

Now, we consider the deformation of the specimen after heating, $t > t_1$. By equating H_N from Eq. (5.3.3) to $S_3\cos\xi$ and applying Eq. (5.4.4), we obtain the following defect intensity for $t > t_1$

$$\psi_N = S_3^2\cos^2\xi + B_T\Delta T\exp(-p_T(t-t_1)) - S_P^2(T_2). \tag{5.4.12}$$

Equations (4.32.1) and (5.4.12) give the following formula for the irreversible strain intensity rate:

$$r\dot{\varphi}_N = K(T_2)\left\{S_3^2\cos^2\xi - S_P^2(T_2) + \left[1 - \frac{p_T}{K(T_2)}\right]B_T\Delta T\exp(-p_T(t-t_1))\right\}. \tag{5.4.13}$$

It is easy to see that a sufficient condition $\dot{\varphi}_N$ that can be positive is $p_T < K(T_2)$.

The thermal stress parameter (5.4.4) decreases with time and, consequently, planes distance (5.3.3) grows meaning the irreversible straining is terminated, i.e., $d\varphi_N = 0$. On the other hand, if $d\varphi_N = 0$, defects relaxation occurs, which is governed by Eq. (4.32.3) giving the movements of tangent planes in the directions of the origin of coordinates. These two concurrent processes occur simultaneously. The solution of Eq. (4.32.3) is

$$\psi_N = \left[S_3^2 \cos^2 \xi + I_T(t_*) - S_P^2(T_2)\right] \exp[-K(T_2)(t - t_*)], \quad (5.4.14)$$

where $I_T(t_*)$ is determined by Eq. (5.4.4) at $t = t_*$, t_* is the instant when certain plane terminates producing irreversible strain (i.e., this plane is not at the endpoint of stress vector).

Let us study the plane distance governed by Eqs. (5.3.3) and (5.4.14) as

$$H_N^2 = S_P^2(T_2) + \left[S_3^2 \cos^2 \xi + I_T(t_*) - S_P^2(T_2)\right] \exp[-K(T_2)(t - t_*)] - I_T(t). \quad (5.4.15)$$

Let time t exceed the instant t_* by infinitesimal magnitude dt, $dt = t - t_*$. It is clear that the thermal integral increment due to dt is $dI_T = I_T(t) - I_T(t_*)$. By approximating $\exp[-K(T_2)(t - t_*)]$ in (5.4.15) by a Tailor series approximation (discarding degrees greater than 2), we get

$$H_N^2 = S_3^2 \cos^2 \xi - dI_T - \left[S_3^2 \cos^2 \xi + I_T(t) - S_P^2(T_2)\right] K(T_2) dt. \quad (5.4.16)$$

To determine the boundary angles β_3 and λ_3 providing that the stress vector reaches tangent planes (region **III** in Fig. 5.5), one must carry out a standard procedure, to equate the distance H_N in (5.14.16) to $S_3 \cos \xi$. As a result, we obtain that

$$\cos^2 \xi_3 = \sin^2 \beta_3 \cos^2 \lambda_3 = \frac{1}{S_3^2} \left[S_P^2(T_2) - I_T - \frac{1}{K(T_2)} \frac{dI_T}{dt} \right]. \quad (5.4.17)$$

At $\lambda = 0$ we have

$$\sin^2 \beta_3 = \frac{1}{S_3^2} \left[S_P^2(T_2) - \left(1 - \frac{p_T}{K(T_2)} \right) I_T \right]. \quad (5.4.18)$$

It is easy to see that $\dot{\varphi}_N$ from Eq. (5.4.13) at $\cos \xi = \cos \xi_3$ is zero. Therefore, Eq. (5.4.17) determines boundary planes located on the endpoint of stress vector on which $\dot{\varphi}_N = 0$. In terms of designations applied in Eq. (5.4.17), Eq. (5.14.13) takes the form

$$r\dot{\varphi}_N = K(T_2)S_3^2 \sin^2\beta_3 \left(\frac{\sin^2\beta\cos^2\lambda}{\sin^2\beta_3} - 1 \right). \qquad (5.4.19)$$

By making use of the above formula and Eq. (3.9.14), we calculate the temperature after-effect strain at $t > t_1$ as

$$\dot{e}_3^P = \frac{\pi K(T_2)S_3^2 \sin^2\beta_3}{3r} F(\sin\beta_3), \qquad (5.4.20)$$

The total temperature after-effect strain (at $t = t_1$ and $t > t_1$) is

$$e_3^P = \Delta e_3^P + \int_{t_1}^{t} \dot{e}_3^P \, dt, \qquad (5.4.21)$$

where Δe_3^P is given by Eq. (5.4.11).

According to Eqs. (5.4.18) and (5.4.20), strain rate $\dot{e}_3^P$ decreases with time and reaches its minimum value as $t \to \infty$:

$$\sin\beta_3 = \frac{S_P(T_2)}{S_3}, \quad \dot{e}_3^P = \frac{\pi}{6r} K(T_2) S_P^2(T_2) F\left(\frac{S_P(T_2)}{S_3} \right). \qquad (5.4.22)$$

The $\dot{e}_3^P$ from the above formula gives the value of steady-state creep rate at constant temperature T_2.

The diagram $e_3^P(t)$ constructed by Eq. (5.4.21) is shown in Fig. 5.4a by segment M_1M_2 and curve 2; the portion OM_1 is the part of steady-state creep diagram at lower temperature T_1 (line 1) prior to the temperature change. As seen from Fig. 5.4a, the curve 2 approximates the dotted line, whose tangent corresponds to the steady-state creep at constant temperature T_2.

5.5 Temperature After-Effect at Cooling

Similar to the previous section, we consider the case when a specimen first deforms under a time-independent shear stress, τ_{xz}, at a constant temperature T_2, i.e., a steady-state creep deformation occurs. Further, let the temperature suddenly decrease at $t = t_2$ to the value of T_1 and further remain unchangeable (Fig. 5.6b). We wish, again, to model the after-effect strain but now at the negative temperature change.

Prior to the temperature change and at the point of cooling, the tangent planes distance is

$$H_N = \sqrt{S_P^2(T_2) + \psi_N} \qquad t < t_2 \tag{5.5.1}$$

$$H_N = \sqrt{S_P^2(T_1) + \psi_N - B_T \Delta T} \qquad t = t_2 \tag{5.5.2}$$

where $B_T \Delta T$ is the value of thermal stress parameter I_T at the temperature drop by $\Delta T = T_2 - T_1$; I_T for $t > t_2$ is

$$I_T = B_T \Delta T \exp(-p_T(t - t_2)). \tag{5.5.3}$$

Since the creep limit of material grows due to cooling, $S_P(T_1) > S_P(T_2)$. On the other hand, the negative term $B_T \Delta T$ appears in the radicand of Eq. (5.5.2) and, consequently, identical values of the defects intensity ψ_N can give the both smaller and greater value of the right-hand side in Eq. (5.5.2) than that in Eq. (5.5.1).

1. Consider the case when

$$S_P^2(T_1) - B_T \Delta T < S_P^2(T_2). \tag{5.5.4}$$

Now, the identical defects intensities lead to the right-hand side in Eq. (5.5.2) that is smaller than that in Eq. (5.5.1), i.e., we come to the situation as in Sec. 5.4. Namely, the defects intensity acquires a finite increment, $\Delta\psi_N$, due to sudden temperature drop and, similarly to Sec. 5.4, the increment $\Delta\psi_N$ produces the finite increment in creep strain Δe_3^P. For $T_2(t) = const$ as $t > t_2$, the time-dependent behavior of the function $e_3^P(t)$ is governed by Eqs. (5.4.3), (5.4.8), (5.4.18), (5.4.11), and (5.4.20) if to switch the role of T_1 and T_2 and t_1 and t_2.

The possible creep diagrams after the temperature drop are shown in Fig. 5.6a, where the straight line 1 is the isothermal steady-state creep diagram at the temperature T_2. Prior to the temperature change ($t < t_2$), the plot of $e_3^P(t)$ lies on line 1. If there were no after-effect, the strain vs. time diagram at $t > t_2$ would coincide with line 2 which is the steady-state creep diagram at the lower temperature T_1. At the presence of the after-effect the diagram is shown by line 3, which approximates the dotted line parallel to line 2.

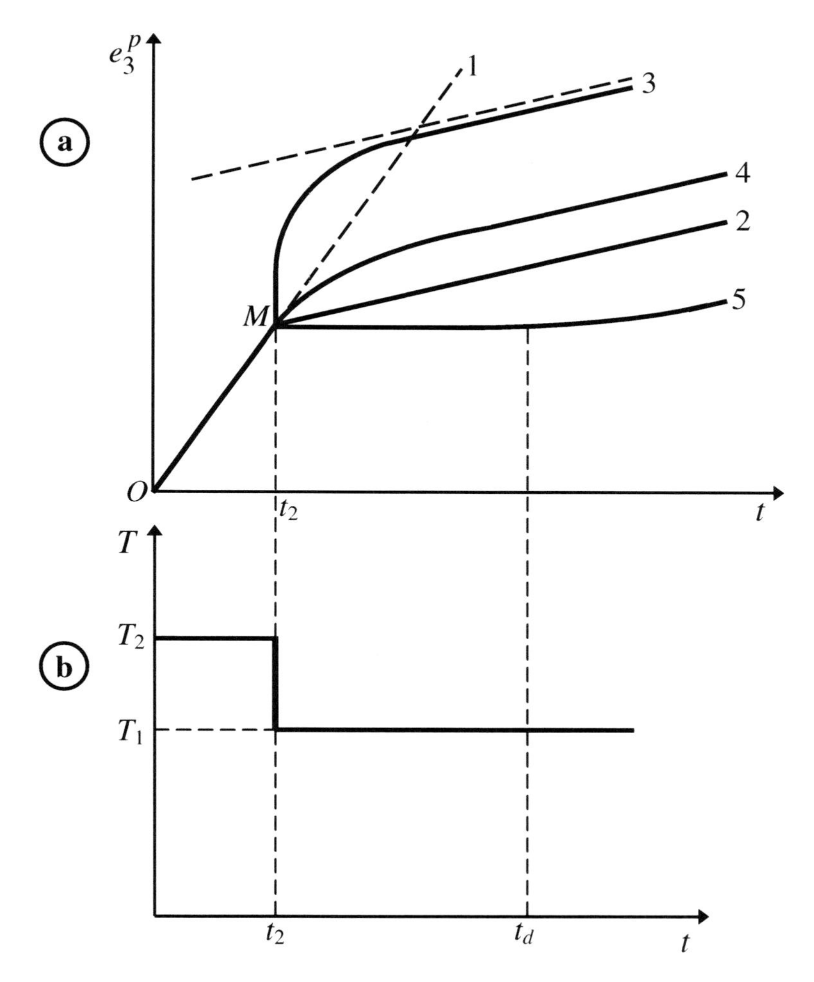

Fig. 5.6 Creep diagrams (a) at sudden temperature decrease (b)

It is easy to show that the after-effect strain at cooling is smaller than that at heating because the creep limit S_P depends on temperature. If the interval of temperature changes is such that the creep limit change is immaterial, the after-effect strains are identical at both heating and cooling.

The fact that upon both heating and cooling the loaded specimen acquires the after-effect strain increments of the same sign contradict to the generally accepted idea that a temperature increase leads to the increase of specimen size and vice verse.

2. Further, consider the case if the radicands in Eqs. (5.5.1) and (5.5.2) are equal to each other,

$$S_P^2(T_1) - B_T \Delta T = S_P^2(T_2), \tag{5.5.5}$$

meaning that the defects intensity ψ_N changes with time in a continuous way, i.e., it does not experience the step-wise increment due to the temperature drop at $t = t_2$. This follows from Eqs. (5.5.1), (5.5.2), and (5.5.5). Hence, $\Delta e_3^P = 0$ and the $e_3^P(t)$ function behaves like the creep diagram marked by 4 in Fig. 5.6a.

3. Finally, we consider the case if

$$S_P^2(T_1) - B_T \Delta T > S_P^2(T_2), \tag{5.5.6}$$

i.e., the right-hand side in Eq. (5.5.2) is greater than that in Eq. (5.5.1). The condition (5.5.6) means that tangent planes move away from the origin of coordinates by a finite magnitude, i.e., they are not located on the endpoint of stress vector $\vec{\mathbf{S}}$ any more. Therefore, the irreversible straining is terminated as $t > t_2$. However, if the irreversible strain intensity does not develop, the defects relaxation governed by Eq. (4.32.3) takes place. The solution of Eq. (4.32.3) at the initial condition similar to (5.4.2) can be written as

$$\psi_N = \left[S_3^2 \cos^2 \xi - S_P^2(T_2)\right] \exp[-K(T_1)(t - t_2)]. \tag{5.5.7}$$

Since the temperature stress parameter (5.3.1) after the temperature change takes the form of Eq. (5.4.4), the planes that produce irreversible strain prior to the temperature change are at a distance of

$$H_N = \left\{S_P^2(T_1) + \left[S_3^2 \cos^2 \xi - S_P^2(T_2)\right] \exp[-K(T_1)(t - t_2)] - B_T \Delta T \exp[-p_T(t - t_2)]\right\}^{1/2} \tag{5.5.8}$$

from the origin of coordinates for $t > t_2$. The right-hand side in the above formula decreases with time meaning that the tangent planes move toward the origin of coordinates. Let us designate by t_d the instant when the plane with coordinates $\beta = \pi/2$ and $\lambda = 0$ is on the endpoint of stress vector $\vec{\mathbf{S}}$, or, in other words, the distance H_N at $\beta = \pi/2$, $\lambda = 0$ is equal to S_3. Equating the right-hand side in Eq. part (5.5.8) to S_3, we obtain

$$S_P^2(T_1) + \left[S_3^2 - S_P^2(T_2)\right] \exp[-K(T_1)(t_d - t_2)] - B_T \Delta T \exp[-p_T(t_d - t_2)] = S_3^2. \tag{5.5.9}$$

The interval of time $t_d - t_2$ is called the creep delay. For $t > t_d$, the irreversible straining develops not only on the plane with coordinates $\beta = \pi/2$

and $\lambda = 0$ alone, but also does on some set of planes as well. Then, instead of Eq. (5.5.8), we have

$$H_N = \left\{S_P^2(T_1) + \left[S_3^2 \cos^2 \xi - S_P^2(T_2)\right]\exp[-K(T_1)(t-t_2)] - B_T \Delta T \exp[-p_T(t-t_2)] + \psi_{1N}\right\}^{1/2}, \quad (5.5.10)$$

where ψ_{1N} is the defects intensity due to irreversible strain developed at $t > t_d$. From this formula and the equality $H_N = S_3 \cos\xi$, the defects intensity ψ_{1N} takes the form

$$\psi_{1N} = S_3^2 \cos^2 \xi - S_P^2(T_1) + B_T \Delta T \exp[-p_T(t-t_2)] - \left[S_3^2 \cos^2 \xi - S_P^2(T_2)\right]\exp[-K(T_1)(t-t_2)]. \quad (5.5.11)$$

The creep strain intensity rate is related to the defects intensity ψ_{1N} by Eq. (4.32.1):

$$r\dot{\varphi}_N = K(T_1) S_3^2 \left(\cos^2 \xi - \sin^2 \beta_3\right), \ \sin^2 \beta_3 = \frac{S_P(T_1)}{S_3^2} - \frac{B_T \Delta T}{S_3^2}\left(1 - \frac{p_T}{K(T_1)}\right)\exp[-p_T(t-t_2)]. \quad (5.5.12)$$

The strain vector rate component is determined by the integrating over $\dot{\varphi}_N$ in Eq. (3.9.14):

$$\dot{e}_3^P = \frac{\pi}{3r} K(T_1) S_3^2 \sin^2 \beta_3 F(\sin \beta_3) \quad (5.5.13)$$

and the strain vector component is

$$e_3^P = \frac{\pi}{3r} K(T_1) \int_{t_d}^{t} S_3^2 \sin^2 \beta_3 F(\sin \beta_3) dt. \quad (5.5.14)$$

Equation (5.5.14) gives the curve marked by 5 in Fig. 5.6a.

5.6 Temperature Strengthening

The temperature after-effect considered above is not the only effect observed during a non-isothermal creep. So-called temperature strengthening is recorded in experiments [23,82] when the temperature behaves as shown in Fig. 5.7b. First, $T = T_2$ for $t \in [0, t_0)$, further the lower temperature prevails, T_1, for the time-period $t_1 - t_0$ and, at $t = t_1$, the temperature returns to its initial value T_2 remaining unaltered for $t > t_1$. The experiments give that under some conditions:

$$\dot{e}_3^1 < \dot{e}_3^0, \quad (5.6.1)$$

where $\dot{e}_3^0$ and $\dot{e}_3^1$ is the stationary creep rate of specimen at $T = T_2$ as $t < t_0$ and $t > t_1$, respectively, and the difference between $\dot{e}_3^0$ and $\dot{e}_3^1$ can take considerable values.

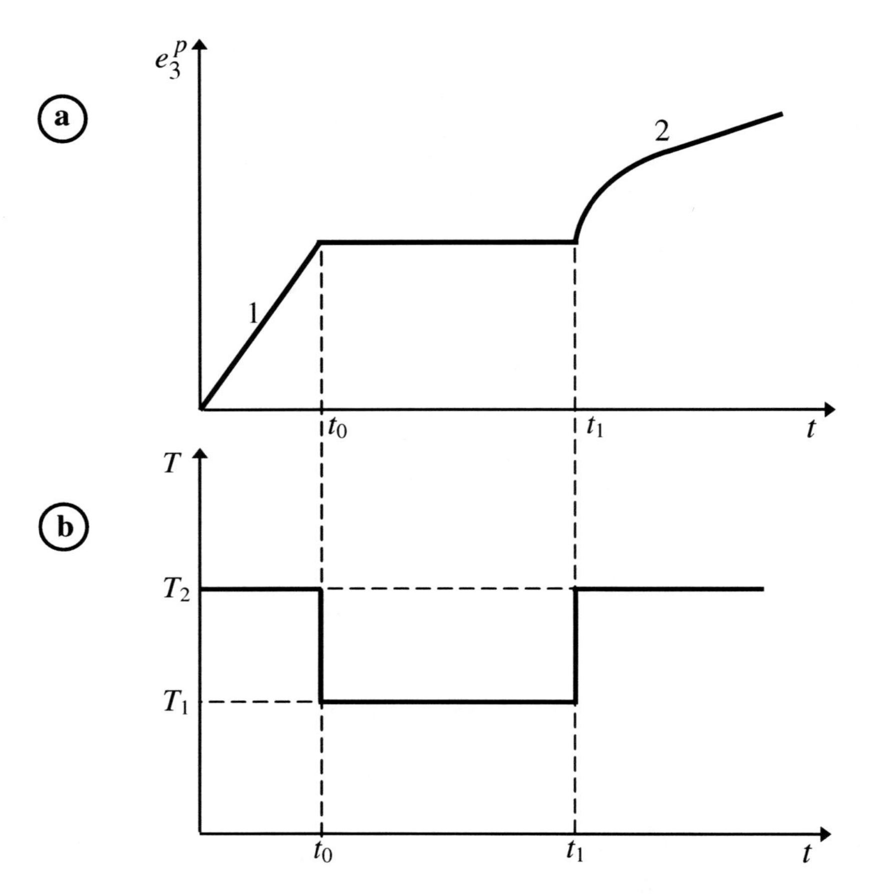

Fig. 5.7 The effect of temperature strengthening

Therefore, the low-temporary period can result in the strengthening effect (5.6.1) that is called the temperature strengthening. The creep diagram at the presence of the temperature strengthening is illustrated by Fig. 5.7a. The line 1 is the isothermal creep diagram at the higher temperature, and the curve 2 is the creep diagram at the presence of strengthening. In Fig. 5.7b, the value of lower temperature T_1 is such that the strain for $t \in (t_0, t_1)$ does not occur at all, or, in other words, the creep delay spreads to the whole lower-temperature period. As one can see from Fig. 5.7a, the tangent of curve 2 on its stationary portion is less than that of 1. The decrease in creep rate at higher temperature reduces with the number of temperature jumps and, eventually, terminates.

The temperature strengthening effect has been recorded in the experiments [23,82] on the metals having the body center cubic structure such as nickel, copper, silver, and brass. This effect is not observed in materials with other structures, they experience only the temperature after-effect.

The main condition for the arising of temperature strengthening is the lower temperature of cycle T_1 to be less than a critical one, T_c, which is about $0.4 \div 0.5T_m$, where T_m is the melting point of material. If $T_1 > T_c$, the temperature strengthening is absent.

Experiments [23] show that the temperature strengthening depends on:

1) the magnitude of temperature jump, $\Delta T = T_2 - T_1$, independent of the values of T_2 and T_1 subject to the restriction that $T_1 < T_c$. Even an insignificant violation of this inequality (only by 10-15°C) results in the absence of strengthening;

2) cooling rate: The temperature strengthening is manifested only at a sufficiently fast cooling;

3) the duration of lower-temperature period, $t_1 - t_0$; the strengthening grows with the increase of this period;

4) the magnitude of acting stress.

The mathematical model of the temperature strengthening can be found in the work [131].

Summarizing the results of Secs. 5.1–5.6, one can see that the temperature after-effect and strengthening result in opposite effects, assisting and hindering creep strains, respectively. If the lower temperature of cycle is larger than the critical value, the strengthening is absent and creep diagrams are qualitatively identical and shown by curve 1 in Fig. 5.3 (independent of the type of lattice). As one can see, the temperature after-effect strain is larger than isothermal creep strain. If there are no thermal effects, the strain develops along line 4 in Fig. 5.3 whose portions are parallel to the linear parts of isothermal diagrams 2 and 3 at higher and lower temperature, respectively. As seen, the diagram 1 is above the broken line 4 and the distance between them increases with the number of temperature cycles.

In the case if $T_1 < T_c$, the creep diagrams of metals having body center cubic structure and other cell structures are considerably different. For non-BCC structures the mutual arrangements of the diagrams 1 and 4 is as such the same as in Fig. 5.3 because of the absence of temperature strengthening. However, the creep diagram of BCC-metals due to temperature strengthening is below line 4 in Fig. 5.3 that is constructed for the case of the absence of thermal effects.

Summarizing, abrupt temperature changes can considerably influence upon the irreversible straining of metals resulting in both the increase and reduction of creep deformation.

5.7 Influence of Atomic Irradiation Upon Plastic and Creep Deformation of Metals[3]

Let us proceed to the mathematical description of the laws of irreversible deformation in radiation field.

Due to nuclear power finds expanding applications, it is necessary to investigate the influence of irradiation upon the deformation properties of metals. The structural members of nuclear reactor are exposed to irradiation by high-powered neutron-, proton-, and electron-flux, produced by radioactive decay.

The macro effect of radiation will be discussed below. Now, we consider an example illustrating the scale of problem. The yield limit of iron, for example, subjected to irradiation by neutron-flux of intensity 10^{12} *particles/cm*^{2} *per second* for one week grows two-fold. Therefore, we deal not with the change in mechanical property of about 3-5 %, but of 100 %.

The basic influence of irradiation upon metals consists in the following. High-energy neutrons, penetrating into the crystal lattice of solid, knock-out atoms from their places. As a result, so-called Frenkel pairs or Frenkel defects ("interstitial atom + vacancy") nucleate. The Frenkel defect forms when an atom leaves its place in the lattice (creating a vacancy), and lodges nearby in the crystal (becoming an interstitial). To displace an atom of such metal as, e.g., copper, and to form the Frenkel pair, the energy of about 20 eV is needed. Therefore, the penetrated neutron with energy of the order of 10^6 eV knocks out thousands of atoms from their places. The distance between knocked-out atoms produced by a single neutron is large enough as the neutron is neutral and it does not experience attractive forces from nuclei; the neutron propagation can be stopped only if it collides with atomic nucleus. Knocked-out atoms can initiate the displacements of other atoms. To imagine the scale of the given phenomenon we refer to data [47]. About 5 % of the atoms of metal taking the dose by radiation of intensity 10^{20} neutron/cm^2 for one year are knocked. In uranium elements, about 5 % of atoms are knocked out for one day.

Together with the Frenkel pairs, other radiation-induced defects arise; they are not considered here. For us, the facts that (i) so-called radiation defects arise and (ii) they are of different nature than those due to deformation are of major importance.

The numerous experiments testify [47] the following properties of radiation defects:

1. The number of radiation defects is proportional to the irradiation time in a short-term irradiation.
2. The saturation effect is observed in long-term irradiation when the number of radiation defects takes a stationary value.

---

[3] Secs. 5.7-5.12 are written by M. Rusinko.

3. The number of radiation defects strongly depends on a radiation temperature, it grows with the homology temperature Θ if $0 \leq \Theta \leq 0.5$, reaches the maximum approximately at $\Theta = 0.5$, and decreases as a function of Θ for $\Theta > 0.5$.

4. If to terminate an irradiation, the number of radiation defects decreases with time and, eventualy, they almost completely disappear for the following reasons [109]: (i) if the number of interstitial atoms is larger than that in equilibrium condition, they start to fill vacancies and (ii) the part of interstitial atoms and vacancies is attracted by dislocations on grain boundaries.

Similarly to any distortion of lattice, radiation-induced defects hinder [110] the movements of dislocations which are one of the basic mechanism of irreversible deformation. On the other hand, blocked dislocations attract the part of interstitial atoms and vacancies that results in the dislocations climb to adjacent slip planes and get around obstacles to their motion, allowing further deformation to occur. Thus, radiation defects assist glide-plus-climb deformation mechanism which is one of the basic creep mechanisms.

Consider, further, how radiation affects a plastic and creep deformation.

(1) Radiation leads to the increase in the yield limit of material and the decrease in plastic deformation. This statement is true for the overwhelming majority of metals irrespective of their crystal structure. Qualitative stress-strain diagrams in tension are shown in Fig. 5.8, where the stress-strain diagram in radiation and an ordinary stress-strain diagram are marked by 1 and 2, respectively. Consequently, to produce some amount of plastic deformation in irradiation field, the greater stress is needed to apply than in the case of ordinary loading.

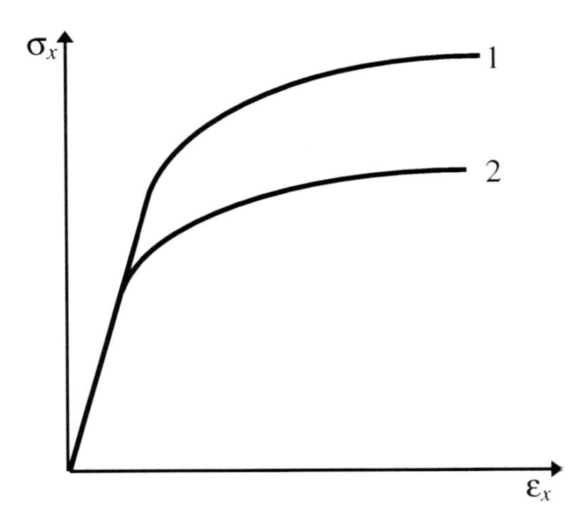

Fig. 5.8 Stress-strain diagram in uniaxial tension under irradiation (1) and without irradiation (2)

(2) Radiation assists a creep deformation; the radiation creep is considerably larger than that without radiation. As specified in Sec. 4.36, steady-state creep is absent for the temperature range $0 < \Theta < 0.3$. In contrast to this, the radiation creep is observed on the whole temperature interval, $0 < \Theta < 1$.

If to denote creep strain components developed in radiation field by ε_{ij}^{p} and ordinary (thermal) creep strain components by ε_{ij}^{pT}, we will call their deference as,

$$\varepsilon_{ij}^{p} - \varepsilon_{ij}^{pT} = \varepsilon_{ij}^{pp} \tag{5.7.1}$$

the radiation creep components.

The diagrams of both the thermal and radiation creep have unsteady-state, steady-state, and tertiary portions. Numerous experimental data give the following dependence of steady-state radiation creep rate on temperature. The steady-state radiation creep rate grows with temperature for $0 \le \Theta \le 0.5$ and has a maximum for $\Theta = 0.5$. As $0.5 < \Theta \le 1$, the function $\dot{\varepsilon}_{ij}^{pp}(\Theta)$ decreases and $\dot{\varepsilon}_{ij}^{pp}(1) = 0$. As seen from Fig. 5.9, a plot of $\dot{\varepsilon}_{ij}^{pp}$ as a function of Θ yields the form of semicircle.

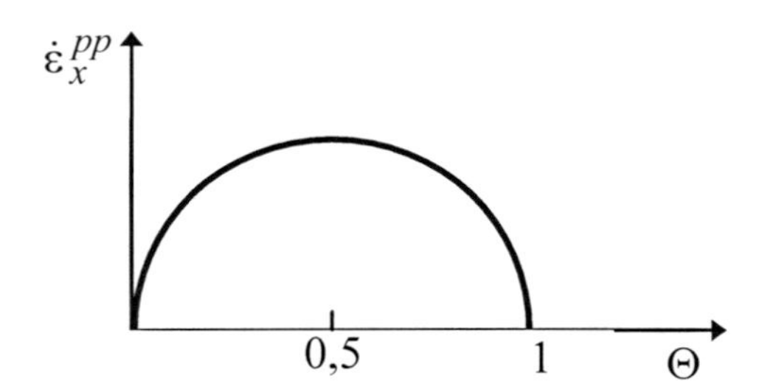

Fig. 5.9 *Radiation steady-state creep rate* vs. *homology temperature* plot

Therefore, summarizing, one can see that the thermal and radiation creep have a different dependence of temperature. The steady-state thermal creep does not occur for $0 < \Theta < 0.3$ at all and grows monotonically at temperatures in excess of a homologous temperature of 0.3, whereas the radiation steady-state creep occurs on the whole range of homology temperature, $0 < \Theta < 1$, however, it is not of a monotonic character like the thermal creep. As an example, Fig. 5.10 illustrates experimental "$\dot{\varepsilon}_{x}^{Pp} - T$" (lines 1 and 1') and "$\dot{\varepsilon}_{x}^{p} - T$" (lines 2 and 2') plots obtained in creep tests on specimens of aluminum in tension. The lines 1' and 2' correspond to the tensile stress of 9.8 MPa and line 1 and 2 – 14.7 MPa. The specimens were exposed to the long-term time-independent irradiation of density $\Phi = 2.5 \cdot 10^{15}$ fluence. As seen from Fig. 5.10, there is the temperature-diapason where the thermal creep does not develop, whereas the radiation creep accumulates. At significant temperatures, the thermal creep prevails [161].

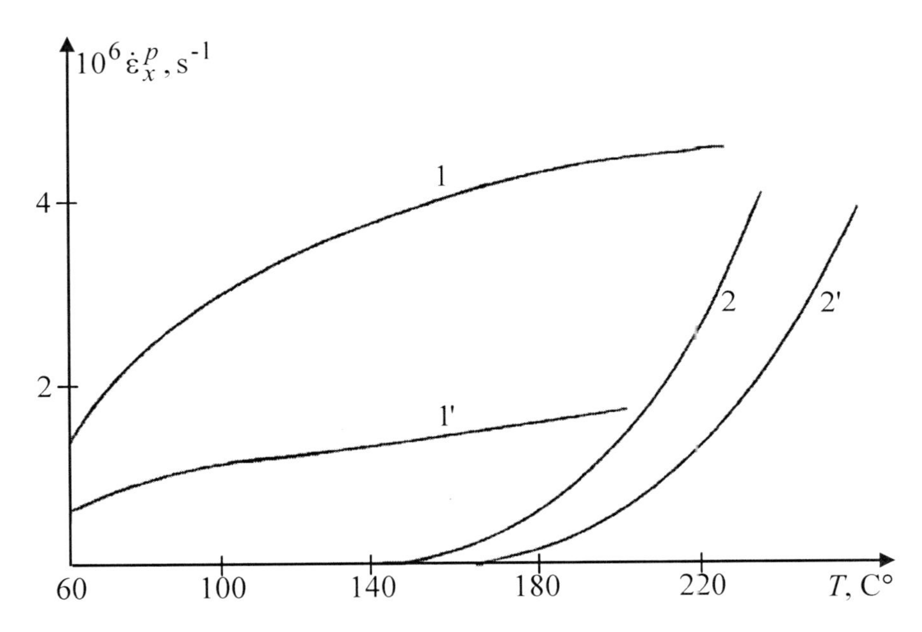

Fig. 5.10 Radiation (1 and 1') and ordinary (2 and 2') steady-state creep rate as a function of temperature

Another experimental facts are that *steady-state-creep-rate* vs. *stress* dependence is different for the thermal and radiation creep. Function $\dot{\varepsilon}_{ij}^{pT}(\sigma_{ij})$ governed by Eqs. (4.1.1) and (4.1.2) is of nonlinear character whereas $\dot{\varepsilon}_{ij}^{pp}$ are proportional to σ_{ij} and the proportionality holds for a wide stress-range.

5.8 Mathematical Theory of the Irreversible Deformation in Irradiation

At present, there is the huge number of works devoted to the influence of radiation on the properties of material. The most of them report the results of tests carried out on atomic-molecular level. They concern, for example, with the following issues: (i) the quantity of primary and secondary Frenkel pairs dependently on irradiation-energy and -time, temperature, and stress; (ii) the behavior of knocked out atoms, their diffusion and interaction with dislocations; (iii) dislocation-climbs activated by the diffusion of interstitial(knocked-out) atoms; (iv) the Frenkel pair annihilation, the formation of more complicated defects than Frenkel pair; (v) radiation defects stability, etc.

The great number of theoretical works is published as well. They are devoted to the study of radiation creep resulted from the motion of dislocations and other defects of crystal lattice. In these works, for example, the following processes, magnitudes, and phenomena are considered and modeled: (i) the time required for dislocation to climb around obstacles; (ii) dislocation free propagation length; (iii) how effectively point (radiation) defects are absorbed by dislocations, the interaction between a dislocation core and radiation defects, etc.

These works, both experimental and theoretical ones, are of unquestionable importance, due to they elucidate how a radiation affects plastic and creep straining. However, we have to note the fact, which does not detract from their merits, that they are incapable of predicting a plastic/creep behavior in radiation filed even for a single crystal grain, to say nothing of a polycrystalline aggregate.

Therefore, to model the influence of radiation upon plastic/creep straining, one need to apply such a theory that depicts the essence of this influence – radiation induces the defects of crystal lattice, radiation defects, which, together with defects due to irreversible straining, determine the deformation properties of material. We carry out the modeling of radiation effects in terms of the generalized synthetic theory formulated in Sec. 4.32 for the following reasons:

1. The basic mechanism of plastic or creep straining in radiation (as well as at the absence of radiation) is slips within crystal grains. Hence, the concept of slip and its modification, the synthetic theory, are applicable.

2. The purpose of researches is to establish the laws of both plastic and creep straining. The generalized synthetic model describes either forms of irreversible deformation.

3. In terms of the synthetic theory, the strain-hardening is determined through defects intensity. The radiation-induced imperfections of crystal lattices, radiation defects, are readily taken into account because of the characteristic of defects appears in the constitutive relationship of the synthetic theory.

We introduce a new notion, the radiation defects intensity, ψ_p, into the relationships of synthetic model; ψ_p is an average measure of defects caused by irradiation,. We suppose that ψ_p is of isotropic nature, i.e., independent of angles α, β, and λ, because of radiation is assumed to be identical in all micro-volumes of body. This assumption is allowable due to the neutron-energy is so large (above 10^{10}eV) that propagate through a body with almost unaltered energy (it is clear that this depends on the body size).

As stated above, a radiation causes the formation of defects; on the other hand, they relax with time after the termination of radiation. According to the Bailey principle, the relaxation rate is proportional to the number of defects. Therefore, we propose the following equation for ψ_p:

$$d\psi_p = c\Phi dt - K_3\psi_p dt\,; \qquad (5.8.1)$$

the form of K_3 will be considered below (Sec 5.12), c is constant, and Φ is the radiation density (fluence), which is defined as the number of particles (e.g., neutrons) that intersect a unit area in a unit time. Its units are $\mathrm{m}^{-2}\cdot\sec^{-1}$ (number of particles per meter squared per second). Further throughout, we assume that the density Φ is a given function of time, i.e., the irradiation regime is known.

As stated in the previous section, the Frenkel pair can arise not only due to neutron bombardment, but also because of the bombardment by other particles (protons, electrons, etc). In terms of the synthetic model, we do not distinguish between types of particles; different values of c in (5.8.1) correspond to different types of bombarding particles.

The solution of Eq. (5.8.1) for initial condition $\psi_p = 0$ at $t = 0$ and at stationary irradiation ($\Phi(t) = cons$) is

$$\psi_p = \frac{c\Phi}{K_3}\left(1 - \exp(-K_3 t)\right). \qquad (5.8.2)$$

At the small values of $K_3 t$ Eq. (5.8.2) becomes

$$\psi_p = c\Phi t\,. \qquad (5.8.3)$$

Therefore, in short-term radiation, the defects intensity is proportional to the radiation time that agrees with experiments (point 1 in Sec. 5.7). For the case of long-term radiation, $t \to \infty$, Eq. (5.8.2) gives that

$$\psi_p = \frac{c\Phi}{K_3}. \qquad (5.8.4)$$

Therefore, the number of defects does not tend to infinity with time, but is bound by the maximal value (5.8.4). The finite value of ψ_p at $t \to \infty$ is in accord with experimental data from the point 2 in Sec. 5.7.

The radiation defects relaxation at $\Phi = 0$ that follows the radiation lasting the length of time t_1 is governed by Eq. (5.8.1),

$$d\psi_p = -K_3\psi_p dt\,, \qquad (5.8.5)$$

whose solution at the initial condition

$$\psi_p = \frac{c\Phi}{K_3}\left(1 - \exp(-K_3 t_1)\right) \tag{5.8.6}$$

is

$$\psi_p = \frac{c\Phi}{K_3}\left(1 - \exp(-K_3 t_1)\right)\exp\left(-K_3(t - t_1)\right). \tag{5.8.7}$$

One can see that ψ_p vanishes that satisfies the point 4 of experimental data in Sec. 5.7.

Since the radiation defects intensity ψ_p depends on temperature in the same way as steady-state radiation creep rate does, function K_3 (we will show below (Sec. 5.12) that the function K_3 does not depend on stress) can be presented as

$$K_3 = \frac{B_p}{\Theta(1-\Theta)}, \tag{5.8.8}$$

where B_p is the model constant. Indeed, $K_3 \to \infty$ as $\Theta \to 0$ and $\Theta \to 1$, and , i.e., $\psi_p = 0$; the function K_3 has a minimum for $\Theta = 0.5$ meaning that ψ_p takes its maximal value at $\Theta = 0.5$. The same behavior of $\dot{\varepsilon}_x^{pp}$ as a function of homology temperature is shown in Fig. 5.9.

Equations (5.8.1) and (5.8.8) determine the radiation defects intensity. Since $\Phi = \Phi(t)$ is a given function of time, we consider ψ_p as a known function of time as well.

Let us designate through ψ_N a total defects intensity, the sum of the radiation and deformation defects intensities. Then $\psi_N - \psi_p$ gives the deformation defects intensity induced by irreversible straining. Now, Eq. (4.32.1), which holds for the deformation defects, takes the form:

$$d\left(\psi_N - \psi_p\right) = r d\varphi_N - K\left(\psi_N - \psi_p\right)dt, \tag{5.8.9}$$

where K is the function determining the deformation-defects relaxation rate; we will set its form in Sec. 5.12. It must be noted that K and K_3 are different functions (though both of them regulates defects relaxation rate) as the deformation and radiation defects are of different nature and, consequently, their relaxations are governed by different laws. Let us also note that the function K in (5.8.9) cannot be reduced to the form of Eq. (4.36.1) obtained at the absence of radiation.

Equations (4.32.4) or (4.32.5) determines a tangent plane distance, the condition for the producing of irreversible strain, Eq. (4.32.6), and Eq. (3.9.4) driving macro irreversible strain vector components remains unchangeable. The relation between tangent plane distances with opposite normals $\vec{\mathbf{N}}$ and $-\vec{\mathbf{N}}$ (Eqs. (4.32.7) or (4.32.8)), the displacement of a plane on the endpoint of stress vector induces the displacement of the plane with normal vector $-\vec{\mathbf{N}}$, remains valid as well. In view of radiation defects, Eq. (4.32.9) is replaced by the following

$$\psi_{-N} - \psi_p = -(\psi_N - \psi_p) \tag{5.8.10}$$

and

$$\psi_{-N} = -\psi_N + 2\psi_p. \tag{5.8.11}$$

Taking into account that the radiation defects intensity is of isotropic nature, i.e., independent of angles α, β, and λ, we have

$$\psi_{-N} = \psi_N = \psi_p, \tag{5.8.12}$$

Since a radiation induces the arising of crystal lattice defects, it is natural to suppose that the local peak stresses also increase. Therefore, in place of Eq. (4.7.1), now we propose the following formula:

$$I_N = B(1 + c_1\psi_p)\int_0^t \frac{d\vec{\mathbf{S}}}{ds} \cdot \vec{\mathbf{N}} Q(t-s)ds. \tag{5.8.13}$$

Equations (3.9.4), (4.32.4)–(4.32.8), (5.8.1), and (5.8.8)–(5.8.13) are the basic relationships of the generalized synthetic theory that describe the influence of radiation upon the plastic/creep strain of metals.

5.9 Creep and Yield Criteria

Let a body be exposed to radiation resulting in the formation of radiation-induced defects whose intensity ψ_p is given by Eq. (5.8.1) and assumed to be a known function of time.

Since, prior to the start of irreversible straining, there is no deformation defects in a body, we have $\psi_N = \psi_p$. In addition, since the creep limit is defined at the yield limit with infinitesimal loading rate the integral of non-homogeneity I_N is zero. Therefore, Eq. (4.32.4), which determines tangent plane distances, H_N, in the Ilyushin stress-deviator space, takes the form

$$H_N = S_P + \psi_p, \tag{5.9.1}$$

where S_P is the creep limit of metal at the absence of radiation. The right-hand side in Eq. (5.9.1) defines the creep limit of metal in radiation, S_{Pp}:

$$S_{Pp} = S_P + \psi_p. \tag{5.9.2}$$

If the distance H_N is given by Eq. (4.32.5), we have

$$S_{Pp} = \sqrt{S_P^2 + \psi_p} \tag{5.9.3}$$

instead of (5.9.2). Let us note that the units of ψ_p in Eqs. (5.9.2) and (5.9.3) are different. If to accept Eq. (5.9.2), the equation of creep surface in Ilyushin's three-dimensional subspace becomes as

$$S_1^2 + S_2^2 + S_3^2 = \left(S_P + \psi_p\right)^2. \tag{5.9.4}$$

Now, let us determine the yield limit of metal in radiation. The condition for the onset of plastic deformation is the stress vector $\vec{\mathbf{S}}$ reaches first tangent plane. In contrast to Eq. (5.9.2), when determining yield limit, the integral of non-homogeneity, I_N, is non-zero and must be taken into account. As an example, consider the case of pure shear when the stress vector $\vec{\mathbf{S}}(0,0,S_3)$ is directed along S_3-axis. Therefore, the first plane tangential to the sphere (5.9.4) to be on the endpoint of the stress vector is perpendicular to S_3-axis ($\beta = \pi/2$). In addition, the fact that the plane is tangential to the creep surface in ***three***-dimensional subspace immediately implies that $\lambda = 0$. Therefore, by designating the yield limit in radiation as S_{Sp}, the condition $S_3 = S_{Sp} = H_N(\beta = \pi/2, \lambda = 0)$ and Eq. (4.32.4) give that

$$S_P + \frac{Bv}{p}\left(1 - \exp(-pt_S)\right)\left(1 + c_1\psi_p\right) + \psi_p = S_{Sp}. \tag{5.9.5}$$

In arriving at the result (5.9.5) we have taken a time-independent value of loading rate, v. Eq. (5.9.5) contains two unknown quantities, the yield limit S_{Sp} and the time, t_S, needed for stress to become equal to S_{Sp}. It is obvious that $S_{Sp} = vt_S$ meaning that (5.9.5) is a transcendental equation for one variable, S_{Sp}. Equation (5.9.5) is obtained for the case of pure shear, but it is also valid for any proportional loading in three-dimensional subspace.

Yield limit vs. *loading rate* diagram is shown in Fig. 5.11 by line 1. If $\Phi = 0$ and $\psi_p = 0$, the yield limit is smaller than that in radiation (line 2 in Fig. 5.11).

Both diagrams have horizontal asymptote corresponding to the case $v \to \infty$. If $v \to \infty$, $t_S \to 0$ and Eq. (5.9.5) in which the term $\exp(-pt_s)$ is expressed by a Tailor series approximation,

$$S_P + BS_{Sp}\left(1 + c_1\psi_p\right) + \psi_p = S_{Sp},$$

gives that

$$S_{Sp} = \frac{S_P + \psi_p}{1 - B\left(1 + c_1\psi_p\right)}, \quad (v \to \infty). \tag{5.9.6}$$

In particular, the yield limit S_S at $\psi_p = 0$, is

$$S_S = \frac{S_P}{1 - B}, \quad (v \to \infty). \tag{5.9.7}$$

The difference between S_{Sp} and S_S,

$$S_{Sp} - S_S = \frac{\psi_p\left[1 - B\left(1 - c_1 S_p\right)\right]}{(1 - B)\left[1 - B\left(1 + c_1\psi_p\right)\right]}, \tag{5.9.8}$$

is shown by segment M_1M_2 in Fig. 5.11. As seen from Eq. (5.9.2), the radiation defects constitute the difference between the creep limits S_{Pp} and S_P, which lies on the axis of ordinates in Fig. 5.11 $(v = 0)$. Hence, $S_{Sp} - S_S > S_{Pp} - S_P$; this follows from Eqs. (5.9.8) and (5.9.2).

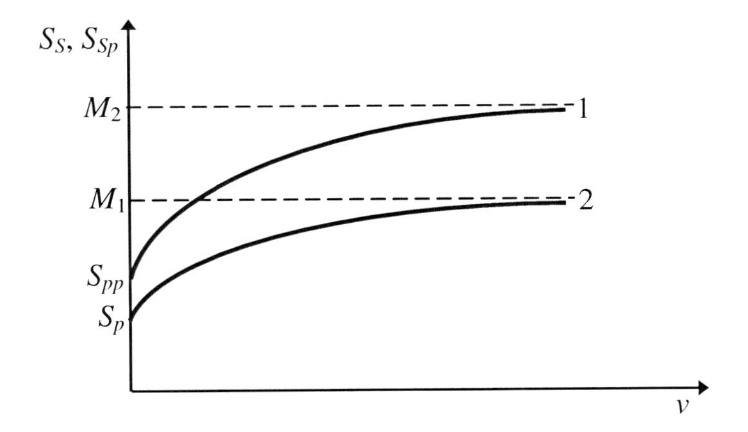

Fig. 5.11 Radiation induced (1) against ordinary (2) yield limit diagram

5.10 Plastic Deformation in Radiation

Consider the case when a specimen loaded in pure shear, $\tau_{xz} = S_3/\sqrt{2}$, is exposed to irradiation so that radiation defects of intensity ψ_p arises.

First let us find the plastic deformation due to loading rate $v = \dot{S}_3$ which is assumed time-independent. Tangent plane distances can be found from Eq. (4.32.4) which, together with Eq. (5.9.5) for S_{Sp}, takes the form

$$H_N = S_P + I(1 + c_1\psi_p)\cos\xi + \psi_N, \quad \cos\xi = \sin\beta\cos\lambda, \qquad (5.10.1)$$

where I is given by Eq. (4.6.1), ψ_N is the total defects intensity (radiation + deformation defects). If $S_3 > S_{Sp}$, an appropriate set of planes "produces" plastic deformation. This means that these planes are on the endpoint of stress vector and Eq. (4.32.6), $H_N = \vec{\mathbf{S}} \cdot \vec{\mathbf{N}} = S_3\cos\xi$, holds and gives the positive values of defects intensity in Eq. (5.10.1):

$$\psi_N = \left[S_3 - I(1 + c_1\psi_p)\right]\cos\xi - S_P. \qquad (5.10.2)$$

Let us write down the above formula in another form

$$\psi_N = S_{Pp}\left(\frac{\cos\xi}{a} - 1\right) + \psi_p, \quad a = \frac{S_{Pp}}{S_3 - I(1 + c_1\psi_p)}, \quad S_{Pp} = S_P + \psi_p. \qquad (5.10.3)$$

As before (Sec. 4.33), when determining plastic strain components, the second term in Eq. (5.8.9) may be neglected in comparison to the first one because it is manifested only at long-term loads (for 100 or 1000 hours). Therefore, as we study the development of plastic strain when loading lasts several minutes, instead of Eq. (5.8.9), we have

$$d(\psi_N - \psi_p) = r\,d\varphi_N, \quad (\psi_N \ge \psi_p) \qquad (5.10.4)$$

and

$$r\varphi_N = \psi_N - \psi_p. \qquad (5.10.5)$$

Hence,

$$r\varphi_N = S_{Pp}\left(\frac{\cos\xi}{a} - 1\right). \qquad (5.10.6)$$

The plastic strain is produced by those planes for which $\psi_N \geq \psi_P$. By employing this condition we also arrive at Eqs. (3.9.12), (3.10.7), (3.10.9), and (5.10.3) that determine the range of angles α, β,and λ where $\psi_N \geq \psi_P$.

Since the intensities (5.10.6) and (4.9.6) are, pre se, of identical form, plastic strain vector component e_3^S is determined as follows

$$e_3^S = \frac{\pi S_{Pp}}{3r} F(a), \tag{5.10.7}$$

where the function F is given by Eq. (3.10.14). The obtained formula is readily generalized to the case of arbitrary proportional loading. Doing the same transformations as in Secs. 3.11 and 4.11, we obtain

$$e_k^S = \frac{\pi S_{Pp}}{3r} \frac{S_k}{S} F(a), \tag{5.10.8}$$

where S is the module of stress vector, the argument a is determined by Eq. (5.10.3), and parameter I, which appears in the formula for a, is given by Eqs. (4.5.20)–(4.5.22). Equation (5.10.8) determines the plastic strain vector components of irradiated body in simple loading. In particular, for the case of uniaxial tension ($\varepsilon_x^S = \sqrt{2}\, e_1^S / \sqrt{3}$, $S_1 = S = \sqrt{2}\, \sigma_x / \sqrt{3}$), in view of elastic components, we have

$$\varepsilon_x = \frac{\sigma_x}{E} + \frac{\sqrt{2}\left(S_P + \psi_p\right)}{3\sqrt{3}r} F(a) \quad S_P = \sqrt{2}\tau_P = \frac{\sqrt{2}\sigma_P}{\sqrt{3}} \tag{5.10.9}$$

where σ_P and τ_P are the creep limits of material in tension and torsion, respectively, in the absence of radiation.

Equations (5.10.9) and (5.10.3) describe the increase in the yield limit and the radiation embrittlement of material (Fig. 5.8). By the notion "radiation embrittlement" we mean that a stress applied to a body under irradiation and without it induces smaller strains in the radiation field.. The qualitative stress-strain diagrams in uniaxial tension in radiation (curve 1) and at $\Phi = 0$ (curve 2), respectively, corresponding to Eq. (5.10.9) are shown in Fig. 5.8.

It must be noted that the generalized synthetic theory is not capable of modeling the case when a sharp yield point arises on stress-strain diagram of radiated specimen, whereas it is absent in ordinary loading.

Equation (5.10.8) or (5.10.9) is obtained if the planes distance H_N is given by Eq. (4.32.4). It is easy to derive appropriate formulae for strain components if H_N is determined by Eq. (4.32.5).

It is easy to see that Eq. (5.10.8) expresses the law of deviator proportionality meaning that the generalized theory gives the existence of universal $\tau_0 - \gamma_0$ curve in radiation.

If to the convert the Ilyushin variables S_k and e_k^S to strain tensor components (by making use of Eqs. (1.7.10) and (1.7.12)), Eq. (5.10.8) yields the form

$$\varepsilon_{ij}^S = \frac{\pi(S_P + \psi_p)}{\sqrt{6}r} \frac{\bar{\sigma}_{ij}}{\tau_0} F(a). \tag{5.10.10}$$

If to write down the deviator proportionality in the standard form, Eq. (1.2.4) or (1.2.5), we arrive at Eq. (1.5.39). By comparing Eq. (1.5.39) to Eq. (5.10.10), we obtain that

$$\frac{1}{G_S} - \frac{1}{G} = \frac{\pi\sqrt{2}(S_P + \psi_p)}{\sqrt{3}r\tau_0} F(a). \tag{5.10.11}$$

This formula shows how the plastic shear modulus G_S is related to the elastic modulus G and function F. Equation (5.10.10) or (5.10.11) allows us to construct the stress-strain diagram in radiation if we have that without radiation. Indeed, to do this, one needs to multiply ordinary plastic strains by the value of $1 + \psi_p / \sigma_P$, while the function F must be calculated for the value of a determined not by formula $a = S_P / (S_3 - I)$, but by Eq. (5.10.3).

5.11 Creep Deformation in Radiation

Let us start with unsteady-state creep at pure shear. The loading regime is the same as that described in Sec. 4.23, namely, the stress vector component S_3 grows with constant speed v from zero to S_M, $S_M > S_S$, at time t_M, i.e., $S_3 = S_M = vt_M$, and $S_3(t) = S_M = const$ for $t \geq t_M$ (see Fig. 4.8a). The plastic strain vector deformation e_M^S at $t = t_M$ is determined by Eq. (5.10.7):

$$e_M^S = \frac{\pi S_{Pp}}{3r} F(a_M) \quad a_M = \frac{S_{Pp}}{S_M - (1 + c_1\psi_p)I_M}. \tag{5.11.1}$$

Since the integral of non-homogeneity I decreases with time as $t > t_M$, the argument a from Eq. (5.10.3) also decreases and, consequently, the strain grows with time. The difference between strains for $t > t_M$ and at $t = t_M$ gives the creep strain component. Therefore, we arrive at Eq. (4.23.5),

$$e_3^P = \frac{\pi S_{Pp}}{3r}\left(F(a) - F(a_M)\right), \tag{5.11.2}$$

where the argument a is given by Eq. (5.10.3) which contains the integral I relaxing with time by Eq. (4.6.2) or (4.6.4).

The obtained formula determines the total (thermal and radiation) creep strain component at pure shear. If radiation defects intensity is zero, we will obtain thermal creep strain component as

$$e_3^T = \frac{\pi S_P}{3r}\left(F(a_T) - F\left(a_{M_T}\right)\right), \tag{5.11.3}$$

where a_T and a_{M_T} are given by Eqs. (5.10.3) and (5.11.1), respectively, if to let $\psi_p = 0$. The radiation part of creep, according to (5.7.1), can be written as

$$e_3^{pp} = \frac{\pi}{3r}\left\{S_{Pp}\left[F(a) - F(a_M)\right] - S_P\left[F(a_T) - F\left(a_{M_T}\right)\right]\right\}. \tag{5.11.4}$$

To show how the integral of non-homogeneity exerts its influence upon the radiation strain intensity, consider the following computations. According to Eq. (5.10.5), the irreversible strain intensity for $t > t_M$ is

$$r\varphi_N = \left[S_M - \left(1 + c_1\psi_p\right)I\right]\cos\xi - S_{Pp}. \tag{5.11.5}$$

In particular, at $t = t_M$ we have

$$r\varphi_{N_M} = \left[S_M - \left(1 + c_1\psi_p\right)I_M\right]\cos\xi - S_{Pp}. \tag{5.11.6}$$

The creep strain intensity i.e., the difference between $r\varphi_N$ and $r\varphi_{N_M}$ is

$$r\varphi_N^p = (I_M - I)\left(1 + c\psi_p\right)\cos\xi \tag{5.11.7}$$

and its thermal ($\psi_p = 0$) and radiation parts are

$$r\varphi_N^{pT} = (I_M - I)\cos\xi, \tag{5.11.8}$$

$$r\varphi_N^{pp} = c(I_M - I)\psi_p\cos\xi, \tag{5.11.9}$$

respectively.

The formulae (5.11.7)–(5.11.9) illustrate that the driving force of unsteady-state is the local peak stresses, their relaxation (the reduction of integral I) induce the unsteady-state creep. In addition, these formulae give the relations between the complete, thermal, and radiation unsteady creep strain intensities.

It is easy to generalize Eq. (5.11.2), which determines the unsteady creep strain vector component in pure shear, to the case of simple loading. Let stress vector components, S_1, S_2, and S_3, vary in proportion to a common parameter with constant speeds. At some instant, t_M, the module of stress vector, S, reaches the value of $S = S_M = vt_M$ and further remains unaltered with time. By analogy with Eq. (4.24.1) creep strain vector components are obtained as

$$e_i^p = \frac{\pi S_{Pp}}{3r}\left(F(a) - F(a_M)\right)\frac{S_i}{S}. \tag{5.11.10}$$

If to convert strain- and stress-vector component, into strain- and stress-deviator-tensor components, we obtain

$$\varepsilon_{ij}^p = \frac{\pi S_{Pp}}{\sqrt{6}r}\left(F(a) - F(a_M)\right)\frac{\bar{\sigma}_{ij}}{\tau_0}, \tag{5.11.11}$$

where ε_{ij}^p are the unsteady-sate creep strain tensor components in radiation and $\bar{\sigma}_{ij}$ the stress-deviator-tensor components.

Let us proceed to the mathematical description of steady-state creep in radiation. Since the integral of non-homogeneity I_N is zero in steady-state creep, Eq. (5.10.2) gives that the defect intensity is of constant value (at the fixed angles β and λ). Assuming that irradiation defects intensity (given by, e.g., Eq. (5.8.4)) is also constant, the steady-state creep strain intensity rate in pure shear is

$$r\dot{\varphi}_N = K\left(S_3 \cos\xi - S_{Pp}\right). \tag{5.11.12}$$

By integrating $\dot{\varphi}_N$ in Eq. (3.9.4) we obtain the following steady-state creep strain vector rate

$$\dot{e}_3^p = \frac{\pi K S_{Pp}}{3r} F\left(\frac{S_{Pp}}{S_3}\right). \tag{5.11.13}$$

For the case of arbitrary state of stress we have

$$\dot{e}_i^p = \frac{\pi K S_{Pp}}{3r}\frac{S_i}{S} F\left(\frac{S_{Pp}}{S}\right), \tag{5.11.14}$$

or

$$\dot{\varepsilon}_{ij}^{p} = \frac{\pi K S_{Pp}}{\sqrt{6}r} \frac{\overline{\sigma}_{ij}}{\tau_0} F\left(\frac{S_{Pp}}{S}\right), \tag{5.11.15}$$

where $\overline{\sigma}_{ij}$ are the stress deviator tensor components.

5.12 The Determination of Function *K*

The function K in thermal creep is determined by Eqs. (4.36.1), (4.36.6), (4.36.10), and (4.36.12). From these formulas it follows that steady-state creep is absent if homology temperature is less than 0.3. Consequently, they are inapplicable to radiation creep as the function K appears in Eq. (5.11.14) as a factor at $\dot{e}_i^p$ which is non-zero for $0 < \Theta \leq 0.3$. We introduce a new function K in the following form

$$K = K_T + \psi_p K_p, \tag{5.12.1}$$

where K_T is the deformation defects relaxation rate related to the number of deformation defects; it is determined by Eqs. (4.36.1), (4.36.6), (4.36.10), (4.36.12), and (4.37.5). As for a new function, K_p, we make use of the fact that, according to Eq. (5.7.1), the radiation creeps is the difference between complete and thermal creep. Therefore, steady-state radiation creep rate in pure shear can be written as

$$\dot{e}_3^{Pp} = \frac{\pi}{3r}\left[KS_{Pp}F\left(\frac{S_{Pp}}{S_3}\right) - K_T S_P F\left(\frac{S_P}{S_3}\right)\right] \tag{5.12.2}$$

Let us note that the radiation or thermal creep develops if the module of stress vector $\vec{\mathbf{S}}$ is larger than the creep limit in radiation S_{Pp} or thermal creep limit S_P, respectively; in addition, $S_{Pp} > S_P$. If $S_P < S_3 < S_{Pp}$, Eqs. (5.11.4) and (5.12.2) are inapplicable as the first term in (5.12.2) is zero. They hold true if S_3 leads to positive strain values in (5.12.2).

Let us simplify the formula (5.12.2) if the component S_3 is much larger than the creep limit meaning that the argument a of function F takes a small value. In this case, according to (3.10.14), function F can be expressed in the following approximate form:

$$F = \frac{\pi}{2a}, \tag{5.12.3}$$

i.e.,

$$F\left(\frac{S_{Pp}}{S_3}\right)=\frac{\pi S_3}{2S_{Pp}}, \quad F\left(\frac{S_P}{S_3}\right)=\frac{\pi S_3}{2S_P}. \tag{5.12.4}$$

Consequently, Eqs. (5.12.1), (5.12.2), and (5.12.4) gives that

$$\dot{e}_3^{Pp}=\frac{\pi^2}{6r}\psi_p K_p S_3. \tag{5.12.5}$$

As follows from experiments presented in Sec. 5.7, the steady-state radiation creep rate depends linearly on stresses. Since, in the right-hand side of Eq. (5.12.5), the function K_p is multiplied by the stress component S_3 of power 1, we conclude that K_p does not depend on stress. For the same reason, the function K_3 appearing in the expression for radiation defects ψ_p does not depend on stress. The multiplier K_p does not depend on temperature as the relation between the temperature and the radiation creep rate is regulated by Eqs. (5.8.4) and (5.8.8) for ψ_p and K_3, respectively. Summarizing, it may be inferred that K_p is a stress- and temperature-independent quantity, i.e., $K_p = const$. Therefore, now, Eq. (5.12.1) is readily used.

In works [160,161], the comparison of theoretical results obtained by Eqs. (5.11.11) and (5.11.15) to experimental ones are presented, *radiation creep rate* vs. *flux density, radiation time, stress and temperature* diagrams are considered. As seen from these works, the satisfactory agreement between the analytical and experimental results is observed.

5.13 Phase Transformations

The following sections are concerned with the modeling of irreversible deformation that accompanies/induces phase transitions of metals.

Let us start with the definition of phase transition (PT). There exists a temperature-diapason within which materials undergo so-called phase transitions (transformations). PTs do not occur by the long-range diffusion of atoms but rather by some form of cooperative, homogeneous movement of many atoms that results in a change in crystal structure. These movements are small, usually less than the interatomic distances, and the atoms maintain their relative positions. The PT temperature interval is different for different metals and alloys. For example, iron has body-centered cubic (BCC) structure at temperature below 910°C, face centered cubic (FCC) structure for temperature range from 910°C to 1391°C; iron structure above 1392°C returns to BCC structure. The phase transformation in tin starts at the temperature of 18°C; titanium, whose melting point is 1660°C, has hexagonal close packed (HPC) structure below 882°C and BCC structure at higher temperature. Following [83], a low-temperature phase is called martensite or martensitic phase and a high-temperature phase – austenite or austenitic phase. Austenite→martensite transformation is referred to as direct transformation and martensite→austenite transformation is called the reverse transformation.

Phase transformations of most metals are not accompanied by noticeable deformation. However, there exist such alloys whose phase transformations are accompanied by significant non-elastic deformation. So-called Shape Memory Alloys (SMAs) belong to them. The shape memory alloy (also known as a smart alloy, memory metal, or muscle wire) is an alloy that "remembers" its shape, and can be returned to initial shape by applying heat to the alloy. A typical representative of SMA is nickel-titanium (Ni-Ti) alloy (metallide with approximately equal Ni- and Ti-concentrations). In temperature interval from -150 to +200°C, this alloy undergoes phase transformations, the character and sequence of which depends on the component concentration and experimental details. In nickel-titanium alloy Ni-Ti (~45% Ni) only martensite→austenite transition upon heating and austenite→martensite transition upon cooling occur, whereas Ni–Ti (~55 % Ni) undergoes so-called two-step phase transformations, upon heating, the martensitic phase first transforms into an intermediate structure and then the austenitic phase starts to grow. There is a situation such that a heat-induced intermediate phase does not nucleate, whereas it arises upon cooling.

Another important feature of martensite/austenite transformations is their reversibility. This means that the kinetics and atom rearrangements in direct and reverse transitions are identical. Further, during both cooling and heating, PT can be terminated at any stage and redirected in the opposite direction. In constant temperature, the structure of load-free material remains unchangeable and there is a balance between the martensite/austenite phase and parent matrix.

Martensite/austenite fraction in a local region (grain) of metal matrix can be expressed through a phase character function, Φ, $\Phi = 1$ for martensite, $\Phi = 0$ for austenite, and $0 < \Phi < 1$ for the case when the region contains both martensitic and austenitic phases. A plot of Φ as a function of temperature, T, in a load-free body is shown in Fig. 5.12 [83]. The behavior of the $\Phi(T)$ is as follows:

A) Heating of fully martensitic material (solid-line arrows) $\Phi = 1$ up to point A_s, $T = A_s'$. Beginning from point A_s, the martensite fraction decreases along the straight line $A_s A_f$ and for $T \geq A_f$ the material is fully austenitic, $\Phi = 0$.

B) Cooling of fully austenitic material (dotted-line arrows). $\Phi = 0$ up to point M_s. For $M_f' \leq T \leq M_s$ (segment $M_s M_f$), the function Φ increases meaning that the martensite grows. Finally, when the martensite finish temperature, $T = M_f'$, is reached, the transformation is complete.

Furthermore, if to terminate the austenite reaction during heating at, e.g., point M_1 on line $A_s A_f$, and start cooling, the austenite fraction does not change until point M_2 on line $M_s M_f$ is reached; upon further cooling, the martensite fraction growth is presented by segment $M_2 M_f$. Similar to this, the transition from cooling to heating at point M_2 on line $M_s M_f$ initiates the return transformation along trajectory $M_2 M_1 A_f$. The hysteresis in Fig. 5.12 practically does not depend on temperature rate. For Ni–Ti (~51 % Ni) alloy, we have [83] $M_f' = 290K$, $M_s = 340K$, $A_f = 460K$, $A_s' = 410K$, so

$$A_s' - M_f' = A_f - M_s, \tag{5.13.1}$$

i.e., the straight lines $A_s A_f$ and $M_f M_s$ are parallel to each other.

Let us note, that such a simple scheme as in Fig. 5.12 does not embrace all the variety of experimental data. More detailed and wide information about PTs can be found in the monograph [83]. We assume that austenite-martensite transformations are as in Fig. 5.12, and temperatures M_s, M_f', A_s', and A_f for a load-free material is suggested to be known and satisfy the condition (5.13.1).

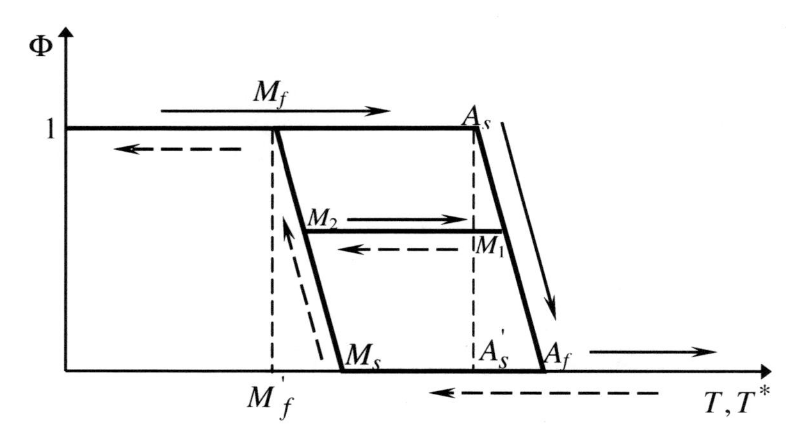

Fig. 5.12 Phase character function Φ change with (effective)temperature

We can distinguish between two kinds of PTs, the PT of the first and second kind. Without entering into detail, the transition of the first kind develops by the propagation of interface between the transformed and parent material. The transformed structure arises because of jump-wise changes in lattice parameters due to subtle but rapid rearrangement of atomic positions. Such a process is schematically shown in Fig. 5.13. Figure 5.13a illustrates an initial crystal lattice, the "cubic" lattice of parent material with lattice parameter of l_1. Figure 5.13b demonstrates the intermediate state of the crystal structure, there is the parent phase to the left of the border M_1M_2, while the transformed phase grows the right of M_1M_2. The propagation of interface M_1M_2 from left to right means that the parent and transformed phase fractions decrease and increase, respectively. Figure 5.13c shows the final product of the transformation with lattice parameters of l_2 and l_3. For the case of first-kind-PT, it is possible to introduce the phase character function Φ which gives the martensite/austenite fractions (Fig. 5.12).

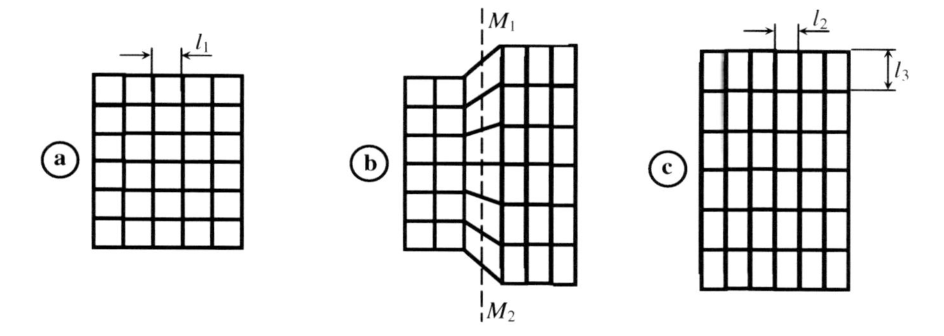

Fig. 5.13 The evolution of PT of the first kind: **a** – initial lattice; **b** – intermediate stage, **c** – the transformation completion

Consider the case of the second-kind-PT. In contrast to the first-kind-PT, the changes in lattice parameter are gradual and occur throughout the parent matrix over the whole temperature-change period. This process is shown schematically in Fig. 5.14. It is clear that the phase character function Φ is not applicable for the second-kind-PT. Further throughout, we will consider only the PTs of the first kind.

As it is known from experiments, the PT of material can be induced by temperature changes as described above, or by the loading/unloading of body. It turns out that the loading of body results in the same effect as a cooling. If to load an austenitic specimen, a martensitic phase starts to form. On the contrary, the unloading is equivalent to a heating. In any case, the reason of direct/reverse phase transitions is that PT-induced structural changes are energetically favorable.

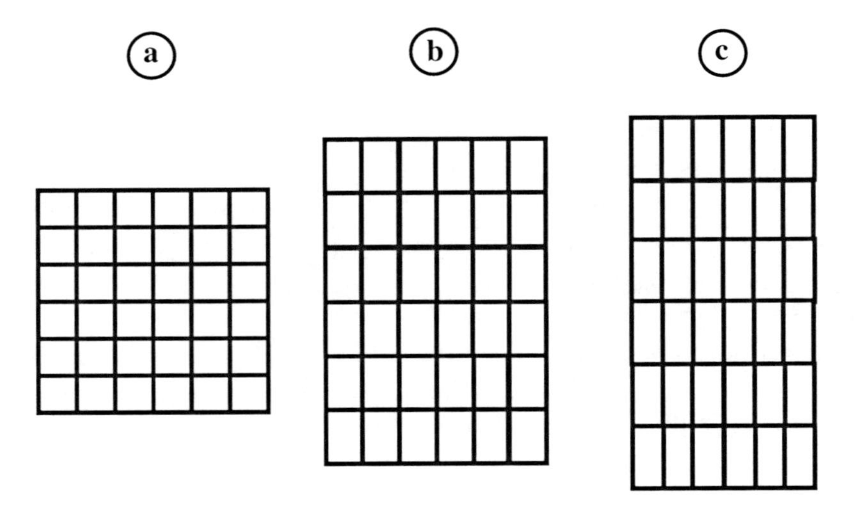

Fig. 5.14 The evolution of PT of the second kind: **a** – initial lattice; **b** – intermediate stage, **c** – the transformation completion

5.14 Stress-Strain Diagram under Phase Transformations

The fact that phase transformations can be accompanied by a significant non-elastic deformation (Kurdjumov and Khandros [70]) excited an active interest in this phenomenon. Martensite transformation is the basic reason for the occurrence of pseudo-elasticity, shape-memory effect, and transformation plasticity which find expanding applications in engineering.

1. ***Pseudo-elasticity.*** Consider the deformation of specimen loaded in tension at the temperature close to the martensite start temperature; its stress-strain diagram is shown schematically in Fig. 5.15. The initial portion of the diagram, segment OM_1, obeys Hooke's law. Beyond point M_1, the specimen experiences intensive flow (portion M_1M_2) caused by the formation of martensite. After the transformation is completed, beginning from point M_2, the material deforms elastically again until the yield limit σ_S is achieved (point M_3); the further increase in stress induces an usual plastic deformation. The unloading from the point lying on the portion M_1M_2 leads to the complete or partial, so-called pseudo-elastic, recover of deformation accompanied, as a rule, with a significant hysteresis. Since an unloading is equivalent to the temperature increase, martensitic phase, which has arisen at tensile, transforms into austenitic phase. It is this fact that explains the reduction of phase-induced strain in unloading. The phase-induced strain in loading is referred to as pseudo-elasticity, which is represented by the portion M_1M_2 in Fig. 5.15.

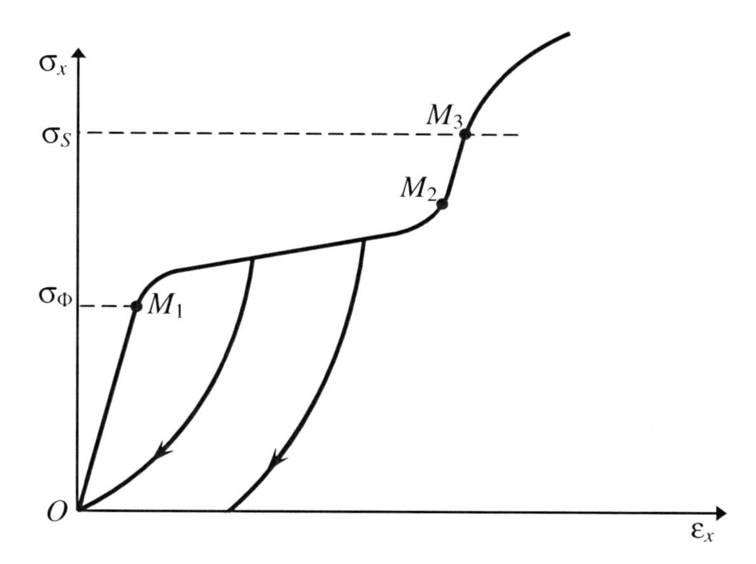

Fig. 5.15 Stress-strain diagram showing pseudo-elastic deformation (section M_1M_2); pseudo-elastic deformation starts arising at stress σ_Φ

2. ***Shape-memory effect.*** Let a specimen be heated/cooled to such a temperature so that two phases are present. Further, the specimen is subjected to (a) loading in tension, (b) unloading, and (c) heating in the unloaded state. The stress-strain and strain-temperature diagram of the specimen is shown in Fig. 5.16. Segment OM_1 gives a pure elastic strain of the specimen. Along the portion M_1M_2, PT-induced strain is occurred, pseudo-elasticity; by the PT the martensitic transformation is meant. The portion of the diagram $OM_1M_2M_4$ in Fig. 5.16 is identical to the diagram in Fig. 5.15. After the complete unloading (point M_4) and heating (along portion M_4M_5 the strain remains unaltered), the intensive decrease in strain (portion M_5M_6) occurs that is caused by austenitic transformation due to the temperature increase.

The phenomenon when a specimen can be returned to its initial shape after being deformed, by applying heat, is called the shape-memory effect. The portion M_5M_6 in Fig. 5.16 displays this effect.

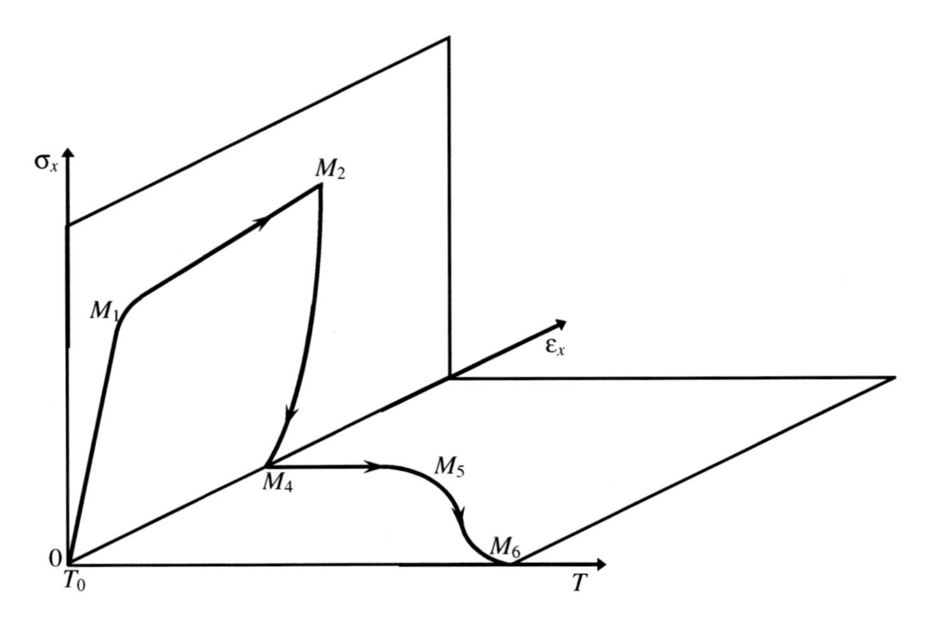

Fig. 5.16 Strain-strain diagram in tension showing pseudo-elastic deformation (M_1M_2) and effect of the shape memory (M_5M_6)

3. ***Transformation plasticity.*** To demonstrate the transformation plasticity the following procedures must be carried out:

(i) a specimen is stretched by tensile stress σ_x to deformation ε_{x0}; the produced deformation can be of two forms: (a) plastic deformation of martensitic specimen (b) pseudo-elastic deformation induced by the martensite transformation of two-phase specimen (portion M_1M_2 in Fig. 5.16).

(ii) the sample heating under constant σ_x;

(iii) the sample cooling under constant σ_x.

Figure 5.17 illustrates *deformation* vs. *temperature* diagram of Ni-Ti alloys under constant loading. Upon heating, the deformation ε_{x0} disappears (portion M_1M_2) and upon the following cooling it is recovered (portion M_3M_4). Upon heating, the deformation decreases due to austenite transition, whereas the relative elongation is recovered because martensite transition occurs upon cooling. The PT-induced deformation of loaded specimen under cooling (portion M_3M_4 in Fig. 5.17) is called the *transformation plasticity*. If the initial deformation ε_{x0} is of pseudo-elastic nature, the portion M_1M_2 represents the shape-memory effect. As seen from Fig. 5.17 the material remembers both the low- and the high-temperature shape. As a result, the strain-trajectory $M_1M_2M_3M_4M_1$ can be reproduced any number of times.

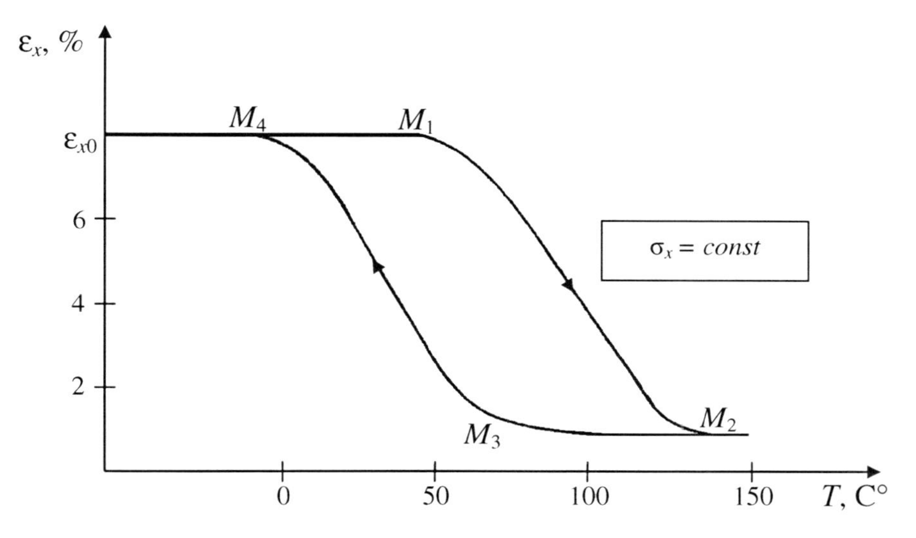

Fig. 5.17 Transformation plasticity

The effect described above can be also obtained if the deformation ε_{x0} is the plastic deformation produced in austenite phase. To observe the transformation plasticity, the specimen first should be cooled and then heated.

Together with the properties considered above, SMAs possess other properties as well. As one can see, the straining of SMAs is far from being standard. Classical theories of plasticity are inapplicable for the modeling of PT-induced strain due to the driving force of deformation under PT is not external loading alone, but the forces of metal-physics and chemical nature as well.

It is worthwhile to note once more that the explanations of different PT effects presented above are strongly simplified. The reader is referred to the monographs [30,83,114,169] for a more detailed and wide analysis of PTs.

5.15 Effective Temperature

The martensite fraction in a local volume of load-free body is determined by the phase character function Φ, the Φ-T diagram has the form of parallelogram $M_f A_s A_f M_s$ in Fig. 5.12 and the temperatures M_f, A_s,... are material constants. However, if the body is subjected to loading, the Φ-T dependence becomes more complex. Namely, the values of temperatures M_f, A_s,... are strongly influenced by the values of acting stress, i.e., they are not material constants any more. At the same time, it turns out that there exists so-called effective temperature (T^*) such that the function $\Phi(T^*)$ under loading is of the same form as $\Phi(T)$ in a load-free case. In other words, the phase character retains its form if to replace T by T^* on the axis of abscissas. This means that the laws of austenite/martensite transformations described in Sec. 5.13 remain valid under loading as well. Thus, the temperatures M_f, A_s,... are again considered as material constants independent of acting stress.

To determine the effective temperature, we rewrite the Clausius-Clapeyron equation giving the slope of the co-existence curve, dp/dT, on a pressure-temperature (p-T) diagram,

$$\frac{dp}{dT} = \frac{q}{\Delta VT}, \tag{5.15.1}$$

in terms of deformation and stress as [83]

$$dT = -\frac{TD_{ij}}{q} d\sigma_{ij}. \tag{5.15.2}$$

Here D_{ij} are the crystal deformation components related to crystallographic axes for a complete phase transformation when the function Φ changes from zero to unit, σ_{ij} are stress components in the local volume related to the crystallographic axes; q is the phase transformation thermal effect. The components D_{ij} and q are the known physical characteristics of phase transformation for a given alloy. If the crystal grain undergoes the phase transformation as shown in Fig. 5.13, then

$$D_{11} = \frac{l_2 - l_1}{l_1}, \quad D_{22} = \frac{l_3 - l_1}{l_1}, \tag{5.15.3}$$

where crystal axes 1 and 2 are horizontal and vertical, respectively. It turns out that the lattice parameters (interatomic distance) l_1, l_2, and l_3 are related to each other as $l_2 + l_3 = 2l_1$. Therefore, the relation

$$D_{11} + D_{22} = 0 \tag{5.15.4}$$

expresses the fact that phase transformations usually are not accompanied by volume changes.

It is clear that in a load-free state $T^* = T$ that allows us to find the effective temperature by integrating in Eq. (5.15.2):

$$\int_{T}^{T^*} \frac{dT}{T} = -\int_{0}^{\sigma_{ij}} \frac{D_{ij}}{q} d\sigma_{ij} . \quad (5.15.5)$$

As in the right-hand side of this formula the D_{ij} and q are constants, we have

$$T^* = T \exp\left(-D_{ij}\sigma_{ij}/q\right). \quad (5.15.6)$$

If to expand the T^* in Eq. (5.15.6) by its Taylor series expansion, dropping the quadratic term, we obtain

$$T^* = T\left(1 - D_{ij}\sigma_{ij}/q\right). \quad (5.15.7)$$

This formula determines the $T^* = T^*\left(T, \sigma_{ij}\right)$-dependence and shows that the effective temperature decreases with the application of loading, whereas the following unloading results in the increase of T^*, i.e., Eq. (5.15.7) is in full accordance with what discussed in Sec. 5.14.

5.16 Mathematical Model of Deformation under Phase Transformations

As SMAs are finding ever-widening applications, the problem of the prediction of their behavior upon the application of thermal and/or mechanical loadings arises. This has resulted in a number of analytical models to describe the deformation of SMAs [78]. The V.A. Lichachov and V.G. Malinin monographs [83] should be emphasized as some principal statements are taken from them to develop the model of PT-induced deformation in terms of the synthetic theory offered in the K.M. Rusynko, I.M. Holyboroda, and A.G. Shandrivsky works [42,43,158]. The basic statements of this theory are the following.

First of all, a dislocation shape change is neglected as a PT-induced deformation occurs under stresses below the yield limit (see, for example, Fig. 5.15). Further, we do not take into account creep deformation either because loading, unloading, and heating/cooling of body last relatively for a short time.

The approach used here is based on the two-level theory of an irreversible deformation. Let a direct (martensitic) or reverse (austenitic) transformation occur in a local volume (grain) of body which progresses according to the diagram in Fig. 5.12. The martensite formation rate is determined as

$$\dot{\Phi} = -\frac{\dot{T}^*}{M_s - M_f}, \quad (5.16.1)$$

where T^* is the effective temperature (5.15.7), and the martensite is formed at the reduction of T^* for the temperature range $M_f' \le T^* \le M_s$. Therefore, Eq. (5.16.1) holds at

$$\dot{T}^* < 0, \quad M_f' < T^* < M_s. \tag{5.16.2}$$

The decrease in the martensite fraction, occurring at the effective temperature-decrease range $A_s' < T^* < A_f$ is

$$\dot{\Phi} = -\frac{\dot{T}^*}{A_f - A_s'}, \tag{5.16.3}$$

i.e., Eq. (5.16.3) is applicable at

$$\dot{T}^* > 0, \quad A_s' \le T^* \le A_f. \tag{5.16.4}$$

As $M_s - M_f' = A_f - A_s'$, Eqs. (5.16.1) and (5.16.3) are identical.

The effective temperature (5.15.7) depends on stress components acting in the local volume. Similarly to the concept of slip, we assume that the state of stress in a local volume is identical to the macro-state of stress. Therefore, the components σ_{ij} in Eq. (5.15.7) are known as components of a given macro-stress and, consequently, the effective temperature T^* is known as well. In addition, the martensite fraction can be calculated by Eq. (5.16.1) as the temperatures M_s and M_f' are known to be physical constants of material.

Now we proceed to the determination of phase-transformation induced strain components generated in local volume, ε_{ij}^{Φ}. In most cases, it is possible to expresses the $\dot{\varepsilon}_{ij}^{\Phi}$ components through the PT-induced lattice distortion D_{ij} (Sec. 5.15) and martensitic transformation rate $\dot{\Phi}$ as [83]:

$$\dot{\varepsilon}_{ij}^{\Phi} = A_{\Phi} D_{ij} \dot{\Phi}, \tag{5.16.5}$$

where A_{Φ} is the factor that takes into account that the PT-induced deformation may not be the same as the deformation of crystal lattice. However, $A_{\Phi} = 1$ is often the case. If the deformation of crystal grain is characterized by the tensor D_{ij} at the complete transformation, the average deformation of the grain at partial transformation, according to Fig. 5.13 is determined by $D_{ij}\Phi$. It is this statement that forms the basis of Eq. (5.16.5) giving $\dot{\varepsilon}_{ij}^{\Phi}$ components both at direct ($\dot{\Phi} > 0$) and reverse ($\dot{\Phi} < 0$) transformation.

We take Eq. (5.16.5) as the basic relationship for the determination of PT-induced strain rate components (further throughout, we will also use the phase deformation notion); to calculate by Eq. (5.16.1) the velocity of martensite fraction formation $\dot{\Phi}$, we need to differentiate Eq. (5.15.7) with respect to time. As a result, Eq. (5.16.5) yields the form

$$\dot{\varepsilon}_{ij}^{\Phi} = -\frac{A_{\Phi} D_{ij}}{M_s - M_f{}'} \left[\dot{T} \left(1 - \frac{D_{kl} \sigma_{kl}}{q} \right) - T \frac{D_{kl} \dot{\sigma}_{kl}}{q} \right] \tag{5.16.6}$$

and is valid for both the condition (5.16.2) and (5.16.4); all the quantities in the right-hand side of this formula are known. Thus, the law of the progress of PT-induced strains in local volume, on micro-level, has been found. The following problem is to establish the rule for the determination of macro-strains, i.e., the rule for the summation of microstrains $\dot{\varepsilon}_{ij}^{\Phi}$ from Eq. (5.16.6).

In contrast to plastic deformation, phase deformation is realized not by shears (see, for example Fig. 5.13). At the same time, in most cases it is not accompanied by volume change. On the other hand, as well known, a deformation of any kind can be obtained as shears (slips) along several (maximum 5) slip systems. Therefore, PT-induced macro-deformation can be determined in terms of the concept of slip or synthetic theory. Similarly to Chapter 2, if to designate through $\vec{\mathbf{n}}$ the vector normal to a slip plane and through $\vec{\mathbf{l}}$ the vector which lies on this plane and gives the slip direction, Eq. (5.16.6) gives shear strain components in the $\vec{\mathbf{n}}$ - $\vec{\mathbf{l}}$ slip system as

$$\dot{\gamma}_{nl}^{\Phi} = -\frac{A_{\Phi} D}{M_s - M_f{}'} \left[\dot{T} \left(1 - \frac{D \tau_{nl}}{q} \right) - T \frac{D \dot{\tau}_{nl}}{q} \right], \tag{5.16.7}$$

where τ_{nl} is the resolve shear stress, D is a reduced deformation in crystal grain under complete phase transformation. The quantity D is the known characteristic of phase transformation for a given crystal lattice. If, for example, the transformation occurs as in Fig. 5.13,

$$D = \frac{D_{22} - D_{11}}{2}, \tag{5.16.8}$$

where D_{11} and D_{22} are the relative elongations of lattice in horizontal and vertical directions. Equation (5.16.8), together with Eq. (5.15.3), takes the form

$$D = \frac{l_3 - l_2}{2l_1} = \frac{l_3 - l_2}{l_3 + l_2}, \tag{5.16.9}$$

where interatomic distances l_2 and l_3 are shown in Fig. 5.13.

In terms of the synthetic theory, each slip system is represented by tangent plane with normal vector $\vec{\mathbf{N}}$ in the Ilyushin five-dimension space and the resolved shear stress τ_{nl} is replaced by scalar product $\vec{\mathbf{S}} \cdot \vec{\mathbf{N}}$, where $\vec{\mathbf{S}}$ is the stress vector. Similar to Sec. 3.4, we designate through $\vec{\mathbf{e}}_0^{\Phi}$ the irreversible (phase) micro-strain vector caused by the displacement of single tangent plane, $\vec{\mathbf{e}}_0^{\Phi}$ is co-directed with the plane normal vector $\vec{\mathbf{N}}$. We propose to determine $\dot{\vec{\mathbf{e}}}_0^{\Phi}$, which is assumed to be co-directed with $\vec{\mathbf{N}}$ as well, as

$$\dot{\vec{\mathbf{e}}}_0^{\Phi} = \dot{\varphi}_N \vec{\mathbf{N}} \qquad r\dot{\varphi}_N = -\dot{T}\left(1 - D_1 \vec{\mathbf{S}} \cdot \vec{\mathbf{N}}\right) + D_1 T \dot{\vec{\mathbf{S}}} \cdot \vec{\mathbf{N}}, \tag{5.16.10}$$

where

$$r = \frac{M_s - M_f{}'}{A_\Phi D}, \quad D_1 = \frac{D}{q}, \tag{5.16.11}$$

and in arriving at the result (5.16.10) we have used relation (5.16.7). Although the magnitude of the deformation $\vec{\mathbf{e}}_0^{\Phi}$ is a finite quantity relative to the grain size, its contribution into macro-deformation is infinitesimal as the number of tangent planes (slip systems) is suggested to be infinitely large.

The macro phase-deformation rate components are

$$d\dot{e}_i^{\Phi} = \dot{\varphi}_N m_i \cos\lambda dV \qquad dV = \cos\beta d\alpha d\beta d\lambda \tag{5.16.12}$$

Therefore, in a similar fashion to Eq. (3.9.14), the macro non-elastic PT-induced deformation rate components are

$$\dot{e}_i^{\Phi} = \iiint_{\alpha\,\beta\,\lambda} m_i \cos\beta \cos\lambda \dot{\varphi}_N d\alpha d\beta d\lambda, \tag{5.16.13}$$

where $\dot{\varphi}_N$ is given by Eq. (5.16.10).

Equations (5.16.10) and (5.16.13) are similar to those proposed in [83], they are based on the suggestions that (i) martensitic transformation induces deformation co-directed with acting stresses (ii), austenitic transformation induces the recovery of deformation. The difference between the model of phase deformation based on the synthetic theory and that proposed in [83] in terms of the slip concept consists in that the synthetic theory (i) is based on the more precise equation, Eq. (5.15.7), for the effective temperature and (ii) gives much more compact relations than in terms of the slip concept, see Chapters 2 and 3.

Taking in view Eqs. (3.9.14) and (5.16.13), we can infer that both plastic and phase strain rates are described by the formulae of identical form. Since the driving forces of plastic and phase deformations are different, its natural that different strain intensities appear in Eqs. (3.9.14) and (5.16.13).

5.17 PT-Deformation under Heating or Cooling

Upon the application of both thermal and mechanical loadings, local volumes (grains) of specimen undergo deformations induced by phase-transformations of different intensity; in some of them the phase reactions does not occur at all. In terms of the synthetic model, a local volume of a body is represented by tangent plane with normal vector $\vec{\mathbf{N}}$. Therefore, first of all, it is necessary to formulate the general rule that determines the orientations of $\vec{\mathbf{N}}$ in which phase micro-deformations progress at a given instant. The reduction of effective temperature T^* in the range $M_f' < T^* < M_s$ (Fig. 5.12) leads to the martensite phase which arises by inducing strain co-directed with the action of load; the temperature increase in the range $A_s' < T^* < A_f$ causes the recover of the strain. Therefore, PT-induced strain develops at a given instant and in a given direction $\vec{\mathbf{N}}$ if

$$\dot{\varphi}_N > 0 \text{ for } \dot{T}^* < 0 \left(M_f' < T^* < M_s\right), \tag{5.17.1}$$

or

$$\dot{\varphi}_N < 0 \text{ for } \dot{T}^* > 0 \left(A_s' < T^* < A_f\right), \tag{5.17.2}$$

where $\dot{\varphi}_N$ is determined by Eq. (5.16.10) whose right-hand side can have different signs and different directions $\vec{\mathbf{N}}$. If the sign of function $\dot{\varphi}_N$ from Eq. (5.16.10) does not satisfy the condition (5.17.1) or (5.17.2) for some directions, the phase deformation does not acquire an increment in these directions at the given instant.

The criterion (5.17.1) or (5.17.2) differs from that established in Chapters 3 and 4, according to which a plane produces an irreversible strain only if $\dot{\varphi}_N > 0$.

There are also other differences between the determination of irreversible strain due to load and phase transitions:

1. In terms of the synthetic model applied to the description of PTs, there is no concept of yield/creep surface. Therefore, planes, which are capable of producing shape changes, fill completely the Ilyushin three-dimensional subspace. Similarly to Chapters 3 and 4, any plane is set by its normal vector $\vec{\mathbf{m}}$ (by angles α and β; Fig. 3.3) and the distance between the plane and the origin of coordinates is

characterized by angle λ; $\lambda = 0$ and $\lambda = \pi/2$ give a plane with zero and infinitely large distance, respectively.

2. Planes do not move.

3. Now the requirement that a tangent plane produces irreversible strain if only it moves on the endpoint of stress vector is not valid.

Consider a load-free specimen of SMA at temperature below the martensite finish temperature, $T < M_f{}'$ (Fig. 5.12), i.e., the specimen is fully martensitic. Upon heating to a temperature from the range $A_s{}' < T < A_f$, the austenite phase starts to form. By making use of Eq. (5.17.2), we wish to determine the macro-deformation caused by austenitic transformation. The condition that the specimen is load free and Eq. (5.15.7) give that $T = T^*$ leading to that the phase strain rate intensity from (5.16.10) is

$$r\dot{\varphi}_N = -\dot{T}. \tag{5.17.3}$$

The $\dot{\varphi}_N$ from the above formula satisfies the condition (5.17.2) and does not depend on angles α, β, and λ meaning that the austenite phase forms in all local volumes simultaneously, i.e., angles α, β, and λ vary in the range

$$0 < \alpha < 2\pi, \quad -\frac{\pi}{2} < \beta < \frac{\pi}{2}, \quad 0 < \lambda < \frac{\pi}{2}. \tag{5.17.4}$$

Therefore, phase-strain rate vector components from Eq. (5.16.13) take the form

$$\dot{e}_k^{\Phi} = -\frac{\dot{T}}{r} \int_0^{2\pi} d\alpha \int_{-\pi/2}^{\pi/2} m_k \cos\beta d\beta \int_0^{\pi/2} \cos\lambda d\lambda. \tag{5.17.5}$$

As one can see, the integral over λ equal to unity for all three components $\dot{e}_k$. Inserting m_k from Eq. (3.5.4) into Eq. (5.17.5), we obtain

$$\begin{aligned}
\dot{e}_1^{\Phi} &= -\frac{\dot{T}}{r} \int_0^{2\pi} \cos\alpha d\alpha \int_{-\pi/2}^{\pi/2} \cos^2\beta d\beta \\
\dot{e}_2^{\Phi} &= -\frac{\dot{T}}{r} \int_0^{2\pi} \sin\alpha d\alpha \int_{-\pi/2}^{\pi/2} \cos^2\beta d\beta \\
\dot{e}_3^{\Phi} &= -\frac{\dot{T}}{2r} \int_0^{2\pi} d\alpha \int_{-\pi/2}^{\pi/2} \sin 2\beta d\beta
\end{aligned} \tag{5.17.6}$$

and integrating that gives

$$\dot{e}_1^{\Phi} = \dot{e}_2^{\Phi} = \dot{e}_3^{\Phi} = 0 \qquad (5.17.7)$$

and $e_1^{\Phi} = e_2^{\Phi} = e_3^{\Phi} = 0$.

Thus, the heating/cooling of load-free body does not induce phase macro-deformation that is in a full agreement with experimental data. The reason is that micro-deformations occurring in micro-volumes (slip systems) cancel one another (see Eq. 5.17.3 where $\dot{\varphi}_N + \dot{\varphi}_{-N} = 0$) thereby producing no macro-deformation.

5.18 Mathematical Description of Pseudo-elasticity

Consider the following procedure. A load-free fully austenitic sample is under temperature T_0, $M_s < T_0$. Further, holding the temperature T_0 constant, the specimen is subjected to an arbitrary proportional loading. Let us consider the case when a thin-walled pipe is subjected to torsion $\left(S_3 = \sqrt{2}\tau_{xz} > 0\right)$. We wish to determine the PT-induced deformation of the pipe in terms of the synthetic theory.

Equation (5.15.7) gives the effective temperature T^* as

$$T^* = T_0\left(1 - D_1 \vec{\mathbf{S}} \cdot \vec{\mathbf{N}}\right). \qquad (5.18.1)$$

In torsion, we have

$$\vec{\mathbf{S}} \cdot \vec{\mathbf{N}} = S_3 \cos\xi, \quad \cos\xi = \sin\beta\cos\lambda. \qquad (5.18.2)$$

Hence, Eqs. (5.18.1) and (5.18.2) give that

$$T^* = T_0(1 - D_1 S_3 \cos\xi) \qquad (5.18.3)$$

meaning that the loading of body at constant temperature is equivalent to the decrease in effective temperature, and this reduction varies from plane to plane with different values of angles $\beta > 0$ and λ. T^* is smallest for the plane with $\beta = \pi/2$ and $\lambda = 0$:

$$T_{\min}^* = T_0(1 - D_1 S_3). \qquad (5.18.4)$$

If $T_{\min}^*$ decrease to the value of M_s, the martensite phase starts to grow in the plane with $\beta = \pi/2$ and $\lambda = 0$. Equating $T_{\min}^*$ to M_s, we obtain the equation for the stress inducing pseudo-elastic strain – phase flow limit S_{Φ} (Fig. 5.15) – in the following form:

$$S_{\Phi} = \frac{1}{D_1}\left(1 - \frac{M_s}{T_0}\right). \tag{5.18.5}$$

As seen from the above formula, the phase flow limit grows with the temperature of body T_0. If $T_0 = M_s$, $S_{\Phi} = 0$ and the diagram 5.15 becomes nonlinear at once from the origin of coordinates.

The further increase in component S_3 leads to that the effective temperature becomes smaller than M_s meaning the martensitic phase starts to form. The formation of martensite induces strains whose intensity $\dot{\varphi}_N$ is determined by Eqs. (5.16.10) and (5.18.2) as

$$r\dot{\varphi}_N = D_1 T_0 \dot{S}_3 \cos\xi . \tag{5.18.6}$$

As seen form (5.18.6), the conditions (5.17.1) and $\dot{\varphi}_N \geq 0$ is satisfied for $0 < \lambda < \pi/2$ and $0 < \beta < \pi/2$. On the other hand, the inequality $T^* \leq M_s$ requires that

$$\frac{1}{S_3 D_1}\left(1 - \frac{M_s}{T_0}\right) \leq \cos\xi \leq 1 \tag{5.18.7}$$

obtained by equating T^* to M_s in Eq. (5.18.3). With account of (5.18.5), the above inequality can be rewritten as

$$\frac{S_{\Phi}}{S_3} \leq \cos\xi \leq 1 . \tag{5.18.8}$$

The set of angles β and λ satisfying the condition (5.18.8) is shown by shaded area in Fig. 5.18a. Therefore, the integration limits in Eq. (5.16.13) are

$$0 \leq \alpha \leq 2\pi, \quad \beta_1 \leq \beta \leq \pi/2, \quad 0 \leq \lambda \leq \lambda_1$$
$$\cos\lambda_1 = \frac{S_{\Phi}}{S_3 \sin\beta} = \frac{\sin\beta_1}{\sin\beta}, \quad \sin\beta_1 = \frac{S_{\Phi}}{S_3} \tag{5.18.9}$$

The macro-strain rate vector components induced by the martensite transformation, by making use of Eq. (5.16.13), acquire the form

$$\dot{e}_k^{\Phi} = \frac{D_1 T_0 \dot{S}_3}{2r} \int_0^{2\pi} d\alpha \int_{\beta_1}^{\pi/2} m_k \sin 2\beta d\beta \int_0^{\lambda_1} \cos^2 \lambda d\lambda . \tag{5.18.10}$$

Integrating in Eq. (5.18.10) at $k=1$ and $k=2$ (the direction cosines m_k are determined by Eq. (3.5.14)) gives that $\dot{e}_1^{\Phi}=\dot{e}_2^{\Phi}=0$. $\dot{e}_3^{\Phi}$ is

$$\dot{e}_3^{\Phi}=\frac{2\pi D_1 T_0 \dot{S}_3}{r}\int_{\beta_1}^{\pi/2}\sin^2\beta\cos\beta d\beta\int_0^{\lambda_1}\cos^2\lambda d\lambda \tag{5.18.11}$$

and integrating in (5.18.11) gives

$$\dot{e}_3^{\Phi}=\frac{\pi D_1 T_0}{3r}Z_1\left(\frac{S_\Phi}{S_3}\right)\dot{S}_3 \quad Z_1(x)=\arccos x+x\sqrt{1-x^2}-2x^3\ln\frac{1+\sqrt{1-x^2}}{x}. \tag{5.18.12}$$

Further we integrate over S_3 in limits from S_Φ to S_3 in the above formula. As a result, we get

$$e_3^{\Phi}=\frac{\pi D_1 T_0}{3r}\int_{S_\Phi}^{S_3}Z_1\left(\frac{S_\Phi}{S_3}\right)dS_3=\frac{\pi D_1 T_0 S_\Phi}{3r}F\left(\frac{S_\Phi}{S_3}\right), \tag{5.18.13}$$

where the function F is determined by Eq. (3.10.14). According to Eqs. (5.16.11) and (5.18.5), we rewrite Eq. (5.18.13) as

$$e_3^{\Phi}=\frac{\pi A_f D(T_0-M_s)}{3(M_s-M_f')}F\left(\frac{S_\Phi}{S_3}\right). \tag{5.18.14}$$

The obtained formula holds if the effective temperature (5.18.3) is less than M_f', i.e., the martensite transformation progresses. Otherwise, according to Fig. 5.12, the direct martensite transformation is finished ($\Phi=1$) at $T^*=M_f'$ and, consequently, the transformation-induced deformation is terminated. Equating the right-hand side of Eq. (5.18.4) to M_f', we obtain the equation for the boundary value of S_3, and S_f, at which Eq. (5.18.13) or (5.18.14) is applicable:

$$S_\Phi\le S_3\le S_f \qquad S_f=\frac{1}{D_1}\left(1-\frac{M_f'}{T_0}\right). \tag{5.18.15}$$

If the component S_3 exceeds the found value of S_f, we can split $\beta-\lambda$ plane into three parts (Fig. 5.18b): $\Phi=0$ meaning that PT-induced strain is not yet in progress; $0<\Phi<1$ – the strain is occurring; $\Phi=1$ – the strain has been finished.

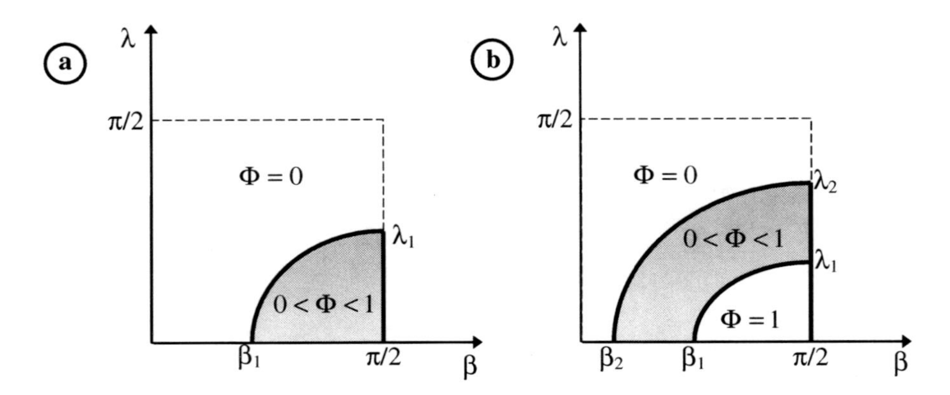

Fig. 5.18 The ranges of angles β and λ where phase transformation occurs

Equation (5.18.13) can be obtained in another, simpler, way. By integrating in Eq. (5.18.6) with respect to time, we obtain

$$r\varphi_N = D_1 T_0 \cos\xi\left(S_3 - C\right), \tag{5.18.16}$$

where C is the value of stress that induces phase deformation at a given point (with coordinates β and λ) within the region **I** in Fig. 5.18a:

$$C = \frac{S_\Phi}{\cos\xi}. \tag{5.18.17}$$

Therefore,

$$r\varphi_N = D_1 T_0 \left(S_3 \cos\xi - S_\Phi\right) = D_1 T_0 S_\Phi \left(\frac{\sin\beta\cos\lambda}{\sin\beta_1} - 1\right), \quad \sin\beta_1 = \frac{S_\Phi}{S_3} \tag{5.18.18}$$

We faced strain intensities similar to (5.18.18) many times (see for example Eqs. (5.4.17) and (5.4.18)) and, consequently, we can infer that we arrive at the result of Eq. (5.18.13).

5.19 Modeling of Pseudo-elasticity at Unloading

Consider, as in the previous Section, the case when a fully austenitic sample at constant temperature T_0 is subjected to torsion ($S_3 > 0$) that causes martensite transformation. After the component S_3 reaches its maximal value, S_0, unloading takes place:

$$S_3 = S_0 - \Delta S. \tag{5.19.1}$$

Our goal is to determine the phase deformation due to the unloading at constant temperature T_0. Let us assume that the minimum value of effective temperature over angles β and λ before unloading ($S_3 = S_0$), $T_*{}^*$, is such that $M_f{}' < T_*{}^* < M_s$ (Fig. 5.19), i.e., the martensite transformation has been not completed, $\Phi < 1$. The effective temperature during unloading,

$$T^* = T_0\left[1 - D_1\left(S_0 - \Delta S\right)\cos\xi\right], \tag{5.19.2}$$

grows with the increase in ΔS at $\cos\xi > 0$. According to the basic statement of the synthetic model, the austenite phase starts to nucleate if the effective temperature T^* for the plane with $\cos\xi = 1$, i.e., $\beta = \pi/2$ and $\lambda = 0$, reaches the value corresponding to point M_1 in Fig. 5.19:

$$T^* = T_*{}^* + A_f - M_s, \tag{5.19.3}$$

where $A_f - M_s$ is the base of parallelogram. The value of $T_*{}^*$ can be expressed through T^* from Eq. (5.19.2) at $\Delta S = 0$ and $\cos\xi = 1$ as

$$T_*{}^* = T_0\left(1 - D_1 S_0\right) \quad \text{and} \quad T^* = T_0\left(1 - D_1 S_0\right) + A_f - M_s \tag{5.19.4}$$

The value of stress-decrease when the effective temperature (5.19.3) is reached, ΔS_*, can be determined by equating to each other the values of T^* from Eq. (5.19.2) at $\cos\xi = 1$ and Eq. (5.19.4)

$$\Delta S_* = \frac{A_f - M_s}{T_0 D_1}. \tag{5.19.5}$$

When $\Delta S = \Delta S_*$, the austenite transformation starts to develop for the plane with coordinates $\beta = \pi/2$ and $\lambda = 0$. At the same time, the effective temperature (5.19.2) at $\Delta S = \Delta S_*$ is incapable of inducing the austenitic-transformation-induced deformation for planes (slip systems) with $\beta < \pi/2$ and $\lambda > 0$. The progressing of austenitic transformation for planes with $\beta < \pi/2$ and $\lambda > 0$ is possible if $\Delta S > \Delta S_*$.

Summarizing, an elastic unloading takes place at $\Delta S \le \Delta S_*$ and the austenite-reaction-induced-deformation occurs at $\Delta S > \Delta S_*$. Let us find the value of ΔS needed for the inducing of phase deformation at arbitrary angles β and λ. The austenite induced deformation starts to develop if the difference between effective temperature calculated by Eq. (5.19.2) at $\Delta S > 0$ and $\Delta S = 0$,

$$T^*_{\Delta S=0} = T_0\left(1 - D_1 S_0 \cos\xi\right), \tag{5.19.6}$$

is equal to the value of segment $A_f - M_s$:

$$T_0 D_1 \Delta S \cos\xi = A_f - M_s. \tag{5.19.7}$$

This means that the austenite induced deformation starts at point M_2 shown in Fig.5.19. Eq. (5.19.7) shows that $\cos\xi$ lies within the range

$$\frac{A_f - M_s}{T_0 D_1 \Delta S} \le \cos\xi \le 1 \tag{5.19.8}$$

meaning that the domain of angles β and λ is the same as in Fig. 5.18a.

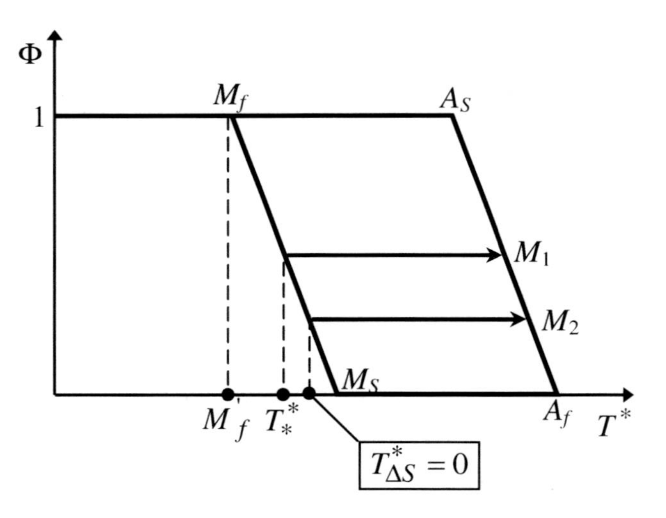

Fig. 5.19 Phase character function Φ and effective temperature $T_*{}^*$ in pseudo-elastic deformation

According to Eq. (5.16.10), the austenite-reaction-induced strain intensity rate can be written as

$$r\dot{\varphi}_N = -D_1 T_0 \left|\dot{S}_3\right| \cos\xi\,, \quad \dot{\mathbf{S}} \cdot \vec{\mathbf{N}} = -\left|\dot{S}_3\right| \cos\xi \tag{5.19.9}$$

By integrating of Eq. (5.19.9) with respect to time, together with Eqs. (5.19.5), (5.19.7) and (5.19.8), we get

$$r\varphi_N = -D_1 T_0 \left(\Delta S \cos\xi - \Delta S_*\right), \tag{5.19.10}$$

or

$$r\varphi_N = -D_1 T_0 \Delta S_* \left(\frac{\sin\beta \cos\lambda}{\sin\beta_1} - 1\right), \quad \sin\beta_1 = \frac{\Delta S_*}{\Delta S}. \tag{5.19.11}$$

By integrating of φ_N from the formula above in Eq. (3.9.14), we obtain the phase deformation in unloading as

$$e_3^{\Phi} = -\frac{\pi}{3r} T_0 D_1 \Delta S_* F\left(\frac{\Delta S_*}{\Delta S}\right). \tag{5.19.12}$$

The total strain, due to loading and unloading, is obtained as the sum of the right-hand sides of Eqs. (5.18.13) and (5.19.8):

$$e_3^{\Phi} = \frac{\pi T_0 D_1 S_{\Phi}}{3r} F\left(\frac{S_{\Phi}}{S_0}\right) - \frac{\pi\left(A_f - M_s\right)}{3r} F\left(\frac{A_f - M_s}{T_0 D_1 \Delta S}\right), \tag{5.19.13}$$

$$\Delta S \geq \Delta S_*.$$

5.20 Modeling of Transformation Plasticity

In this Section, we deal with the strain of a specimen of SMA upon cooling under constant loading which is called the transformation plasticity (Sec. 5.14)[4]. Consider a load-free, fully-austenitic specimen at $T_0 > M_s$. Then, the specimen is subjected to torsion so that PT-induced deformation does not start meaning that the effective temperature (5.18.3) does not drop below M_s during loading. Further, the shear stress is held constant and the temperature decreases. The temperature decrease leads to the formation of martensite phase which induces the phase deformation.

The effective temperature, according to Eq. (5.18.1), is

$$T^* = T\left(1 - D_1 S_3 \cos\xi\right). \tag{5.20.1}$$

[4] We restrict ourselves to the determination of PT-induced strain upon cooling only.

As $\dot{S}_3 = 0$, the phase strain rate intensity (5.16.10) takes the form

$$r\dot{\varphi}_N = -\dot{T}\left(1 - D_1 S_3 \cos\xi\right). \tag{5.20.2}$$

According to the problem formulation, we have $\dot{T} < 0$ implying that $\dot{\varphi}_N > 0$ and $\dot{T}^* < 0$. Now, we wish to determine the martensite start temperature, T_s. Equating the minimum value of T^* (over β and λ) to M_s, we obtain the equation for the T_s as a function of S_3 in the following form

$$T_s = \frac{M_s}{1 - D_1 S_3}. \tag{5.20.3}$$

In Eq. (5.20.3), the term $D_1 S_3$ is assumed to be less then unity.

If $T^* < T_s$, the martensite grows for the set of plane (slip systems) which is determined by the equating of T^* to M_s:

$$\begin{gathered} 0 \le \alpha \le 2\pi, \quad \beta_1 \le \beta \le \pi/2, \quad 0 \le \lambda \le \lambda_1, \\ \cos\lambda_1 = \frac{\sin\beta_1}{\sin\beta}, \quad \sin\beta_1 = \frac{1}{D_1 S_3}\left(1 - \frac{M_s}{T}\right). \end{gathered} \tag{5.20.4}$$

The obtained ranges (5.20.4) shown in Fig. 5.18a give the diapasons of angles α, β, and λ within which Eq. (5.20.2) holds true. The variant $\Phi = 1$ in some planes, i.e., the martensite transformation is completed, is also possible (Fig. 5.18b), which depends on temperature *T*. We, however, will further work under the condition that $T > M_s$.

Let us integrate Eq. (5.20.2) with respect to time:

$$r\varphi_N = -T\left(1 - D_1 S_3 \cos\xi\right) + C, \tag{5.20.5}$$

where *C* is the integration constant. The constant *C* is determined from the condition that the range (5.20.4) for $\dot{\varphi}_N$ must be the same for φ_N, i.e., $C = M_s$. Therefore,

$$r\varphi_N = -\left(T - M_s\right) + T D_1 S_3 \cos\xi, \tag{5.20.6}$$

Phase strain component, e_3^{Φ}, according to Eqs. (3.9.4) and (5.20.6), can be written as

$$e_3^{\Phi} = -\frac{\pi(T - M_s)}{r} \int_{\beta_1}^{\pi/2} \sin 2\beta d\beta \int_0^{\lambda_1} \cos\lambda d\lambda + \frac{2\pi T D_1 S_3}{r} \int_{\beta_1}^{\pi/2} \sin^2\beta \cos\beta d\beta \int_0^{\lambda_1} \cos^2\lambda d\lambda \cdot \quad (5.20.7)$$

By integrating in (5.20.7), we obtain

$$e_3^{\Phi} = \frac{\pi}{3r} T D_1 S_3 Z_1(\sin\beta_1)$$

$$Z_1(x) = \arccos x - 2x\sqrt{1-x^2} + x^3 \ln\frac{1+\sqrt{1-x^2}}{x}. \quad (5.20.8)$$

In works [42,43,158], the comparison of model results in experimental data concerning the PT-induced deformation, pseudo-elasticity, and transformation plasticity are presented. As seen from these works the results obtained in terms of the synthetic theory are in satisfactory agreement with experiments.

5.21 Conclusions

1. In Secs. 5.1–5.6, the peculiarities of non-isothermal creep have been considered. It turns out that the creep diagram under periodic step-wise temperature changes considerably which differs from standard ones at $T = const$; the *temperature after-effect* and *temperature strengthening* phenomena are observed. In this case, classical theories of creep are inapplicable as they give strains by an order of magnitude smaller than experimental data. On the other hand, results obtained in terms of the generalized synthetic theory are in satisfactory agreement with experiments.

2. To model the creep strain under the condition of step-wise thermal loading, the thermal stress parameter I_T, which depends on a temperature-change history, is introduced into the $H(\psi)$ relation thereby expanding to a very considerable extent the physical base of synthetic model. The parameter I_T is the average measure of the second-kind microstresses arising in a body due to temperature-jump-changes.

3. It is worthwhile to note that the integral of non-homogeneity I_N (introduced in Chapter 4) and the thermal stress parameter I_T in different ways affect the deformation properties of a material, I_N expresses strain hardening, whereas I_T assists irreversible straining.

4. Sections 5.7–5.12 concern the modeling of plastic/creep deformation of a body subjected to irradiation. For this purpose, we have introduced the measure of radiation-induced defects into the constitutive equation of synthetic theory. These defects are set as a function of the intensity, duration, and regime of irradiation.

The sum of the radiation induced and deformation defects govern the deformation properties of body.

5. Irradiation influences upon a creep and plastic deformation in different ways. As generally accepted, the radiation decreases plastic deformation but assists creep deformation. In this context, as in Sec. 4.17, plastic and creep straining may seem to be modeled by different theories. However, this approach is unacceptable. Indeed, the fact that radiation affects in different way plastic and creep deforming is manifested if only the acting stress is considerably above the creep limit. If stress is slightly greater than creep limit, the radiation-induced creep strains is smaller than radiation-free creep strains meaning that radiation affects in identical way both plastic and creep deformation.

6. In Secs. 5.13-5.20, the phase-transformation-induced strains of the bodies of shape-memory alloys are studied. The stress-strain diagrams of such alloys are of so unusual (non-classical) form that it seems that their analytical description is possible in terms of only very specific and extremely complicated theory. The non-elastic deformation of SMA can be readily described in terms of the synthetic theory.

7. Phase deformations are induced by the direct or reverse, martensitic or austenitic, transformations. The driving force of martensite/austenite transitions constitutes a considerable problem for many sciences such as metallurgy, thermodynamics, and chemistry. In the monograph, this problem is solved by the establishment of the phenomenological law of martensite/austenitic formation based on the transformation parallelogram $M_sM_fA_sA_f$ (Fig. 5.12).

8. As it is known, since the martensite/austenite fraction, which is a function of the effective temperature, varies from one local(micro) volume (e.g. crystal grains) to another, the contributions of PT-induced micro strains in the macro strain are non-uniform. This fact is displayed in Secs. 5.15–5.20, where different planes (micro volumes) have different effective temperatures meaning different martensite fractions. Here, the advantage of two-level models is manifested in full measure; the synthetic theory is capable of describing such unusual diagrams as in Fig. 5.15–5.18.

5.22 Conclusions

The synthetic theory of irreversible deformation is capable of modeling a number of the effects and phenomena which cannot be described in terms of any classical theories. This fact makes this theory a perspective one.

At simple loading, synthetic theory leads to the fulfillment of the deviator proportionality. One of the most remarkable facts is that a corner point arises on loading surface during loading. This allows us to describe the peculiarities of irreversible straining due to linear piece-wise loading paths. The kinetics of loading surface is not prescribed *a priori*, as in terms of flow plasticity theories with, e.g., kinematic or isotropic hardening, but is fully determined by loading regime. This statement results from the fact that the yield/loading surface is constructed as the inner-envelop surface to the planes tangential to this surface.

During loading, stress vector translates the set of the planes thereby transforming the surface. Consequently, in terms of the synthetic theory, it is enough to set plane distances to follow the loading surface transformations and irreversible strain developments. Although the plane distance is set in a really simple fashion (see, e.g., Eq. (4.32.4), or (4.32.5)), a number of problems – the modeling of plastic deformation, loading-rate effects, unsteady-state creep both in loading and partial unloading, irreversible strain induced by non-force factors (temperature after-effect, radiation induced deformation, PT-induced deformation) – has been solved.

Beyond doubt, a theory which embraces such broad class of problems must possess an extended physical basis, meaning that real mechanisms inducing/ accompanying irreversible deformations must be reflected in constitutive equations.

At present, physicists have investigated in detail the mechanisms of irreversible deformation on both atomic and crystal grain scale. Dislocation motion induces that the parts of grain slip past each other along their close-packed directions. The result is a permanent change of shape within the crystal grain and plastic deformation. If to suggest that the development of irreversible strain caused by a single dislocation has been studied more or less completely, the interaction between dislocation groups and conglomerations is far from being investigated. At the same time considerable collective forces arise in these dislocation formations. Furthermore, if to model an irreversible strain on the grain scale, one can see that it varies from one grain to other; this follows at least from grains different orientations. As a result, other considerable forces, which originate from the interaction between the grains, arise. Since the law of straining is established through the stresses acting in a body, in view of the forces specified above, there is an extremely difficult problem to find real stresses exerting influence upon the body. Let us note that as the orientation of grains is of random character, these real stresses are random as well.

Therefore, to find real forces acting in a body is practically impossible. As a consequence, to take them into account, one needs to apply some simplifying hypotheses. In this context, it is important to stress that the simplifying hypotheses must be adequate to the accuracy adopted in a given model.

Let us explain the above stated by an example of the concept of slip in terms of which an irreversible deformation is calculated as the summation of slips occurring on some set of slip systems (in terms of the synthetic theory, tangent planes represent the slip systems). Consequently, the slip concept reflects the essence of irreversible straining in a real body, dislocation movements induce shears(slips) within crystals. On the other hand, together with the dislocation movements, there are other mechanisms producing irreversible strain such as twinning, point defects diffusion, grain boundary flow, and grain-boundary migration. And now the following two possibilities arise. The first of them is to formulate how each of the mechanism specified above influences upon irreversible straining. The second possibility consists in that we do not detail the nature of plastic shears in a real body. In this case, we deal with the notion of plastic slip intensity, which represents the average measurement of all the slips occurring in body. The second way seems to be less physical, but it is more reliable as not requiring superfluously detailed and sometimes even erroneous hypotheses.

As an example, consider the modeling of, e.g., of temperature after-effect. This type of deformation is caused by different reasons such as the dispersion of elastic modulus and thermal expansion coefficient above crystal grains as functions of temperature, *etc*. It would be possible to detail which of these reasons play the main part in temperature after-effect, but we do not need to do it. The important point here is only that the specified dispersions due to step-wise temperature changes induce additional (thermal) stresses which in turn induce additional (temperature after-effect) deformations.

This raises the question of great importance and difficulty, in what extents it is expedient to develop the physical basis of model. In other words, what knowledge of nuclear physics, the physics of solid and, other adjacent sciences should be taken into account? The answer to this question is not unequivocal and each researcher chooses priorities himself. A mathematical model must be capable of describing a considered phenomenon by relationships as simple as possible.

References

1. Andrade, E.: The viscous flow in metals and allied phenomena. Proc. R. Soc. London 84, 1–12 (1910)
2. Andrusik, J., Rusinko, K.: The deformation of loading surface. Izv. AN SSSR, Mekh. Tverd. Tela 3, 98–103 (1979) (in Russian)
3. Andrusik, J., Rusinko, K.: Plastic strain of work-hardening materials under loading in three-dimensional subspace of five-dimensional stress-deviator space. Izv. RAN (Russian Academy of Sciences), Mekh. Tverd. Tela 2, 92–101 (1993) (in Russian)
4. Andrusik, J., Rusinko, K.: A new approach to the phenomenological interpretation of the concept of slippage. J. Math. Sci. 63, 355–357 (1993)
5. Andrusik, J., Rusinko, K.: Plastic deformation of hardening materials under cyclic loading. Izv. RAN (Russian Academy of Sciences), Mekh. Tverd. Tela 3, 76–85 (1999) (in Russian)
6. Annin, B.: The Contemporary Models of Plastic Solids Novosibirsk (1975) (in Russian)
7. Annin, B., Zsigalkin, V.: Materials Behavior at Complex Loading, Novosibirsk (1999) (in Russian)
8. Batdorf, S., Budiansky, B.: A mathematical theory of plasticity based on the concept of slip, NACA, Technical Note, 871 (1949)
9. Batdorf, S., Budiansky, B.: Polyaxial stress-strain relation on strain hardening metal. J. Appl. Mech. 21, 323–326 (1954)
10. Béda, G., Kozák, I., Verhás, J.: Kontinuum mechanika. Budapest (1986) (in Hungarian)
11. Betten, J.: Creep Mechanics. Springer, Heidelberg (2005)
12. Bolshanina, M.: Hardening and recovery as the basic phenomenon of plastic deformation. Izv. AN SSSR, Physics 14, 223–231 (1950) (in Russian)
13. Bolshanina, M., Panin, V.: The latent energy of deformation. Solids Physics Researches, Moscow, 193–233 (1957) (in Russian)
14. Bridgman, P.: Effects of high hydrostatic pressure on the plastic properties of metals. Mod. Phys. 17, 3 (1945)
15. Budiansky, B.: A reassessment of deformation theories of plasticity. J. Appl. Mech. 26, 259–264 (1959)
16. Budiansky, B., Dow, N., Peters, R., Shepherd, R.: Experimental studies of polyaxial stress-strain laws of plasticity. Proc. Nat. Congr. Appl. Mech., 503–512 (1951/1952)
17. Chakrabarty, J.: Applied Plasticity. Springer, Heidelberg (2000)
18. Chen, W., Han, D.: Plasticity for structural engineers. Springer, Heidelberg (1988)
19. Cicala, P.: Sulle deformatione plastiche, Acc. Dei Lincei, Rendicenti, 5 (1950)
20. Cicala, P.: On the plastic deformation. Atti Acad. Naz. Lincei, Rendicenti, Mat. E Natur. 8, 583–586 (1950)

21. Cottrell, A.: Dislocations and Plastic Flow in Crystals. Oxford University Press, London (1953)
22. Curpal, I.: Stress concentration around circle hole in nonlinear elastic continuum. J Appl. Mech. 8, 45–50 (1962)
23. Davidenkov, N., Lichachov, V.: Irreversible Strain under Cyclic Heating, Moscow, Leningrad (1962) (in Russian)
24. Davis, E.: Increase of stress with permanent strain and stress-strain relations in the plastic state for copper under combined stresses. J. Appl. Mech. 10, 187–196 (1943)
25. Davis, E.: Yielding and fracture of medium-carbonate steel under combined stress. J. Appl. Mech. 12, 13–24 (1945)
26. Drucker, D.: A more fundamental approach to plastic stress-strain relations. In: Proc. 1st. U.S. Nat. Congr. Appl. Mech., vol. 1, pp. 487–491 (1951)
27. Drucker, D., Prager, W., Greenberg, H.: Extended limit design theorems for continuous media. Quart. Appl. Math. 9, 381 (1952)
28. Drucker, D.: A definition of stable inelastic material. J. Appl. Mech. 26, 101–106 (1959)
29. Drucker, D., Palgen, L.: On the stress-strain relations suitable for cyclic and other loading. J. Appl. Mech. 48, 479 (1981)
30. Duerig, T., Melton, K., Stöckel, D., Wayman, C.: Engineering Aspects of Shape Memory alloys. Butterworth Heinemann, London (1990)
31. Fastov, N.: Energy of distorted crystalline grid. Problemy Metallovedenija i Fiziki Metallov 4, 377–387 (1955) (in Russian)
32. Feigen, M.: Inelastic behavior under combined tension and torsion. In: Proc. 2nd US Nat. Congr. of Appl. Mech., pp. 469–476 (1954)
33. Garofalo, F.: Fundamentals of Creep and Creep-Rupture in Metals. MacMillan, New York (1965)
34. Ginsztler, J., Skelton, R.: Component Reliability under Creep-fatigue Conditions. Springer, Heidelberg (1998)
35. Gusenkov, A.: Properties of Cyclic Stress-Strain Diagrams at Room Temperature, Nauka, Moscow (1967) (in Russian)
36. Haasen, P., Kelly, A.: A yield phenomenon in face-centered cubic single crystals. Acta Metallurgica 5, 192–199 (1957)
37. Handelman, G., Lin, C., Prager, W.: On the mechanical behavior of metals in the strain-hardening rage. Quart. Appl. Math., 397–407 (1947)
38. Hencky, H.: Zur theorie plastischer deformationen und der hierdurch im material herforgerufenen nacshpannungen. Z. Ang. Mech. 4, 323–334 (1924)
39. Hill, R.: The Mathematical Theory of Plasticity. Oxford University Press, New York (1950)
40. Hodge, P.: A general theory of piecewise linear plasticity based on maximum shear. J. Mech. Phys. Solids 5, 242 (1975)
41. Hodge, P.: A piecewise linear theory of for an initially isotropic material in plane stress. J. Mech. Sci. 22, 21 (1980)
42. Holyboroda, I., Rusinko, K.: Derivation of a universal relation between tangential stress and shear strain intensities in describing reversible martensitic deformation within the framework of a synthetic model. J. Appl. Mech. and Tech. Phys. 37, 447–453 (1996)
43. Holyboroda, I., Rusinko, K., Tanaka, K.: Description of an Fe-based shape memory alloy thermomechanical behavior in terms of the synthetic model. J. Comp. Mater. Sci. 13, 218–226 (1999)

44. Honeycombe, R.: The Plastic Deformation of Metals. Edward Arnold, London (1984)
45. Hutchinson, J.: Elasto-plastic behavior of polycrystalline and composites. Proc. Roy. Soc. London Ser. A 319, 247 (1970)
46. Hutchinson, J.: Plastic Buckling. Advances in Applied Mechanics 14, 67–144 (1974)
47. Ibragimov, S., Kirsanov, V., Pjatiletov, J.: The Radiation Defects of Metals and Alloys, Moscow (1985) (in Russian)
48. Ilyushin, A.: Plasticity, Moscow (1963) (in Russian)
49. Ishlinsky, A.: A general theory of plasticity with linear strain-hardening. Ukrain. Mat. Zh. 6, 314–325 (1954) (in Russian)
50. Ivanova, G.: Creep of EI-437B alloy at variable temperature. Izv. AN SSSR, Tekhn. Nauki 4, 98–99 (1958) (in Russian)
51. Ivlev, D., Bikovcev, G.: Theory of Hardening Plastic Solid. Moscow (1971) (in Russian)
52. Joshimure, J.: Comment on the slip theory of Batdorf and Budiansky. Bull. ISME 1(2), 109–113 (1958)
53. Kachanov, A.: Foundations of the Theory of Plasticity. North-Holland, Amsterdam (1971)
54. Kadashevitch, I., Novozhilov, V.: The theory of plasticity that takes into account residual microstresses. Journal of Applied Mathematics and Mechanics 22, 104–118 (1959)
55. Kadashevich, J., Novozhilov, V.: Micro-Stresses in Materials, Leningrad (1990) (in Russian)
56. Kalatinec, A., Parhomenko, J., Rusinko, K.: The analytical and experimental study of alternating loading. Izv. AN SSSR, Mekh. Tverd. Tela 23, 722–731 (1959) (in Russian)
57. Kaliszky, S.: Plasticity. Theory and Engineering Applications. Elsevier, Amsterdam (1989)
58. Kelly, A.: Strong Solids. Oxford University Press, London (1973)
59. Kennedy, A.: Processes of Creep and Fatigue in Metals. Oliver and Boyd, Edinburg (1962)
60. Kliushnikov, V.: On plasticity laws for work-hardening materials. Journal of Applied Mathematics and Mechanics 22, 129–160 (1958)
61. Kliushnikov, V.: New concepts in plasticity and deformation theory. Journal of Applied Mathematics and Mechanics 23, 1030–1042 (1959)
62. Kliushnikov, V.: Surface of loading and tolerance at its experimental determination. Dokl. AN SSSR 121, 299–300 (1975) (in Russian)
63. Kliushnikov, V.: Theory of plasticity, its contemporary state and development perspectives. Izv. AN SSSR, Mekh. Tverd. Tela 2, 102–116 (1993)
64. Koiter, W.: Stress-strain relations, uniqueness and variational theorems for elastic-plastic materials with a singular yield surface. Quart. Appl. Math. 11, 350–354 (1953)
65. Koiter, W.: General Theorems. In: Sneddon, I., Hill, R. (eds.) Progress in Solid Mechanics, vol. I. North-Holland, Amsterdam (1960)
66. Kuksa, L.: Comparative investigations of the inhomogeneity of elastic and plastic deformation of metals. Strength of Materials 18, 349–354 (1986)
67. Kuksa, L., Lebedev, A., Koval'chuk, B.: Laws of distribution of microscopic strains in two-phase polycrystalline alloys under simple and complex loading. Strength of Materials 18, 1–5 (1986)

68. Kunejev, V., Rusinko, K.: Orthogonal additional-loading of materials with different behavior in tension and compression. Izv. AN SSSR, Mekh. Tverd. Tela 5, 90–98 (1971) (in Russian)
69. Kurdiumov, G.: X-ray investigation of the distortions of crystalline grid. In: Probl. Metallovedenija i Fiziki Metallov, Moscow, vol. 4, pp. 339–359 (1955)
70. Kurdiumov, G., Handros, L.: Thermal-elastic equilibrium under martensite transformation. Dokl. AN SSSR 74, 211–214 (1949) (in Russian)
71. Lebedev, A.: Equivalence criteria under conditions of creep in a complex stressed state. Strength of Materials 2, 340–343 (1970)
72. Lebedev, A., Koval'chuk, B., Lamashevskii, V.: Poinsson's ratio for carbon steel and gray iron at normal and low temperatures. Strength of Materials 3, 292–297 (1971)
73. Lenskij, V.: Experimental verifications of the basic postulates of the theory of elastic-plastic deformation. In: Voprosi Teorij Plastichnosti, Moscow, pp. 58–82 (1961) (in Russian)
74. Leonov, M.: Elements of the analytical theory of plasticity. Dokl. AN SSSR 205, 303–306 (1972) (in Russian)
75. Leonov, M.: Deformation and Fracture Mechanics, Frunze (1981) (in Russian)
76. Leonov, M., Shvajko, J.: Complex plane deformation. Dokl. AN SSSR 159, 1007–1010 (1964) (in Russian)
77. Lévy, M.: Mémoire sur les équations générales des mouvements intérieurs des corps solids ductiles au dela des limites, où l'élasticité pourrait les ramener à leur premier état. In: Compt. Rend. Acad. Sci., Paris, vol. 70, pp. 1323–1325 (1870)
78. Liang, C., Rogers, C.: One dimensional thermomechanical constitutive relations for shape memory materials. In: Dyn. and Mater. Conf., Washington, D.C, pp. 16–28 (1990)
79. Lin, T.: Analysis of elastic and plastic strains of a face centered cubic crystal. J. Mech. Phys. Solids 5, 143 (1957)
80. Lin, T.: Physical theory of plasticity. Adv. Appl. Mech. 11, 255–311 (1971)
81. Lin, T.: Physical theory of plasticity and creep. J. Engng. Mater. Technol., Trans. ASME 106, 290 (1984)
82. Lichachov, V., Malygin, G.: Temperature aftereffect. Prochnost Metallov i Splavov, Leningrad, 98–99 (1958) (in Russian)
83. Lichachov, V., Malinin, V.: Structure-analytical theory of strength, St. Peterburg (1993) (in Russian)
84. Lode, W.: Versuche über den einfluss der mittleren hauptspannun auf das fliessen der metalle eisen, kupfer und nickel. Z. Physik. 36, 913–939 (1926)
85. Lode, W.: The influence of the mean principal stress upon the flow of meta. In: Book The theory of plasticity, Moscow (1948) (in Russian)
86. McLean, D.: Grain Boundaries in Metals. Oxford University Press, London (1957)
87. McLean, D.: Mechanical Properties of Metals and Alloys. John Wiley, New York (1977)
88. Malinin, V.: Strain-loading rate relation. Polzuchest Tverdogo Tela, Frunze, pp. 52–63 (1974) (in Russian)
89. Malmejster, A., Tamuzs, V., Teters, G.: Strength of Polymer Materials, Riga (1972) (in Russian)
90. Oding, J. (ed.): Machinery encyclopedia-reference, Moscow, vol. 3 (1947) (in Russian)
91. Mises, R.: Mechanik der festen körper im plastisch-deformablen zustang. Göttingen Nachr. Math. Phys. K. 1, 582–592 (1913)

92. Mises, R.: Mechanik der plastischen formänderung von kristallen. Z. Ang. Math. Mech. 8, 161–185 (1928)
93. Myshljajev, M.: Creep of polygonal structures. In: Nesovershenstva kristallicheskogo stroenija i martensitnyje prevraschenija, Moscow, pp. 194–234 (1972) (in Russian)
94. Myshljajev, M., Hannanov, S., Hodos, I.: The structure of dislocation boundary in deformed germanium. Izv. AN SSSR, Phys. 38, 2389–2402 (1974) (in Russian)
95. Mitrochin, N., Jagn, J.: On the systematic character of deviations from the laws of plasticity. Dokl. AN SSSR 135, 796–799 (1960) (in Russian)
96. Nádai, Á.: Der bildsame Zustand der Werkstoffe. Springer, Berlin (1927)
97. Nádai, Á.: Theory of Flow and Fracture of Solids, vol. 1. McGraw-Hill, New York (1950)
98. Nádai, Á.: Theory of Flow and Fracture of Solids, vol. 1. McGraw-Hill, New York (1963)
99. Naghdi, P., Rowles, D.: An experimental study of biaxial stress-strain relations in plasticity. J. Mech-Phys. Solids 3, 63–80 (1954)
100. Namestnikov, V.: Creep under alternating loading in complex stress-state. In: Izv. AN SSSR, Tekhn. Nauki, vol. 10 (1957) (in Russian)
101. Namestnikov, V., Chvostunkov, A.: Creep of duralumin under constant and alternating loading. Prikl. Mekh. i Tekhn. Fiz. 4 (1960) (in Russian)
102. Novozsilov, V.: On the relation between stresses and strains in non-linear elastic continuum. Prikl. Matem. i Mekh. 15, 183–194 (1951) (in Russian)
103. Novozsilov, V.: On the physical significance of stress invariants. Prikl. Matem. i Mekh. 5, 617–619 (1952) (in Russian)
104. Oding, J., Ivanova, V., Burdugskij, V., Geminov, V.: Theory of the Creep and Strength of Metals, Moscow (1959) (in Russian)
105. Odquist, F.: Mathematical Theory of Creep and Creep Rupture. University Press, Oxford (1966)
106. Ohashi, Y., Kawai, M., Mamose, T.: Effects of prior plasticity on subsequent creep of type 316 stainless steel at elevated temperature. AOIM, D 108, 99–111 (1986)
107. Osipiuk, W.: Zastosovanie teorii poslizgow do opisu pelzania wstecsnego. Rozpravki inzynierskie 27, 52–57 (1991) (in Polish)
108. Paul, B.: Mathematical criteria of plastic flow and fracture. In: Liebovitz, H. (ed.) Fracture, New York, London, vol. 2, pp. 336–520 (1968)
109. Pisarenko, G., Kiselevskij, V.: The Strength and Plasticity of Materials due to Radiation Flow, Kiev (1979) (in Russian)
110. Pjatiletov, J.: The contribution of slip into radiation creep of metals. The Physics of Metals and Metallography 50 (1980)
111. Polianski, J., Rusinko, K.: Deviation of strain diagram from straight line at unloading. J. Trybology 1, 56–59 (1999)
112. Polianski, J., Rusinko, K.: Plastic deformation and creep of light alloys. Proc. Ternopil State University 1, 112–116 (2000) (in Russian)
113. Polyans'kyi, Y., Rusynko, K.: Synthesized theory of plastic deformation depending on the loading rate. Materials Science 36, 42–47 (2000)
114. Porter, D., Easterling, K.: Phase transformations in metals and alloys. Chapman & Hall, Boca Raton (1992)
115. Prager, W.: Strain hardening under combined stresses. J. Appl. Phys. 16, 837–840 (1945)
116. Prager, W.: A new method of analyzing stress and strain in work-hardening plastic solids. J. Appl. Mech. 23, 493 (1956)

117. Prager, W.: An Introduction to Plasticity. Addison-Wesley, Reading (1959)
118. Prager, W., Hodge, P.: Theory of Perfectly Plastic Solids. Wiley, New York (1951)
119. Prager, W., Hodge, P.: On the laws of plasticity for hardening materials. J. PMM 22 (1958)
120. Prandtl, L.: Spanmungsverteilung in plastischen körpern. In: Proc. 1st Int. Conf. Appl. Mech., Delft, pp. 43–54 (1924)
121. Rabotnov, Y.: Strength of Materials, Moscow (1962) (in Russian)
122. Rabotnov, Y.: Creep Problems in Structural Members. North-Holland, Amsterdam (1966)
123. Rabotnov, Y.: Deformed Solid Mechanics, Moscow (1979) (in Russian)
124. Reuss, E.: Berucksichtigung der elastichen formänderungen in der plasizitätstheorie. Z. Angew. Math. Mech. 10, 266 (1930)
125. Rivlin, R.: Further remarks on the stress-deformation relations for isotropic materials. Journal of Applied Mechanics 27(6), 303–308 (1955)
126. Rivlin, R.: Non-Linear Continuum Theories in Mechanics and Physics and their Applications. In: Centro Internacionale Mathematico Estivo (C.I.M.E.), Edizione Cremonese, Roma (1970)
127. Rogozin, I.: On the slip theory. Continuum Dynamics, Novosibirsk 4, 148–153 (1970) (in Russian)
128. Rovinskij, B.: On the problem of stress relaxation mechanism in metals. Izv. AN SSSR, Tekhn. Nauki, 67–74 (1954) (in Russian)
129. Rovinskij, B., Liutcau, V.: Oriented micro-stresses relaxation. J. Tekhn. Phys. 27, 345–350 (1957) (in Russian)
130. Rovinskij, B., Sinajskij, V.: On the influence of strain rate upon residual oriented-micro-stresses. Izv. AN SSSR, Tekhn. Nauki. 3, 159–160 (1962) (in Russian)
131. Rusinko, A.: Creep with temperature hardening. Materials Science 33, 813–817 (1997)
132. Rusinko, A.: Mathematical description of ultrasonic softening of metals within the framework of synthetic theory. Materials Science 37, 671–676 (1999)
133. Rusinko, A.: Analytic dependence of the rate of stationary creep of metals on the level of plastic prestrain. Strength of Metals 34, 381–389 (2002)
134. Rusinko, A.: Effect of preliminary mechanical and thermal treatment on the unsteady creep of metals. Materials Science 38, 824–832 (2002)
135. Rusinko, A.: Influence of preliminary mechanical and thermal treatment on the steady-state creep of metals. Materials Science 40, 223–231 (2004)
136. Rusinko, A.: Analytic description of the effect of duration of the procedure of annealing performed after deformation on the steady-state creep of metals. Materials Science 41, 280–283 (2005)
137. Ruszinko, E.: The Influence of Preliminary Mechanical-thermal Treatment on the Plastic and Creep Deformation of Turbine Disks. J. Meccanica 43 (2008)
138. Rusinko, A., Ginsztler, J., Dévényi, L.: Analytic Description of the Formation of Pores Under Conditions of Steady-state Creep of Metals. Strength of Materials 39, 74–79 (2007)
139. Rusinko, A., Rusinko, K.: The mathematical description of temperature aftereffect, Zeszyty naukove politechniki Bialostockej. Mechanika 112, 145–150 (1997)
140. Rusinko, A., Rusinko, K.: Synthetic theory of irreversible deformation in the context of fundamental bases of plasticity. Int. J. Mech. Mater. 41, 106–120 (2009)
141. Rusinko, K.: The generalization of Cicala's formula. Izv. AN SSSR, Mekh. Tverdogo Tela 6, 37–44 (1971) (in Russian)

142. Rusinko, K.: Theory of Plasticity and Unsteady Creep, Lvov (1981) (in Russian)
143. Rusinko, K.: The Peculiarities of Non-Elastic Deformation of Solids, Lvov (1986) (in Russian)
144. Rusinko, K., Andrusik, J.: The orthogonal break of loading path. In: Prochnost Materialov i Elementov Konstrukcii, Kiev, pp. 223–228 (1986) (in Russian)
145. Rusinko, K., Basaraba, D.: Creep under thermal cyclic action. Materials Science 20, 387–391 (1984)
146. Rusinko, K., Blinov, E.: Theoretical analysis of deformation under two-stage loading. International Applied Mechanics 7, 1220–1225 (1971)
147. Rusinko, K., Blinov, E.: Analytic investigation of the stress-strain relation for an arbitrary loading trajectory. Mechanics of Composite Materials 7, 874–878 (1971)
148. Rusinko, K., Chormonov, M.: Stress concentration at circle hole in semi-fragile plate. In: Koncentracija Napryazhenij, Kiev, vol. 3, pp. 134–139 (1969) (in Russian)
149. Rusinko, K., Gazda, M.: Locally macroscopic investigation of creep in metals. International Applied Mechanics 16, 1036–1041 (1980)
150. Rusinko, K., Holiboroda, I.: Irreversible deformation in a cubic metal on temperature change. Materials Science 25, 125–129 (1989)
151. Rusinko, K., Ivasyuk, V.: Elements of the mathematical theory of inelastic deformation. Materials Science 16, 400–405 (1981)
152. Rusinko, K., Yablochko, Y.: Plastic deformation of an anisotropic body. Strength of Materials 9, 384–389 (1977)
153. Rusinko, K., Kalatinets, A.: Complicated alternating deformation of elastoplastic bodies. International Applied Mechanics 12, 433–438 (1976)
154. Rusinko, K., Kalatinets, A., Dreval, S.: The concept of slip in the theory of plasticity. International Applied Mechanics 10, 1–14 (1974)
155. Rusinko, K., Malinin, V.: Deformation of a solid body taking into account the time. International Applied Mechanics 11, 124–129 (1975)
156. Rusinko, K., Mulyar, R.: Plastic deformation at elevated temperatures. Strength of Materials 9, 150–154 (1977)
157. Rusinko, K., Panova, L.: Dependence of plastic deformation on temperature and loading rate. Materials Science 13, 427–431 (1978)
158. Rusynko, K., Shandrivs'kyi, A.: Irreversible deformation in the course of a martensite transformation. Materials Science 31, 786–789 (1996)
159. Rusinko, K., Shlyakhov, S.: Plastic deformation under a generalized proportional loading. Journal of Applied Mechanics and Technical Physics 22, 283–286 (1981)
160. Rusinko, K., Sypa, O.: Macroscopic investigation of the radiation creep of metals. Strength of Materials 22, 364–367 (1990)
161. Rusinko, M.: Radiation hardening. J. Fiz.-Khim Mekh. Mater. 6, 119–120 (1997) (in Russian)
162. Saint-Venant, B.: Mémoir sur l'établissement des équations différentielles de mouvements intérieurs opérés dans les corps solides ductiles au delá des limites où l'élasticité pourrait les ramener à leur premier état. Compt. Rend. 70, 473–480 (1870)
163. Salli, A.: Creep of Metals and Alloys, Moscow (1953) (in Russian)
164. Salli, A.: The contemporary stage of knowledge about the creep of metals. In: Uspekhi Fiziki Metallov, Moscow, pp. 157–221 (1960) (in Russian)
165. Sanders, I.: Plastic stress-strain relations based on linear loading function. In: Proc. 2nd U.S. Nat. Congr. Appl. Mech., pp. 455–460 (1954)
166. Sarrak, V., Shubin, V.: On the relaxation of local stresses in iron. J. The Physics of Metals and Metallography 25, 522–528 (1968) (in Russian)

167. Sarrak, V., Shubin, V., Entin, R.: The inhomogeneous distribution of internal stresses and the tendency to fragile fracture. J. The Physics of Metals and Metallography 2, 143–149 (1970) (in Russian)
168. Savin, G.: Stress Distribution Around Holes, Kiev (1968) (in Russian)
169. Shimizu, K., Tadaki, T.: Shape Memory Alloys. In: Funakubo, H. (ed.), Gordon and Breach Science Publishers, New York (1987)
170. Schmid, E., Boas, W.: Kristallplastizität. Springer, Berlin (1935)
171. Shvaiko, M.: Correctness of the Theories of Plasticity Taking into Account the Mutual Influence of the Mechanisms of Inelastic Deformation. Materials Science 37, 199–209 (2001)
172. Shvaiko, M.: Slip theory and its application. Materials Science 38, 590–605 (2002)
173. Smirnov, B., Shpejzman, V., Privalova, T., Samojlova, T.: Boundaries migrations at the steady-state creep of crystals. Fizika Tverdogo Tela 18, 2432–2434 (1976) (in Russian)
174. Sosnin, O.: On the problem of creep potential. Izv. AN SSSR, Mekh. Tverdogo Tela 5 (1971) (in Russian)
175. Suvorova, Y.: Delayed yield in steels (review). Journal of Applied Mechanics and Technical Physics 9, 270–274 (1968)
176. Sveshnikova, V.: On the plastic deformation of hardening metals. Izv. AN SSSR, Tekhn. Nauki 1, 155–161 (1956) (in Russian)
177. Teters, G.: The Combined Loading and Stability of Shells of Polymer Materials, Riga (1969) (in Russian)
178. Usikov, M., Utevskij, L.: Microscopic studies of the dislocation structure of nickel and nickel-alloys. In: Problemi Metallovedenija i Fizika Metallov, Moscow, pp. 77–100 (1964) (in Russian)
179. Vavakin, A., Viktorov, V., Mechanikova, I., Mohel, A., Stepanov, L.: The study of influence of temporal effects upon the plastic strain in steel under combined loading, Preprint AN SSSR, No. 235, Moscow (1984) (in Russian)
180. Viktorov, V., Shapiro, G.: Dynamic stress-strain diagrams in tension of metals at medium strain rates. Eng. J. Mekh. Tverd. Tela 2, 184–187 (1968) (in Russian)
181. Volkov, S.: Statistical theory of strength, Moscow, Sverdlovsk (1960) (in Russian)
182. Zarubin, V., Poliakov, A.: The model of anisotropy plastic deformation of polycrystalline materials. In: Termicheskije Napryazhenija v Elementach Konstrukcij, vol. 3 (1970) (in Russian)
183. Zarubin, V., Poliakov, A.: The influence of plastic deformation upon creep. In: Termicheskije Napryazhenija v Elementach Konstrukcij, vol. 15 (1975) (in Russian)
184. Zhigalkin, V.: Character of strain-hardening of a ductile material. Strength of Materials 12, 176–181 (1980)
185. Zhigalkin, V., Lindin, G.: A model for an anisotropic elastic state. Journal of Applied Mechanics and Technical Physics 18, 139–143 (1977)
186. Zsukov, A.: The plastic deformation of steel under combined loading. Izv. AN SSSR, Tekhn. Nauki 11, 53–61 (1954) (in Russian)
187. Zsukov, A.: The theory of plasticity of isotropy materials under combined loading. Izv. AN SSSR, Tekhn. Nauki 8, 84–92 (1955) (in Russian)

CPSIA information can be obtained at www.ICGtesting.com
Printed in the USA
LVOW012128230911

247657LV00004B/49/P

9 783642 212123